BIOFUELS
Biotechnology, Chemistry, and Sustainable Development

DAVID M. MOUSDALE

CRC Press
Taylor & Francis Group
Boca Raton   London   New York

CRC Press is an imprint of the
Taylor & Francis Group, an **informa** business

CRC Press
Taylor & Francis Group
6000 Broken Sound Parkway NW, Suite 300
Boca Raton, FL 33487-2742

Library of Congress Cataloging-in-Publication Data

Mousdale, David M.
  Biofuels : biotechnology, chemistry, and sustainable development / David M.
Mousdale.
    p. ; cm.
  CRC title.
  Includes bibliographical references and index.
  ISBN-13: 978-1-4200-5124-7 (hardcover : alk. paper)
  ISBN-10: 1-4200-5124-5 (hardcover : alk. paper)
    1. Alcohol as fuel. 2. Biomass energy. 3. Lignocellulose--Biotechnology. I. Title.
  [DNLM: 1. Biochemistry--methods. 2. Ethanol--chemistry. 3. Biotechnology.
4. Conservation of Natural Resources. 5. Energy-Generating Resources. 6.
Lignin--chemistry. QD 305.A4 M932b 2008]

  TP358.M68 2008
  662'.6692--dc22
                                                    2007049887

Visit the Taylor & Francis Web site at
http://www.taylorandfrancis.com

and the CRC Press Web site at
http://www.crcpress.com

Contents

Preface

When will the oil run out? Various estimates put this anywhere from 20 years from now to more than a century in the future. The shortfall in energy might eventually be made up by developments in nuclear fusion, fuel cells, and solar technologies, but what can substitute for gasoline and diesel in all the internal combustion engine-powered vehicles that will continue to be built worldwide until then? And what will stand in for petrochemicals as sources of building blocks for the extensive range of "synthetics" that became indispensable during the twentieth century?

Cellulose — in particular, cellulose in "lignocellulosic biomass" — embodies a great dream of the bioorganic chemist, that of harnessing the enormous power of nature as the renewable source for all the chemicals needed in a modern, bioscience-based economy.[1] From that perspective, the future is not one of petroleum crackers and industrial landscapes filled with the hardware of synthetic organic chemistry, but a more ecofriendly one of microbes and plant and animal cells purpose-dedicated to the large-scale production of antibiotics and blockbuster drugs, of monomers for new biodegradable plastics, for aromas, fragrances, and taste stimulators, and of some (if not all) of the novel compounds required for the arrival of nanotechnologies based on biological systems. Glucose is the key starting point that, once liberated from cellulosic and related plant polymers, can — with the multiplicity of known and hypothesized biochemical pathways in easily cultivatable organisms — yield a far greater multiplicity of both simple and complex chiral and macromolecular chemical entities than can feasibly be manufactured in the traditional test tube or reactor vessel.

A particular subset of the microbes used for fermentations and biotransformations is those capable of producing ethyl alcohol — ethanol, "alcohol," *the* alcohol whose use has both aided and devastated human social and economic life at various times in the past nine millennia. Any major brewer with an international "footprint" and each microbrewery set up to diversify beer or wine production in contention with those far-reaching corporations use biotechnologies derived from ancient times, but that expertise is also implicit in the use of ethanol as a serious competitor to gasoline in automobile engines. Hence, the second vision of bioorganic chemists has begun to crystallize; unlocking the vast chemical larder and workshop of natural microbes and plants has required the contributions of microbiologists, microbial physiologists, biochemists, molecular biologists, and chemical, biochemical, and metabolic engineers to invent the technologies required for industrial-scale production of "bioethanol."

The first modern social and economic "experiment" with biofuels — that in Brazil — used the glucose present (as sucrose) in cane sugar to provide a readily available and renewable source of readily fermentable material. The dramatic rise in oil prices in 1973 prompted the Brazilian government to offer tax advantages to those who would power their cars with ethanol as a fuel component; by 1988, 90% of the cars on Brazilian roads could use (to varying extents) ethanol, but the collapse in oil prices then posed serious problems for the use of sugar-derived ethanol. Since then, cars have evolved to incorporate "dual-fuel" engines that can react to fluctuations in the market price of oil, Brazilian ethanol production has risen to more than 16 million liters/year, and by 2006, filling up with ethanol fuel mixes in Brazil cost up to 40% less than gasoline.

Sugarcane thrives in the equatorial climate of Brazil. Further north, in the midwestern United States, corn (maize, *Zea mays*) is a major monoculture crop; corn accumulates starch that can, after hydrolysis to glucose, serve as the substrate for ethanol fermentation. Unlike Brazil, where environmentalists now question the destruction of the Amazonian rain forest to make way for large plantations of sugarcane and soya beans, the Midwest is a mature and established ecosystem with high yields of corn. Cornstarch is a more expensive carbon substrate for bioethanol production, but with tax incentives and oil prices rising dramatically again, the production of ethanol for fuel has become a significant industry. Individual corn-based ethanol production plants have been constructed in North America to produce up to 1 million liters/day, and in China 120,000 liters/day, whereas sugarcane molasses-based facilities have been sited in Africa and elsewhere.[2]

In July 2006, the authoritative journal *Nature Biotechnology* published a cluster of commentaries and articles, as well as a two-page editorial that, perhaps uniquely, directed its scientific readership to consult a highly relevant article ("Ethanol Frenzy") in *Bloomberg Markets*. Much of the discussion centered on the economic viability of fuel ethanol production in the face of fluctuating oil prices, which have inhibited the development of biofuels more than once in the last half century.[3] But does bioethanol production consume more energy than it yields?[4] This argument has raged for years; the contributors to *Nature Biotechnology* were evidently aware of the controversy but drew no firm conclusions. Earlier in 2006, a detailed model-based survey of the economics of corn-derived ethanol production processes concluded that they were viable but that the large-scale use of cellulosic inputs would better meet both energy and environmental goals.[5] Letters to the journal that appeared later in the year reiterated claims that the energy returns on corn ethanol production were so low that its production could only survive if heavily subsidized and, in that scenario, ecological devastation would be inevitable.[6]

Some energy must be expended to produce bioethanol from any source — in much the same way that the pumping of oil from the ground, its shipping around the world, and its refining to produce gasoline involves a relentless chain of energy expenditure. Nevertheless, critics still seek to be persuaded of the overall benefits of fuel ethanol (preferring wind, wave, and hydroelectric sources, as well as hydrogen fuel cells). Meanwhile its advocates cite reduced pollution of the atmosphere, greater use of renewable resources, and erosion of national dependence on oil imports as key factors in the complex overall cost-benefit equation.

To return to the "dream" of cellulose-based chemistry, there is insufficient arable land to sustain crop-based bioethanol production to more than fuel-additive levels worldwide, but cellulosic biomass grows on a massive scale — more than 7×10^{10} tons/year — and much of this is available as agricultural waste ("stalks and stems"), forestry by-products, wastes from the paper industry, and as municipal waste (cardboard, newspapers, etc.).[7] Like starch, cellulose is a polymeric form of glucose; unlike starch, cellulose cannot easily be prepared in a highly purified form from many plant sources. In addition, being a major structural component of plants, cellulose is combined with other polymers of quite different sugar composition (hemicelluloses) and, more importantly, with the more chemically refractive lignin. Sources of lignocellulosic biomass may only contain 55% by weight as fermentable sugars and usually require extensive pretreatment to render them suitable as substrates for any microbial fermentation, but that same mixture of sugars is eminently suitable for the production of structures as complex as aromatic intermediates for the chemical industry.[8]

How practical, therefore, is sourcing lignocellulose for bioethanol production and has biotechnology delivered feasible production platforms, or are major developments still awaited? How competitive is bioethanol without the "special pleading" of tax incentives, state legislation, and (multi)national directives? Ultimately, because the editor of *Nature Biotechnology* noted that, for a few months in 2006, a collection of "A-list" entrepreneurs, venture capitalists, and investment bankers had promised $700 million to ethanol-producing projects, the results of these developments in the real economy may soon refute or confirm the predictions from mathematical models.[9] Fiscal returns, balance sheets, and eco audits will all help to settle the major issues, thus providing an answer to a point made by one of the contributors to the flurry of interest in bioethanol in mid-2006: "biofuels boosters must pursue and promote this conversion to biofuels on its own merits rather than by overhyping the relative political, economic and environmental advantages of biofuels over oil."[10]

Although the production of bioethanol has proved capable of extensive scale up, it may be only the first — and, by no means, the best — of the options offered by the biological sciences. Microbes and plants have far more ingenuity than that deduced from the study of ethanol fermentations. Linking bioethanol production to the synthesis of the bioorganic chemist's palette of chemical feedstocks in "biorefineries" that cascade different types of fermentations, possibly recycling unused inputs and further biotransforming fermentation outputs, may address both financial and environmental problems. Biodiesel (simple alkyl esters of long-chain fatty acids in vegetable oils) is already being perceived as a major fuel source, but further down the technological line, production of hydrogen ("biohydrogen") by light-driven or dark fermentations with a variety of microbes would, as an industrial strategy, be akin to another industrial revolution.[11]

A radically new mind-set and a heightened sense of urgency were introduced in September 2006 when the state of California moved to sue automobile manufacturers over tailpipe emissions adding to atmospheric pollution and global warming. Of the four major arguments adduced in favor of biofuels — long-term availability when fossil fuels become depleted, reduced dependence on oil imports, development of sustainable economies for fuel and transportation needs, and the reduction in greenhouse gas emissions — it is the last of these that has occupied most media attention in the last three years.[12] In October 2006, the first quantitative model of the economic costs of not preventing continued increases in atmospheric CO_2 produced the stark prediction that the costs of simply adapting to the problems posed by global warming (5–20% of annual global GDP by 2050) were markedly higher than those (1% of annual global GDP) required to stabilize atmospheric CO_2.[13] Although developing nations will be particularly hard hit by climate changes, industrialized nations will also suffer economically as, for example, rising sea levels require vastly increased flood defense costs and agricultural systems (in Australia and elsewhere) become marginally productive or collapse entirely.

On a more positive note, the potential market offered to technologies capable of reducing carbon emissions could be worth $500 billion/year by 2050. In other words, while unrestrained increase in greenhouse gas emissions will have severe consequences and risk global economic recession, developing the means to enable a more sustainable global ecosystem would accelerate technological progress and establish major new industrial sectors.

In late 2007, biofueled cars along with electric and hybrid electric–gasoline and (in South America and India) compressed natural gas vehicles represented the only immediately available alternatives to the traditional gasoline/internal combustion engine paradigm. Eventually, electric cars may evolve from a niche market if renewable energy sources expand greatly and, in the longer term, hydrogen fuel cells and solar power (via photovoltaic cells) offer "green" vehicles presently only known as test or concept vehicles. The International Energy Agency estimates that increasing energy demand will require more than $20 trillion of investment before 2030; of that sum, $200 billion will be required for biofuel development and manufacture even if (in the IEA's assessments) the biofuels industry remains a minor contributor to transportation fuels globally.[14] Over the years, the IEA has slowly and grudgingly paid more attention to biofuels, but other international bodies view biofuels (especially the second-generation biofuels derived from biomass sources) as part of the growing family of technically feasible renewable energy sources: together with higher-efficiency aircraft and advanced electric and hybrid vehicles, biomass-derived biofuels are seen as key technologies and practices projected to be in widespread use by 2030 as part of the global effort to mitigate CO_2-associated climate change.[15]

In this highly mobile historical and technological framework, this book aims to analyze in detail the present status and future prospects for biofuels, from ethanol and biodiesel to biotechnological routes to hydrogen ("biohydrogen"). It emphasizes ways biotechnology can improve process economics as well as facilitate sustainable agroindustries and crucial elements of the future bio-based economy, with further innovations required in microbial and plant biotechnology, metabolic engineering, bioreactor design, and the genetic manipulation of new "biomass" species of plants (from softwoods to algae) that may rapidly move up the priority lists of funded research and of white (industrial biotech), blue (marine biotech), and green (environmental biotech) companies.

A landmark publication for alternative fuels was the 1996 publication *Handbook on Bioethanol: Production and Utilization*, edited by Charles E. Wyman of the National Renewable Energy Laboratory (Golden, Colorado). That single-volume, encyclopedic compilation summarized scientific, technological, and economic data and information on biomass-derived ethanol ("bioethanol"). While highlighting both the challenges and opportunities for such a potentially massive production base, the restricted use of the "bio" epithet was unnecessary and one that is now (10 years later) not widely followed.[16] Rather, all biological production routes for ethanol — whether from sugarcane, cornstarch, cellulose ("recycled" materials), lignocellulose ("biomass"), or any other nationally or internationally available plant source — share important features and are converging as individual producers look toward a more efficient utilization of feedstocks; if, for example, sugarcane-derived ethanol facilities begin to exploit the "other" sugars (including lignocellulosic components) present in cane sugar waste for ethanol production rather than only sucrose, does that render the product more "bio" or fully "bioethanol"?

As the first biofuel to emerge into mass production, (bio)ethanol is discussed in chapter 1, the historical sequence being traced briefly from prehistory to the late nineteenth century, the emergence of the petroleum-based automobile industry in the early twentieth century, the intermittent interest since 1900 in ethanol as a fuel, leading to the determined attempts to commercialize ethanol–gasoline blends in Brazil and in the United States after 1973. The narrative then dovetails with that in *Handbook on Bioethanol: Production and Utilization*, when cellulosic and lignocellulosic substrates are considered and when the controversy over calculated energy balances in the production processes for bioethanol, one that continued at least until 2006, is analyzed. Chapters 2, 3, and 4 then cover the biotechnology of ethanol before the economics of bioethanol production are discussed in detail in chapter 5, which considers the questions of minimizing the social and environmental damage that could result from devoting large areas of cultivatable land to producing feedstocks for future biofuels and the sustainability of such new agroindustries.

But are bioethanol and biodiesel (chapter 6) merely transient stopgaps as transportation fuels before more revolutionary developments in fuel cells usher in biohydrogen? Both products now have potential rivals (also discussed in chapter 6). The hydrogen economy is widely seen as providing the only workable solution to meeting global energy supplies and mitigating CO_2 accumulation, and the microbiology of "light" and "dark" biohydrogen processes are covered (along with other equally radical areas of biofuels science) in chapter 7. Finally, in chapter 8, rather than being considered as isolated sources of transportation fuels, the combined production of biofuels and industrial feedstocks to replace eventually dwindling petrochemicals — in "biorefineries" capable of ultimately deriving most, if not all, humanly useful chemicals from photosynthesis and metabolically engineered microbes — rounds the discussion while looking toward attainable future goals for the biotechnologists of energy production in the twenty-first century, who very possibly may be presented with an absolute deadline for success.

For to anticipate the answer to the question that began this preface, there may only be four decades of oil left in the ground. The numerical answer computed for this shorter-term option is approximately 42 years from the present (see Figure 5.13 in chapter 5) — exactly the same as the answer to the ultimate question of the universe (and everything else) presented in the late 1970s by the science fiction writer Douglas Adams (*The Hitchhiker's Guide to the Galaxy*, Pan Books, London). The number is doubly unfortunate: for the world's senior policy makers today, agreement (however timely or belated) on the downward slope of world oil is most likely to occur well after their demise, whereas for the younger members of the global population who might have to face the consequences of inappropriate actions, misguided actions, or inaction, that length of time is unimaginably distant in their own human life cycles.

Four decades is a sufficiently long passage of time for much premier quality scientific research, funding of major programs, and investment of massive amounts of capital in new ventures: the modern biopharmaceutical industry began in the early 1980s from a scattering of research papers and innovation; two decades later, biotech companies like Amgen were dwarfing long-established pharmaceutical multinationals in terms of income stream and intellectual property.

But why (in 2008) write a book? When Jean Ziegler, the United Nations' "independent spokesman on the right to food," described the production of biofuels as a "crime against humanity" and demanded a five-year moratorium on biofuels production so that scientific research could catch up and establish fully the methods for utilizing nonfood crops, he was voicing sentiments that have been gathering like a slowly rising tide for several years.[17] Precisely because the whole topic of biofuels — and especially the diversion of agricultural resources to produce transportation fuels, certainly for industry, but also for private motorists driving vehicles with excellent advertising and finance packages but woefully low energy efficiencies — is so important, social issues inevitably color the science and the application of the derived technology. Since the millennium, and even with rocketing oil prices, media coverage of biofuels has become increasingly negative. Consider the following selection of headlines taken from major media sources with claims to international readerships:

Biofuel: Green Savior or Red Herring? (CNN.com, posted April 2, 2007)

Biofuels: Green Energy or Grim Reaper? (BBC News, London, September 22, 2006)

Scientists Are Taking 2nd Look at Biofuels (*International Herald Tribune*, January 31, 2007)

Green Fuel Threatens a 'Biodiversity Heaven' (*The Times*, London, July 9, 2007)

Biofuel Demand to Push Up Food Prices (*The Guardian*, London, July 5, 2007)

Plantation Ethanol 'Slaves' Freed (*The Independent*, London, July 5, 2007)

The Biofuel Myths (*International Herald Tribune*, July 10, 2007)

Biofuel Gangs Kill for Green Profits (*The Times*, London, June 3, 2007)

Dash for Green Fuel Pushes Up Price of Meat in US (*The Times*, London, April 12, 2007)

The Big Green Fuel Lie (*The Independent*, London, March 5, 2007)

How Biofuels Could Starve the Poor (*Foreign Affairs*, May/June 2007)

Biofuel Plant 'Could Be Anti-Green' (*The Scotsman*, Edinburgh, July 5, 2007)

To Eat … or to Drive? (*The Times*, London, August 25, 2007)

These organizations also carry (or have carried) positive stories about biofuels ("The New Gold Rush: How Farmers Are Set to Fuel America's Future" or "Poison Plant Could Help to Cure the Planet,"[18]) but a more skeptical trend emerged and hardened during 2006 and 2007 as fears of price inflation for staple food crops and other concerns began to crystallize. In the same week in August 2007, New Zealand began its first commercial use of automobile bioethanol, whereas in England, the major long-distance bus operator abandoned its trials of biodiesel, citing environmental damage and unacceptable diversion of food crops as the reasons. On a global ecological basis, plantations for biofuels in tropical regions have begun to be seriously questioned as driving already endangered wildlife species to the edge of oblivion.

Perhaps most damning of all, the "green" credentials of biofuels now face an increasing chorus of disbelief as mathematical modeling erodes the magnitudes of possible benefits of biofuels as factors in attempts to mitigate or even reverse greenhouse gas emissions — at its most dramatic, no biofuel production process may be able to rival the CO_2-absorbing powers of reforestation, returning unneeded croplands to savannah and grasslands.[19] The costs of biofuels escalate, whereas the calculated benefits in reducing greenhouse gas emissions fall.[20] The likely impact of a burgeoning world trade in biofuels — and the subject already of highly vocal complaints about unfair trade practices — on the attainment of environmental goals in the face of economic priorities[21] is beginning to cause political concern, especially in Europe.[22]

But why write a *book*? The Internet age has multiple sources of timely information (including all the above-quoted media stories), regularly updated, and available 24/7. The thousands of available sites offer, however, only fragmentary truths: most are campaigning, selective in the information they offer, focused, funded, targeting, and seeking to persuade audiences or are outlets for the expression of the views and visions of organizations ("interested parties"). Most academic research groups active in biofuels also have agendas: they have intellectual property to sell or license, genetically engineered microbial strains to promote, and results and conclusions to highlight in reviews. This book is an attempt to broaden the discussion, certainly beyond bioethanol and biodiesel, placing biofuels in historical contexts, and expanding the survey to include data, ideas, and bioproducts that have been visited at various times over the last 50 years, a time during which widely volatile oil prices have alternately stimulated and wrecked many programs and initiatives. That half century resulted in a vast library of experience, little of it truly collective (new work always tends to supplant in the biotech mind-set much of what is already in the scientific literature), many claims now irrelevant, but as a body of knowledge, containing valuable concepts sometimes waiting to be rediscovered in times more favorable to bioenergy.

Each chapter contains many references to published articles (both print and electronic); these might best be viewed as akin to Web site links — each offers a potentially large amount of primary information and further links to a nexus of data and ideas. Most of the references cited were peer-reviewed, the remainder edited or with multiple authorships. No source used as a reference requires a personal subscription or purchase — Internet searches reveal many thousands more articles in trade journals and reports downloadable for a credit card payment; rather, the sources itemized can either be found in public, university, or national libraries or are available to download freely. Because the total amount of relevant information is very large, the widest possible quotation basis is required, but (as always with controversial matters) all data and information are subject to widely differing assessments and analyses.

Meanwhile, time passes, and in late 2007, oil prices approached \$100/barrel, and the immediate economic momentum for biofuels shows no signs of slackening. Hard choices remain, however, in the next two decades or, with more optimistic estimates of fossil fuel longevity, sometime before the end of the twenty-first century. Perhaps, the late Douglas Adams had been more of a visionary than anyone fully appreciated when he first dreamed of interstellar transportation systems powered by equal measures of chance and improbability and of an unremarkable, nonprime, two-digit number.

NOTES AND REFERENCES

1. See, for example, the article by Melvin Calvin (who discovered the enzymology of the photosynthetic CO_2 fixation cycle in plants), "Petroleum plantations for fuels and materials" (*Bioscience*, 29, 553, 1979) on "gasoline plants" that produce volatile, highly calorific terpenoids.
2. http://www.vogelbusch.com.
3. Holden, C., Is bioenergy stalled?, *Science*, 227, 1018, 1981.
4. Pimentel, D. and Patzek, T.W., Ethanol production using corn, switchgrass, and wood; biodiesel production using soybean and sunflower, *Nat. Resour. Res.*, 14, 65, 2005.
5. Farrell, A.E., Plevin, R.J., Turner, B.T., Jones, A.D., O'Hare, M., and Kammenet, D.M., Ethanol can contribute to energy and environmental goals, *Science*, 311, 506, 2006.
6. Letters from Cleveland, C.J., Hall, C.A.S., Herendeen, R.A., Kaufmann, R.K., and Patzek, T.W., *Science*, 312, 1746, 2006.
7. Kadam, K.K., Cellulase preparation, in Wyman, C.E. (Ed.), *Handbook on Bioethanol: Production and Utilization*, Taylor & Francis, Washington, DC, 1996, chap. 11.
8. Li, K. and Frost, J.W., Microbial synthesis of 3-dehydroshikimic acid: a comparative analysis of D-xylose, L-arabinose, and D-glucose carbon sources, *Biotechnol. Prog.*, 15, 876, 1999.
9. Bioethanol needs biotech now [editorial], *Nat. Biotechnol.*, 24, 725, July 2006.
10. Herrera, S., Bonkers about biofuels, *Nat. Biotechnol.*, 24, 755, 2006.
11. Vertès, A.L., Inui, M., and Yukawa, H., Implementing biofuels on a global scale, *Nat. Biotechnol.*, 24, 761, 2006.
12. Canola and soya to the rescue, unsigned article in *The Economist*, May 6, 2006.
13. Stern, N., *The Economics of Climate Change*, prepublication edition at http://www.hm-treasury.gov.uk/independent_reviews/stern_review_report.cfm.
14. *World Energy Outlook 2005*, International Energy Agency/Organisation for Economic Co-operation and Development, Paris, 2006.
15. IPCC, *Climate Change 2007: Mitigation Contribution of Working Group III to the Fourth Assessment Report of the Intergovernmental Panel on Climate Change*, Cambridge University Press, Cambridge, UK, 2007, www.ipcc.ch.
16. Leiper, K.A., Schlee, C., Tebble, I., and Stewart, G.G., The fermentation of beet sugar syrup to produce bioethanol, *J. Ins. Brewing*, 112, 122, 2006.
17. Reported in *The Independent*, London, 27 October 2007. The professor of sociology at the University of Geneva also appears seriously behind the times: all the relevant methods have already been thoroughly established, at least in scientific laboratories — see chapters 3 and 4.
18. A newspaper report about *Japtroha* seeds, a candidate for nonfood crop production of biodiesel; ingesting three of the seeds can be lethal.
19. Righelato, R. and Spracklen, D.V., Carbon mitigation by biofuels or by saving and restoring forests?, *Science*, 317, 902, 2007.
20. Biofuels policy costs double, *The Guardian*, London, 10 October, 2007.
21. 2006. –7 Production statistics confirm a strong growth in the EU, but legislation and fair trade improvements are urgently needed to confirm expansion, press release, July 17, 2006, European Biodiesel Board, www.ebb-eu.org.
22. In EU, a shift to foreign sources for "green fuel," *International Herald Tribune*, 29 March, 2006.

Author

David M. Mousdale was educated at Oxford (B.A. in biochemistry, 1974) and Cambridge (Ph.D., 1979). He researched growth control and integration mechanisms in plants and plant cell cultures before turning to enzyme responses to xenobiotics, including the first isolation of a glyphosate-sensitive enzyme from a higher plant.

In the microbial physiology and biochemistry of industrial fermentations, he developed metabolic analysis to analyze changes in producing strains developed by serendipity (i.e., classical strain improvement) or by rational genetic engineering, becoming managing director of beòcarta Ltd. (formerly Bioflux) in 1997. Much of the work of the company initially focused on antibiotics and other secondary metabolites elaborated by *Streptomycetes* but was extended to vitamins and animal cell bioreactors for the manufacture of biopharmaceuticals.

Recent projects have included immunostimulatory polysaccharides of fungal origin, enzyme production for the food industry, enzymes for processing lignocellulose substrates for biorefineries, recycling glycerol from biodiesel manufacture, and the metabolic analysis of marine microbes.

1 Historical Development of Bioethanol as a Fuel

1.1 ETHANOL FROM NEOLITHIC TIMES

There is nothing new about biotechnology. Stated more rigorously, the practical use — if not the formal or intuitive understanding of microbiology — has a very long history, in particular with regard to the production of ethanol (ethyl alcohol). The development of molecular archaeology, that is, the chemical analysis of residues on pottery shards and other artifacts recovered from archaeological strata, has begun to specify discrete chemical compounds as markers for early agricultural, horticultural, and biotechnological activities.[1] Among the remarkable findings of molecular archaeology, put into strict historical context by radiocarbon dating and dendrochronology techniques, as well as archaeobotanical and archaeological approaches, are that

- In western Asia, wine making can be dated as early as 5400–5000 BC at a site in what is today northern Iran and, further south in Iran, at a site from 3500 to 3000 BC.[1]
- In Egypt, predynastic wine production began at approximately 3150 BC, and a royal wine-making industry had been established at the beginning of the Old Kingdom (2700 BC).[2]
- Wild or domesticated grape (*Vitis vinifera* L. subsp. *sylvestris*) can be traced back to before 3000 BC at sites across the western Mediterranean, Egypt, Armenia, and along the valleys of the Tigris and Euphrates rivers. This is similar to the modern distribution of the wild grape (used for 99% of today's wines) from the Adriatic coast, at sites around the Black Sea and southern Caspian Sea, littoral Turkey, the Caucasus and Taurus mountains, Lebanon, and the islands of Cyprus and Crete.[3]
- Partial DNA sequence data identify a yeast similar to the modern *Saccharomyces cerevisiae* as the biological agent used for the production of wine, beer, and bread in Ancient Egypt, ca. 3150 BC.[2]

The occurrence of *V. vinifera* in regions in or bordering on the Fertile Crescent that stretched from Egypt though the western Mediterranean and to the lower reaches of the Tigris and Euphrates is crucial to the understanding of Neolithic wine making. When ripe, grapes supply not only abundant sugar but also other nutrients (organic and inorganic) necessary for rapid microbial fermentations as well as the causative yeasts themselves — usually as "passengers" on the skins of the fruit. Simply crushing ("pressing") grapes initiates the fermentation process, which, in unstirred vessels (i.e., in conditions that soon deplete oxygen levels), produces ethanol at 5–10% by volume (approximately, 50–100 g/l).

In China, molecular archaeological methodologies such as mass spectroscopy and Fourier transform infrared spectrometry have placed "wine" (i.e., a fermented mixture of rice, honey, and grape, as well as, possibly, other fruit) as being produced in an early Neolithic site in Henan Province from 6500 to 7000 BC.[4] Geographically, China lies well outside the accepted natural range of the Eurasian *V. vinifera* grape but is home to many other natural types of grape. Worldwide, the earliest known examples of wine making, separated by more than 2,000 km and occurring between 7000 and 9000 years ago, were probably independent events, perhaps an example on the social scale of the "convergent evolution" well known in biological systems at the genetic level.

The epithet "earliest" is, however, likely to be limited by what physical evidence remains. Before domestication of cereals and the first permanent settlements of *Homo sapiens*, there was a long but unrecorded (except, perhaps, in folk memory) history of hunter-gatherer societies. Grapes have, in some botanical form or other, probably been present in temperate climates for 50 million if not 500 million years.[3] It would seem entirely possible, therefore, that such nomadic "tribes" — which included shamans and/or observant protoscientists — had noted, sampled, and replicated natural fermentations but left nothing for the modern archaeologist to excavate, record, and date. The presently estimated span of wine making during the last 9000 years of human history is probably only a minimum value.

Grape wines, beers from cereals (einkorn wheat, one of the "founder plants" in the Neolithic revolution in agriculture was domesticated in southeastern Turkey, ca. 8000 BC), and alcoholic drinks made from honey, dates, and other fruits grown in the Fertile Crescent are likely to have had ethanol concentrations below 10% by volume. The concentration of the ethanol in such liquids by distillation results in a wide spectrum of potable beverages known collectively as "spirits." The evolution of this chemical technology follows a surprisingly long timeline:[5,6]

- Chinese texts from ca. 1000 BC warn against overindulgence in distilled spirits.
- Whisky (or whiskey) was widely known in Ireland by the time of the Norman invasion of 1170–1172.
- Arnold de Villeneuve, a French chemist, wrote the first treatise on distillation, ca. 1310.
- A comprehensive text on distilling was published in Frankfurt-am-Main (Germany) in 1556.
- The production of brandies by the distillation of grape wines became widespread in France in the seventeenth century.
- The first recorded production of grain spirits in North America was that by the director general of the colony of New Netherland in 1640 (on Staten Island).
- In 1779, 1,152 stills had been registered in Ireland — this number had fallen drastically to 246 by 1790 as illicit "moonshine" pot stills flourished.
- In 1826, a continuously operating still was patented by Robert Stein of Clackmannanshire, Scotland.
- The twin-column distillation apparatus devised by the Irishman Aeneas Coffey was accepted by the Bureau of Excise of the United Kingdom in 1830; this apparatus, with many variations and improvements to the basic design, continues to yield high-proof ethanol (94–96% by volume).

Distillation yields "95% alcohol," a binary azeotrope (a mixture with a constant composition) with a boiling point of 78.15°C. "Absolute" alcohol, prepared by the physical removal of the residual water, has the empirical formula C_2H_6O and molecular weight of 46.07; it is a clear and colorless liquid with a boiling point of 78.5°C and a density (at 20°C) of 0.789 g/mL. Absolute alcohol absorbs water vapor rapidly from the air and is entirely miscible with liquid water. As a chemical known to alchemists and medicinal chemists in Europe and Asia, it found many uses as a solvent for materials insoluble or poorly soluble in water, more recently as a topical antiseptic, and (although pharmacologically highly difficult to dose accurately) as a general anesthetic. For the explicit topic of this volume, however, its key property is its inflammability: absolute alcohol has a flash point of 13°C.[7]

By 1905, ethanol was emerging as the fuel of choice for automobiles among engineers and motorists,* opinion being heavily swayed by fears about oil scarcity, rising gasoline prices, and the monopolistic practices of Standard Oil.[8] Henry Ford planned to use ethanol as the primary fuel for his Model T (introduced in 1908) but soon opted for the less expensive alternative of gasoline, price competition between ethanol and gasoline having proved crucial. The removal of excise duty from denatured ethanol (effective January 1, 1907) came too late to stimulate investment in large-scale ethanol production and develop a distribution infrastructure in what was to prove a narrow window of opportunity for fuel ethanol.[8]

Ford was not alone in considering a variety of possible fuels for internal combustion engines. Rudolf Diesel (who obtained his patent in 1893) developed the first prototypes of the high-compression, thermally efficient engine that still bears his name, with powdered coal in mind (a commodity that was both cheap and readily available in nineteenth-century Germany). Via kerosene, he later arrived at the use of crude oil fractions, the marked variability of which later caused immense practical difficulties in the initial commercialization of diesel engines.[9] The modern oil industry had, in effect, already begun in Titusville, Pennsylvania, in the summer of 1859, with a drilled extraction rate of 30 barrels a day, equivalent to a daily income of $600.[10] By 1888, Tsarist Russia had allowed Western European entrepreneurs to open up oil fields in Baku (in modern Azerbaijan) with a productive capacity of 50,000 barrels a day. On January 10, 1901, the Spindletop well in Texas began gushing, reaching a maximum flow of 62,000 barrels a day. Immediately before the outbreak of World War I, the main oil-producing countries could achieve outputs of more than 51 million tons/year, or 1 million barrels a day. In 1902, 20,000 vehicles drove along American roads, but this number had reached more than a million by 1912. These changes were highly welcome to oil producers, including (at least, until its forced breakup in 1911) the Standard Oil conglomerate: kerosene intended for lighting domestic homes had been a major use of oil but, from the turn of the century, electricity had increasingly become both available and preferable (or fashionable). The rapid growth in demand for gasoline was a vast new market for J.D. Rockefeller's "lost" oil companies.

Greatly aiding the industry's change of tack was the dominance of U.S. domestic production of oil: in 1913, the oil produced in the United States amounted to more

---

* The Automobile Club of America sponsored a competition for alcohol-powered vehicles in 1906.

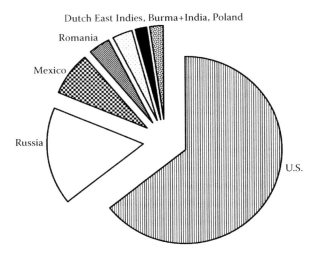

FIGURE 1.1 Geographical breakdown of world oil production in 1913. (Data from Tugendhat and Hamilton.[10])

than 60% of the worldwide total (figure 1.1). The proximity within national boundaries of the world's largest production line for automobiles (in Detroit) and oil refining capacities firmly cast the die for the remainder of the twentieth century and led to the emergence of oil exploration, extraction, and processing, and the related petrochemical industry as the dominant features of the interlinked global energy and industrial feedstock markets.

Nevertheless, Henry Ford continued his interest in alternative fuels, sponsoring conferences on the industrial uses of agricultural mass products (grain, soybeans, etc.) in 1935–1937; the Model A was often equipped with an adjustable carburetor designed to allow the use of gasoline, alcohol, or a mixture of the two.[11]

1.2   ETHANOL AND AUTOMOBILES, FROM HENRY FORD TO BRAZIL

Many commentators state that the Oil Crisis of 1973, after the Yom Kippur War, catalyzed the interest in and then sustained the development of biofuels on the national and international stages. This is an overly simplistic analysis. The following words were spoken by Senator Hubert Humphrey in May 1973, some five months before war in the Middle East broke out:[12]

> I have called these hearings because ... we are concerned about what is going on with gasoline; indeed, the entire problem of energy and what is called the fuel crisis. Gas prices are already increasing sharply and, according to what we hear, they may go much higher. ... We were saved from a catastrophe in the Midwest — Wisconsin, Iowa and Minnesota — and in other parts of the country, by the forces of nature and divine providence. We had one of the mildest winters in the past 25 years, and had it not been for the unusually warm weather, we would have had to close schools and factories, we would have had to shut down railroads, and we would have had to limit our use of electrical power.

Security of oil supplies and the pressures of price inflation have, since the 1970s, been major issues that continue to the present day.

Even a cursory glance at figure 1.1 will show how disadvantaged were the German, Austro-Hungarian, and Ottoman empires in comparison with the Allied powers in World War I, especially after the entry of the United States in 1917, with only Polish and some Romanian oil fields beyond the vagaries of naval blockade and interception; the ingenuity of the German chemical industry was severely stretched by the effort to substitute imports (including fuel oils) by innovations with synthetic, ersatz products. Since then, and throughout the twentieth and early twenty-first centuries, any state entering into global or regional wars faces the same strategic imperatives: how to ensure continued oil supplies and how (if possible) to control access to them. From the naval blockades of 1914 to the air strikes of the 2006 Hezbollah-Israel conflict, oil refineries and storage tanks are to be targeted, sea-lanes interdicted, and, if possible, foreign oil fields secured by invasion. In those 90 years, wars and economic depressions often demanded attempts to substitute ethanol for gasoline. In the 1920s and 1930s, several countries (Argentina, Australia, Cuba, Japan, New Zealand, the Philippines, South Africa, and Sweden) used ethanol blends in gasoline; alcohol-fueled vehicles became predominant in Germany during World War II and, by 1944, the U.S. Army had developed a nascent biomass-derived alcohol industry.[11] Such programs were, however, mostly of a contingency (or emergency) nature, highly subsidized, and, once oil began flowing in increasingly large amounts after 1945, generally abandoned.

In the decade immediately preceding 1973, the United States had lost its dominance of world oil production (figure 1.2). Other major players were expanding (e.g., the Middle East reached 30% of world oil production) and new producers were appearing: Africa (Libya, Algeria, and Nigeria) already produced 13% of world oil.[13] Allowing for inflation, world oil prices slowly decreased throughout the 1960s (figure 1.3). At the time, this was perceived as a "natural" response to increasing oil production, especially with relative newcomers such as Libya and Nigeria contributing significantly; global production after World War II followed an exponential rate of increase (figure 1.2). Political changes (especially those in Libya) and a growing cooperation between oil-producing states in the Organization of Petroleum Exporting Countries (OPEC) and the Organization of Arab Petroleum Exporting Countries (OAPEC) led to new agreements between oil producers and oil companies being negotiated in Tehran (Iran) and Tripoli (Libya) in 1970 and 1971, which reversed the real oil price erosion.

Then, Libya and Kuwait began to significantly reduce oil output in a structured, deliberate manner. In Libya, average production was reduced from a peak of 3.6 million barrels/day before June 1970 to approximately 2.2 million barrels/day in 1972 and early 1973; the Kuwaiti government enforced a ceiling of 3 million barrels/day in early 1972, shifting down from peak production of 3.8 million barrels/day.[10] Structural imbalances in the global supply of oil had by that time become apparent because of short- and medium-term causes:

- High demand for oil exceeded predictions in 1970.
- The continued closure of the Suez Canal after the 1967 war between Israel and Egypt was confounded by a shortage of tanker tonnage for the much longer voyage around South Africa.

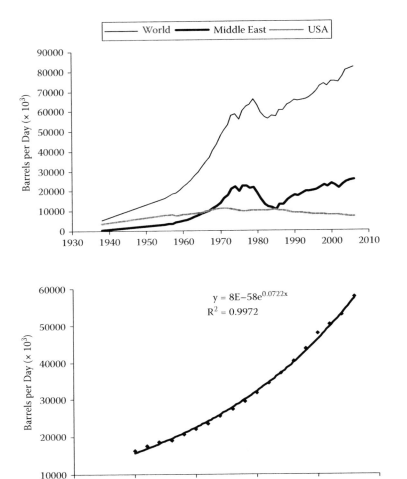

FIGURE 1.2 Oil production. (Data from *BP Statistical Review of World Energy.*[20])

- Accidental damage resulted in the prolonged closure of the pipeline carrying oil from Saudi Arabia to the Mediterranean.
- Supply and demand became very much more closely matched, imposing acute pressures on shipping and refinery kinetics; the estimated spare capacity in crude oil shrank from 7 million barrels/day in 1965 to less than 0.5 million barrels/day in early 1973.

A rapid response to the outbreak of war in October 1973 continued the politically motivated reduction in crude oil output: OAPEC proposed with immediate effect to cut back output by 5% with a further 5% each month until a settlement in accord with United Nations resolutions was effected. In addition, the Gulf States of OPEC, together with Iran, imposed unilateral price rises of up to 100%. The immediate effect on world oil prices was severe (figure 1.3). More importantly, however, the effect was

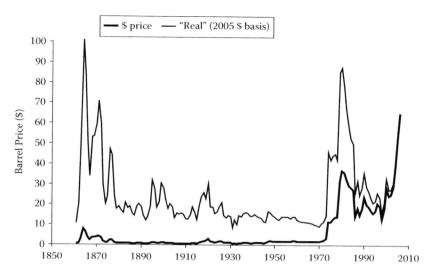

FIGURE 1.3   Historical oil price. (Data from *BP Statistical Review of World Energy*.[20])

not transitory: although prices decreased from the initial peaks in 1973–1974, prices began a second wave of rapid increase in 1979 after the Iranian revolution, to reach a new maximum in 1981. From more than $50 a barrel in 1981, prices then confounded industry analysts again, despite the subsequent conflict between Iran and Iraq, and crashed down to $20 by the late 1980s, but for over a decade real oil prices had been continuously threefold more (or greater) than those paid in 1970. Although not reaching the real prices recorded in the 1860s during the American Civil War (when industrialization was a new phenomenon for most of the world), the oil price inflation between 1973 and 1981 represented a markedly different scenario from any experienced during the twentieth century — in dollar or real terms — despite world wars and major depressions (figure 1.3).

Across the industrially developed states of the Organisation for Economic Cooperation and Development (OECD) — the United States, Japan, Germany, France, United Kingdom, Italy, and Canada — while the real price of imported crude oil had decreased between 1960 and 1973 by an average of 1%/annum, the inflation-adjusted price increased by 24.5%/annum between 1973 and 1980; the result was that the oil crisis soon developed into a deep economic crisis even in those economically and technically advanced OECD nations.[14] Because gasoline prices were "buffered" by the (frequently high) taxes included in the at-pump prices in the OECD countries, gasoline prices to motorists increased by only two- to threefold between 1970 and 1980, whereas crude oil prices rose by more than eightfold; in contrast, industrial and domestic oil prices increased by approximately fivefold.[14]

Furthermore, viewed from the perspective of 1973, the future for oil supplies to net oil importers was highly problematic. Although known oil reserves amounted to 88×10^9 tons, more than 55% of these lay in the Middle East, and mostly in OAPEC countries (figure 1.4). In the days of the then-Cold War, the Soviet Union (USSR), Eastern Europe, and China accounted for only 16.3% of world oil production but

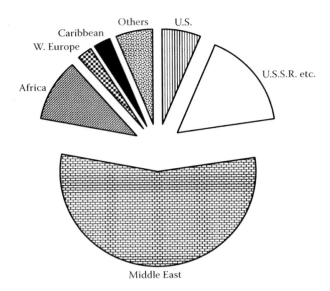

FIGURE 1.4   Known oil reserves at the end of 1973. (Data from *BP Statistical Review of the World Oil Industry, 1973.*[15])

were net exporters of both crude oil and oil products, whereas the United States had become a net importer of both (figure 1.5). In the United States, oil represented 47% of total primary energy consumption.[15] In other OECD countries, the dependence on oil was even more marked: 64% in Western Europe and 80% in Japan. The developed economies of the OECD countries responded to the oil price "shocks" of the 1970s by becoming more oil-efficient: while total OECD gross domestic product (GDP) increased by 19% between 1973 and 1980, total oil imports fell by 14%, and the oil used to produce each unit of GDP fell by 20% — to offset the reduced use of oil, however, coal and (especially) nuclear energy source utilization increased greatly.[16] Energy conservation became a priority ("energy-demand management" measures), and technologies for the improved efficiency of energy use were much developed, advertised, and retrofitted to both domestic and industrial premises. "Fuel switching" was much less obvious in the strategies adopted by OECD countries. While the substitution of gasoline for road transport by alcohol, liquefied gas, and so forth was widely advocated, by 1980, Canada was unique in having adopted a comprehensive policy (the "off oil conversion programme") covering all aspects of oil use and providing oil reduction targets as well as financial incentives.

For an "emerging" economy like Brazil's, the economic dislocation posed by sustained oil price rises was potentially catastrophic. In November 1973, Brazil relied on imports for more than 80% of the country's oil consumption; in the course of the following year, the total import bill rose from $6.2 billion to $12.6 billion, and the trade balance collapsed (figure 1.6). For the preceding decade, the Brazilian economy had enjoyed high growth rates (figure 1.6). Industrialization had proceeded well, and the inflation rate had reached its lowest level since the 1950s.[17] The Brazilian government opted against economic stagnation; rather, it aimed to pay

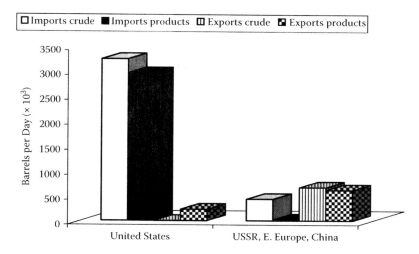

FIGURE 1.5   Oil imports/exports 1973. (Data from *BP Statistical Review of the World Oil Industry, 1973.*[15])

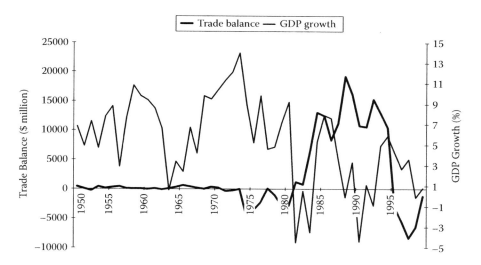

FIGURE 1.6   Brazil's economy 1950–1999. (Data from Baer.[17])

for the higher oil bills by achieving continued growth. To meet the challenges of energy costs, the Second National Development Plan (1975–1979) decreed the rapid expansion of indigenous energy infrastructure (hydroelectricity) as well as nuclear power and alcohol production as a major means of import substituting for gasoline.

In the next decades, some of these macroeconomic targets were successfully realized. Growth rates were generally positive after 1973, and historically massive positive trade balances were recorded between 1981 and 1994. The counterindicators were, however, renewed high rates of inflation (reaching >100%/annum by 1980) and a spiral of international debt to fund developmental programs that made Brazil the third world's largest debtor nation and resulted in a debt crisis in the early 1980s. Arguments continue concerning the perceived beneficial and detrimental effects of the costs of developmental programs on political, social, and environmental indices in Brazil.[17]

Cane sugar was the key substrate and input for Brazil's national fuel alcohol program. Sucrose production from sugarcane (*Saccharum* sp.) in Brazil has a long history, from its days as a colony of Portugal. Brazil had become the world's leading sugar supplier by the early seventeenth century, but sugar production was based initially on slave labor and remained (even in the twentieth century) inefficient. This, however, represented a potential for rapid growth after 1975 because large monoculture plantations had been long established in the coastal regions of the northeast and southeast of the country. Expansion of cultivated land was greatly encouraged for the "modern" export crops — sugarcane, cotton, rice, corn, soybeans, and wheat — at the expense of the more traditional crops, including manioc, bananas, peanuts, and coffee. Sugarcane cultivation increased by 143% between 1970 and 1989 when expressed as land use, but production increased by 229% as Brazil's historically low use of fertilizer began to be reversed.[17]

Brazil is also the southernmost producer of rum as an alcoholic spirit, but *cachaça* is the oldest and most widely consumed national spirit beverage, with a yearly production of *ca.* 1.3 billion liters.[18] The primary fermentation for *cachaça* uses sugarcane juice, and large industrial plants had been established after the end of World War II; a variety of yeasts had been developed, suitable for continuous or discontinuous fermentations, the former reusing and recycling the yeast cells.[18] Before distillation, the fermentation is (as are all traditional potable alcohol processes) allowed to become quiescent, the yeast cells settling and then being removed (along with other residual solids) by, in technologically more advanced facilities, centrifugation; batch ("pot still") and continuous distillation are both used, and final alcohol concentrations are in the 38 to 48% range (by volume). Predating the oil crises of the 1970s and 1980s, the first moves toward using cane sugar as a substrate for industrial ethanol production independent of beverages dated from 1930, when the Sugar and Alcohol Institute (Instituto do Açúcar e do Álcool) was set up; in 1931, a decree imposed the compulsory addition of 5% ethanol to gasoline, and the blending was increased to 10% in 1936. Four decades of experience had, therefore, been garnered in Brazil before fuel substitution became a priority on the political agenda.[19]

The final element in Brazil's developing strategy to produce "gasohol" was, ironically, petroleum itself. Brazil had produced oil at a low rate from at least 1955, but the offshore deposits discovered by the state-owned company PETROBRÁS were so large that by 1998 domestic oil production equaled 69% of domestic consumption.[17] Production continued to increase (figure 1.7), and by 2005, Brazil had become a significant global producer, accounting for 2.2% of world oil production, equivalent to that of the United Kingdom, considerably higher than either Malaysia or India (both 0.9%) and approaching half that of China (4.6%).[20] Indigenous refining capacity also increased during the 1970s and again after 1996 (figure 1.7). The ability to produce alcohol as a fuel or (when mixed with gasoline) as a fuel additive became — if need be, at an unquantified ecological cost (chapter 5, section 5.5.3) — an ongoing feature of Brazilian economic life.

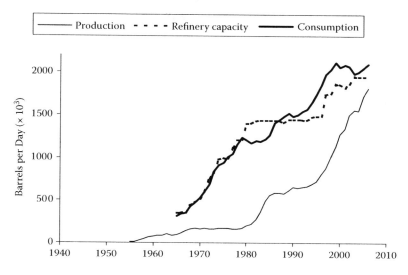

FIGURE 1.7   The Brazilian oil economy up to 2006. (Data from *BP Statistical Review of World Energy*.[20])

1.3   ETHANOL AS A TRANSPORTATION FUEL AND ADDITIVE: ECONOMICS AND ACHIEVEMENTS

As a volatile chemical compound viewed as a gasoline substitute, pure ethanol has one major drawback. Internal combustion engines burn fuels; ethanol, in comparison with the typical hydrocarbon components of refined oils, is more oxygenated, and its combustion in oxygen generates less energy compared with either a pure hydrocarbon or a typical gasoline (table 1.1). This is not mitigated by the higher density of ethanol because liquid volumes are dispensed volumetrically and higher weights in fuel tanks represent higher loads in moving vehicles; a gallon of ethanol contains, therefore, only 70% of the energy capacity of a gallon of gasoline.[11,21] A review of the relative merits of alternative fuels in 1996 pointed out that ethanol not only had a higher octane number (leading to higher engine efficiencies) but also generated an increased volume of combustion products (gases) per energy unit burned; these factors in optimized ethanol engines significantly eroded the differential advantages of gasoline.[21] Similar arguments could not be extended to a comparison between ethanol and diesel fuel, and ethanol had only 58 to 59% of the energy (net heat of combustion) of the latter.[21]

The high miscibility of ethanol and refined oil products allows a more conservative option, that is, the use of low-ethanol additions to standard gasoline (e.g., E10: 90% gasoline, 10% ethanol) and requires no modifications to standard gasoline-burning vehicles. Dedicated ethanol-fueled cars were, however, the initial favorite of the Brazilian Alcohol Program (PROÁLCOOL); sales of alcohol-powered vehicles reached 96% of total sales in 1980 and more than 4 million such vehicles were estimated to be in the alcohol "fleet" by 1989.[22] Such high market penetration was not, however, maintained, and sales of alcohol-powered vehicles had almost ceased by 1996 (figure 1.8). The major reason for this reversal of fortune for ethanol-fueled

TABLE 1.1
Energy Parameters for Ethanol, Isooctane, Gasoline, and Diesel

	Ethanol	Isooctane	Gasoline	Diesel
Density, lb/gal	6.6	5.8	6.25	7.05
Net heat of combustion, Btu ($\times 10^3$)/gal	75.7–76.0	110.5–119.1	109.0–119.0	128.7–130
Octane number (mean of research and motor octane numbers)	104.5		90.5	
Octane number (research octane number)	106	100		

Source: Data from Cheremisinoff[11] and Bailey.[21]

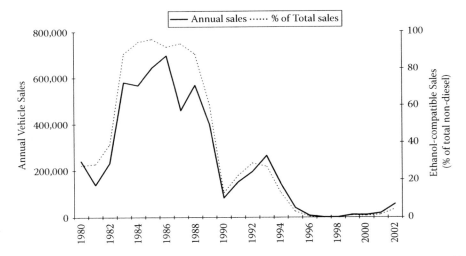

FIGURE 1.8   Ethanol-compatible vehicles in Brazil, 1980–1996. (Data from ANFAVEA[24] and Melges de Andrade et al.[22])

vehicles was the collapse in oil prices during the late 1980s and 1990s — by 1998, the real price of crude oil was very similar to that before November 1973 (figure 1.3). Ethanol production from sugarcane in Brazil increased from a low and declining production level in early 1972, by nearly 20-fold by 1986, and then continued to increase (although at a greatly reduced rate) until 1998 (figure 1.9). The government responded to the novel "crisis" of the competing ethanol-gasoline market in several ways[23]:

- In 1990, the Instituto do Açúcar e do Álcool, the body through which governmental policy for ethanol production had been exercised, was abolished.
- In 1993, a law was passed that all gasoline sold in Brazil would have a minimum of 20% ethanol by volume.

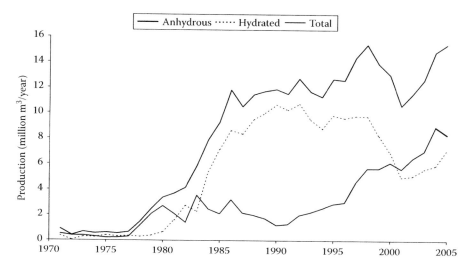

FIGURE 1.9 Ethanol production in Brazil after 1970. (Data from UNICA.[25])

- Prices of sugarcane and ethanol were deregulated as of January 1, 1997.
- Tariffs on sugar exports were abolished in 1997.
- In January 2006, the tax rate for gasoline was set to be 58% higher than that for hydrated ethanol (93% ethanol, 7% water), and tax rates were made advantageous for any blend of gasoline and anhydrous ethanol with ethanol contents of more than 13%.

Brazilian automobile producers introduced truly flexible-fuel vehicles (FFVs) in 2003, with engines capable of being powered by gasoline, 93% aqueous ethanol, or by a blend of gasoline and anhydrous ethanol.[24] In 2004, "flex-fuel" cars sold in Brazil were 16% of the total market, but during 2005, sales of FFVs overtook those of conventional gasoline vehicles (figure 1.10). This was a very "prescient" development as crude oil prices, which had been only slowly increasing during 2003 and early 2004, surged to new dollar highs in 2005 (figure 1.3). Domestic demand for ethanol-containing fuels became so great that the ethanol percentage was reduced from 25% to 20% in March 2006; this occurred despite the increased production of anhydrous ethanol for blending.[25] Brazil had evolved a competitive, consumer-led dual-fuel economy where motorists made rational choices based on the relative prices of gasoline, ethanol, and blends; astute consumers have been observed to buy ethanol only when the pump price is 30% below gasoline blends — equal volumes of ethanol and gasoline are still, as noted above, divergent on their total energy (and, therefore, mileage) equivalents.

Other pertinent statistics collected for Brazil for 2004–2006 are the following:[23]

- In 2004–2005, Brazil was the world's largest producer of ethanol, with 37% of the total, that is, 4.5 billion gallons.
- Brazil exported 15% of its total ethanol production in 2005.

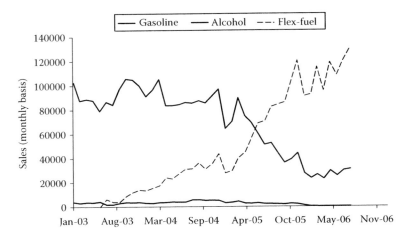

FIGURE 1.10   Sales of flexibly fueled vehicles in Brazil. (Data from ANFAVEA.[24])

- Real prices for ethanol in Brazil decreased by two-thirds between 1973 and 2006.
- São Paolo state became the dominant contributor to national ethanol production and PETROBRÁS began the construction of a 1,000-mile pipeline from the rural interior of the state to the coast for export purposes.

A significant contraindicator is that ethanol-compatible vehicles still remain a minority of the total on Brazilian roads: in 1997, before FFVs became available, ethanol-compatible vehicles were only 21% of the total of ca. 15 million.[22] The introduction of FFVs in 2005 is expected to gradually improve this ratio (figure 1.10).

Another predictable but little emphasized problem is that improvements to sugarcane harvesting methods have lead to the unemployment of 8% of seasonal sugarcane workers.[23] Since 1998, Brazil has restricted the traditional practice of burning sugarcane crops (to eliminate the leaves) before manual harvesting in favor of the mechanical harvesting of green canes.[19] Although far from straightforward (because the lack of burning requires changes in pest management), this change in agricultural practice has contributed to a growing surplus of energy from sugar/alcohol plants as electricity generated on-site and offered to the distribution grids.[19,22]

Any overall cost-benefit of Brazil's 30-year experience of ethanol as a biofuel is inevitably colored by the exact time point at which such an assessment is made. In April 2006, crude oil prices exceeded $70, and this price was exceeded during the summer of 2006, with crude trading briefly at $78/barrel (figure 1.11). Although the emphasis on oil prices may be perceived as one-dimensional,[26] it undeniably focuses attention on real historic events, especially those on a short time scale that may, if not counterbalanced by government action and/or fiscal policies, determine the success of embryonic attempts at oil/gasoline substitution — as evidenced (negatively) by the 1990s in Brazil (figure 1.8). A survey published in 2005 by

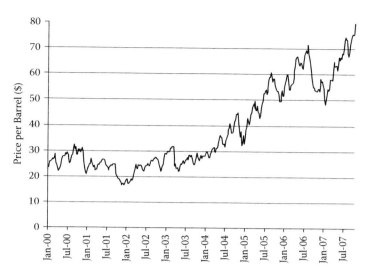

FIGURE 1.11 Crude oil price post-2000. (Data from U.S. Department of Energy, Energy Information Administration.)

Brazilian authors summarized many official statistics and Portuguese-language publications; the major impact factors claimed for fuel ethanol production in Brazil were the following:[27]

- After 1975, fuel ethanol substituted for 240 billion liters of gasoline, equivalent to $56 billion in direct importation costs and $94 billion if costs of international debt servicing are included — after 2004, the severe increases in oil prices clearly acted to augment the benefits of oil substitution (figure 1.11).
- The sugar/ethanol sector presented 3.5% of the gross national product and had a gross turnover of $12 billion, employed (directly and indirectly) 3.6 million people, and contributed $1.5 billion in taxation revenues; approximately half of the total sugarcane grown in Brazil in 2003 was dedicated to ethanol production.
- In 2004, sugarcane production required 5.6 million hectares and represented only 8.6% of the total harvested land, but more than 120 million of low-productivity pasture, natural pastures, and low-density savannas could be dedicated to sugarcane production for ethanol, with a potential ethanol yield of more than 300 billion liters/year.

Ethanol became a major exported commodity from Brazil between 1998 and 2005; exports of ethanol increased by more than 17-fold, whereas sugar exports increased by less than twofold, although price volatility has been evident with both commodities (figure 1.12).[25] As a report for the International Bank for Reconstruction and Development and World Bank (first published in October 2005) noted, average

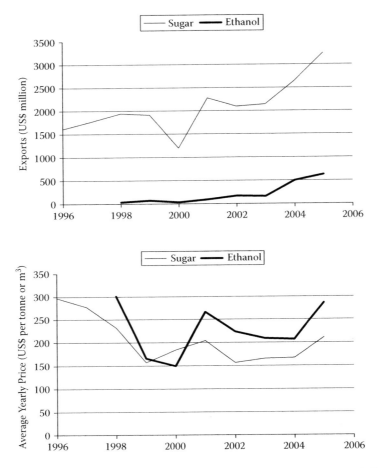

FIGURE 1.12 Exports of sugar and ethanol from Brazil, 1996–2005. (Data from UNICA.[25])

wages in the sugar-ethanol sector are higher than the mean for all sectors in Brazil.[28] As a source of employment, sugarcane ethanol production directly employs more than 1 million people and is far more labor-intensive than the petrochemical industry: 152 times more jobs are estimated to have been created than would have been the case from an equivalent amount of petroleum products.[29]

Despite the apparent vibrancy of ethanol production in Brazil, ethanol use amounts to only 20 to 30% of all liquid fuels sold in Brazil.[27] True levels of subsidies remain difficult to accurately assess; for example, public loans and state-guaranteed private bank loans were estimated to have generated unpaid debts of $2.5 billion to the Banco do Brasil alone by 1997.[28] The ban on diesel-powered cars has also artificially increased fuel prices because diesel prices have been generally lower than gasohol blend prices.[27,28] PROÁLCOOL had invested $11 billion before 2005 but, by that time, could claim to have saved $11 billion in oil imports.[29]

Viewed from the perspectives of fermentation technology and biochemical engineering, ethanol production in Brazil improved after 1975; fermentation productivity (cubic meters of ethanol per cubic meter of fermentation tank capacity volume per day) increased by 130% between 1975 and 2000.[27] This was because of continuous incremental developments and innovations; no reports of radically new fermentor designs in Brazil have been published (although very large fermentors, up to 2 million liters in capacity, are used), and ethanol concentrations in batch fermentations are in the 6 to 12% (v/v) range; the control of bacterial infection of fermentations has been of paramount importance, and selection of robust wild strains of the yeast *S. cerevisiae* has systematized the traditional experience that wild strains frequently overgrow "laboratory" starter cultures.[30] The use of flocculent yeast strains and the adoption of continuous cultivation (chapter 4, section 4.4.1) have also been technologies adopted in Brazil in response to the increased production of sugarcane ethanol.[31] Technical development of downstream technologies have been made in the largest Brazilian provider of distillation plants (Dedini S/A Indústrias de Base, www.dedini.com.br): conventional (bubble cap trays), sieve tray, and azeotropic distillation methods/dehydration (cyclohexane, monoethylene glycol, and molecular sieving) processes operate at more than 800 sites — up from 327 sites before 2000.[31]

On a longer-term basis, genomic analysis of sugarcane promises to identify plant genes for programs to improve sugar plant growth and productivity by genetic engineering.[32,33]

1.4 STARCH AS A CARBON SUBSTRATE FOR BIOETHANOL PRODUCTION

If ethanol production in Brazil exemplified the extrapolation of a mature technology for sugar-based fermentation and subsequent distillation, the development of the second major ethanol fuel market — from corn in the United States — adopted a different approach to alcohol production, adapting and developing that employing starchy seeds in the production of malt and grain spirits (bourbon, rye, whiskey, whisky, etc.). The biological difference from sugar-based ethanol fermentations lies in the carbon substrate, that is, starch glucan polymers (figure 1.13). Historically, seeds and grains have been partially germinated by brewers to generate the enzymes capable of depolymerizing "storage" polysaccharides. With whisky, for example, barley (*Hordeum vulgare* L.) seeds are germinated and specialized cells in the seed produce hydrolytic enzymes for the degradation of polysaccharides, cell walls, and proteins; the "malted" barley can be used as a source of enzyme activities to break down the components of starch in cooked cereals (e.g., maize [*Zea mays* L.]) solubilized in sequential hot-water extractions (which are combined before the yeast cells are added) but not sterilized so as to maintain the enzyme activities into the fermentation stage.[34,35] Starch is usually a mixture of linear (amylose) and branched (amylopectin) polyglucans. For starch hydrolysis, the key enzyme is α-amylase, active on α-1,4 but not α-1,6 linkages (in amylopectin); consequently, amylose is broken down to maltose and maltotriose and (on prolonged incubation) to free glucose and maltose, but amylopectin is only reduced to a mixture of maltose, glucose, and

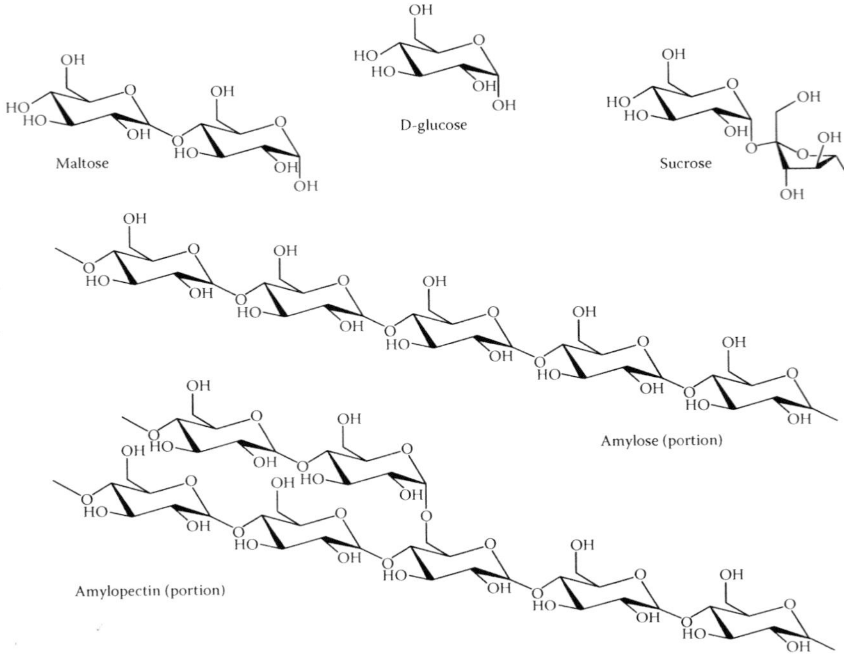

FIGURE 1.13   Chemical structures of glucose, disaccharides, and components of starch.

oligosaccharides containing α-1,6-linked glucose residues, thus limiting the amount of fermentable sugars liberated (figure 1.14). Cereal-based ethanol production plants use the same biochemical operations but replace malted grains with α-amylase and other polysaccharide-degrading enzymes added as purified products.

For much of the twentieth century, ethanol production as a feedstock in the formation of a large number of chemical intermediates and products was dominated in the United States by synthetic routes from ethylene as a product of the petrochemical industry, reaching 8.8×10^5 tonnes/year in 1970.[36] The oil price shocks of the early 1970s certainly focused attention on ethanol as an "extender" to gasoline, but a mix of legislation and economic initiatives starting in the 1970s was required to engender a large-scale bioprocessing industry; in particular, three federal environmental regulations were important:[37,38]

- The 1970 Clean Air Act (amended in 1977 and 1990) began the requirement for cleaner burning gasoline and (eventually) the mandatory inclusion of "oxygenates," that is, oxygen-rich additives.
- The 1988 Alternative Motor Fuels Act promoted the development of ethanol and other alternative fuels and alternative-fuel vehicles (AFVs).
- The 1992 Energy Policy Act defined a broad range of alternative fuels but, more urgently, required that the federal vehicle fleet include an increasing number of AFVs and that they be powered by domestically produced alternative fuels.

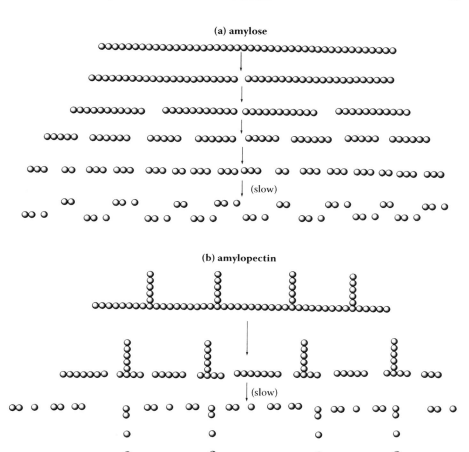

FIGURE 1.14 Prolonged degradation of starch components by amylase.

Although ethanol was always a good oxygenate candidate for gasoline, the compound first approved by the Environmental Protection Agency was methyl tertiary butyl ether (MTBE), a petrochemical industry product. Use of MTBE increased until 1999, but reports then appeared of environmental pollution incidents caused by MTBE spillage; state bans on MTBE came into force during 2002,[39] and its consumption began to decline (figure 1.15). In the Midwest, ethanol was by then established as a corn-derived, value-added product; when the tide turned against MTBE use, ethanol production increased rapidly after showing little sustained growth for most of the 1990s (figure 1.16). California, New York, and Connecticut switched from MTBE to ethanol in 2004; after 2006, with many refiners discontinuing MTBE use, U.S. ethanol demand was expected to expand considerably.[40] In the seven years after January 1999, the number of ethanol refineries in the United States nearly doubled, and production capacity increased by 2.5-fold (figure 1.17). In 2005, the United States became the largest ethanol producer nation; Brazil and the United States accounted for 70% of global production, and apart from China, India,

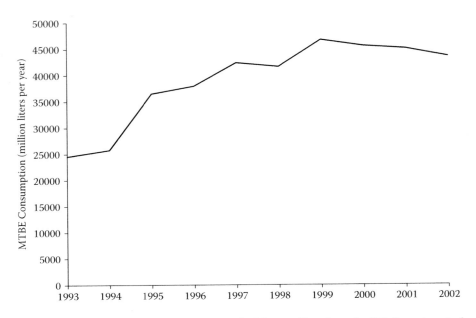

FIGURE 1.15   MTBE consumption in the United States. (Data from the U.S. Department of Energy, Energy Information Administration.)

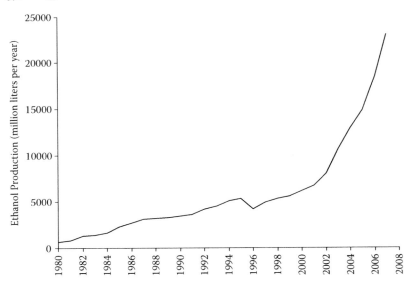

FIGURE 1.16   Ethanol production in the United States. (Data from the Renewable Fuels Association, including a projected figure for 2007.)

France, and Russia, no other nation accounted for more than 1% of the total ethanol produced. To further underline the perceived contribution of renewable fuels to national energy use, the 2005 Energy Policy Act created a Renewable Fuels Standard that envisioned renewable fuel use increasing from 4 billion gallons/year in 2006 to

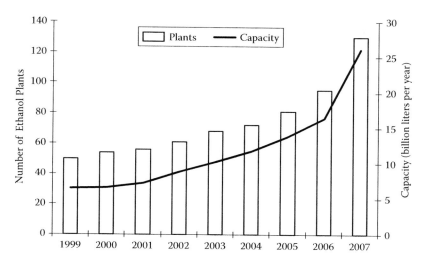

FIGURE 1.17 Growth of U.S. ethanol production January 1999–July 2007. (Data from the Renewable Fuels Association.)

7.5 billion gallons/year in 2012. This implies a further expansion of ethanol production because of the dominant position of E85 (85% ethanol, 15% gasoline) vehicles in the AFV and hybrid-fuel marketplace (figure 1.18).

Fuel ethanol production in the United States has been almost exclusively from corn, although sorghum (*Sorghum bicolor* L.), barley, wheat, cheese whey, and brewery waste have made small contributions. A detailed study of sugar sources for ethanol production concluded that only sugarcane molasses offered competitive feedstock and processing costs to established corn-based technologies (figure 1.19), although annual capital cost investments could be comparable for corn, sugarcane, and sugarbeet molasses and juice as rival feedstocks.[41] Corn ethanol production developed from wet milling of corn; data compiled in the mid-1990s indicates that more than 70% of the large ethanol facilities then used wet milling.[38] Wet milling, schematized in figure 1.20, produces four important liquid or solid by-products:

- Corn steep liquor (a lactic acid bacterial fermentation product, starting from ca. 5% of the total dry weight of the grain extracted with warm water, with uses in the fermentation industry as a nitrogen source)[42]
- Corn oil (with industrial and domestic markets)
- Corn gluten feed (a low-value animal feed)
- Corn gluten meal (a higher-value, high-protein animal feed)

Together with the possibility of collecting CO_2 from the fermentation step as a saleable commodity, this multiplicity of products gave wet milling flexibility in times of variable input and output prices, although requiring a higher initial capital investment.[38] Other sources of flexibility and variation in the wet milling procedure arise at the starch processing stage; while α-amylase is used to liquefy the starch, saccharification (using glucoamylase) can be differently controlled, at one extreme producing

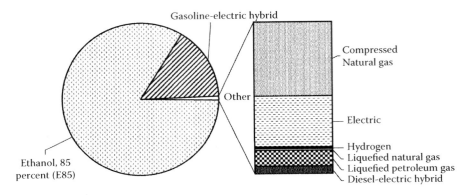

FIGURE 1.18 Alternative- and hybrid-fuel vehicles. (Data from U.S. Department of Energy, Energy Information Administration.)

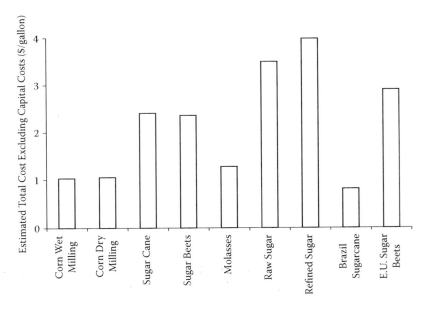

FIGURE 1.19 Estimated ethanol production costs. (Data from Shapouri et al.[41])

a high-glucose, low-solids substrate for fermentation and at the other producing a low glucose concentration but which is continually replenished during the fermentation by the ongoing activity of the glucoamylase in the broth.

In contrast to wet milling, dry milling produces only CO_2 and distillers dried grains with solubles (DDGS) as by-products but has become the favored approach for corn ethanol production because of lower start-up costs.[43] Dry milling should conserve more of the nutrients for yeast growth in the fermentation step — in particular, nitrogenous inputs (free amino acids, peptides, and protein), inorganic

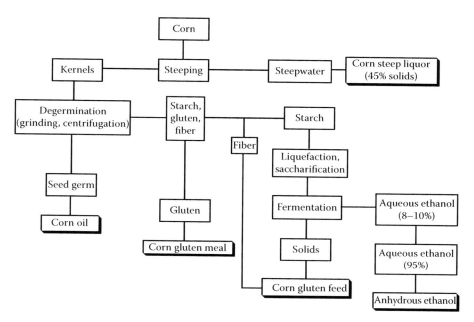

FIGURE 1.20   Outline of corn wet milling and ethanol production.

and organic phosphates, and some other inorganic ions (potassium, sodium, magnesium, etc.) — but this has little, if any, impact on overall process economics (table 1.2). A detailed account of a dry milling process was published by Alltech in 2004.[44] The scheme in figure 1.21 is a simplified derivative of the information provided then as a representative example of the complete bioprocess for ethanol and DDGS.

Unlike Brazilian sucrose-based ethanol, corn-derived ethanol has been technology-driven, especially in the field of enzymes and improved yeast strains with high ethanol tolerance and may be (or become) capable of yielding up to 23% by volume of ethanol in batch fermentations within 60 hours.[44,45] Typical commercially available enzymes used liberate the sugars present in starches. Their properties are summarized in table 1.3. Innovations in biocatalysts and fermentation engineering for corn ethanol facilities are covered at greater length in chapter 3. The availability of enzyme preparations with increasingly high activities for starch degradation to maltooligosaccharides and glucose has been complemented by the use of proteases that can degrade corn kernel proteins to liberate amino acids and peptides to accelerate the early growth of yeast cells in the fermentor; protein digestion also aids the access of amylases to difficult-to-digest starch residues, thus enhancing overall process efficiency and starch to ethanol conversion. table 1.4 contains indicative patents and patent applications awarded or filed since 2003 for corn ethanol technologies.

As the multiplicity of U.S. corn ethanol producers has increased, the relative contributions of large and small facilities have shifted: in 1996, Archer Daniels Midland accounted for more than 70% of the total ethanol production, but by late

TABLE 1.2
Estimated Ethanol Production Costs ($/Gallon) from Corn Milling Technologies

	Wet milling 2005[a]	Dry milling 2005[a]	Dry milling model data[b] 40 mgy[c]	Dry milling model data[b] 80 mgy[c]
Feedstock costs	0.712	0.707	0.877	0.840
By-product credits	0.411	0.223	0.309	0.286
Net feedstock costs	0.301	0.484	0.568	0.554
		Operating costs		
Electricity	0.061	0.058	0.040	0.039
Fuels	0.145	0.211	0.160	0.112
Waste management	0.031	0.007		
Water	0.015	0.003	0.004	0.004
Enzymes	0.067	0.042	0.040	0.040
Yeast	0.031	0.005	0.010	0.006
Chemicals	0.055	0.036	0.010	0.013
Denaturant	0.059	0.054	0.072	0.062
Maintenance	0.088	0.062	0.020	0.052
Labor	0.093	0.058	0.010	0.020
Adminstrative	0.055	0.042		
Other	0.000	0.004		
Total variable cost	1.002	1.065	0.934	0.902

[a] Shapouri et al., 2006[41]
[b] Dale and Tyner, 2006[43]
[c] mgy = million gallons per year output

2006, this had fallen to just 21%.[38,46] Although the largest 4 producers still account for 42%, 8 smaller companies each claim 1 to 2% of the total capacity (figure 1.22). The mix of producers includes local initiatives and farmer-owned facilities, and production is heavily concentrated in the Midwest (to minimize transportation costs for raw materials) but with existing and planned expansion in states from Georgia to Oregon. In September 2006, ethanol production capacity in the United States amounted to 5 billion gallons/year, with a further 3 billion gallons/year under construction.[46]

Presently, ethanol blends commercially available are the 10% (E10) and 85% (E85) versions. The 2004 Volumetric Ethanol Excise Tax Credit made E85 eligible for a 51 cent/gallon tax break; various states (including Pennsylvania, Maine, Minnesota, and Kansas) levy lower taxes on E85 to compensate for the lower mileage with this fuel. In Hawaii, the tax rate positively discriminates in favor of E85.[47] The 2005 Energy Policy Act established tax credits for the installation of a clean-fuel infrastructure, and state income tax credits for installing E85 fueling equipment have been introduced. FFVs capable of using standard gasoline or E85 began to appear in 1995–1998 (Ford), and since then, Daimler Chrysler, General Motors, Isuzu, Lincoln, Mazda, Mercedes Benz, Mercury, and Nissan have introduced

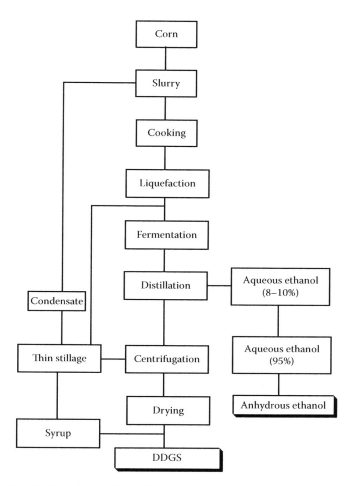

FIGURE 1.21 Outline of corn dry milling and ethanol production.

models as FFVs.[47] Usage of ethanol blends is highest in California — 46% of total U.S. consumption.[46]

Outside North America, construction of the first bioethanol facility in Europe to utilize corn as the feedstock commenced in May 2006 in France; AB Bioenergy France aims to begin production in 2007. The parent company Abengoa Bioenergy (www.abengoa.com) operates three facilities in Spain, producing 5,550 million liters of ethanol a year from wheat and barley grain. A plant in Norrköping, Sweden, began producing 50 million liters of ethanol annually from wheat in 2001; the product is blended with conventional imported gasoline at up to 5% by volume. These and other representative bioethanol facilities in Europe and Asia are listed in table 1.5. Similar industrial plants, to use a variety of agricultural feedstocks, are presently planned or under construction in Turkey, Bulgaria, Romania, El Salvador, Colombia, and elsewhere.

TABLE 1.3

Typical Enzymes for Fuel Ethanol Production from Cereals

Manufacturer and enzyme	Type of enzyme	Use	Properties
		Novozymes	
BAN® (Thermozyme®)	α-amylase	Starch liquefaction	
Termamyl®	α-amylase	Starch liquefaction	Heat stable
Liquozyme®	α-amylase	Starch liquefaction	Heat stable, broad pH tolerance, low calcium requirement
Viscozyme®	α-amylase	Starch liquefaction	Optimized for wheat, barley, and rye mashes
Spirizyme®	Glucoamylase	Saccharification	Heat stable
Alcalase®	Protease	Fermentation	
		Genencor International	
Spezyme®	α-amylase	Starch liquefaction	Heat stable
Distillase®	Glucoamylase	Saccharification	
G-Zyme®	Glucoamylase	Saccharification	Also added pre-saccharification
STARGEN™	α-amylase + glucoamylase	Saccharification and fermentation	Enzyme blend
FERMGEN™	Protease	Fermentation	
Fermenzyme®	Glucoamylase + protease	Saccharification and fermentation	Enzyme blend
		Alltech[a]	
Allcoholase I™	α-amylase	Starch liquefaction	
High T™	α-amylase	Starch liquefaction	Heat stable
Allcoholase II™	Glucoamylase	Saccharification	

[a] Now marketed by Enzyme Technology

1.5   THE PROMISE OF LIGNOCELLULOSIC BIOMASS

Sucrose functions in plants as a highly water-soluble and readily transported product of carbon fixation in leaves, although it can also accumulate in storage organs (e.g., sugarbeets). Starch is mainly a storage polymer in, for example, cereal grains. Cellulose, on the other hand, is essentially a structural polymer in plants (figure 1.23), highly insoluble, organized into crystalline macroscopic fibers, mixed with other polysaccharides (e.g., hemicelluloses), and protected from enzymic attack in native woods by the physical presence of lignin (figure 1.24). Lignins are polyphenolic polymers generated by enzyme-catalyzed free radical reactions from phenylpropanoid alcohols. Unlike nucleic acids and proteins, they have no informational content but are neither inert to enzyme-catalyzed degradation nor incapable of being converted (by hydrogenolysis or oxidative breakdown) to useful chemical intermediates, for example, in the manufacture of synthetic resins, perfume, dyes, and pharmaceuticals.[48] As wood chemicals,

TABLE 1.4

Patent Applications and Patents Awarded for Corn Ethanol Technologies

Date of filing or award	Title	Applicant/assignee	Patent/application
January 21, 2003	Process for producing ethanol	ZeaChem, Inc., Golden, CO	US 6509180
October 1, 2004	Improved process for the preparation of ethanol from cereals	Etea S.r.l., Savigliano, Italy	EP 1 536 016 A!
March 9, 2004	Method for producing fermentation-based products from high oil corn	Renesson LLC, Bannockburn, IL	US 6703227
May 25, 2004	Fermentation-based products from corn and method	Renesson LLC, Bannockburn, IL	US 6740508
October 20, 2005	Methods and systems for producing ethanol using raw starch and fractionation	Bruin and Associates, Inc., Sioux Falls, SD	US 2005/0233030 A1
October 27, 2005	Continuous process for producing ethanol using raw starch	Bruin and Associates, Inc., Sioux Falls, SD	US 2005/0239181 A1
February 23, 2006	Removal of fiber from grain products including distillers dried grains with solubles	R. Srinivasan and V. Singh	US 2006/0040024 A1
May 2, 2006	Heterologous expression of an *Aspergillus kawachi* acid-stable alpha amylase …	Genencor International, Palo Alto, CA	US 7037704
July 11, 2006	Process for producing ethanol from corn dry milling	ZeaChem, Inc., Golden, CO	US 7074603
August 1, 2006	Method for producing fermentation-based products from corn	Renesson LLC, Deerfield, IL	US 7083954
September 5, 2006	Alcohol production using sonication	UltraForce Technology LLC, Ames, IA	US 7101691
November 16, 2006	Hybrid enzymes	Novozymes A/S, Bagsvaerd, Denmark	US 2006/0257984 A1

extracted celluloses are, however, far more widely known, especially in the paper industry and also (as acetylated, nitrated, and other derivatives) find applications as varied as components of explosives, cigarette filters, cosmetics, and medical products such as gauze and bandages. Among tree species, hardwoods and softwoods differ in their compositions, hardwoods having (perhaps, paradoxically) less lignin (figure 1.25). Detailed data for more than 120 tree species lists cellulose contents as high as 57% (and as low as 38%), with lignin in the 17 to 37% range (by weight).[48] This plasticity of biomass chemical composition suggests that plant breeding programs and genetic technologies can accelerate the evolution of "bioenergy" plants as novel cultivars.

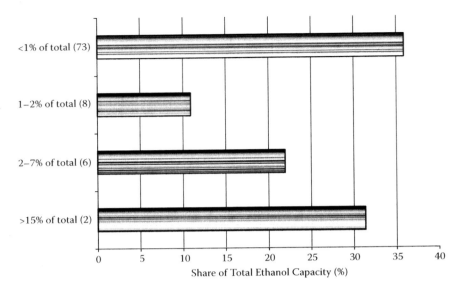

FIGURE 1.22   Contributors to U.S. fuel ethanol production. (Data from the Renewable Fuel Association.)

TABLE 1.5
Industrial Sites for Bioethanol Production from Cereals and Sugar in Europe and Asia

Manufacturer	Location	Substrate	Capacity (million liters per year)
Agroethanol AB	Norrkoping, Sweden	Wheat grain	50 (planned expansion to 200 in 2008)
Biowanze SA	Wanze, Belgium	Wheat grain, sugar beet	300 (by 2008)
Suedzucker Bioethanol GmbH	Zeitz, Germany	Sugar beet	270
AB Energy France	Lacq, France	Corn	150 (by 2008)
Abengoa Bioenergia	3 sites in Spain	Wheat and barley grain	550
HSB Agro Industries Ltd.	Ringus, Rajasthan, India	Rice and sorghum grain	11
Jilin Fuel Ethanol Co. Ltd.	Jilin, China	Corn	900
Harbin Winery	Harbin, China	Corn	4

Hemicelluloses are a diverse group of polysaccharides, with different plant species elaborating structures with two or three types of sugar (sometimes further modified by O-methylation or O-acetylation) and a sugar acid; the major sugar monomers are the pentoses xylose and arabinose and the hexoses glucose, galactose, and mannose (figure 1.23). Most plant species contain xylans (1,4-linked polymers of xylose); in addition, hardwood and softwood trees contain copolymers of glucose and mannose (glucomannans) — larchwoods are unusual in having a core polymer of galactose. Table 1.6 summarizes compositional data based on sugar

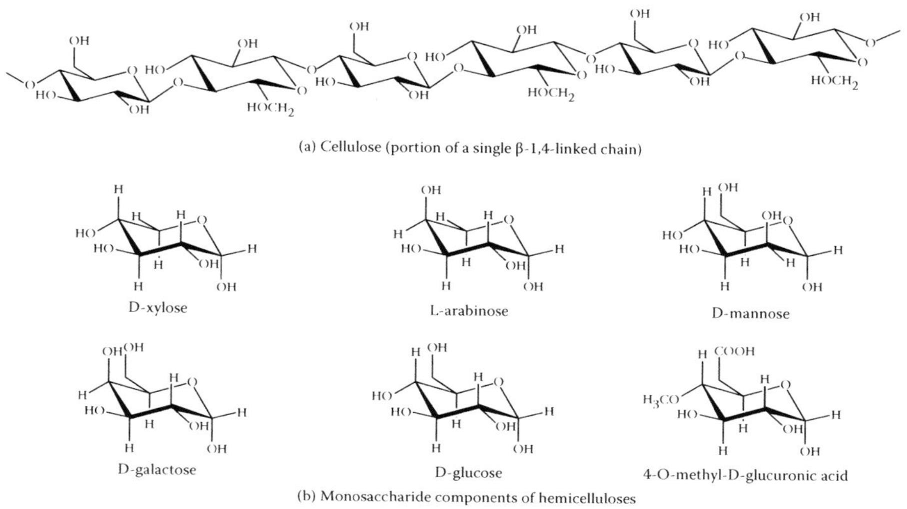

(a) Cellulose (portion of a single β-1,4-linked chain)

D-xylose

L-arabinose

D-mannose

D-galactose

D-glucose

4-O-methyl-D-glucuronic acid

(b) Monosaccharide components of hemicelluloses

FIGURE 1.23   Chemical structures of cellulose and sugar components of hemicelluloses.

type, taken from analyses presented by the National Renewable Energy Laboratory (Golden, Colorado) for a range of tree species, paper products for recycling, cereals, and grasses.[49]

In principle, a bioprocess for producing ethanol from a lignocellulosic substrate could be modeled on those developed for cornstarch (figures 1.20 and 1.21). If only cellulosic glucose is considered as a substrate, the essential stages are

- Milling/grinding of the plant material to reduce particle size
- Chemical and/or physical pretreatment of the plant material to increase the exposure of the cellulose to enzyme (cellulase) attack
- Separation of soluble sugars and oligosaccharides
- Addition of either cellulase or a microorganism capable of secreting active cellulase and utilizing the released sugars for ethanol production by fermentation (simultaneous saccharification and fermentation) or direct microbial conversion

If the hemicellulosic sugars are also to be utilized, then either hemicellulases need to be added or a mixture of organisms used in cofermentations or sequential fermentations. As more of the total potential substrate is included in the fermentation step, the biology inevitably becomes more complex — more so if variable feedstocks are to be used during the year or growing season — as the total available carbohydrate input to the biological fermentation step will alter significantly (table 1.6).

In 1996, several years' experience with pilot plants worldwide using either enzyme conversion or acid-catalyzed hydrolysis of candidate cellulosic feedstocks inspired the prediction that technologies for the conversion of lignocellulosic biomass to ethanol would be rapidly commercialized.[50] A decade on, "generic" technologies have failed to emerge on large-scale production sites; in April 2004, Iogen

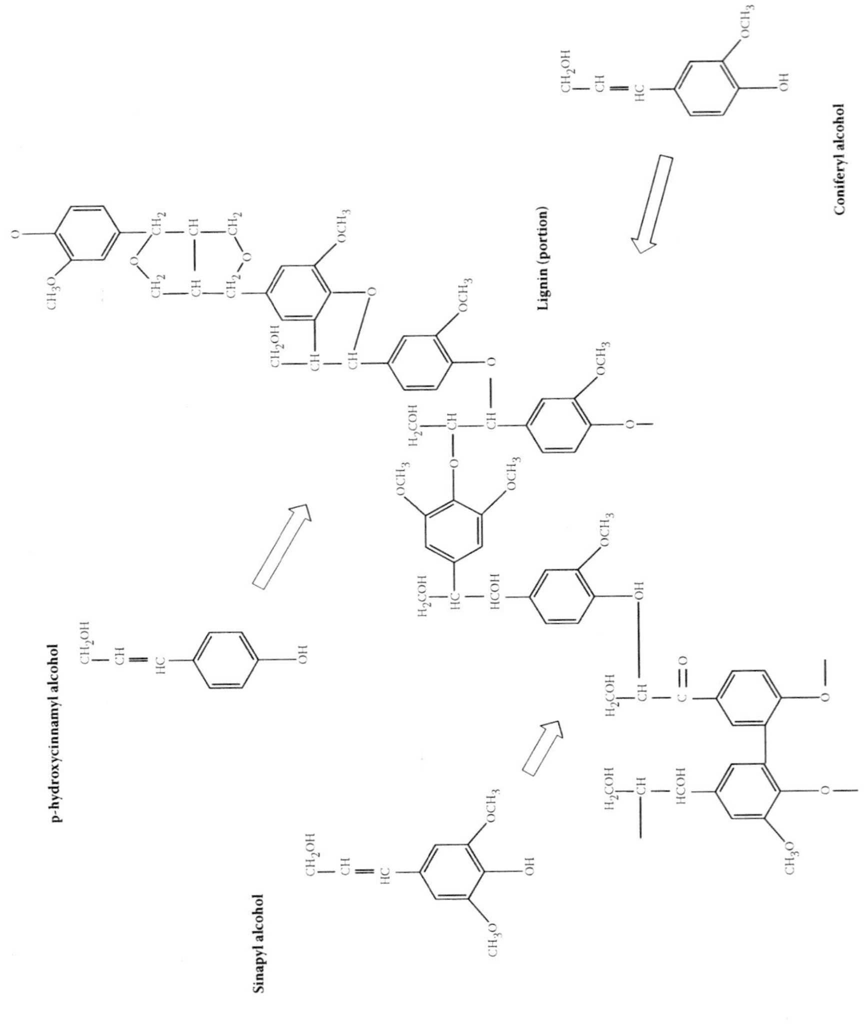

FIGURE 1.24 Outline of lignin biosynthesis.

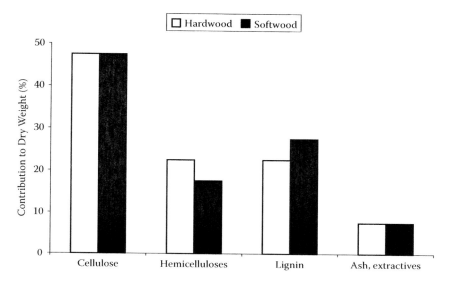

FIGURE 1.25   Chemical composition of wood. (Data from Sudo et al.[48])

TABLE 1.6

Compositional Analyses of Tree Species, Paper Recyclates, Cereal Wastes, and Grasses (% Dry Weight Basis)

	Hexose sugars			Pentose sugars		Lignin	Water and alcohol extractives	Ash
	Glucans	Galactans	Mannans	Xylans	Arabinans			
Tree species[a]	46.2	1.0	3.3	16.0	2.4	24.2	4.0	0.8
Paper recyclates[b]	62.3		6.7	10.4				20.7
Cereal wastes[c]	37.4	0.9	0.4	20.5	2.3	20.5	8.2	7.6
Cane sugar bagasse[d]	40.2	0.7	0.3	21.1	1.9	25.2	4.4	4.0
Grasses[e]	31.8	1.1	0.2	18.3	2.8	17.3	18.1	6.3

[a] Mean of 18 softwood and hardwood species
[b] Municipal solid waste and office paper
[c] Corn stover, wheat straw and rice straw
[d] Cane sugar after removal of cane juice
[e] Four species including switchgrass

(Ottawa, Ontario, Canada) opened a demonstration facility capable of processing 40 tonnes of feedstock/day and producing 3 million liters of ethanol annually from wheat, oat, and barley straw, corn cobs, and corn stalk.[51] Iogen was founded in 1974 and has received CAN$91.8 million in research funding from the government of Canada, Petro-Canada, and Shell Global Solutions International BV. Iogen and its

European partners are studying the feasibility of producing cellulosic ethanol in Germany and, in May 2006, attracted CAN$30 million from Wall Street investors Goldman Sachs.[52–54] The unique status of Iogen indicated that the technical hurdles to be overcome for cellulosic materials were considerably higher than for corn-starch, and it is highly likely that major advances in biotechnology will be required to tailor enzymes for the fledgling industry as well as providing novel biocatalysts for fermentative steps and optimizing plant species as "energy crops."[55] Significantly, Iogen is also an industrial producer of enzymes used in textiles, pulp and paper, and animal feed.*

Since 2006, a pilot facility in Jennings, Louisiana, has operated for the production of cellulosic ethanol; in February 2007, construction commenced on an adjacent demonstration facility designed to utilize regionally available feedstocks, including sugarcane bagasse, and the technology has also been licensed to Japanese companies to develop a project facility at Osaka to produce 1.3 million liters of ethanol annually from demolition wood waste (www.verenium.com). After developing facilities to produce more than a million tons of ethanol from corn and wheat grain by 1995, Chinese scientists at Shandong University devised a bioethanol production process from corn cob, and a 50,000 ton/year plant for ethanol and xylose-derived products is planned to be constructed at Yucheng.[56]

The "promise" of lignocellulosic bioethanol remains quantitatively persuasive. Estimates of land area available for biomass energy crops and of the utilization of wood industry, agricultural, and municipal solid waste total 1.3–2.3 billion tons of cellulosic biomass as potential annual inputs to bioethanol production, potentially equivalent to a biofuel supply matching 30–50% of current U.S. gasoline consumption.[37,57] In stark contrast, even if all U.S. corn production were to be dedicated to ethanol, only 12% of the gasoline demand would be met.[58] Data from Canada show similar scenarios, i.e., the total 2004 demand for fuel ethanol was met from 2,025 million liters of wheat, barley, corn, and potatoes, but the available nonfood crop supplies then already amounted to nearly 11,500 million liters as corn stover, straw, wood residues, and forest residues.[59]

The 2005 Energy Policy Act (http://www.ferc.gov) continued the influential role of legislation on renewable energy sources with initiatives to

- Increase cellulosic ethanol production to 250 million gallons/year
- Establish loan guarantees for new facilities
- Create an advanced biofuels technologies program

Continued interest in novel biotechnological solutions to the problems of lignocellulosic bioethanol are highly likely to be maintained over the next decade. The scientific aspects of present developments and future requirements are discussed in chapters 2 and 3.

---

* Six Canadian companies, starting operations between 1981 and 2006, with total capacity of 418 million liters of ethanol per year, mostly from wheat starch, with a further 390 million liters under construction (Canadian Renewable Fuels Strategy, Canadian Renewable Fuels Association, Toronto, Ontario, Canada, April 2006).

1.6 THERMODYNAMIC AND ENVIRONMENTAL ASPECTS OF ETHANOL AS A BIOFUEL

Arguments in favor of bioethanol (and other biofuels) tend to mix some (or all) of four key points, variously worded, and with different degrees of urgency:

- Fossil energy resources are finite and may become seriously depleted before 2050 (the "peak oil" argument — see chapter 5, section 5.6).
- Biofuels avoid dependency on oil imports (the "energy security" argument).
- Biofuels augment sustainable development across the globe, more efficiently utilizing agricultural wastes (the "long-term environmental" argument).
- Biofuels can reduce greenhouse gas and other harmful emissions (the "acute climatic" argument).

Economists and socioeconomists tend to concentrate the first two cases, whereas biological scientists are far more comfortable with the latter two arguments. So far, the present discussion has touched on topics pertinent to the "energy security" problem, but conflicting data and conclusions have been apparent for nearly three decades on how biofuels (and bioethanol in particular) may — or may not — be solutions to the global energy supply and the ecological crisis of global warming caused by greenhouse gas emissions.

1.6.1 NET ENERGY BALANCE

Even before the end of the 1970s, serious scientific discussion had commenced regarding the thermodynamics of fuel ethanol production from biological sources. The first in-depth study that compiled energy expenditures for ethanol production from sugarcane, sorghum, and cassava under Brazilian conditions concluded that the net energy balance (NEB), that is, the ratio between the energy produced (as ethanol) and the total energy consumed (in growing the plants, processing the harvest, and all the various stages of the ethanol production process) was positive.[60] This issue must not be confused with economic price but is one of thermodynamics: does the production of ethanol require a net input when a summation of the inputs is made, including such obviously energetic factors as the heat employed to distill the ethanol from the aqueous fermentation broth and also more subtle energy costs such as those involved in the manufacture of fertilizers and pesticides?

Thermodynamically, energy can be neither created nor destroyed but merely converted from one form into another. Energy is unavoidably expended in the production of ethanol, and the summation can be made across the entire production process including human labor (that must be replenished, i.e., energy goes into supplying the food for the work force), machinery, fuel, seeds, irrigation, and all agrochemicals (many of which are derived from petrochemicals); these can be described as "direct" or "mechanical" as distinct from essential environmental energy inputs such as sunlight.* The output (i.e., ethanol) has a measurable energy yield in internal

---

* Climatic factors including seasonal hours of sunlight, precipitation, average temperature, and others certainly affect crop yield and will, therefore, impact on the economics of ethanol production as the cost of, for example, corn grain as an essential input into the ethanol production process varies from year to year.

combustion engines; the energy inputs I have termed "energy conversion debits" that can be — although they are not unavoidably — fossil fuel consumptions (as diesel, coal, natural gas, etc.). The ratio between energy in the ethanol produced and the energy consumed in the conversion debits is the net energy yield or energy balance.

Four distinct cases can be distinguished with far-reaching implications for macroeconomics and energy policy:

1. Ethanol production has a net yield of energy and has no absolute dependency on nonrenewable energy inputs.
2. Ethanol production has a net yield of energy but has a dependency on nonrenewable energy inputs.
3. Ethanol production has a net loss of energy (a yield of <1) but has no dependency on nonrenewable energy inputs.
4. Ethanol production has a net loss of energy as well as being dependent on nonrenewable energy inputs.

Viewed broadly, the most acceptable conclusion economically as well as to policy makers is case 1. Case 2 would pose problems to the short-term adoption of biofuels, but eventually, there could be no insuperable requirement to either "fund" bioethanol by expending fossil-fuel energy as alternative energy sources become more widely available or to support the required agricultural base with agrochemicals derived exclusively from petrochemicals or other nonrenewable sources (especially if fermentation-derived bioprocesses supplanted purely synthetic routes). Moreover, cases 3 and 4 are not in themselves intrinsically unacceptable — as Sama pointed out nearly 20 years ago, power stations burning oil or coal all show a net energy loss but that is not used as an argument to shut them down.[61] Domestic and industrial devices that are electrically driven require power stations as an interface for energy conversion, with the associated economic and thermodynamic costs. Filling gasoline tanks with sugar, corncobs, or wood chips cannot fuel automobiles; converting these biological substrates to ethanol can — but again at economic and energetic costs. Case 3 and especially case 4 are equivalent to wartime scenarios where fuel is produced "synthetically"; case 3 is one of sustainable, long-term development if nonfossil energy sources such as hydroelectric, geothermal, wind, wave, and solar can be used to meet the net energy costs inherent in bioethanol production.

A review of a broad portfolio of renewable energy technologies in the mid-1970s that included methane production from algae and livestock waste as well as sugarcane, cassava, timber, and straw concluded that only sugarcane-derived ethanol was capable of yielding a net energy gain.[62] Data from early Brazilian and U.S. studies of sugarcane-derived ethanol are summarized in table 1.7. It was evident even in these early studies that the NEB was highly influenced by the utilization of bagasse, that is, the combustion of the waste product to supply steam generation and other energy requirements on-site or (although not considered at that time) as a saleable power commodity to the broader community. The energy balance was, however, much lower than the comparative quoted case of gasoline produced from oil extracted in the Gulf of Mexico oil fields, that is, 6:1.[63]

TABLE 1.7
Energy Balances for Ethanol Production from Sugarcane

Publication	Energy conversion debits[a]			Energy outputs[b]			Energy balance
	Sugarcane growing	Ethanol production	Total	Ethanol	Residue	Total	
			Da Silva et al., 1978				
Ethanol only considered	4138	10814	14952	18747		18747	1.25
All bagasse converted to steam	4138	10814	14952	18747	17500	36247	2.42
			Hopkinson and Day, 1980				
Ethanol only considered	8500	10800	19300	18400		18400	0.95
All bagasse converted to steam	8500	10800	19300	18400	17200	35600	1.84

[a,b] 10^6 kcal per hectare per year

Several reports of the technical feasibility of ethanol production from corn had been published by 1978. Researchers from the University of Illinois, Urbana, collected the data and developed a mathematical model that extended the analysis to include mileage estimates and measurements for 10% ethanol blends with gasoline.[64] From the data, the overall energy balance could be computed to be small (approximately 1.01); even attaining the energy break-even point required the efficient utilization of wastes for on-site energy generation and the sale and agricultural use of by-products counted as energy outputs (see figures 1.21 and 1.22). Few data were available at that time for mileages achieved with gasohol blends; if positive advantages could be gained, this would have translated into significantly increased net energy gains in what was, in effect, the first testing of a "well-to-wheel" model for fuel ethanol.[64]

Subsequently, both the energetics of corn- and sucrose-derived bioethanol processes have been the subjects of scrutiny and increasingly elaborate data acquisition and modeling exercises. Economic analyses agree that the production of ethanol from sugarcane is energetically favorable with net energy gains that rival or equal those for the production of gasoline from crude oil in subterranean deposits.[65,66] For corn-derived fuel alcohol, however, the energy position is highly ambiguous. By 2002, that is, before the steep rise in world oil prices (see figures 1.3 and 1.11) — a factor that might easily have exercised undue weight on the issue — conflicting estimates had appeared, but the published accounts during two decades displayed very different methodological approaches and quantitative assumptions. Even as crucial a parameter as the energy content of ethanol used in the computations varied in the range 74,680 to 84,100 Btu/gallon — the choice of lower or higher heat values (measured by reference to water or steam, respectively) is also required for the energy inputs, and if the choice is used consistently in the calculations, there will be no effect on the net energy gained or lost. Ten of the studies were discussed in a 2002 review.[67–76] Table 1.8 collects the data, adding to the comparison an extra publication from 2001.[77] The obvious spread of energy balance values has generated a sustained argument over the choice of relevant input parameters, but a more pertinent conclusion is that none of the balances greatly exceeded 1.00 (the arithmetical average is 1.08); it could have been concluded before 2002, therefore, that only a highly efficient production process (including the maximal utilization of whatever by-products were generated) could deliver a net energy gain, although changes in agricultural practices, higher crop yields, increased fermentation productivity, and others might all be anticipated to contribute to a gradual trend of piecemeal improvement.

Since 2002, the polemics have continued, if anything, with an increased impression of advocacy and counteradvocacy. Among the peer-reviewed journal articles, the following considered NEB values (on a similar mathematical basis to that used in tables 1.7 and 1.8), but with various assumptions and calculations for energy input data:

- Patzek[78] — 0.92 (with no by-product energy credits allowed).
- Pimentel and Patzek[79] — 0.85 (after adjustment for the energy in by-products).
- Dias de Oliveira et al.[80] — 1.10 (compared with 3.7 for sugar-derived ethanol in Brazil).
- Farrell et al.[81] — 1.20.
- Hill et al.[58] — 1.25, although this depended mostly on counting the energy represented by the DDGS by-product (Figure 1.21) into the calculations for NEB and NEB ratio.

TABLE 1.8
Energy Balances for Ethanol Production from Corn

Source	Energy balance HHV[a]	Energy balance LHV[b]
Ho, 1989[67]		0.96
Marland and Turhollow, 1990[68]	1.25	
Pimentel, 1991[69]		0.74
Keeny and DeLuca, 1992[70]		0.92
Lorentz and Morris, 1995[71]	1.38	
Shapouri, Duffield and Graboski, 1995[72]	1.20	
Agriculture and Agri-Food Canada, 1999[73]		1.32
Wang, Saricks and Santini, 1999[74]		1.33
Pimentel, 2001[75]		0.74
Shapouri, Duffield and Wang, 2002[76]	1.27	
Berthiaume et al., 2001[77]	0.79	

a   High heat value for ethanol (83,961 Btu per gallon)
b   Low heat value for ethanol (76,000 Btu per gallon)

The continued failure to demonstrate overall NEBs much more than 1.00 is again striking. This has provided an impetus to define better metrics, including decreased fossil fuel usage and greenhouse gas emissions in the assessments.[81] Put into the perspective of the historical measures of energy balance from the oil industry corn-derived ethanol remains relatively inefficient.[82-84] Assuming an NEB of 1.2, the notional expenditure of 5 units of bioethanol would be required to generate each net unit. This argument against ethanol as a biofuel is, of course, far less persuasive if all the energy needs for bioethanol production are met by renewable sources — this was highly speculative in the 1970s, and energy inputs to biofuels have (implicitly or explicitly) assumed a fossil fuel basis. With an energy input defined as "nonrenewable," Hammerschlag normalized data from six studies to compute a range of "energy return on investment," that is, total product energy divided by the nonrenewable energy to its manufacture, of 0.84–1.65.[86] For an economy like Brazil's, where hydroelectricity is the single largest source of power generation (64% in 2002), renewable energy is a practical option, but for the world at large (where hydroelectricity and all other renewable energy sources account for less than 10% of total power generation), the dependency on coal, oil, and natural gas is likely to remain for some decades.[87]

In comparison with corn ethanol, biomass-derived ethanol has received less attention. Estimates of the NEB range from the "pessimistic" (0.69 from switchgrass [*Panicum virgatum*], 0.64 from wood) to the "optimistic" (2.0).[79,81] Hammerschlag suggested a range for cellulosic ethanol of 4.40–6.61 but ignored one much lower value.[86] The Greenhouse gases, Regulated Emissions and Energy use in Transportation (GREET) model developed by the Argonne National Laboratory (Argonne, Illinois) also predicts large savings in petroleum and fossil fuels for ethanol produced from switchgrass as the favored candidate lignocellulosic feedstock.[88] Technologies for bioethanol manufactured from nonfood crops appear, therefore, to warrant further attention as energy-yielding processes capable of narrowing the considerable gap in energy gain and expenditure that exists between corn-derived ethanol production and the oil industry.

Why the large discrepancy between corn-derived and lignocellulosic ethanol? Intuitively, lignocellulosic substrates are more intractable (and, therefore, more costly to process) than are starches. As all practitioners of mathematical modeling appreciate, the answer lies in the assumptions used to generate the modeling. Specifically, modeling methodologies for cellulosic ethanol processes assume that electricity is generated on-site from the combustion of components of biomass feedstocks not used for fermentation (or that are unused after the fermentation step) in combined heat and power plants; this was built into the earliest detailed models of biomass ethanol production in studies undertaken by the Argonne National Laboratory and by the National Renewable Energy, Oak Ridge National, and the Pacific Northwest Laboratories, and the results are pivotal not only for energy balances but also for reductions in pollutant emissions.[81,88,89] This methodology approximates energy use in biomass (cellulose) ethanol plants to that in sugar ethanol plants in Brazil (see table 1.7).

In Europe, where bioethanol production is much less developed than in North America and Brazil, attention has focused on wheat starch and beet sugar. Studies of energy balances with wheat have consistently shown negative energy balances (averaging 0.74), whereas values for sugarbeet range from 0.71 to 1.36 but average 1.02, although projections made by the International Energy Agency suggest that both feedstocks will show positive energy balances as combined fertilizer and pesticide usage drops and biotechnological conversion efficiencies improve.[90]

The question of deriving metrics adequate to accurately compare fuel alcohol and gasoline production processes also continues to exercise the imagination and inventiveness, particularly of European analysts. Portuguese analysts defined energy renewability efficiency (ErenEf) as:

$$\text{ErenEf} = (\text{FEC} - E_{in.\ fossil.\ prim}) \times 100\ /\ \text{FEC}$$

where FEC is the fuel energy content and $E_{in.\ fossil.\ prim}$ represents the unavoidable fossil energy input required for the production of the biofuel. Under French conditions, sugarbeet-derived ethanol was renewable even without taking into account any coproduct credits but was maximal with an allocation of the energy inputs based on the mass of the ethanol and coproducts (ErenEf = 37%), whereas wheat grain-derived ethanol was entirely dependent on this allocation calculated for the DDGS (ErenEf = 48%).[91] These values were equivalent to positive NEBs of 1.59 (sugarbeet) and 1.92 (wheat). Conversion of the ethanol from either source to ethyl tertiary butyl ether by reaction with petroleum-derived isobutylene results in a product superior as an oxygenate to MTBE, but the energy gains calculated for the fuel ethanol were almost entirely lost if this extra synthetic step was included.

Accepting that the fossil fuel requirement of corn-derived ethanol approximates the net energy value of the product, Nielsen and Wenzel then advanced the argument that gasoline requires an equal amount of energy as fossil fuels in its production — the net fossil fuel saving when counting in the gasoline saved when corn ethanol is combusted in an automobile therefore generates a fossil energy saving 90% of that of the fuel alcohol.[92] No supporting data were quoted for this assertion, that is that gasoline required such a high outlay of fossil fuel, and other analysts markedly disagree.[82–84] A commentator from the Netherlands argued that the opportunity costs for crop production must be taken into account, that is, that the energy costs for corn- and biomass-derived ethanol cannot include those implicit in the generation of the fermentation substrate unless the land for their production is otherwise left idle.[93] This is again contentious, as no arable land must be gainfully farmed; as a citizen of a European Union state, the Dutch author would have been aware that land deliberately left idle even has a monetary value: the so-called "set-aside" provisions of the Common Agricultural Policy aim to financially encourage farmers not to overproduce agricultural surpluses, which otherwise would require a large financial outlay for their storage and disposal (see chapter 5, section 5.2.3). Even with this economic adjustment, the production of bioethanol was not a net energy process with either switchgrass or wood as the biomass input.

When analysts tacitly assume that all (or most) of the energy required for biofuel production is inevitably derived from fossil fuels, this is equivalent to the

International Energy Agency's Reference Scenario for future energy demands, that is, that up until 2030, coal, gas, and oil will be required for 75% of the world's power generation, with nuclear, hydro, biomass, and other renewables accounting for the remaining 25%.[94] The agency's alternative scenario for 2030 postulates fossil fuel usage for power generation decreasing to 65% of the total. In the future, therefore, the fossil fuel inputs and requirements for corn- and biomass-derived ethanol may significantly decrease, thus affecting the quantitative assessment of (at least) the fossil fuel energy balance. As an interim measure, however, defining the crucial net energy parameter as the ratio between the energy retrieved from ethanol and the fossil fuel energy inputs involved in its production, the "fossil energy ratio" or "bioenergy ratio"[66] gives:

$$(E_{net} + E_{coproduct})/(E_A + E_B + E_C + E_D)$$

where E_A is the fossil fuel energy required for the production of the plant inputs, E_B is the fossil fuel energy required during crop growth and harvesting, E_C is the fossil fuel required during transport of the harvested crop, and E_D is that expended during the conversion process.

On this basis, different industrial processes could result in widely divergent bioenergy ratios (figure 1.26). The most obvious inconsistency is that between two molasses-derived ethanol processes, one (in India) was derived from the case of a distillery fully integrated into a sugar mill, where excess low-pressure steam was used for ethanol distillation, whereas a South African example was for a distillery distant from sugar mills and reliant on coal and grid electricity for its energy needs.[66] In terms of energy, corn and corn stover and wheat and wheat straw were all inferior to Brazilian sugarcane, whereas Indian bagasse was a biomass source that gave a high result.

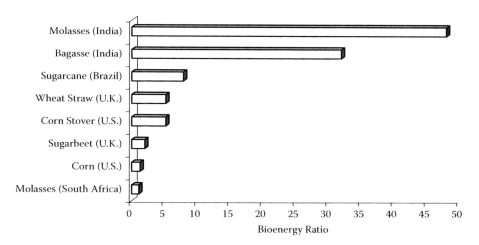

FIGURE 1.26 Ratio of ethanol energy content to fossil fuel energy input for ethanol production systems. (Data from von Blottnitz and Curran.[66])

1.6.2 EFFECTS ON EMISSIONS OF GREENHOUSE GASES AND OTHER POLLUTANTS

Direct comparisons of gas emissions resulting from the combustion of anhydrous ethanol, ethanol-gasoline blends, and gasoline are straightforward to perform but are poor indicators of the overall consequences of substituting ethanol for gasoline. Instead, from the early 1990s, full fuel-cycle analyses were performed to estimate gas emissions (projected beyond 2000) throughout the production process for bio-ethanol to gauge the direct and indirect consequences of gasoline replacement, including

- Changes in land use and the replacement of native species by energy crops
- Agricultural practices and the potential for utilizing agricultural wastes
- Materials manufacture and the construction of facilities
- Bioethanol production
- Transportation of feedstocks and bioethanol
- Fuel usage per mile driven

With the three major greenhouse gases, carbon dioxide (CO_2), methane (CH_4), and nitrous oxide (N_2O), a major difference between corn-derived and cellulosic ethanol was immediately apparent.[89] Figure 1.27 compares the projected CO_2 emissions for a range of fuels; noncarbon-based fuels (i.e., electric vehicles powered using electricity generated by nuclear and solar options) were superior, whereas corn ethanol showed no net advantage. More recent estimates place corn ethanol production as giving modest reductions in greenhouse gas emissions, 12–14%.[58,81] The GREET model of the Argonne National Laboratory indicates steeper reductions (figure 1.28).[95]

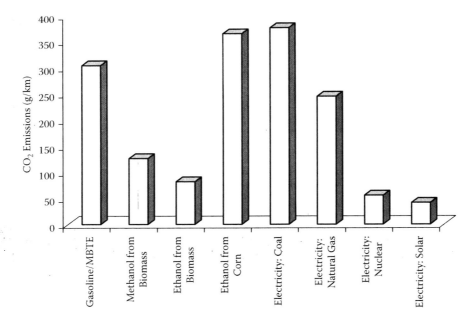

FIGURE 1.27   Total fuel cycle carbon dioxide emissions. (Data from Bergeron.[89])

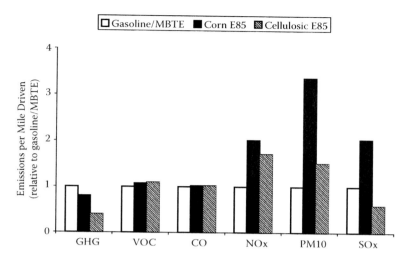

FIGURE 1.28 Total fuel cycle emissions. (Data from Wu et al.[95])

An unavoidable complication in any such calculation, however, is that they assume that harvested crops are from land already under cultivation; converting "native" ecosystems to biofuel production would inevitably alter the natural carbon balance and reduce the potential savings in greenhouse gas emissions.[58]

As with energy balance computations, including combined heat and power systems into cellulosic ethanol production displaces fossil fuel use and accounts for the reductions in greenhouse gas emissions.[81,88] In some corn ethanol production scenarios, the use of fossil fuels such as coal to power ethanol distillation plants and diesel to enable the long-distance transportation of corn can actually increase total greenhouse gas emissions.[81,96,97] If the NEB is less than 1, the total ethanol production cycle would also increase net greenhouse emissions unless renewable energy sources rapidly supplanted fossil fuels in power generation.[78,97,98] Corn stover ethanol can, if its production process shares facilities with grain production, exhibit lower life cycle emissions than switchgrass-derived ethanol in projections up to 2010.[99]

For other priority pollutants, bioethanol production (even from cellulosic resources) and the use of bioethanol as a transportation fuel have ambiguous effects: increased nitric oxides, particulate matter, carbon monoxide, and volatile organic carbon but decreased sulfur oxides when measured in total in both urban and rural locations (figure 1.28). Combined fuel cycle and vehicle life cycle analyses also show major reductions in greenhouse gases but not in other emissions with cellulosic ethanol as an alternative fuel.[100]

For production sources other than sugarcane, therefore, the actual reductions in greenhouse gas emissions resulting from the adoption of biofuels may be considerably less than anticipated. One estimate for corn ethanol concluded that, by itself, fuel alcohol use in the United States would struggle to reduce transportation-dependent emissions by more than 10%.[101] This study used the following data and arguments:

- For U.S. gasoline consumption of 460×10^9 L/year, corn ethanol replaced 0.8% of this while using 1% of the total cropland.
- To replace 10% of the gasoline consumption, corn ethanol would need to be produced on 12% of the total U.S. cropland.
- Corn ethanol only avoids 25% of the CO_2 emissions of the substituted gasoline emissions when the fossil fuel-dependent energy consumed to grow and process the corn is taken into account.
- Offsetting 10% of the CO_2 emissions from gasoline consumption would require a fourfold higher production of corn ethanol, that is, from 48% of U.S. cropland.
- "A challenge for providing transportation fuels is to be able to substitute a joule of energy in harvestable biomass for a joule of primary energy in fossil fuel and to do this without significant fossil energy consumption. If a joule for joule consumption can be achieved, dedicated biomass plantations may be able to displace some 15% of the CO_2 emissions expected from all uses of fossil fuels globally by 2030. For biofuels to replace more than about 15% of fossil-fuel CO_2 emissions in 2030 will require a more rapid improvement in the energy efficiency of the global economy than is apparent from past trends in efficiency."[101]

This pessimistic — or realistic — view of the quantitative impacts of biofuels on greenhouse gas scenarios in the twenty-first century was strongly reinforced by a mid-2007 study of alternatives to biofuels, specifically conserving and (if possible) extending natural forests, savannahs, and grasslands: compared with even the best biomass-to-biofuel case, forestation of an equivalent area of land was calculated to sequester at least twice the amount of CO_2 during a 30-year period than the emissions avoided by biofuel use.[102] As the authors concluded:

> If the prime object of policy on biofuels is mitigation of carbon dioxide-driven global warming, policy makers may be better advised in the short term (30 years or so) to focus on increasing the efficiency of fossil fuel use, to conserve the existing forests and savannahs, and to restore natural forest and grassland habitats on cropland that is not needed for food.

Present national policies have, however, more diverse "prime object" aims and goals, including those of ensuring continuity of supply of affordable transportation fuels and (increasingly important on the geopolitical stage) establishing some degree of "energy independence." Nevertheless, the assessment of potential savings in greenhouse gas emissions remains a powerful argument for the adoption of a considered, balanced raft of biofuels options.

1.7   ETHANOL AS A FIRST-GENERATION BIOFUEL: PRESENT STATUS AND FUTURE PROSPECTS

The developments of sugar-derived ethanol as a major transportation fuel in Brazil and corn-derived ethanol as a niche (but rapidly growing) market in the United States were conditioned by different mixes of economic and environmental imperatives from the early 1970s to the present day. Together and separately, they have been criticized as unsuitable for sustainable alternatives to gasoline in the absence of tax incentives; highly integrated production processes, with maximum use of coproducts and high degrees of energy efficiency, are mandated to achieve full economic competitiveness. Although technically proven as a technology, the scope of ethanol production from food crops (primarily sugar and corn) is limited by agricultural and geographical factors, and only cellulosic sources offer the quantitative availability to significantly substitute for gasoline on a national or global basis. Even for that goal, however, bioethanol is only one player in a diverse repertoire of strategic and tactical ploys and mechanisms — from carbon capture and storage to hydrogen-utilizing fuel cells — with which to change the oil economy and reduce dependence on fossil fuels.[103]

Both corn- and sugarcane-derived ethanol have — unarguably but not irreversibly — emerged as industrial realities.[104] Even detailed questions about their NEBs and the implications of their use in mitigating greenhouse gas emissions have become unavoidable on Web site discussion forums. A major article in the October issue of *National Geographic* magazine accepted the marginal energy gain in corn ethanol production and the disappointingly small reduction of total CO_2 emissions in the total production/use cycle but noted the much clearer benefits of sugarcane ethanol.[105] Two quotes encapsulate the ambiguous reactions that are increasingly evident as the public debate over biofuels spreads away from scientific interest groups to embrace environmentalist and other "lobbies":

It's easy to lose faith in biofuels if corn ethanol is all you know,

and

If alcohol is a "clean" fuel, the process of making it is very dirty ... especially the burning of cane and the exploitation of cane workers.

All forms of ethanol produced from plant substrates are, however, best viewed as "first-generation" biofuels, operating within parameters determined by (in varying allocated orders): the internal combustion engine, mass personal transport, and the global oil industry.[105] Together with biodiesel (i.e., chemically esterified forms of the fatty acids present in vegetable oils — see chapter 6, section 6.1), they have proven potential for extending the availability of gasoline blends and are affordable within the budgets accepted by consumers in the late twentieth and early twenty-first centuries while helping to reduce greenhouse gas emissions and (with some technologies, at least) other atmospheric pollutants. It is, however, likely that different countries and supranational economic groupings (e.g., the European Union) will value differently energy security, economic price, and ecological factors in the face of fluctuating oil prices and uncertain mid- to long-term availabilities of fossil fuels.[106] Fermentation products other than ethanol (e.g., glycerol, butanol) and their thermochemical conversion to synthesis gas mixtures to power electricity generation are second-generation technologies, whereas biohydrogen, fuel cells, and microbial fuel cells represent more radical options with longer lead times.[107,108]

Sugar- and corn-derived ethanol production processes were simply extrapolated from preexisting technologies, and to varying extents, they reflect the limitations of "add-on" manufacturing strategies.[109,110] As one example of the *National Geographic* writer's anticipated "breakthrough or two," lignocellulosic ethanol only emerged as a commercial biofuel reality in 2004, and biotechnology is crucial to its establishment as a major industry.[105,111] For that reason, and because of the uniquely global potential of this nonfood sector of the bioethanol supply chain, the microbial biotechnology of cellulosic ethanol will now be considered at length, to provide not only a snapshot of commercially relevant contemporary science but also to extrapolate existing trends in the development of the scientific base.

REFERENCES

1. McGovern, P.E., *Ancient Wines: The Search for the Origins of Viniculture*, Princeton University Press, Princeton, NJ, 2003, chap. 3.
2. McGovern, P.E., *Ancient Wines: The Search for the Origins of Viniculture*, Princeton University Press, Princeton, NJ, 2003, chap. 5.
3. McGovern, P.E., *Ancient Wines: The Search for the Origins of Viniculture*, Princeton University Press, Princeton, NJ, 2003, chap. 1.
4. McGovern, P.E. et al., Fermented beverages of pre- and proto-historic China, *Proc. Nat. Acad. Sci., USA*, 101, 17593, 2004.
5. Rose, A.H., History and scientific basis of alcoholic beverage production, in Rose, A.H. (Ed.), *Economic Microbiology. Volume 1. Alcoholic Beverages*, Academic Press, London, 1977, chap. 1.
6. Lyons, T.P. and Rose, A.H., Whisky, in Rose, A.H. (Ed.), *Economic Microbiology. Volume 1. Alcoholic Beverages*, Academic Press, London, 1977, chap. 10.
7. Budavari, S. (Ed.), *The Merck Index*, 11th ed., Merck, Rahway, NJ, 1989, 594.
8. McCarthy, T., The coming wonder? Foresight and early concerns about the automobile, *Environ. History*, 6, 46, 2001.
9. Thomas, D.E., *Diesel: Technology and Society in Industrial Germany*, University of Alabama Press, Tuscaloosa, 1987, chap. 5.
10. Tugendhat, C. and Hamilton, A., *Oil: The Biggest Business*, Eyre Methuen, London, 1975, chap. 1–7.
11. Cheremisinoff, N.P., *Gasohol for Energy Production*, Ann Arbor Science Publishers, Ann Arbor, MI, 1979, chap. 6.
12. *The Gasoline and Fuel Oil Shortage*, Hearings before the Subcommittee on Consumer Economics of the Joint Economic Committee, Congress of the United States, Ninety-third Congress, U.S. Government Printing Office, Washington, DC, 1973, p. 1.
13. *Oil, Gas and Coal Supply Outlook*, International Energy Agency, Paris, 1995, chap. 2.
14. *World Energy Outlook*, International Energy Agency, Paris, 1982, chap. 1–2.
15. *BP Statistical Review of the World Oil Industry 1973*, British Petroleum, London, 1974.
16. *World Energy Outlook*, International Energy Agency, Paris, 1982, chap. 3.
17. Baer, W., *The Brazilian Economy: Growth and Development*, 5th ed., Praeger Publishers, Westport, CT, 2001, chap. 6.
18. Faria, J.B., Franco, D.W., and Piggott, J.R., The quality challenge: cachaça for export in the 21st century, in Bryce, J.H. and Stewart, G.G. (Eds.), *Distilled Spirits: Tradition and Innovation*, Nottingham University Press, Nottingham, 2004, chap. 30.
19. Braunbeck, O., Macedo, I., and Cortez, L.A.B., Modernizing cane sugar production to enhance the biomass base in Brazil, in Silveira, S. (Ed.), *Bioenergy — Realizing the Potential*, Elsevier, Amsterdam, 2005, chap. 6.

20. *BP Statistical Review of World Energy*, British Petroleum, London, 2006.
21. Bailey, B.K., Performance of ethanol as a transportation fuel, in Wyman, C.E. (Ed.), *Handbook on Bioethanol: Production and Utilization*, Taylor & Francis, Washington, DC, 1996, chap. 3.
22. Melges de Andrade, A., Morata de Andrade, C.A., and Bodinaud, J.A., Biomass energy use in Latin America: focus on Brazil, in *Biomass Energy: Data, Analysis and Trends*, International Energy Agency, Paris, 1998, 87.
23. Martines-Filho, J., Burnquist, H.L., and Vian, C.E.F., Bioenergy and the rise of sugarcane-based ethanol in Brazil, *Choices*, 21(2), 91, 2006.
24. ANFAVEA (Associação Nacional dos Fabricantes de Veículos Automotores), http://www.anfavea.com.br.
25. UNICA (União da Agroindústria Canvieira do São Paolo), http://www.unica.com.br.
26. Herrera, S., Bonkers about biofuels, *Nat. Biotechnol.*, 24, 755, 2006.
27. Moreira, J.R., Noguiera, L.A.H., and Parente, V., Biofuels for transport, development, and climate change: lessons from Brazil, in Bradley, R. and Baumert, K.A. (Eds.), *Growing in the Greenhouse: Protecting the Climate by Putting Development First*, World Resources Institute, Washington, DC, 2005, chap. 3.
28. Kojima, M. and Johnson, T., *Potential for Biofuels for Transport in Developing Countries*, International Bank for Reconstruction and Development/World Bank, Washington, DC, 2005, chap. 2.
29. Pessoa, A. et al., Perspectives on bioenergy and biotechnology in Brazil, *Appl. Biochem. Biotechnol.*, 121–124, 59, 2005.
30. Amorim, H.V., Basso, L.C., and Lopes, M.L., Evolution of ethanol fermentation in Brazil, in Bryce, J.H. and Stewart, G.G. (Eds.), *Distilled Spirits: Tradition and Innovation*, Nottingham University Press, Nottingham, 2004, chap. 20.
31. Zanin, G.M. et al., Brazilian bioethanol program, *Appl. Biochem. Biotechnol.*, 84–86, 1147, 2000.
32. Telles, G.P. et al., Bioinformatics of the sugarcane EST project, *Genet. Mol. Biol.*, 24, 8, 2001.
33. Vincentz, M. et al., Evaluation of monocot and eudicot divergence using the sugarcane transcriptome, *Plant Physiol.*, 134, 951, 2004.
34. MacLeod, A.M., The physiology of malting, in Pollock, J.R.A. (Ed.), *Brewing Science*, Volume 1, Academic Press, London, 1979, chap. 2.
35. Bryce, J.H. et al., Optimising the fermentability of wort in a distillery — the role of limit dextrinase, in Bryce, J.H. and Stewart, G.G. (Eds.), *Distilled Spirits: Tradition and Innovation*, Nottingham University Press, Nottingham, 2004, chap. 11.
36. Waddams, A.L., *Chemicals from Petroleum. An Introductory Survey*, 4th ed., John Murray, London, 1978, chap. 8.
37. Putsche, V. and Sandor, D., Strategic, economic, and environmental issues for transportation fuels, in Wyman, C.E. (Ed.), *Handbook on Bioethanol: Production and Utilization*, Taylor & Francis, Washington, DC, 1996, chap. 2.
38. Elander, R.T. and Putsche, V.L., Ethanol from corn: technology and economics, in Wyman, C.E. (Ed.), *Handbook on Bioethanol: Production and Utilization*, Taylor & Francis, Washington, DC, 1996, chap. 15.
39. Energy Information Administration, U.S. Department of Energy, http://www.eia.doe.gov.
40. *Ethanol Industry Outlook*, Renewable Fuels Association, Washington, DC, 2006.
41. Shapouri, H., Salassi, M., and Fairbanks, J.N., *The Economic Feasibility of Ethanol Production from Sugar in the United States*, U.S. Department of Agriculture, Washington, DC, July 2006.
42. Cejka, A., Preparation of media, in Brauer, H. (Ed.), *Biotechnology*, Volume 2, VCH Verlagsgesellschaft, Weinheim, 1985, chap. 26.
43. Dale, R.T. and Tyner, W.E., Economic and technical analysis of ethanol dry milling: model description, Purdue University Staff Paper #06-04, West Lafayette, IN, April 2006.

44. Lyons, T.P., Fuel ethanol: the current global situation. Lessons for the beverage industry, in Bryce, J.H. and Stewart, G.G. (Eds.), *Distilled Spirits: Tradition and Innovation*, Nottingham University Press, Nottingham, 2004, chap. 19.
45. Schubert, C., Can biofuels finally take center stage?, *Nat. Biotechnol.*, 24, 777, 2006.
46. Renewable Fuels Association, Washington, DC, http://www.ethanolrfa.org.
47. National Ethanol Vehicle Coalition, Jefferson City, MO, http://www.e85fuel.com.
48. Sudo, S., Takahashi, F., and Takeuchi, M., Chemical properties of biomass, in Kitani, O. and Hall, C.W. (Eds.), *Biomass Handbook*, Gordon and Breach, New York, 1989, chap. 5.3.
49. Wiselogel, A., Tyson, S., and Johnson, D., Biomass feedstock resources and composition, in Wyman, C.E. (Ed.), *Handbook on Bioethanol: Production and Utilization*, Taylor & Francis, Washington, DC, 1996, chap. 6.
50. Schell, D. and Duff, B., Review of pilot plant programs for bioethanol conversion, in Wyman, C.E. (Ed.), *Handbook on Bioethanol: Production and Utilization*, Taylor & Francis, Washington, DC, 1996, chap. 17.
51. Iogen Corporation, Ottawa, Canada, http://iogen.ca.
52. Iogen Corporation, Ottawa, Canada, press release 21 April, 2004.
53. Iogen Corporation, Ottawa, Canada, press release 8 January, 2006.
54. Iogen Corporation, Ottawa, Canada, press release 1 May, 2006.
55. Bioethanol needs biotech now, editorial in *Nat. Biotechnol.*, 24, 725, July 2006.
56. Yinbo, Q. et al., Studies on cellulosic ethanol production for sustainable supply of liquid fuel in China, *Biotechnol. J.*, 1, 1235, 2006.
57. Perlack, R.D. et al., *Biomass as a feedstock for a bioenergy and bioproducts industry: the technical feasiblity of a billion-ton annual supply*, U.S. Department of Energy, April 2005 – available at http.//www.osti.gov/bridge.
58. Hill, J. et al., Environmental, economic, and energetic costs and benefits of biodiesel and ethanol biofuels, *Proc. Natl. Acad. Sci., USA*, 103, 11206, 2006.
59. *Economic, Financial, Social Analysis and Public Policies for Fuel Ethanol Phase 1*, Natural Resources Canada, Ottawa, November 2004.
60. da Silva, J.G. et al., Energy balance for ethyl alcohol production from crops, *Science*, 201, 903, 1978.
61. Sama, D.A., Net energy balance, ethanol from biomass, in Kitani, O. and Hall, C.W. (Eds.), *Biomass Handbook*, Gordon and Breach, New York, 1989, chap. 3.2.10.
62. Lewis, C., Energy relationships of fuel from biomass, *Proc. Biochem.*, 11 (part 9, November), 29, 1976.
63. Hopkinson, C.S. and Day, J.W., Net energy analysis of alcohol production from sugarcane, *Science*, 207, 302, 1980.
64. Chambers, R.S. et al., Gasohol: does it or doesn't it produce positive net energy?, *Science*, 206, 789, 1979.
65. Geller, H.S., Ethanol fuel from sugar cane in Brazil, *Ann. Rev. Energy*, 10, 135, 1985.
66. von Blottnitz, H. and Curran, M.A., A review of assessments conducted on bio-ethanol as a transportation fuel from a net energy, greenhouse gas, and environmental life-cycle perspective, *J. Cleaner Prod.*, 15, 607, 2007.
67. Ho, S.P., Global warming impact of ethanol versus gasoline, presented at 1989 National Conference *Clean Air Issues and America's Motor Fuel Business*, Washington, DC, October 1989.
68. Marland, G. and Turhollow, A.F., CO_2 emissions from the production and combustion of fuel ethanol from corn, Oak Ridge National Laboratory, Oak Ridge, TN, Environmental Sciences Division, no. 3301, U.S. Department of Energy, May 1990.
69. Pimentel, D., Ethanol fuels: energy security, economics, and the environment, *J. Agri. Environ. Ethics*, 4, 1, 1991.

70. Keeney, D.R. and DeLuca, T. H, Biomass as an energy source for the Midwestern U.S., *Am. J. Altern. Agri.*, 7, 137, 1992.
71. Lorenz, D. and Morris, D., *How Much Energy Does it Take to Make a Gallon of Ethanol?* Institute for Local Self-Reliance, Washington, DC, May 1995.
72. Shapouri, H., Duffield, J.A., and Graboski, M.S., *Estimating the Net Energy Balance of Corn Ethanol*, Economic Research Service, AER-721, U.S. Department of Agriculture, Washington, DC, 1995.
73. Henderson, S., *Assessment of Net Emissions of Greenhouse Gases From Ethanol-Gasoline Blends in Southern Ontario*, Agriculture and Agri-Food Canada, 2000.
74. Wang, M., Saricks, C., and Santini, D., *Effects of Fuel Ethanol Use on Fuel-Cycle Energy and Greenhouse Gas Emissions*, U.S. Department of Energy, Argonne National Laboratory, Center for Transportation Research, Argonne, IL, 1999.
75. Pimentel, D., The limits of biomass utilization, in *Encyclopedia of Physical Science and Technology*, 3rd ed., Academic Press, New York, 2001, pp. 159–171.
76. Shapouri, H., Duffield, J.A., and Wang, M., *The Energy Balance of Corn Ethanol: An Update*, Agricultural Economic Report No. 8134, U.S. Department of Agriculture, Economic Research Service, Office of the Chief Economist, Office of Energy Policy and New Uses, Washington, DC, 2002.
77. Berthiaume, R., Bouchard, C., and Rosen, M.A., Exergetic evaluation of the renewability of a biofuel, *Exergy Int. J.*, 1, 256, 2001.
78. Patzek, T.W., Thermodynamics of the corn-ethanol biofuel cycle, *Crit. Rev. Plant. Sci.*, 23, 519, 2004.
79. Pimentel, D. and Patzek, T.W., Ethanol production using corn, switchgrass, and wood; biodiesel production using soybean and sunflower, *Nat. Resour. Res.*, 14, 65, 2005.
80. Dias de Oliveira, M.E., Vaughan, B.E., and Rykiel, E.J., Ethanol as a fuel: carbon dioxide balances, and ecological footprint, *Bioscience*, 55, 593, 2005.
81. Farrell, A.E. et al., Ethanol can contribute to energy and environmental goals, *Science*, 311, 506, 2006.
82. Cleveland, C.J., Hall, C.A.S., and Herendeen, R.A., Letter, *Science*, 312, 1746, 2006.
83. Hagens, N., Costanza, R., and Mulder, K., Letter, *Science*, 312, 1746, 2006.
84. Kaufmann, R.K., Letter, *Science*, 312, 1747, 2006.
85. Patzek, T.W., Letter, *Science*, 312, 1747, 2006.
86. Hammerschlag, R., Ethanol's energy return on investment: a survey of the literature 1990–present, *Environ. Sci. Technol.*, 40, 1744, 2006.
87. *World Energy Outlook 2004*, International Energy Agency, Paris, 2005, Annex A.
88. Wu, M., Wu, Y., and Wang, M., Energy and emission benefits of alternative transportation liquid fuels derived from switchgrass: a fuel life cycle assessment, *Biotechnol. Prog.*, 22, 1012, 2006.
89. Bergeron, P., Environmental impacts of bioethanol, in Wyman, C.E. (Ed.), *Handbook on Bioethanol: Production and Utilization*, Taylor & Francis, Washington, DC, 1996, chap. 5.
90. Henke, J.M., Klepper, G., and Schmitz, N., Tax exemption for biofuels in Germany: is bioethanol really an option for climate policy?, *Energy*, 30, 2617, 2005.
91. Malça, J. and Friere, F., Renewability and life-cycle energy efficiency of bioethanol and bio-ethyl tertiary butyl ether (bioETBE): assessing the implications of allocation, *Energy*, 31, 3362, 2006.
92. Nielsen, P.H., and Wenzel, H., *Environmental Assessment of Ethanol Produced from Corn Starch and Used as an Alternative to Conventional Gasoline for Car Driving*, Institute for Product Development, Technical University of Denmark, June 2005.
93. Wesseler, J., Opportunities (costs) matter: a comment on Pimentel and Patzek "Ethanol production using corn, switchgrass, and wood; biodiesel production using soybean and sunflower," *Energy Policy*, 35, 1414, 2007.

94. *World Energy Outlook,* International Energy Agency, Paris, 2006, Annex A.
95. Wu, M., Wu, Y., and Wang, M., Mobility chains analysis of technologies for passenger cars and light-duty vehicles fueled with biofuels: application of the GREET model to the role of biomass in America's energy future (RBAEF) project, Center for Transportation Research, Argonne National Laboratory, Argonne, IL, May 2005.
96. Wang, M., Wu, M., and Huo, H., Life-cycle energy and greenhouse gas emission impacts of different corn ethanol plant types, *Environ. Res. Lett.,* 2, 024001, 2007.
97. Patzek, T.W. et al., Ethanol from corn: clean renewable fuel for the future, or drain on our resources and pockets?, *Environ. Dev. Sustain.,* 7, 319, 2005.
98. Pimentel, D., Patzek, T., and Cecil, G., Ethanol production: energy, economic, and environmental losses, *Rev. Environ. Contam. Toxicol.,* 189, 25, 2007.
99. Spatari, S., Zhang, Y., and MacLean, H.L., Life cycle assessment of switchgrass- and corn stover-derived ethanol-fueled automobiles, *Environ. Sci. Technol.,* 39, 9750, 2005.
100. Lave, L. et al., Life-cycle analysis of alternative automobile fuel/propulsion technologies, *Environ. Sci. Technol.,* 34, 3598, 2000.
101. Kheshgi, H.S., Prince, R.C., and Marland, G., The potential of biomass fuels in the context of global climate change: focus on transportation fuels, *Ann. Rev. Energy Environ.,* 25, 199, 2000.
102. Righelato, R. and Spracklen, D.V., Carbon mitigation by biofuels or by saving and restoring forests?, *Science,* 317, 902, 2007.
103. Prugh, T., Flavin, C., and Sawin, J.L., Changing the oil economy, in Starke, L. (Ed.), *State of the World 2005,* Earthscan, London, 2005, chap. 6.
104. Pearce, F., Fuels gold: big risks of the biofuel revolution, *New Scientist,* 23 September 2006, 36.
105. Bourne, J.K., Green dreams. Making fuel from crops could be good for the planet — after a breakthrough or two, *National Geographic,* October 2007, 38–59.
106. Lorenz, P. and Zinke, H., White biotechnology: differences in US and EU approaches?, *Trends Biotechnol.,* 23, 570, 2005.
107. Soares, R.R., Simonetti, D.A., and Dumesic, J.A., Glycerol as a source for fuels and chemicals by low-temperature catalytic processing, *Angew. Chem. Int. Ed.,* 45, 3982, 2006.
108. Rabaey, K. and Verstraete, W., Microbial fuel cells: novel biotechnology for energy generation, *Trends Biotechnol.,* 23, 291, 2005.
109. Gray, K.A., Zhao, L., and Emptage, M., Bioethanol, *Curr. Opin. Chem. Biol.,* 10, 141, 2006.
110. Gong, C.S. et al., Ethanol production from renewable resources, *Adv. Biochem. Eng. Biotechnol.,* 65, 207, 1999.
111. Hahn-Hägerdal, B., et al., Bio-ethanol — the fuel of tomorrow from the residues of today, *Trends Biotechnol.,* 24, 549, 2006.

2 Chemistry, Biochemistry, and Microbiology of Lignocellulosic Biomass

2.1 BIOMASS AS AN ENERGY SOURCE: TRADITIONAL AND MODERN VIEWS

Biomass energy in its traditional sense is vegetation (mostly woody plants but also sun-dried grasses) and, extrapolating further up the food chain, animal manure, combusted as a direct source of heat for cooking and heating. For commercial purposes, wood was also the major energy substrate before the rapid development of coal extraction in the late eighteenth and nineteenth centuries ushered in the Industrial Revolution. Even in the early twenty-first century, traditional biomass still accounts for 7% of the total global energy demand, amounting to 765 million tonnes of oil equivalents (Mtoe) in 2002, and this is projected to increase to 907 Mtoe by 2030.[1] Especially in Sub-Saharan Africa, much of this primary energy demand is unsustainable, as population growth outstrips the biological capacity of increasingly drought- and crisis-damaged ecosystems to replace continuous harvestings of firewood.

Biomass has, however, a much more modern face in the form of substrates for power generation, especially in combined heat and power production in OECD regions. For example, biomass-based electricity was 14% in Finland and 3% in Austria in 2002, and as discussed in chapter 1, the burning of sugarcane bagasse in Brazil is a significant energy source, 3% of national electricity in 2002.[1] The use of biomass for power generation is increasingly attractive as a decentralized mechanism of supplying electricity locally or for isolated communities as well as cofiring with coal to reduce CO_2 emissions. In contrast, even taken together, hydroelectricity, solar, geothermal, wind, tide, and wave energy may account for 4% or less of total global energy demand by 2030 (figure 2.1).

Thermal conversion (combustion, burning) of lignocellulosic biomass is an ancient but inefficient means of liberating the energy content of the biological material. Compared with solid and liquid fossil fuels, traditionally used biomass has only 0.33–0.50 of their energy densities (e.g., as expressed by the higher heat value).[2] The higher the water content, the lower the energy density becomes and the more difficult is the task of extracting the total calorific equivalent, as the gas-phase flames are relatively cool. In a well-oxygenated process, the final products of biomass combustion are CO_2, water, and ash (i.e., the inorganic components and salts); intermediate reactions, however, proceed by a complex group of compounds including carbon monoxide (CO), molecular hydrogen, and a wide range of

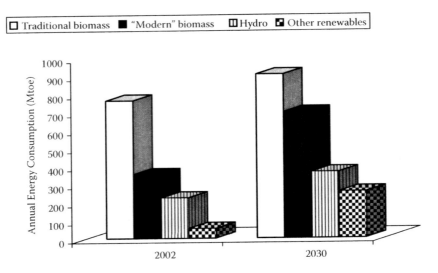

FIGURE 2.1 Predicted global renewable energy consumption trends. (Data from *World Energy Outlook 2004.*[1])

oxygenated hydrocarbons. It is only at temperatures above 1100°C that the thermal conversion steps are efficient and only CO is a significant waste product of incomplete combustion (although even this can be minimized above a threshold residence time in the zone of highest temperatures); below 700°C, the use of catalytic conversion is essential to avoid serious air pollution with gases and volatilized material including tars.[2] With catalytic combustors in residential wood burners designed to high engineering standards, overall efficiencies can be 80% of the theoretical maximum and particulate emissions as low as 1 g/kg.

Direct combustion is, however, only one of three thermochemical options for biomass utilization. Gasification (incomplete combustion) yields different mixes of products depending on the conditions used:[2,3]

- With pure oxygen as the combustant, a producer (or synthesis) gas with a high CO content is produced.
- The use of air rather than oxygen reduces the heating value because nitrogen dilutes the mixture of gases.
- If water is present and high temperatures are reached, hydrogen may also be formed, but excess water tends to result in high CO_2 concentrations and greatly reduces the heating value of the gaseous product.

Producer/synthesis gas resulting from gasification technologies generally has a low heating value (4–10 MJ/L) and is best suited to in situ power heat and/or generation. The third thermochemical method, pyrolysis (i.e., heating in the absence of air or oxygen) can be an efficient means of generating a gas high in hydrogen and CO but can also yield charcoal, a material with many (sometimes ancient) industrial uses.

Biomass utilization has generally been regarded as a low-technology solution for renewable energy and slow to generate public enthusiasm or investor funding for projects in most OECD countries. Nevertheless, energy analysis calculations have shown that combustion, pyrolytic, and gasification technologies are competitive in their energy conversion efficiencies with biotechnological production of fuel ethanol or biogas, that is, methane (figure 2.2).[4] Such biomass-utilizing facilities are highly suitable for inputs that can be broadly described as "waste"; these materials include sawmill residues and forestry, herbaceous agriculture, and construction/demolition waste that might otherwise be simply burned off or dumped in landfill sites. The appropriateness of thermochemical energy production is well illustrated by the commercialization of the SilvaGas process, originally demonstrated in a 10-ton/day plant at the Battelle Memorial Institute (Columbus, Ohio) and subsequently at a 200-ton/day plant in Vermont. The company Biomass Gas & Electric, LLC (Atlanta, Georgia, www.biggreenenergy.com) announced an agreement in October 2006 to construct a 35-MW plant capable of supplying electricity to 8% of residential properties in Tallahassee, Florida.[5] By 2010, this plant will convert 750 tons/day of timber, yard trimmings, and clean construction debris to a medium-energy producer gas by a patented advanced gasification process. The product can substitute for natural gas in most industrial applications, and the process can, therefore, either generate electricity in situ in gas turbines or be injected into the natural gas distribution network. A second (20-MW) facility is under construction northeast

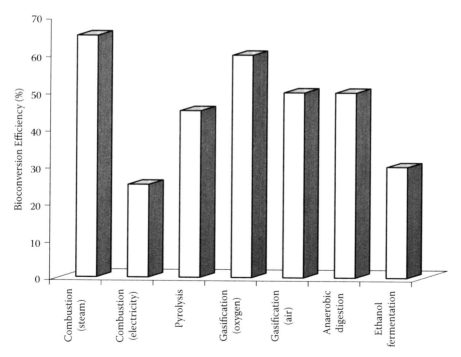

FIGURE 2.2   Bioconversion efficiencies of thermochemical and microbiological processes for biomass. (Data from Lewis.[4])

of Atlanta, adjacent to a landfill site that will divert construction and woody plant material — compared with untreated material, the SilvaGas process reduces solid waste by up to 99%.

To underline the contemporary nature of biomass fuels, U.S. Patent 7,241,321 awarded to Ecoem, LLC (Greenwich, Connecticut) in July 2007 describes an innovative procedure for the production of biomass fuel briquettes: finely ground wood chips, bark, sawdust, or wood charcoal powder can, after drying, be mixed with a vegetable oil to produce a material capable of being pressed into briquettes or ingots. This "organically clean biomass fuel" provides a clean burning, nontoxic fuel, presumably a superior fuel for domestic heating (as well as for igniting the charcoal on barbecues) and offers an alternative route for using vegetable oils as biofuels (chapter 6, section 6.1).

"Biomass" extends well beyond the collection of firewood. Between the late 1960s and the mid-1980s, a program in the United States explored seaweeds as rivals to terrestrial biomass plants because seaweeds have high growth rates, probably more than 100 dry tons/hectare/year, that is, comparable with the highest rates determined for sugarcane and such invasive/problem plant species as water hyacinth.[6] The ocean margins are filled with such highly productive species, including the following:

- The giant kelp (*Macrocystis pyrifera*), whose massive fronds can be repeatedly harvested and, when managed, a "plantation" might require reseeding only every five to ten years, if at all
- Other kelps, including the oriental kelp (*Laminaria japonica*), already harvested in tens of thousand of tons annually as food in China and Japan (both of which countries actively have considered this species as a possible biomass source)
- Other brown algae, including *Sargassum* spp., some of which harbor nitrogen-fixing bacteria and require little artificial fertilizer application
- Red algae such as *Gracilaria tikvahiae*, once considered as a possible summer crop in Long Island Sound and capable of growth without any attachment in closed or semiclosed cultures
- Green algae, many of which have very high biomass production rates

A feature common to many of these species is that anaerobic digestion with methane-forming bacteria is particularly efficient at converting their biomass to a combustible, widely used fuel gas.[6]

Research in China and Japan has continued until the present day, exploring the potential for the development of "marine bioreactors."[7] Genetic transformation techniques have also been adapted from the established body of knowledge acquired with terrestrial plants for indoor cultivation systems with such transgenic kelp.

2.2   "SLOW COMBUSTION" — MICROBIAL BIOENERGETICS

Biomass — as described in the preceding section — has a calorific value that can be at least partly "captured" in other, more immediately useful forms of energy. Two of the examples included in figure 2.2 were fermentations, biological processes involving microbial bioenergetics in the chemical transformations. "Fermentation" has come a long way semantically since its coining by Louis Pasteur as *life without air (la vie sans l'air)*.[8] Pasteur was referring to yeast cells and how they altered their respiratory functions when air (or oxygen) became limiting. The modern use of the term is best understood by considering how microbes oxidize (or metabolize or combust) a substrate such as glucose released by the hydrolysis of, for example, starch or cellulose (figures 1.13 and 1.23). The complete biological oxidation of glucose can be written as

$$C_6H_{12}O_6 + 6O_2 \rightarrow 6CO_2 + 6H_2O$$

As written, this chemical transformation is of little biological use to a population of microbial cells (yeast, fungi, or bacteria) because it implies only the generation of heat (metabolic heat) by the cells and that all the carbon is evolved as CO_2 and cannot be utilized in cell replication. In a real microbiological scenario, glucose represents a valuable organic carbon supply as well as an energy source; part of that carbon is transformed by biochemical reactions inside the cells into the materials for new cells (before cell division), and much of the free energy that is released by the energetically favorable (exothermic) oxidation of the remainder of the glucose is used to drive the energetically unfavorable (endothermic) reactions of biosynthetic pathways; only a small portion of the total energy available is "wasted" as heat by biochemical reactions having mechanisms with thermodynamic efficiencies less than 100%:

$$nC_6H_{12}O_6 + mO_2 \rightarrow \text{cells} + pCO_2 + qH_2O + \text{evolved heat (where } p,q \ll 6n).$$

Such a system is, to the microbial physiologist, simply "growth." If, however, a metabolic output takes the form of a chemical product — especially where that product is commercially desirable or useful — then the process generates cells and product, whereas the CO_2 evolved is (from an industrial or bioprocess standpoint) "waste." In large industrial fermentors (with volumes in excess of 500,000 L), the generation of "metabolic" heat by dense cultures of microbial cells necessitates the expenditure of large amounts of energy to cool the liquid mass and, in turn, stimulates considerable efforts to recycle the "waste" heat on the industrial plant.[9] Microbial processes producing antibiotics, enzymes, amino acids, and other products, recombinant proteins, flavors, and pharmaceutical active ingredients are generally termed "fermentations" although large volumes of oxygen-bearing air are supplied (via heavy-duty compressors) at high rates to the vessels so that microaerobic or anaerobic conditions are deliberately prevented, large quantities of glucose (or other carbon and energy sources) are utilized, and commercial products are accumulated as rapidly as the nutritional and physical conditions allow.

The primary fermentation in ethanol production is, on the other hand, a classic fermentation, that is, one where the ability of the cells to absorb glucose and other sugars from the medium (and to complete the early steps in their metabolic oxidation) exceeds the supply of oxygen required to fully complete the oxidative reactions. Under such conditions, yeasts such as *Saccharomyces cerevisiae* will accumulate (mostly) ethanol, whereas other yeasts, fungi, and bacteria produce combinations of the following:[10]

- Alcohols (ethanol, glycerol, n-propanol, n-butanol)
- Acids (formic, acetic, lactic, propionic, butyric)
- Decarboxylated acids (acetoin, acetone, diacetyl, 2,3-butanediol)

All these products (including ethanol) are the products (or intermediates) of a cluster of closely linked metabolic pathways and most of them are formed biosynthetically by reduction, thus regenerating the finite pool of redox carriers — in particular, nicotinamide adenine dinucleotide (NAD) — in their oxidized forms inside the cell (figure 2.3). With the exception of ethanol (a much earlier development), all of these products were also industrially produced by microbial processes that were commercialized in the twentieth century; they represent "overflow" products of metabolism, given the supply of a high-value carbon and energy source such as glucose and an imbalanced nutritional environment where only part of the available carbon and energy can be fully used in cell replication and cell division. In particular, any cell population has, as determined by its precise genetic profile, a finite maximum specific oxygen uptake rate (i.e., grams of O_2 per grams of cells per hour); feeding glucose as the growth-limiting nutrient to induce specific growth rates requiring the cells to exceed this maximum specific oxygen uptake rate causes the accumulation of "overflow" metabolites such as acetic and formic acids (figure 2.3).[11] Ethanol production by S. cerevisiae, however, exhibits a highly important relationship: even under aerobic conditions (where the capacity to fully metabolize, or combust, glucose is not compromised by the lack of O_2), feeding glucose at an increasingly rapid rate eventually swamps the ability of the cells to both grow and respire the glucose, that is, to provide the energy required to support growth and all the required anabolic, biosynthetic reactions. Below this crucial rate of glucose supply (most readily demonstrated in continuous cultures where the rate of glucose entry, and therefore, growth, is determined by the experimenter without any limitation of O_2 or other nutrient supply), no ethanol is formed;* above this threshold, ethanol is accumulated, the "Crabtree effect" (see section 3.1.1), which is so vital to the long historical use of "wine" and other yeasts for ethanol production:[10]

$$nC_6H_{12}O_6 + mO_2 \rightarrow cells + pCO_2 + qH_2O + rC_2H_5OH + evolved\ heat$$

Much has been written quasi-philosophically (and from a distinctly anthropocentric viewpoint) on how the central metabolic pathways in microbes and plants have evolved and on their various thermodynamic inefficiencies.[13] The single most pertinent point is, however, very straightforward: in a natural environment, there is no selective advantage in microbes having the capability of extracting the full thermodynamic energy in foodstuffs; the problem is that of kinetically utilizing (or using up) any available food source as rapidly as possible, thereby extracting as much energy

---

* Ethanol is formed if a culture growing below the crucial rate of glucose entry is transferred to anaerobic conditions; the faster the subcritical rate of feeding, the faster ethanol is formed after transfer to a "fermentation" environment.[12]

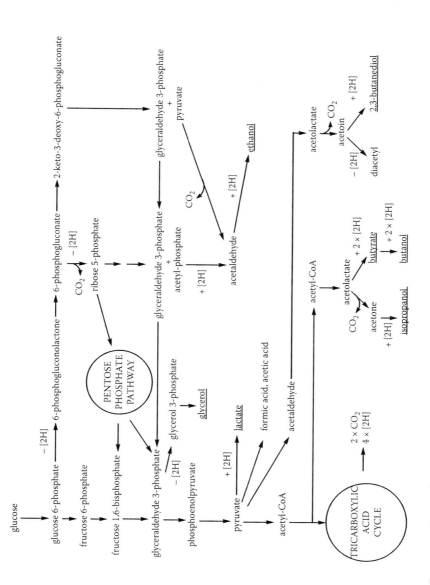

FIGURE 2.3 Enzyme pathways for glucose metabolism in microbial fermentations (+[2H], −[2H]) represent reductive and oxidative steps, respectively.

and nutritional benefit in the shortest possible time in periods of alternating "feast and famine."[14] Many microbes are adventitious feeders, being capable of adapting even to extremes of oxygen availability in aerobic and anaerobic lifestyles. It is merely a human description that overspill products such as ethanol, acetic and lactic acids, and solvents such as acetone are "useful"; in a natural ecosystem, each might be the food-stuff for successive members of microbial consortia and lower and higher plants — in that sense of the term, nature is never wasteful. Axenic cultures of microbes in shake flasks and industrial fermentors are (in the same way as fields of monoculture crops such as soybean and sugarcane) "unnatural" because they are the products of modern agricultural practice and industrial microbiology and offer the economic advantages and potential environmental drawbacks of both of those twentieth-century activities.

Once the growth rate in a fermentation of glucose by an ethanol-forming organism becomes very low, and if glucose continues to be supplied (or is present in excess), the biochemistry of the process can be written minimally:

$$C_6H_{12}O_6 \rightarrow 2CO_2 + 2C_2H_5OH$$

This represents a fully fermentative state where no glucose is respired and no fermentation side products are formed: for every molecule of glucose consumed, two molecules of ethanol are formed; for every 180 g of glucose consumed, 92 g of ethanol are formed, or 51.1 g of ethanol are produced per 100 g of glucose utilized. Any measurable growth or side-product formation will reduce the conversion efficiency of glucose to ethanol from this theoretical maximum.

Microbial energetics and biochemistry can, therefore, predict maximum yields of ethanol and other fermentation products from the sugars available in cane juices, starch, and cellulose (figure 2.3). How closely this theoretical maximum is reached is a function of fermentation design, control, engineering, and the biochemistry of the microorganism used for ethanol production. If a lignocellulosic substrate is to be used, however, the range of biochemical pathways required increases greatly both in scope and complexity as more sugars (pentoses as well as hexoses, plus sugar acids such as glucuronic acid) are potentially made available; this introduces substantially more variables into the putative industrial process, and preceding even this, there is the major hurdle of physically and chemically/enzymically processing biomass materials to yield — on an acceptable mass and time basis — a mixture of fermentable sugar substrates.

2.3 STRUCTURAL AND INDUSTRIAL CHEMISTRY OF LIGNOCELLULOSIC BIOMASS

2.3.1 LIGNOCELLULOSE AS A CHEMICAL RESOURCE

Cellulose, hemicellulose, and lignin are the polymers that provide the structural rigidity in higher plants that grow vertically from a few centimeters to tens of meters — the giant redwood (*Sequoiadendron giganteum*) reaches up to 300 ft (90 m) in height. Although a multitude of microorganisms can elaborate enzymes to degrade cellulose and hemicelluloses, the success of the lignocellulose architecture in the global eco-system is such that it was only with the advent of *Homo sapiens* with flint and (later) metal axes that the domination of deciduous and coniferous forests (especially in the Northern Hemisphere) was seriously challenged.

At the cellular level, plants derive their remarkable resilience to physical and microbial weathering and attack from having evolved the means to greatly thicken their cell walls, using cellulose in linear polymers of high molecular weight (500,000–1,500,000) that are overlapped and aggregated into macroscopic fibers.[15] Linear strands of cellulose have a close molecular arrangement in fibrillar bundles that are sufficiently regular to have X-ray diffraction patterns characteristic of "crystals." This not only augments the structural cohesion but also limits access by water-soluble components and enzymes, and native cellulose is essentially insoluble in water.

Hemicelluloses are diverse in both sugar components and structure, with polymeric molecular weights below 50,000.[15] The heterogeneity of lignins is even greater; any estimate of molecular weight is highly dependent on the method used for extraction and solubilization, and average molecular weight distributions may be less than 10,000.[16] Lignin and hemicelluloses may form chemically linked complexes that bind water-soluble hemicelluloses into a three-dimensional array, cemented together by lignin, that sheaths the cellulose microfibrils and protects them from enzymic and chemical degradation.[15,17,18]

2.3.2 Physical and Chemical Pretreatment of Lignocellulosic Biomass

Given the refractory nature of native lignocelluloses, it is not surprising that chemical processing techniques using acids or alkalis and elevated temperatures have been essential for their use as industrial materials. Conventionally, the starting point has been feedstock material such as wood chips, sawdust, and chopped stalks and stems from herbaceous plants.[19] Mechanical size reduction is unavoidable and, therefore, has both economic and energy costs unless fragmented waste or by-products (e.g., sawdust) is the starting material.[20]

Diverse techniques have been explored and described for the pretreatment of size-reduced biomass materials with the aim of producing substrates that can be more rapidly and efficiently hydrolyzed — by either chemical or biological (enzymic) means — to yield mixtures of fermentable sugars. Physical and thermochemical methods described in the literature are summarized in table 2.1. These approaches have in common the use of conditions and procedures to greatly increase the surface area to which aqueous reactants and/or enzymes have access, in particular, the percentage of the major cellulosic materials that are opened up to attack and thereby reduced to glucose and oligosaccharides on hydrolysis within feasible time limits in batch or continuous processes.

Milling has been little favored because the fibrous nature of lignocellulosic materials requires lengthy processing times and unacceptably high energy inputs; only compression milling has been taken to a testing scale beyond the laboratory. Nevertheless, several studies have concluded that milling can greatly increase the susceptibility to enzymic depolymerization of cellulose.[21] Irradiation with gamma rays and electron beams was a research topic from the 1950s to the 1980s; fragmentation of polysaccharides and lignin was demonstrated to increase the rates of hydrolysis of

TABLE 2.1
Physical and Thermochemical Pretreatments of Lignocellulosic Biomass

Physical method	Pilot plant use?	Thermochemical method	Pilot plant use?[a]
Milling		Steaming	+
dry	−	Steaming/explosion	+
wet	−		
vibratory ball	−	Wet-heat expansion	−
compression	+	Dry-heat expansion	−
vibro energy	−		
colloid	−	Hydrothermolysis	−
Irradiation			
gamma	−	Pyrolysis	−
electron beam	−		
microwave	−		

a Data from Hsu[19] and Thygesen et al.[24]

cellulose when subsequently treated with acids or enzymes, but contradictory results, differential responses when using different wood species, and high investment costs meant that no irradiation technique progressed to pilot-scale evaluation.

Both milling and irradiation give single product streams with only minor degradation of lignocellulosic polymers. Thermochemical methods, in particular those using steam explosion,* can result in extensive degradation of hemicelluloses.[22] Potentially, therefore, a twin-product stream process can be devised by separating solid and liquid phases, the former containing the bulk of the cellulose and the latter the pentose and hexose components of hemicelluloses, although these may be predominantly present in oligosaccharides.[23] At temperatures close to 200°C, even short (10-minute) pretreatment times have major impacts on surface area and enzyme accessibility (figure 2.4). Lignin–carbohydrate bonds are disrupted, some of the lignin is depolymerized, and much of the morphological coherence of the lignified plant cell wall is destroyed.[22] In addition, aqueous extraction at elevated temperatures removes much of the inorganic salts — this is of particular importance with feedstocks such as wheat straw whose combustion (or combustion with coal, etc.) is impeded by their high salt content and the consequent corrosion problems.[24,25]

Chemical pretreatment methods have usually implied hydrolytic techniques using acids and alkalis, although oxidizing agents have also been considered (table 2.2). The use of such chemical reactants introduces a much higher degree of polysaccharide breakdown and greater opportunities for separately utilizing the various potential substrates in lignocellulosic materials. In fact, chemical fractionation procedures for plant cell walls have often been described and have been of inestimable value in the separation and structural elucidation of the cell wall polymers in plant cells and plant organs.[26] With wheat straw, for example, sequential treatments (figure 2.5) with

---

* Steam explosion was originally patented for fiberboard production in 1926. In the 1970s, batch and continuous processes were developed by the Iotech Corporation and the Stake Corporation, respectively, in Canada.

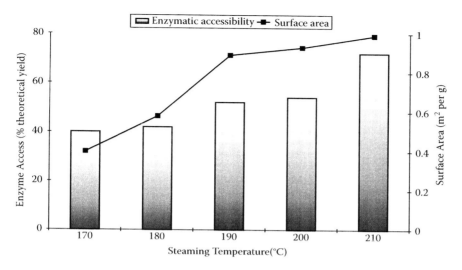

FIGURE 2.4   Efficacy of steaming pretreatments with birch wood. (Data from Puls et al.[23])

TABLE 2.2
Chemical Pretreatment Methods for Lignocellulosic Biomass

Method	Principle	Pilot plant use?[a]
Acids	Hemicellulose solubilization	
dilute sulfuric		−
dilute hydrochloric		−
dilute nitric		−
dilute phosphoric		−
steaming with sulfuric acid impregnation		+
steam explosion/sulfuric acid impregnation		−
steam explosion/sulfur dioxide		−
steam explosion/carbon dioxide		−
Alkalis	Delignification + hemicellulose removal	
sodium hydroxide		−
sodium hydroxide + peroxide		−
steam explosion/sodium hydroxide		−
aqueous ammonia		−
calcium hydroxide		−
Solvents	Delignification	
methanol		−
ethanol		−
acetone		−

[a]   Data from Hsu.[19]

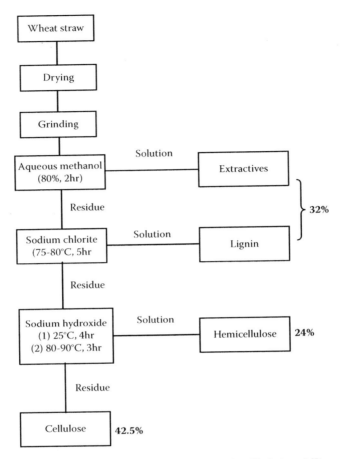

FIGURE 2.5 Chemical fractionation of wheat straw. (After Chahal et al.[27])

an aqueous methanol, sodium chlorite, and alkali yield distinct extractives, lignin, cellulose, and hemicellulose fractions.[27] Similarly, sequential treatments with alkali (lignin removal) and dilute acid (hemicellulose hydrolysis) to leave a highly enriched cellulosic residue have been devised for a variety of feedstocks including switchgrass, corn cob, and aspen woodchip.[28] Because the usual intention is to utilize the sugars present in the polysaccharides as fermentation substrates, however, the developments for bioindustrial applications have invariably focused on faster, simpler, and more advanced engineering options, including some that have been progressed to the pilot-plant scale. Pretreatments involving acids (including SO_2 steam explosion) primarily solubilize the hemicellulose component of the feedstock; the use of organic solvents and alkalis tends, on the other hand, to cosolubilize lignin and hemicelluloses. As with thermochemical methods, the product streams can be separated into liquid and solid (cellulosic) phases; if no separation is included in the process, inevitably, a complex mixture of hexoses and pentoses will be carried forward to the fermentation step.

Combinations of physical, thermochemical, and chemical pretreatments have often been advocated to maximize cellulose digestibility by subsequent chemical or enzymic treatments; this usually involves higher capital and processing costs, and the potential economic benefits of increased substrate accessibility have seldom been assessed in detail. Table 2.3 presents a historical sequence of pilot plants in North America, Japan, New Zealand, Europe, and Scandinavia developed for the processing of lignocellulosic feedstocks to illustrate different approaches to pretreatment choices and cellulose processing; not all of these initiatives included fermentation steps producing ethanol, but every one of them was the result of intentions to generate sugar solutions suitable for subsequent fermentative treatments. Different biomass feedstocks may require different technologies for optimized upstream processing; for example, ammonia-based pretreatments (ammonia fiber explosion and ammonia-recycled percolation) are more effective with agricultural residues (including corn stover and corn straw) than with woody materials.[28] Hardwoods yield higher degrees of saccharification after steam explosion than do softwoods.[22] An organic base, n-butylamine, has been recommended for pretreatment of rice straw on the evidence of efficient delignification, highly enhanced cellulose hydrolysis by cellulase, the ease of recovery of the amine, and the almost complete reprecipitation of the solubilized lignin when the butylamine is removed by distillation.[29,30]

Many accounts of pretreatment optimizations have been published; over a decade ago, a report commissioned by the Energy Research and Development Corporation of Australia estimated that "several thousand" articles dealt with physical and chemical pretreatment methods for lignocellulosics (including by-products from the paper and pulping industries).[31] There are several reasons for this sustained research effort, including the large number of lignocellulosics of potential industrial use, the multiplicity of pretreatment methods (tables 2.1 and 2.2) singly and in combinations, uncoordinated funding from national and international agencies, and the various scales, from the laboratory bench up to demonstration units with the capacity of processing two tonnes/hr of feedstocks (table 2.3). A multiauthor review in 2005 of four thermochemical methods, two pretreatments with acid, two with ammonia, and one with lime (calcium hydroxide) as an alkali — all described as "promising technologies" — concluded that although all nine approaches gave positive outcomes on increasing accessible surface area and solubilizing hemicelluloses and although all but one altered lignin structure, only five could reduce the lignin content, and only two (the ammonia-based methods) decrystallized cellulose.[32] Exceptions and caveats were, however, noted; for example, ammonia fiber/freeze explosion worked well with herbaceous plants and agricultural residues and moderately well with hardwoods, but poorly with softwoods.[20]

Detailed comparisons of different pretreatment methods in controlled, side-by-side studies of multiple technologies using single feedstocks are very rare. A collaboration between the National Renewable Energy Laboratory and six universities in the United States compared ammonia explosion, aqueous ammonia recycle, controlled pH, dilute acid, flow-through with compressed hot water and lime approaches to prepare corn stover for subsequent biological conversion to sugars; material balances and energy balances were estimated for the processes, and the digestibilities of the solids were assessed by a standardized cellulase procedure.[33–39] With this feedstock (already a major "waste" product resulting from the corn ethanol industry), all six pretreatment options resulted in high yields of glucose from cellulose by subsequent treatment with cellulase; in addition, the use of high-pH methods offered potential

TABLE 2.3
Pilot Plants Developed for the Saccharification of Lignocellulosic Biomass

Pilot plant process	Decade	Feedstock	Pretreatment method	Cellulose hydrolysis
U.S. Forest Products Laboratory	1940s	Douglas fir	Steaming	Dilute sulfuric acid
Tennessee Valley Authority	1950s	Wood chips	Steam/dilute sulfuric acid	Dilute sulfuric acid
Iotech (Canada)	1970s	Wood	Steam explosion	Cellulase
American Can Co.	1980s	Sawdust, newspaper, straw, bagasse	Steam/dilute sulfuric acid	Dilute sulfuric acid
New York University	1980s	Wood, recycled newspaper pulp, etc.	Steam/dilute sulfuric acid	Dilute sulfuric acid
New Zealand Wood Hydrolysis Process	1980s	Softwood, hardwood chips and sawdust	Steaming	Dilute sulfuric acid
Solar Energy Research Institute[a]	1980s	Mixed hardwood flour	Steam/dilute sulfuric acid	Dilute sulfuric acid
Research Association for Petroleum Alternatives Development (Japan)	1980s	Bagasse, rice straw	Alkali	Cellulase
New Energy and Industrial Technology Development Organization (Japan)	1980s	Wood chips	Steam explosion	Cellulase-producer[b]
Voest-Alpine Biomass Technology Center (Austria)	1980s	Wheat straw, beech bark, etc.	Cooking	Cellulase-producer
GeoProducts Corporation/ University of California	1980s	Wood chips	Steam/dilute sulfuric acid	Dilute sulfuric acid
Institut Francais du Petrole	1990s	Corn cob	Steam explosion	Cellulase
DOE/National Renewable Energy Laboratory	1990s	Woody and herbaceous crops, etc.	Dilute acid hydrolysis	Cellulase
Denini S/A Industrias de Base (Brazil)	1990s	Bagasse	Organic solvent	Dilute sulfuric acid
NERL/Purdue University/ Aventine Renewable Energy, Inc.	2000s	Corn fiber	pH-controlled hydrothermolysis	Cellulase
Riso National Laboratory (Denmark)	2000s	Wheat straw	Hydrothermolysis	Cellulase
SEKAB E-technology, Ornskoldsvik (Sweden)	2000s	Softwoods	Dilute acid hydrolysis	Dilute acid or cellulase
Iogen Corporation demonstration facility (Canada)	2000s	Wheat straw	Steam explosion/ dilute sulfuric acid	Cellulase

[a] Now National Renewable Energy Laboratory

[b] Direct microbial conversion by cellulase-secreting ethanol producer

for reducing cellulase amounts required in cellulose hydrolysis.[40] Differences were, however, observed in the kinetics of sugar release that were sufficient to influence the choice of process, enzymes, and fermentative organisms. This conclusion was foreshadowed by Swedish research on steam pretreatment of fast-growing willow (*Salix*) with or without SO_2 impregnation that showed that, while glucose yields of more than 90% and overall xylose yields of more than 80% could be obtained both with and without SO_2, the most favorable pretreatment conditions for the separate yields of glucose and xylose were closest when using SO_2-impregnated wood chips.[41] To a large extent, therefore, all pretreatment strategies are likely to include a partial compromise because of the very different susceptibilities to hydrolytic breakdown and solubilization of cellulose and hemicelluloses; highly efficient industrial solutions will require biotechnological approaches to provide fermenting organisms capable of using both hexoses and pentoses and both monomeric and oligomeric (and possibly polymeric) carbohydrates (this is discussed in detail in chapter 3). Even for a single choice of pretreatment method, variation in the biological material (the feedstock) will inevitably occur in, for example, the water content that will either necessitate a flexible technology or extra cost outlay to standardize and micromanage the inflow of biomass material.[42]

The most recent development in pretreatment technologies has been the demonstration that microcrystalline cellulose can be readily solubilized and recovered using a class of chemicals called "ionic liquids." These are salts that are liquids at room temperature and stable up to 300°C; their extreme nonvolatility would also have minimal environmental impact.[43] With one such ionic liquid (an *n*-butyl-methylimidazolium chloride), cellulose could be solubilized by comparatively short (<3-hr) treatments at 300°C, the cellulose could then be recovered by the addition of "antisolvents" such as water, methanol, and ethanol, and the resulting cellulose was 50-fold more susceptible to enzyme-catalyzed hydrolysis as compared with untreated cellulose.[42] Chinese researchers have shown that ionic liquids can successfully pretreat materials such as wheat straw; full commercialization still requires economic synthesis routes and toxicological assessments.[44,45]

2.3.3 BIOLOGICAL PRETREATMENTS

In contrast to (thermo)chemical pretreatments, the use of microbial degradation of lignin to increase feedstock digestibility has several advantages:[19]

- Energy inputs are low.
- Hardware demands are modest.
- No environmentally damaging waste products are generated.
- Hazardous chemicals and conditions are avoided.

All of these features have associated economic cost savings. Against this, the need for lengthy pretreatment times and the degradation of polysaccharides (thus reducing the total fermentable substrate) have acted to keep interest in biological preprocessing of lignocellulosic materials firmly in the laboratory. A careful choice of organism (usually a wood-rotting fungus) or a mixture of suitable organisms can, however, ensure a high degree of specificity of lignin removal.[47,48] Extrapolations of this approach could involve either the preprocessing of in situ agricultural areas for local production facilities or the sequential use of the biomass feedstock first as a substrate for edible mushroom production before further use of the partially depleted material by enzymic hydrolysis to liberate sugars from polysaccharides.

2.3.4 ACID HYDROLYSIS TO SACCHARIFY PRETREATED LIGNOCELLULOSIC BIOMASS

Historically, the use of dilute acid hydrolysis predated enzymic hydrolysis as a methodology used for cellulose processing beyond the laboratory stage of development (table 2.3). In the former Soviet Union, large-scale processes for single-cell protein* as animal feeds using acid-hydrolyzed woody materials were developed.[49] More recently, highly engineered reactors have been devised and investigated for the efficient hydrolysis of lignocellulosic biomass with dilute sulfuric acid, including[50]

- Batch reactors operating at temperatures up to 220°C
- Plug-flow reactors, that is, flow-through reactors in which liquid and solid phases travel at the same velocity and reduce the residence time at high temperature (up to 230°C)
- Percolation reactors, including two-stage reverse-flow and countercurrent geometries

Hydrolysis efficiencies can now rival those in enzymic (cellulase) hydrolyses with the advantages that none of the feedstock need be dedicated to support enzyme production and very low acid concentrations used at high temperatures may be economically competitive with enzyme-based approaches.

Two-stage processes employ mild hydrolysis conditions (e.g., 0.7% sulfuric acid, 190°C) to recover pentose sugars efficiently, whereas the more acid-resistant cellulose requires a second stage at higher temperature (e.g., 215°C); sugars are recovered from both stages for subsequent fermentation steps.[51] Concentrated (30–70%) sulfuric acid hydrolysis can be performed at moderate temperature (40°C) and result in more than 90% recovery of glucose but the procedure is lengthy (2–6 hr) and requires efficient recovery of the acid posttreatment for economic feasibility.[52]

The major drawback remains that of the degradation of hexoses and pentoses to growth-inhibitory products: hydroxymethylfurfural (HMF) from glucose, furfural from xylose, together with acetic acid (figure 2.6). HMF is also known to break down in the presence of water to produce formic acid and other inhibitors of ethanol-producing organisms.[53] In addition, all thermochemical methods of pretreatments suffer, to varying extents, from this problem; even total inhibition of ethanol production in a fermentation step subsequent to biomass presteaming has been observed (figure 2.7).

---

* The conversion of renewable substrates into single-cell protein was the topic in the 1960s and 1970s that most clearly resembles bioethanol production 30–40 years later (see, for example, reference 23). Historically it was the first major failure of industrial biotechnology.

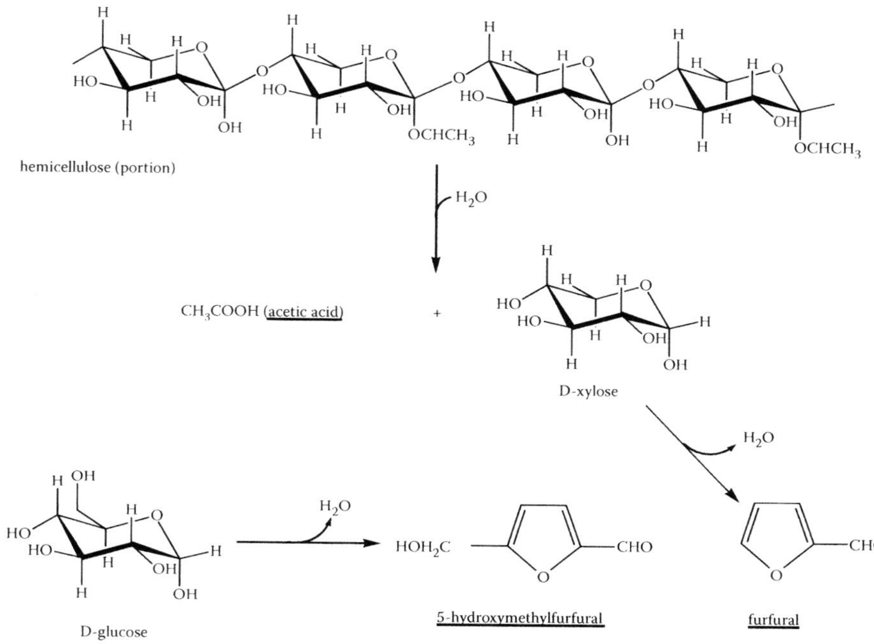

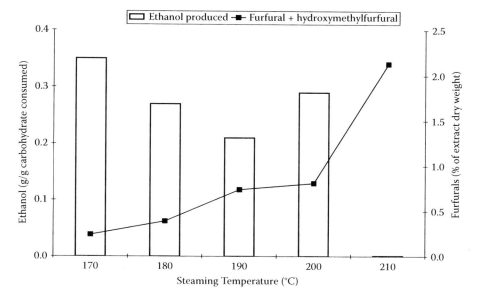

FIGURE 2.6 Chemical degradation of hemicellulose, xylose, and glucose during acid-catalyzed hydrolysis.

FIGURE 2.7 Effects of furfural inhibitors generated by steam pretreatment of wood on ethanol production by *Fusarium oxysporum*. (Data from Puls et al.[23])

Two contrasting views have become apparent for dealing with this: either the growth-inhibitory aldehydes are removed by adsorption or they can be considered to be an additional coproduct stream capable of purification and resale.[54,55]

Bioprocess engineering indicates that simply feeding a cellulosic hydrolysate with high concentrations of furfurals and acetic acid to yeast cells, rather than presenting the full "load" of inhibitors in the batched medium, conditions the microorganism to detoxify and/or metabolize the inhibitory products of sugar degradation.[56,57] A more proactive strategy is to remove the inhibitors by microbiological means, and a U.S. patent details a fungus (*Coniochaeta lignaria*) that can metabolize and detoxify furfural and HMF in agricultural biomass hydrolysates before their saccharification and subsequent use for bioethanol production.[58]

2.4 CELLULASES: BIOCHEMISTRY, MOLECULAR BIOLOGY, AND BIOTECHNOLOGY

2.4.1 ENZYMOLOGY OF CELLULOSE DEGRADATION BY CELLULASES

Enzymic saccharification of pretreated biomass has gradually supplanted acid hydrolysis in pilot plant developments for ethanol production from lignocellulosic substrates (table 2.3). "Cellulase" is a deceptively complex concept, a convenient shorthand term for four enzyme activities and molecular entities, each with their Enzyme Commission (EC) identifying numbers, required for the complete hydrolytic breakdown of macromolecular cellulose to glucose:[59,60]

1. Endoglucanases (1,4-β-D-glucan-4-glucanohydrolases, EC 3.2.1.4) decrease the degree of polymerization of macromolecular cellulose by attacking accessible sites and breaking the linear cellulose chain.
2. Cellodextrinases (1,4-β-D-glucan glucanohydrolases, EC 3.2.1.74) attack the chain ends of the cellulose polymers, liberating glucose.
3. Cellobiohydrolases (1,4-β-D-glucan cellobiohydrolases, EC 3.2.1.91) attack the chain ends of the cellulose polymers, liberating the disaccharide cellobiose, the repeating unit of the linear 1,4-linked polyglucan chain (figure 1.23).
4. Finally, β-glucosidases (EC 3.2.1.21) hydrolyze soluble cellodextrins (1,4-β-D-glucans) and cellobiose to glucose.

Plurals have been used for each discrete enzyme activity because cellulolytic organisms often possess multiple genes and separable enzymically active proteins; for example, the fungus *Hypocrea jecorina** contains two cellobiohydrolases, five endoglucanases, and two β-glucosidases.[61–63]

Cellulases are widely distributed throughout the global biosphere because not only is cellulose the single most abundant polymer, but many organisms have also evolved in widely different habitats to feed on this most abundant of resources. Bacteria and

---

* Historically, this organism (designated until recently as *Trichoderma reesei*) represents the beginning of biotechnological interest in cellulase, as it caused the U.S. Army major equipment and supply problems in the Pacific during World War II by digesting military cotton garments.

fungi produce cellulases in natural environments and while contained in the digestive systems of ruminant animals and wood-decomposing insects (e.g., termites), but insects themselves may also produce cellulases, and other higher life forms — plants and plant pathogenic nematodes — certainly do.[60,64] Higher plants need to reversibly "soften" or irreversibly destroy cell wall structures in defined circumstances as part of normal developmental processes, including plant cell growth, leaf and flower abscission, and fruit ripening; these are highly regulated events in cellular morphology.

The online Swiss-Prot database (consulted in December 2006) listed no fewer than 120 endoglucanases, 22 cellobiohydrolases, 4 cellodextrinases, and 27 β-glucosidases (cellobiases) in its nearly 250,000 entries of proteins fully sequenced at the amino acid level; most of these cellulase components are from bacteria and fungi, but higher plants and the blue mussel (*Mytilus edulis*) are also represented in the collection.[65] The enormous taxonomic diversity of cellulase producers has aroused much speculation; it is likely, for example, that once the ability to produce cellulose had evolved with algae and land plants, cellulase producers arose on separate occasions in different ecological niches; moreover, gene transfer between widely different organisms is thought to occur easily in such densely populated microbial environments as the rumen.[66]

From the biotechnological perspective, fungal and bacterial cellulase producers have been foci of attention as potential industrial sources. More than 60 cellulolytic fungi have been reported, including soft-rot, brown-rot, and white-rot species — the last group includes members that can degrade both cellulose and lignin in wood samples.[67] The penetration of fungal hyphae through the solid growth substrate represented by intact wood results in an enormous surface area of contact between the microbial population and lignocellulosic structures; the release of soluble enzymes then results in an efficient hydrolysis of accessible cellulose as different exo- and endoglucanases attack macroscopic cellulose individually at separate sites, a process often referred to as "synergy" — it is also highly significant that even different members of the same exo- or endoglucanase "family" have different substrate selectivities (table 2.4), thus ensuring a microdiversity among the catalytic population and a maximized capacity to hydrolyze bonds in cellulosic glucans that may have different polymer-polymer interactions.[59] Aerobic bacteria have a similar strategy in that physical adherence to cellulose microfibers is not a prerequisite for cellulose degradation, and a multiplicity of "cellulases" is secreted for maximal cellulose degradation — presently, the most extreme example is a marine bacterium whose extraordinary metabolic versatility is coded by 180 enzymes for polysaccharide hydrolysis, including 13 exo- and endoglucanases, two cellodextrinases, and three cellobiases.[68]

Anaerobic bacteria, however, contain many examples of a quite different biochemical approach: the construction of multienzyme complexes (cellulosomes) on the outer surface of the bacterial cell wall; anaerobic cellulolytics grow optimally when attached to the cellulose substrate, and for some species, this contact is obligatory.[60] The ability of such anaerobic organisms to break down cellulose and to ferment the resulting sugars to a variety of products including ethanol has prompted several investigators to promote them as ideal candidates for ethanol production from lignocellulosic biomass.[69]

The drive to commercialize cellulases — in applications as diverse as the stonewashing of denims, household laundry detergent manufacture, animal feed

TABLE 2.4
Substrate Selectivities of *Trichoderma* Cellulase Components

Substrate	Exoglucanase		Endoglucanase	
	Cellobiohydrolase CBHI	Cellobiohydrolase CBHII	EG1	EGII
Macromolecular				
β-glucan	0	4	5	3
Hydroxyethyl cellulose	0	1	4	2
Carboxymethyl cellulose	1	2	4	5
Crystalline cellulose	4	3	1	1
Amorphous cellulose	1	3	5	3
Small molecule				
Cellobiose	0	0	0	1
p-nitrophenyl glucoside	0	0	0	1
Methylumbelliferyl cellotrioside	0	0	0	1

Source: Data from Tolan and Foody.[70]

Note: Numbers represent relative activity: 0 inactive, 5 maximum activity

production, textile "biopolishing," paper deinking, baking, and fruit juice and beverage processing[70] — has ensured that the biochemistry of the exo- and endoglucanases that attack macromolecular cellulose has been extensively researched. Most of these enzymes share a fundamental molecular architecture comprising two "domains" or "modules": a cellulose-binding region (CBD or CBM) and a catalytic module or core.[71] As more cellulase enzymes have been sequenced at the levels of either amino acids or (almost invariably now) genes, families of conserved polypeptide structures for CBD/CBM have been recognized; they form part of the 34 presently recognized carbohydrate-binding modules collated in a continuously updated database.[72] All proteins in three families (CBM1, CBM5, and CBM10) bind to crystalline cellulose, whereas proteins in the CBM4 and CBM6 families bind to cellulose as well as to xylans and other polysaccharides using different polysaccharide binding sites.[73] Removal of the portion of the cellulase responsible for binding to cellulose reduces cellulase activity with cellulose as the substrate but not with cellodextrins; conversely, the isolated binding domains retain their affinity for cellulose but lack catalytic action.[74] The contribution of cellulose binding to overall cellulase activity has more recently been elegantly demonstrated in a more positive manner: the endoglucanase II from *Hypocrea jecorina* has five amino acid residues in a topographically distinct planar surface in the CBM; selectively altering the amino acids at two positions increased and decreased cellulose binding affinity and cellulose hydrolysis rate in synchrony (table 2.5) and demonstrated that the native enzyme could be catalytically improved by combinatorial mutagenesis.[75]

TABLE 2.5
Bioengineering of Improved Cellulose Binding and Endoglucanase Activity

	Residue 29	Residue 34	Cellulose binding	Cellulose hydrolysis
parent	asparagine	glutamine	100	100
	asparagine	alanine	80	62
	histidine	glutamine	120	115
	valine	glutamine	90	98
	alanine	valine	70	40
	threonine	alanine	150	130

Source: Data from Fukuda et al.[75]

The catalytically active sites in exo- and endoglucanases have different geometries:[63,76,77]

- Cellobiohydrolases have a tunnel-like structure at the active site in which steric hindrance "freezes" the cellulose polymer chain and restricts the hydrolytic reaction to specifically liberating the dimeric cellobiose product.
- In contrast, endoglucanases have more open active sites, and the reaction mechanism is more flexible, resulting in the formation of significant amounts of glucose, cellobiose, and cellotriose as well as random scission of the cellulose polymer chain.

The carboxyl side chains of two acidic amino acids are thought to be directly involved in the chemical mechanism of hydrolysis and cleavage of the β-1,4 glycosidic bond.[77,78]

Joining the separate sites for cellulose binding and cellulose hydrolysis is a linker region, of variable length (6–59 amino acid residues) and with O-glycosylated residues in fungal cellulases.[70,78] The linker appears to optimize the geometry between the catalytic and binding regions of the protein, and the degree of glycosylation (in fungal cellulases) determines the flexibility of the linker peptide; with the CBM temporarily "anchored," the catalytic site hydrolyzes bonds within a linear range set by the linker and then the CBM translocates along the surface of the cellulose to avoid compression or overextension of the linker region in a caterpillar-like motion.[79]

Anaerobic cellulolytic bacteria (*Clostridium, Fibrobacter, Ruminococcus*, etc.) exhibit a very different solution to the multienzyme problem posed by cellulose as an insoluble macrosubstrate. The cellulosome multienzyme complex was first described in 1983 for *Clostridium theromocellum*.[80] The same basic assembly occurs in other species:[81,82]

- Noncatalytic cellulosome integrating protein (Cip) contains cohesin domains that act as receptors for dockerin domains on the catalytic proteins.
- Multiple Cip proteins bind cellobiohydrolases, endoglucanases, hemicellulases, mannanases, pectinase, and other enzymes.
- Hydrophilic modules and cellulose binding modules complete the array.

- The Cip units are anchored to the cell wall via other cohesin domains.
- Large and stable complexes are formed with molecular weights in the range of 2–16 MDa, and polycellulosomes occur with molecular weights up to 100 MDa.

The elaboration of such complex structures is thought to aid the energy-deficient anaerobes by maximizing the uptake of cellodextrins, cellobiose, and glucose by the spatially adjacent bacteria and ensuring a greatly increased binding affinity to the cellulose.[59,60]

The cellulases of cellulolytic anaerobes may also be more catalytically efficient than those of typical aerobes and, in particular, the soluble cellulases secreted by fungi — but this is controversial because enzyme kinetics of the cellulose/cellulase system are problematic: the equations used for soluble enzyme and low-molecular-weight substrates are inadequate to describe the molecular interactions for cellulases "dwarfed" and physically binding to macroscopic and insoluble celluloses.[59] Nevertheless, some results can be interpreted as showing that clostridial cellulases are up to 15-fold more catalytically efficient (based on specific activity measurements, i.e., units of enzyme activity per unit enzyme protein) than are fungal cellulases.[60] Similarly, a comparison of fungal (*Hypocrea*, *Humicola*, and *Irpex* [*Polyporus*]) and aerobic bacterial (*Bacillus*, *Pseudomonas*, and *Thermomonospora*) cellulases noted a higher specific activity of more than two orders of magnitude with the bacterial enzymes.[67]

2.4.2 CELLULASES IN LIGNOCELLULOSIC FEEDSTOCK PROCESSING

Three serious drawbacks have been noted with cellulases — especially of fungal origin — for the efficient saccharification of cellulose and lignocellulosic materials on the industrial and semi-industrial scales. First, cellulases have often been described as being catalytically inferior to other glycosidases. This statement is certainly true when crystalline cellulose is the substrate for cellulase action.[84] When more accessible forms of cellulose are hydrolyzed, the catalytic efficiency increases, but comparison with other glycosidases shows how relatively poor are cellulases even with low-molecular-weight substrates (table 2.6).

Second, cellulases from organisms not normally grown at elevated temperatures have poor stability at incubation temperatures higher than 50°C typically used for cellulose digestion: fungal cellulases show half-life times at 65°C as low as 10 minutes, whereas thermophilic clostridial enzymes may be stable for periods longer than ten days.[84] This is one of the much-explored causes for the rapid decline of hydrolysis rate when cellulase is mixed with cellulose as a substrate.[60] Another possible explanation for this third problem encountered with cellulase-mediated saccharification is the inhibition of cellulases by cellobiose, a major immediate product of cellulase action; a strong product inhibition would be a major drawback, limiting the amounts of soluble sugars that can be produced unless they are rapidly removed; in other words, batch hydrolysis would be inferior to a continuous process with withdrawal of the fermentable sugars or a simultaneous release and utilization of cellobiose would be required. Although such a product inhibition is readily demonstrable

TABLE 2.6

Catalytic Parameters for Fungal Cellulases

Organism	Substrate	kcat (s^{-1})
	Endoglucanases	
Trichoderma reesei	Crystalline cellulose	0.027–0.051
Trichoderma longibrachiatum	Crystalline cellulose	0.05–0.67
Myceliophtora thermophila	Crystalline cellulose	0.013
Trichoderma reesei	Carboxymethyl cellulose	40–60
Trichoderma longibrachiatum	Carboxymethyl cellulose	19–35
Myceliophtora thermophila	Carboxymethyl cellulose	58–140
Glucoamylase	Starch	58
Other glycosidases	Soluble substrates	>100–>1000
	Cellobiohydrolase	
Trichoderma reesei	*p*-nitrophenyl-β-D-lactoside	0.063

Source: Data from Klyosov.[84]

with low-molecular-weight soluble substrates for cellulase, macromolecular cellulose does not show a marked sensitivity to cellobiose.[85] The hydrolysis of cellobiose by β-glucosidase to yield free glucose is itself inhibited by glucose.[86]

The failure to maintain the initial rate of degradation of macromolecular cellulose, however, necessitates an explanation involving the interaction between cellulose and cellulase, and various formulations of a hypothesis have been made in which the "reactivity" of the cellulose decreases during cellulase digestion.[60] A more intuitive line of reasoning is simply that, cellulose being macroscopic and inevitably heterogeneous on an enzyme protein scale, cellulases would attack most rapidly any sites on the cellulose substrate that are, by their very nature, most vulnerable (e.g., with directly accessible regions of the glucan polymer chain); once these sites are cleaved, the active sites remain as intrinsically active as they were initially but most then are restricted to operating at less accessible regions of the available surface. The experimental evidence is contradictory: although scanning electron microscopy has revealed changes in the conformations and packing of cellulose microfibers during prolonged cellulase-catalyzed hydrolysis, "restart" experiments (where the cellulase is removed and fresh enzyme added) show that, at least with crystalline cellulose, no rate-limiting decrease in substrate site accessibility occurs while enzyme action continues.[87,88] Detailed kinetic models have postulated physical hindrance to the movement of exoglucanases along the cellulose and a time variation in the fraction of the β-glucosidic bonds accessible by cellulases bound to the macromolecular cellulose surfaces as factors leading to the failure to maintain hydrolysis rates.[89,90]

2.4.3 MOLECULAR BIOLOGY AND BIOTECHNOLOGY OF CELLULASE PRODUCTION

The consequence of these limitations of cellulases as catalysts for the degradation of cellulose is that large amounts of cellulase have been considered necessary to rapidly process pretreated lignocellulosic substrates, for example, 1.5–3% by weight of the cellulose, thus imposing a high economic cost on cellulose-based ethanol production.[91] From the standpoint of the established enzyme manufacturers, this might have been a welcome and major expansion of their market. The multiplicity of other uses of cellulases meant, however, that competition resulted in massive increases in cellulase fermentation productivity — in the 1980s, space-time yields for cellulases increased by nearly tenfold; between 1972 and 1984, total cellulase production doubled every two years by the selection of strains and the development of fed-batch fermentation systems.[70] Some trends in *Hypocrea jecorina* cellulase productivity and costs were published by the Institut Français du Pétrole from their pilot-plant scale-up work (figure 2.8).[92] By 1995, the productivities in industrial fermentations were probably 400% higher, and process intensification has undoubtedly continued.[93] The development of some of the *H. jecorina* strains from one originally isolated at the U.S. Army Laboratory (Natick, Massachusetts) by physical and chemical mutagenesis treatments has also been described (figure 2.9).[94] From a baseline price of $15/kg in 1990, cellulases have decreased markedly, but they still average three to five times the cost of the much more readily produced (and more enzymically active) amylases.[95]

Major enzyme manufacturers including Genencor (Palo Alto, California), Novozymes (Denmark), Iogen Corporation (Canada), and Rohm (Finland) have long produced cellulases from organisms including *H. jecorina*, *Aspergillus oryzae*, *A. niger*, *Humicola insolens*, and *Penicillium* spp. Different producing organisms yield cellulases with different profiles of cellulase components (figure 2.10), and nonbiofuel markets recognize premium products from particular biological sources with optimum properties for particular applications.[70]

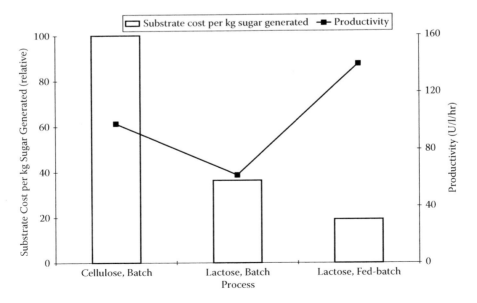

FIGURE 2.8 Process development for fungal cellulase production. (Data from Pourquié et al.[92])

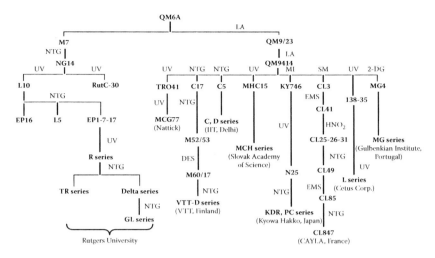

FIGURE 2.9   Mutagenesis and selection of cellulase-hyperproducers of *Trichoderma reesei*. UV, ultraviolet; LA, linear accelerator; NTG, nitrosoguanidine; DES, diethyl sulfate; EMS, ethylmethane sulfate; HNO_2, nitrous acid; MI, monoculture isolation; SM, spontaneous mutation. (After Durand et al.[94])

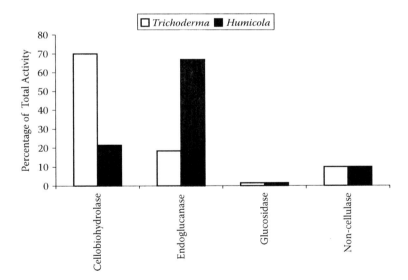

FIGURE 2.10   Differential distribution of cellulase components in cellulases. (Data from Tolan and Foody.[70])

For all microbial producers of cellulase, however, a key area of knowledge is that of the genetic regulation of cellulase synthesis in the life cycle of the organism and — most acutely — when the producing cells are functioning inside the fermentor. For a fungus such as *H. jecorina*, the evidence points to the low-level basal production of cellulase components generating an inducer of rapid cellulase synthesis when cellulose enters the nutritional environment:

*   Deletion of the genes coding for discrete cellulase components prevents the expression of other cellulase genes.[96,97]
*   The general carbon catabolite repressor protein CRE1 represses the transcription of cellulase genes, and a hyperproducing mutant has a *cre1* mutation rendering cellulase production insensitive to glucose.[98,99]
*   Production of cellulases is regulated at the transcriptional level, and two genes encoding transcription factors have been identified.[100–102]
*   The disaccharide sophorose (β-1,2-glucopyranosyl-D-glucose) is a strong inducer of cellulases in *H. jecorina*.[103–105]
*   Cellulases can also catalyze transglycosylation reactions.[106]

Low activities of cellulase are theorized to at least partially degrade cellulose, liberating cellobiose, which is then transglycosylated to sophorose; the inducer stimulates the transcription of cellulase genes, but this is inhibited if glucose accumulates in the environment, that is, there is a triple level of regulation.[107–109] This overall strategy is a typical response of microbes to prevent the "unnecessary" and energy-dependent synthesis and secretion of degradative enzymes if readily utilizable carbon or nitrogen is already present. Augmenting this elaborate mechanism, sophorose also represses β-glycosidase; because sophorose is hydrolyzed by β-glycosidase, this repression acts to maintain sophorose concentrations and thus maximally stimulate cellulase formation.[105] Another disaccharide, gentiobiose (β-1,6-glucopyranosyl-D-glucose), also induces cellulases in *H. jecorina*.[105] In anaerobic cellulolytic bacteria, cellobiose may be the inducer of cellulase, but whether induction occurs in anaerobic bacteria is unclear.[60] In at least one clostridial species, cell wall-attached cellulosomes are formed during growth on cellulose but not on cellobiose; although the cellulase synthesis is determined by carbon catabolite repression, cellulose (or a breakdown product) is a "signal" that can be readily recognized by the cells.[69,110]

Can sophorose be used to increase fungal cellulase expression in fermentations to manufacture the enzyme on a large scale? As a fine chemical, sophorose is orders of magnitude more expensive than glucose,* and its use (even at low concentrations) would be economically unfeasible in the large fermentors (>50,000 l) mandated for commercial enzyme manufacture. Scientists at Genencor discovered, however, that simply treating glucose solutions with *H. jecorina* cellulase could generate sophorose — taking advantage of the transglycosylase activity of β-glucosidase — to augment cellulase expression and production in *H. jecorina* cultures.[111] Lactose is used industrially as a carbon source for cellulase fermentations to bypass the catabolite repression imposed by glucose but adding cellulase-treated glucose increased both cellulase production and the yield of enzyme per unit of sugar consumed (figure 2.11).

---

* The susceptibility of sophorose to β-glucoside-catalyzed degradation precludes the possibility of the disaccharide being used as a recoverable "catalyst" for cellulase expression, although a chemical analog lacking the glycosidic linkage may be more stable and recyclable.

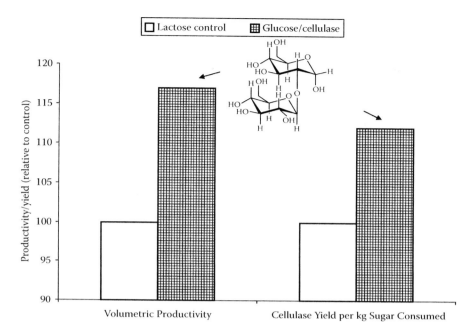

FIGURE 2.11   Productivity of cellulase fermentations in the presence of the inducer sophorose generated from glucose by the transglycosylating action of cellulase. (Data from Mitchinson.[111])

The improving prospects for cellulase usage in lignocellulosic ethanol production has engendered an intense interest in novel sources of cellulases and in cellulase-degrading enzymes with properties better matched to high-intensity cellulose saccharification processes. Both Genencor and Novozymes have demonstrated tangible improvements in the catalytic properties of cellulases, in particular, thermal stability; such enzyme engineering involved site-directed mutagenesis and DNA shuffling (table 2.7). Other discoveries patented by Genencor derived from an extensive and detailed study of gene expression in *H. jecorina* that revealed 12 previously unrecognized enzymes or proteins involved in polysaccharide degradation; some of these novel proteins may not function directly in cellulose hydrolysis but could be involved in the production and secretion of the cellulase complex or be relevant when other polysaccharides serve as growth substrates.[112] The "traditional" and long-established four major components of *H. jecorina* cellulase — two cellobiohydrolases and two endoglucanases — together constitute more than 50% of the total cellular protein produced by the cells under inducing conditions and can reach more than 40 g/l in contemporary industrial strains that are the products of many years of strain development and selection.[112,113] The poor performance of *H. jecorina* as a cellulase producer — once described as the result of nature opting for an organism secreting very large amounts of enzymically incompetent protein rather than choosing an organism elaborating small amounts of highly active enzymes[84] — have engendered many innovative and speculative

TABLE 2.7
Post-2000 Patents and Patent Applications in Cellulase Enzymology and Related Areas

Date, Filing Date	Title	Assignee/Applicant	Patent, Application
1/18/2000	Genetic constructs and genetically modified microbes for enhanced production of β-glucosidase	Iogen Corporation, Canada	US 6,015,703
6/12/2001	Carboxymethyl cellulose from *Thermotoga maritima*	Diversa, San Diego, CA	US 6,245,547
6/26/2002	Polypeptides having cellobiohydrolase I activity …	Novozymes Biotech, Inc., Davis, CA	US 2004/0197890
12/19/2003	Polypeptides having cellobiohydrolase II activity …	Novozymes Biotech, Inc., Davis, CA	US 2006/0053514
4/30/2004	Variants of β-glucosidases	Novozymes Biotech, Inc., Davis, CA	US 2004/0253702
8/25/2004	Variants of glycoside hydrolases	Novozymes Biotech, Inc., Davis, CA	US 2005/0048619
10/26/2004	Cell wall-degrading enzyme variants	Novozymes A/S, Denmark	US 6,808,915
1/28/2005	Polypeptides having cellulolytic enhancing activity …	Novozymes Biotech, Inc., Davis, CA	US 2006/0005279
2/15/2005	Endoglucanases	Novozymes A/S, Denmark	US 6,855,531
9/1/2005	Polypeptides having cellulolytic enhancing activity …	Novozymes Biotech, Inc., Davis, CA	US 2005/0191736
9/29/2005	Polypeptides having β-glucosidase activity …	Novozymes Biotech, Inc., Davis, CA	US 2005/0214920
1/6/2006	Polypeptides having cellobiohydrolase activity …	Novozymes Biotech, Inc., Davis, CA	US 2006/0218671
4/6/2006	Polypeptides having cellobiase activity …	Novozymes Biotech, Inc., Davis, CA	US 2006/0075519
2/28/2003	Cellulase-degrading enzymes of *Aspergillus*	Gielkens et al.	US2004/0001904
3/20/2003	Endoglucanase mutants and mutant hydrolytic depolymerizing enzymes	NERL, Golden, CO	US 2003/0054535
5/22/2003	Thermal tolerant exoglucanase from *Acidothermus cellulyticus*	NERL, Golden, CO	US 2003/0096342
6/12/2003	Thermal tolerant avicelase from *Acidothermus cellulyticus*	NERL, Golden, CO	US 2003/0108988
6/13/2006	Thermal tolerant cellulase from *Acidothermus cellulyticus*	Midwest Research Institute, Kansas City, MO	US 7,059,993
9/20/2005	Method for enhancing cellobiase activity of *Termitomyces clypeatus* using a glycosylation factor	CSIR, New Delhi, India	US 6,946,277

(Continued)

TABLE 2.7   (*Continued*)

12/18/2000	Novel cellulase-producing Actinomycetes ...	Genencor International, Inc., Palo Alto, CA	US 2002/0076792
6/14/2002	Cellulase for use in industrial processes	Genencor International, Inc., Palo Alto, CA	US 2002/0193272
6/26/2003	BGL4 β-glucosidase and nucleic acids encoding the same	Genencor International, Inc., Palo Alto, CA	WO 03/052118
6/26/2003	BGL5 β-glucosidase and nucleic acids encoding the same	Genencor International, Inc., Palo Alto, CA	WO 03/052054
6/26/2003	EGVI endoglucanase and nucleic acids encoding the same	Genencor International, Inc., Palo Alto, CA	WO 03/052057
6/26/2003	EGVII endoglucanase and nucleic acids encoding the same	Genencor International, Inc., Palo Alto, CA	WO 03/052055
6/26/2003	EGVIII endoglucanase and nucleic acids encoding the same	Genencor International, Inc., Palo Alto, CA	WO 03/052056
8/15/2003	Novel variant *Hypocrea jecorina* CBHI cellulases	Genencor International, Inc., Palo Alto, CA	US 2005/0127172
11/5/2003	BGL6 β-glucosidase and nucleic acids encoding the same	Genencor International, Inc., Palo Alto, CA	US 2006/0258554
3/19/2004	Novel CBHI homologs and variant CHBI cellulases	Genencor International, Inc., Palo Alto, CA	US 2005/0054039
1/27/2005	BGL3 β-glucosidase and nucleic acids encoding the same	Genencor International, Inc., Palo Alto, CA	US 2005/0214912
3/23/2005	Exo-endo cellulase fusion protein	Genencor International, Inc., Palo Alto, CA	US 2006/0057672
12/22/2005	Novel variant *Hypocrea jecorina* CBHII cellulases	Genencor International, Inc., Palo Alto, CA	US 2005/0205042
8/22/2006	Variant EGIII-like cellulase compositions	Genencor International, Inc., Palo Alto, CA	US 7,094,588

studies on radical alternatives to the "*Trichoderma* cellulase" paradigm. Proximal to commercially realizable applications are cellulases immobilized on inert carriers that can offer significant cost savings by the repeated use of batches of stabilized enzymes.[114,115] With a commercial β-glucosidase from *Aspergillus niger*, immobilization resulted in two important benefits: greatly improved thermal stability at 65°C and a quite unexpected eightfold increase in maximal enzyme activity at saturating substrate concentration, as well as operational stability during at least six rounds of lignocellulose hydrolysis.[116]

Evidence for an unambiguously novel type of cellulose-binding protein in *H. jecorina* has, however, resulted from the discovery of a family of "swollenins," proteins that bind to macroscopic cellulose and disrupt the structure of the cellulose fibers without any endoglucanase action.[117] This fulfils a prediction made by some of the pioneers of cellulase biotechnology who envisaged that "swelling" factors would be secreted by the fungus to render cellulose more susceptible to cellulase-catalyzed attack.[118] Fusing cellulose-disrupting protein domains with cellulase catalytic domains could generate more powerful artificial exo- and endoglucanases. Whether cellulose-binding domains/modules in known cellulases disrupt cellulose structures remains unclear.[119] Adding a nonionic detergent to steam-pretreated barley greatly increased total polysaccharide saccharification, and the detergent may have been acting partly as a lignocellulose disrupter.[120] An unexpected potential resource for laboratory-based evolution of a new generation of cellulases is the very strong affinity for cellulose exhibited not by a cellulase but by a cellobiose dehydrogenase; nothing has yet been disclosed of attempts to combine this binding activity with cellobiohydrolases.[121]

The relatively low activities of β-glycosidases in fungal cellulase preparations — possibly an unavoidable consequence of the sophorose induction system — have been considered a barrier to the quantitative saccharification of cellulose. Supplementing *H. jecorina* cellulase preparations with β-glucosidase reduces the inhibitory effects caused by the accumulation of cellobiose.[122] Similarly, supplementing cellulase preparations with β-glucosidase eliminated the measured differences in saccharification rates with solvent-pretreated hardwoods.[123] Mixtures of cellulase from different cellulolytic organisms have the advantage of maximally exploiting their native traits.[124] Additionally, such a methodologically flexible approach could optimize cellulase saccharification for a single lignocellulosic feedstock with significant seasonal or yearly variation in its composition or to process differing feedstock materials within a short or medium-length time frame.

Finally, novel sources of cellulases have a barely explored serendipitous potential to increase the efficiency of saccharification; for example, cellulases from "nonstandard" fungi (*Chaetomium thermophilum*, *Thielavia terrestris*, *Thermoascus aurantiacus*, *Corynascus thermophilus*, and *Mycellophthora thermophila*), all thermophiles with high optimum growth temperatures in the range of 45–60°C, improved the sugar yield from steam-pretreated barley straw incubated with a benchmark cellulase/β-glucosidase mix; the experimental enzyme preparations all possessed active endoglucanase activities.[125] "Biogeochemistry" aims to explore the natural diversity of coding sequences available in wild-type DNA.[126] Forest floors are an obvious source of novel microbes and microbial communities adept at recycling lignocelluloses.

2.5 HEMICELLULASES: NEW HORIZONS IN ENERGY BIOTECHNOLOGY

2.5.1 A MULTIPLICITY OF HEMICELLULASES

Mirroring the variety of polysaccharides containing pentoses, hexoses, or both (with or without sugar hydroxyl group modifications) collectively described as hemicelluloses (figure 1.23), hemicellulolytic organisms are known across many species and genera:[15]

- Terrestrial and marine bacteria
- Yeasts and fungi
- Rumen bacteria and protozoa
- Marine algae
- Wood-digesting insects
- Molluscs and crustaceans
- Higher plants (in particular, in germinating seeds)

Table 2.8 summarizes major classes of hemicellulases, their general sites of action, and the released products. Microorganisms capable of degrading hemicelluloses have, however, multiple genes encoding many individual hemicellulases; for example, *Bacillus subtilis* has in its completely sequenced genome at least 16 separate genes for enzymes involved in hemicellulose degradation.[127]

TABLE 2.8
Major Hemicellulases, Their Enzymic Sites of Action and Their Products

Hemicellulase	EC number	Site(s) of action	Released products
Xylanases:			
Endo-β-1,4-xylanase	3.2.1.8	Internal β-1,4-linkages in xylans, L-arabino-D-xylans etc.	Xylose, xylobiose, xylan oligomers, xylan-arabinan oligomers, etc.
Exo-β-1,4-xylosidase	3.2.1.37	External β-1,4-linkages in xylan oligomers, etc.	Xylose
Arabinanases:			
Endoarabinanase	3.2.1.99	Internal α-1,5- and/or α-1,3-linkages in arabinans	Arabinose
α-L-arabinofuranosidase	3.2.1.55	Side chain α-1,2- and/or α-1,3-linkages in xyloarabinans and external α-1,5 linkages in arabinans	Arabinose, xylan oligomers
Mannanases:			
Endo-β-1,4-mannanase	3.2.1.78	Internal β-1,4-linkages in mannans, galactomannans, and glucomannans	Mannose, mannan oligomers, etc.
Exo-β-1,4-mannosidase	3.2.1.25	External β-1,4-linkages in mannan oligomers	Mannose
Galactanases:			
Endo-β-1,4-galactanase	3.2.1.89	Internal β-1,4-linkages in galactans and arabinogalactans	Galactose, galactan oligomers, etc.
α-galactosidase	3.2.1.25	Side chain α-1,6-linkages in galactomannan oligomers	Galactose, mannan oligomers
Other:			
β-glucosidase	3.2.1.21	External β-1,4-linkages in glucomannan oligomers	Glucose, mannan oligomers
α-glucuronidase	3.2.1.139	Side chain 4-O-methyl-α-1,2-linkages in glucuronoxylans	Galactose, mannan oligomers
Esterases:			
Acetyl esterase	3.2.1.6	2- or 3-O acetyl groups on mannan and xylose	Acetic acid, mannose, xylose
Arylesterase	3.2.1.2	3-O-feruoyl/coumaryl-α-L-arabinofuranose side chains	Ferulic, coumaric acids, arabinoxylans

Hemicelluloses comprise both linear and branched heteropolysaccharides. Endoxylanases fragment the xylan backbone, and xylosidases cleave the resulting xylan oligosaccharides into xylose; removal of the side chains is catalyzed by glucuronidases, arabinofuranosidases, and acetylesterases — the action of these enzymes can limit the overall rate of hemicellulose saccharification because endoacting enzymes cannot bind to and cleave xylan polymers close to sites of chain attachment.[128]

Much of the fine detail of hemicellulase catalytic action is beginning to emerge and will be vital for directed molecular evolution of improved hemicellulase biocatalysts.[129,130] Already, however, a thermostable arabinofuranosidase has been identified and shown to have a unique selectivity in being able to degrade both branched and debranched arabinans.[131] Synergistic interactions among different microbial arabinofuranosidases have also been demonstrated to result in a more extensive degradation of wheat arabinoxylan than found with individual enzymes.[132] The activity of biotech companies in patenting novel hemicellulase activities is evident (table 2.9) in exploring hemicellulases from unconventional microbial sources, and deep-sea thermophilic bacteria from the Pacific have been shown to synthesize thermotolerant xylanases, to be active over a wide pH range, and to degrade cereal hemicelluloses.[133]

2.5.2 Hemicellulases in the Processing of Lignocellulosic Biomass

The importance of including hemicellulosic sugars in the conversion of lignocellulosic feedstocks to ethanol to ensure process efficiency and an economic base for biofuel production has often been emphasized.[134,135] Given that hemicellulosic sugars constitute a fermentable resource equal to approximately 50% of the glucose residues present in the cellulose in most plant species (table 1.5), this is an unsurprising conclusion.

Thermochemical and acid-catalyzed pretreatments of lignocellulosic biomass materials extensively degrade hemicelluloses (see above, section 2.3.2); depending

TABLE 2.9
Post-2000 Patents and Patent Applications in Hemicellulase Enzymology

Date, Filing Date	Title	Assignee/Applicant	Patent, Application
12/31/2002	Xyloglucanase from *Malbranchea*	Novozymes A/S, Denmark	US 6,500,658 B2
10/7/2003	Family 5 xyloglucanases	Novozymes A/S, Denmark	US 6,630,340 B2
11/9/2004	Family 44 xyloglucanases	Novozymes A/S, Denmark	US 6,815,192 B2
4/25/2006	Polypeptides having xyloglucanase activity ...	Novozymes, Inc., Palo Alto, CA	US 7,033,811 B2
9/17/2002	Novel recombinant xylanases derived from anaerobic fungi ...	Hseu and Huang	US2004/0053238 A1
6/27/2004	Xylanase from *Trichoderma reesei* ...	Genencor International, Inc., Palo Alto, CA	US 6,768,001 B2
3/11/2004	Novel xylanases and their use	Georis et al.	EP 1 574 567 A1

on the pretreatment method and the feedstock. However, hemicellulose solubilization may be as high as 100% or as low as 10%, and the hemicellulose sugars may be present primarily as monomers (xylose, arabinose, etc.) or as oligomers.[60] A detailed investigation of destarched corn fiber as a starting material, initially extracted with hot water, showed that 75% of the total hemicellulose could be easily extracted; if enzyme preparations generated by growing *Hypocrea jecorina* and *Aspergillus niger* (containing xylanase, β-xylosidase, and feruloyl esterase as well as cellulase) were then used, the recovery of total xylan as xylose increased greatly (from <15% to >50%), and the arabinose yield also improved, although from a much higher starting point of more than 50%; total recoveries of hemicellulosic sugars such as arabinose and xylose reached 80% when supplementations with commercial glucoamylase, β-glucosidase, and feruloyl esterase were included.[136] Similarly, a xylose yield of 88% of the theoretical — as the free monosaccharide xylose — was achieved from pretreated corn stover by supplementing cellulase with xylanase, pectinase, and β-glucosidase activities.[122] Commercial cellulase preparations contain variable but often high activities of hemicellulases; this may have adventitiously contributed to the production of hemicellulose sugars in lignocellulosic materials processed for ethanol production.[137,138]

2.6   LIGNIN-DEGRADING ENZYMES AS AIDS IN SACCHARIFICATION

Cellulases are known to be adsorbed by lignins, thus reducing their catalytic potential for cellulose digestion.[139,140] Steam pretreatment may even increase the potential cellulase-binding properties of lignin.[141] Enzymatic removal of lignin residues from pretreated lignocellulosic materials has received little attention. Laccase (*p*-diphenol:O_2 oxidoreductase, EC 1.10.3.2) catalyzes the cleavage of C–C and C–ether bonds in lignins.[142] Laccase treatment of steam-pretreated softwoods was shown to improve subsequent cellulase-mediated saccharification.[143] The second major class of lignin-degrading enzymes, peroxidases, includes many additional targets for enzyme manufacturers to produce and test as adjuncts to cellulases for the digestion of pretreated lignocellulosic materials.[144]

Coating the surfaces of lignin residues in pretreated lignocellulosic materials by adding a protein before cellulase incubation is an extension of a long-practiced biochemist's technique known as "sacrificial protein" use to protect sensitive enzymes. Adding a commercially available protein, bovine serum albumin, to corn stover pretreated with either dilute sulfuric acid or ammonia fiber explosion and to Douglas fir pretreated by SO_2 steam explosion increases cellulose saccharification and is a promising strategy to reduce enzyme requirements for feedstock processing if a suitable source of low-cost protein is identified.[145]

2.7   COMMERCIAL CHOICES OF LIGNOCELLULOSIC FEEDSTOCKS FOR BIOETHANOL PRODUCTION

The previous sections have itemized the various technologies that have been developed to process lignocellulosic materials to mixtures of glucose, pentose sugars, and oligosaccharides suitable for fermentation by microbes with the production of ethanol. But what commercial and economic forces are presently acting to determine (or limit) the choice of lignocellulosic materials for the first large-scale bioethanol facilities?

The most detailed description so far available for the demonstration plant constructed by the Iogen Corporation in Canada candidly lists the possibilities for starting materials:[146]

- Straws (wheat, barley, etc.) and corn stover as the leading candidates
- Cane bagasse as a localized leading candidate for some tropical locations
- Grass "energy crops" as possible second-generation candidates
- Native forest wood but difficult to process
- Tree farms too expensive because of demands of other markets
- Bark tree waste — cellulose and hemicellulose contents too low
- Sawdust and other mill waste — too expensive because of pulp and paper market demands
- Municipal solid waste and waste paper — too expensive because of paper demand

For a start-up lignocellulosic ethanol facility, there are crucial issues of cost and availability. An industrial plant may require, for example, close to a million tons of feedstock a year, that feedstock should be (for operational stability) as uniform as possible and as free from high levels of toxic impurities and contaminations as possible. Some materials (e.g., wood bark) have compositions that are incompatible with the high yields achieved in starch- and sugar-based materials, and some softwoods demand unattractively high inputs of cellulase.[147] Any lignocellulosic material is subject to some competitive use, and this may dictate cost considerations (table 2.10). Some of these direct competitors are long-established, mature industries, whereas others have unarguably "green" credentials for recycling waste materials or in renewable energy generation in OECD and non-OECD economies.[148] Agricultural waste materials have, in addition, great potential as substrates for the "solid-state" fermentative production of a wide spectrum of fine chemicals, including enzymes, biopesticides, bioinsecticides, and plant growth regulators.[149,150]

The enormous size of the potential supply of lignocellulose is frequently asserted; for example, "Lignocellulose is the most abundant renewable natural resource and substrate available for conversion to fuels. On a worldwide basis, terrestrial plants produce 1.3×10^{10} metric tons (dry weight basis) of wood/year, which is equivalent to 7×10^9 metric tons of coal or about two-thirds of the world's energy requirement."[69]

Some of this "wood," however, represents trees grown (or harvested) as food crops and used as direct domestic or even industrial energy resources, while most is intimately involved in the global carbon cycle and in stabilizing the CO_2 balance in the global ecosystem. Much of what is calculable as available biomass may not, therefore, be commercially harvestable on a short-term basis without the large-scale planting of dedicated "energy crops" such as fast-growing willow trees that are presently planted and harvested in Sweden for burning in district, local, and domestic heating systems.

TABLE 2.10

Competing Uses for Lignocellulosic Biomass Materials Considered for Bioethanol Production

Material	Source	Uses
	Agriculture	
Grain straw, cobs, stalks, husks	Grain harvesting	Animal feed, burning as fuel, composting, soil conditioning
Grain bran	Grain processing	Animal feed
Seeds, peels, stones, rejected fruit	Fruit and vegetable harvesting	Animal feed, fish feed, seeds for oil extraction
Bagasse	Sugar cane industries	Burning as fuel
Sheels, husks, fiver, presscake	Oils and oilseed plants	Animal feed, fertilizer, burning as fuel
	Forestry	
Wood residues, bark, leaves	Logging	Soil conditioning and mulching, burning as fuel
Woodchips, shavings, sawdust	Milling	Pulp and paper, chip and fiber board
Fiber waste, sulfite liquor	Pulping	Use in pulp and board industries as fuel
Paper, cardboard, furniture	Municipal solid waste	Recycling, burning as fuel

Source: Data from Howard et al.[148]

Indeed, these were important issues addressed in a multiauthor projection of biomass options for 2030 to replace 30% of U.S. fuel demands.[151] Ignoring questions of cost and efficiency of bioconversion, this study identified both forestry and agricultural resources for use in biofuels production. Three types of forest resources were quantified:

1. Primary (logging residues, removal of excess biomass in timberland fuel treatments, and fuel wood extracted from forestlands)
2. Secondary (mill residues)
3. Tertiary (urban wood residues from construction/demolition and recycling)

Similarly, lignocellulosic agricultural resources were divided among

1. Primary (crop residues, perennial grasses, and perennial woody crops)
2. Secondary (food/feed processing residues)
3. Tertiary (municipal solid waste recycling)

Approximately 280 million tons (dry weight) of such resources were estimated to be available by the time of the report on an annual basis (figure 2.12).* Augmentation

---

* Much of the report was focused on projections for meeting a much larger biomass supply by combination of altered land use and increased productivity. These topics are covered later, in chapter 5, section 5.4.

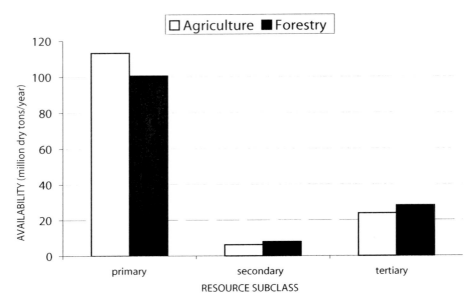

FIGURE 2.12 Estimated availability of biomass resources from agriculture and forestry in the United States for lignocellulosic ethanol production. (Data from Perlack et al.[151])

of this supply could arise from programs to thin native forests strategically so as to reduce fire hazards — in California, for example, more than 750,000 dry tonnes/year could be generated by such activities.[152] Data from Sweden (with its relatively low population density and high degree of forestation) suggest that 25% of the country's gasoline requirements could be substituted by lignocellulose-derived ethanol from existing biomass resources.[153]

Of the two major immediately available sources of lignocellulosic material, however, field crop residues have the distinct advantage of being generated in close proximity to cereal crops intended (partly or entirely) for ethanol production. In August 2005, Abengoa Bioenergy (www.abengoa.com) began constructing the world's first industrial-scale cellulosic bioethanol plant (to use wheat straw as the feedstock) immediately adjacent to its existing 195 million liters/year, Cereal Ethanol Plant (Biocarburantes de Castilla y Leon, BcyL), at Babilfuente, Salamanca, Spain, to dovetail supply trains and technologies. The biomass plant will process more than 25,000 tonnes of wheat straw and other materials to produce 5 million liters of ethanol annually in addition to preparing lignin, pentose sugars, and animal feed products as manufacturing outputs.

As an "energy crop" feedstock for bioethanol production in North America, a grass such as switchgrass (*Panicum virgatum*) has persuasive advantages — economic, social, and agricultural.[154] But the technology for harvesting and processing grasses must be scaled up considerably from that used in, for example, silage fermentation. In the absence of such purpose-dedicated crops, intelligent choices will be mandatory to access sufficiently large supplies of suitable cellulosic feedstocks for start-up facilities.

How much biomass can any nation or region abstract without harming the environment? In a European context, this is an urgent question because any expansion of biomass use for industrial purposes brings in the threat of placing additional pressures on soil and water resources, farmland and forestry biodiversity, and may run counter to extant legislation aiming to encourage environmentally sound farming practices. The European Environment Agency naturally became interested in this issue as projections for large-scale biomass harvesting for bioenergy began to be more ambitious.[155] Taking a cautious view of "harming the environment," that is, with protected forest areas being maintained, residue removal excluded, with no grasslands or olive groves transformed into arable land, and with at least 20% of arable land maintained under environmentally friendly cultivation, the EEA estimated that by 2030 large amounts of biomass could be made available for ambitious bioenergy programs, reaching approximately 300 million tonnes of oil equivalent annually, or 15% of the total primary energy requirements. The calculated increase during the 27 years from 2003 were assumed to be the result of improved agricultural and forestry productivity and liberalization steps leading to more cultivable land being used with higher oil prices and imposed carbon taxes encouraging this expansion. The most revealing aspect of the data presented by the EEA was, however, a ranking of different plant species planted as annual crops with bioenergy as a significant end use: only a mixture of species could ensure that no increased environmental risks were likely, and it is interesting that maize (corn) ranks highest (most potentially damaging) as a monoculture (figure 2.13). The implication was that some of the

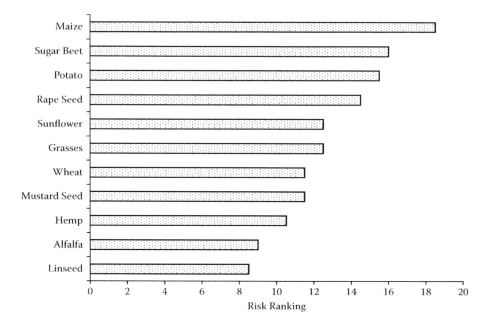

FIGURE 2.13 Environmental risk ranking of annual crops planted for bioenergy production in temperate western regions of Europe: Portugal, northern Spain, France, Belgium, the Netherlands, southern England, and Ireland. (Data from Wiesenthal et al.[155])

favored biofuel crops (maize, sugarbeet, and rapeseed) were in the highest risk category, and careful monitoring of the areas planted with those crops would be advised to minimize environmental damage; grasses were middle ranking purely on the grounds of increased fire risk.

2.8 BIOTECHNOLOGY AND PLATFORM TECHNOLOGIES FOR LIGNOCELLULOSIC ETHANOL

From the perspective of 2007, an important strategic crossroads in the means of providing bioenergy has been reached. A wide complement of technologies has been developed to process realistically available amounts of lignocellulosic materials on at least a semi-industrial scale preparatory to bioethanol production.[156] A crucial component in this is the massively increased production and continued improvement of the "molecular machines" of cellulases (endoglucanases).[157] Chemical (pyrolytic and thermal "cracking") methodologies are being critically developed for biomass substrates to generate synthesis/producer gas both as a direct energy source and as an intermediary stage for "green" chemical refineries.[158] Despite these advances, simple wood-burning stoves for domestic use and furnaces for district heating projects still represent far larger energy gains than does any form of biofuel production.[159]

Nascent industrial facilities for lignocellulosic ethanol are focused on exploiting the supply of cereal crop waste materials — in particular, wheat straw — that together comprise only 30% of the presently available lignocellulosic biomass resources.[151] Worldwide, the industry may have to adapt to a succession of different seasonally available feedstocks, each of which presents unique challenges to pretreatment processing; softwood trees are, for example, dominant contributors to vegetation in Canada, northern Europe, Russia, and Scandinavia, and have both dedicated and passionate advocates and a history of several decades of scientific research; their technoeconomic features and factors are discussed in chapter 4.

Biotechnology must, therefore, demonstrate that multiple carbon sources (hexoses, pentoses, sugar acids, oligosaccharides) can be efficiently converted to ethanol, a proposal that flies in the face of the fermentation industry's tradition of simple (single), highly reproducible carbon inputs. Moreover, biochemical engineering solutions must be found to maximize the value extracted from the processed cellulose, hemicelluloses, and lignin. The former area is that of metabolic engineering, the latter that of bioprocess control; both are considered next as integral parts of evolving models and paradigms of bioethanol production.

REFERENCES

1. *World Energy Outlook 2004*, International Energy Agency, Paris, 2005, chap. 7.
2. Malte, P.C., Combustion, in Kitani, O. and Hall, C.W. (Eds.), *Biomass Handbook*, Gordon and Breach, New York, 1989, chap. 2.3.1.
3. Li, C.-Z., Gasification: a route to clean energy, *Process Safety and Environmental Protection, Trans. IChemeE*, 84, 407, 2006.
4. Lewis, C., Energy analysis of biomass systems, in Kitani, O. and Hall, C.W. (Eds.), *Biomass Handbook*, Gordon and Breach, New York, 1989, chap. 3.4.6.

5. Nelson, A., Florida contract energizes BG&E's strategy, *Atlanta Business* October 20–26, 2006.
6. Gerard, V.A., Seaweeds, in Kitani, O. and Hall, C.W. (Eds.), *Biomass Handbook*, Gordon and Breach, New York, 1989, chap. 1.3.2.
7. Qin, S., Jiang, P., and Tseng, C., Transforming kelp into a marine bioreactor, *Trends Biotechnol.*, 23, 264, 2005.
8. Thimann, K.V., *The Life of Bacteria*, 2nd ed., Macmillan, New York, chap. 5.
9. Curran, J.S., Smith, J., and Holms, W., Heat-and-power in industrial fermentation process, *Applied Energy*, 34, 9, 1989.
10. Moat, A.G. and Foster, J.W., *Microbial Physiology*, 3rd ed., Wiley-Liss, New York, 1995, chap. 1.
11. Eiteman, M.A. and Altman, E., Overcoming acetate in *Escherichia coli* recombinant protein fermentations, *Trends Biotechnol.*, 24, 530, 2006.
12. van Hoek, P., van Dijken, J.P., and Pronk, J.T., Effect of specific growth rate on fermentative capacity of baker's yeast, *Appl. Environ. Microbiol.*, 64, 4226, 1998.
13. Holms, H., Flux analysis and control of the central metabolic pathways in *Escherichia coli*, *FEMS Microbiol. Rev.*, 19, 85, 1996.
14. Koch, A.L., The adaptive responses of *Escherichia coli* to a feast and famine existence, *Adv. Microbiol. Physiol.*, 6, 147, 1971.
15. Singh, A. and Mishra, P., *Microbial Pentose Utilization. Current Applications in Biotechnology, Prog. Ind. Microbiol.*, 33, 1995, chap. 1.
16. Froment, P. and Pla, F., Determinations of average molecular weight distributions of lignin, in Glasser, W.G. and Sarkanen, S. (Eds.), *Lignin. Properties and Materials, ACS Symposium Series 397*, American Chemical Society, Washington, DC, 1989, chap. 10.
17. Koshima, T., Watanabe, T., and Yaku, F., Structure and properties of lignin-carbohydrate complex polymer as an amphipathic substance, in Glasser, W.G. and Sarkanen, S. (Eds.), *Lignin. Properties and Materials, ACS Symposium Series 397*, American Chemical Society, Washington, DC, 1989, chap. 2.
18. Sasaki, T., Cellulase, in Kitani, O. and Hall, C.W. (Eds.), *Biomass Handbook*, Gordon and Breach, New York, 1989, chap. 2.2.2.
19. Hsu, T.-A., Pretreatment of biomass, in Wyman, C.E. (Ed.), *Handbook on Bioethanol: Production and Utilization*, Taylor & Francis, London, 1996, chap. 10.
20. McMillan, J.D., Pretreating lignocellulosic biomass. A review, in *Enzymatic Conversion of Biomass for Fuels Production*, Himmel, M.E., Baker, J.O., and Overend, R.P. (Eds.), *ACS Symposium Series* 566, American Chemical Society, Washington, DC, 1994, chap. 15.
21. Singh, A. and Mishra, P., *Microbial Pentose Utilization. Current Applications in Biotechnology, Prog. Ind. Microbiol.*, 33, 1995, chap. 3.
22. Higuchi, T., Steam explosion of wood, in Kitani, O. and Hall, C.W. (Eds.), *Biomass Handbook*, Gordon and Breach, New York, 1989, chap. 2.5.3.
23. Puls, J. et al., Biotechnical utilization of wood carbohydrates after steaming pretreatment, *Appl. Microbiol. Biotechnol.*, 22, 416, 1985.
24. Thygesen, A. et al., Hydrothermal treatment of wheat straws on pilot plant scale, Proceedings of the world conference and technology exhibition on biomass for energy, industry and climate protection, Rome, Italy, 10-15 May 2004, ETA-Florence, Florence, 2004.
25. Thomsen, M.H. et al., Preliminary results on optimization of pilot scale pretreatment of wheat straw used in coproduction of bioethanol and electricity, *Appl. Biochem. Biotechnol.*, 129–132, 448, 2006.
26. Breet, C. and Waldron, K., *Physiology and Biochemistry of Plant Cell Walls*, Unwin Hyman, London, 1990, chap. 2.
27. Chahal, D.S., Moo-Young, M., and Dhillon, G.S., Bioconversion of wheat straw and wheat straw components into single-cell protein, *Can. J. Microbiol.*, 25, 793, 1979.

28. Gong, C.S. et al., Ethanol production from renewable resources, *Adv. Biochem. Eng. Biotechnol.*, 65, 207, 1999.
29. Tanaka, M. et al., Optimal conditions for pretreatment of rice straw with n-butylamine for enzymatic solubilization, *Appl. Microbiol. Biotechnol.*, 22, 13, 1985.
30. Tanaka, M. et al., Evaluation of effectiveness of pretreating rice straw with n-butylamine for improvement of sugar yield, *Appl. Microbiol. Biotechnol.*, 22, 19, 1985.
31. Brooks, R.B. et al., *Technical Improvements in the Derivation of Ethanol from Lignocellulosic Feedstocks*, Energy Research and Development Corporation, Energy Publications, New South Wales, Australia, 1994.
32. Mosier, N. et al., Features of promising technologies for pretreatment of lignocellulosic biomass, *Bioresour. Technol.*, 96, 673, 2005.
33. Wyman, C.E. et al., Coordinated development of leading biomass pretreatment technologies, *Bioresour. Technol.*, 96, 1959, 2005.
34. Lloyd, T.A. and Wyman, C.E., Combined sugar yields for dilute sulfuric acid pretreatment of corn stover followed by enzymatic hydrolysis of the remaining solids, *Bioresour. Technol.*, 96, 1967, 2005.
35. Liu, C. and Wyman, C.E., Partial flow of compressed-hot water through corn stover to enhance hemicellulose sugar recovery and enzymatic digestibility of cellulose, *Bioresour. Technol.*, 96, 1978, 2005.
36. Mosier, N. et al., Optimization of pH controlled liquid hot water pretreatment of corn, *Bioresour. Technol.*, 96, 1986, 2005.
37. Kim, S. and Holtzapple, M.T., Lime pretreatment and enzymatic hydrolysis of corn stover, *Bioresour. Technol.*, 96, 1994, 2005.
38. Kim, T.H. and Lee, Y.Y., Pretreatment and fractionation of corn stover by ammonia recycle percolation process, *Bioresour. Technol.*, 96, 2007, 2005.
39. Teymouri, F. et al., Optimization of the ammonia fiber explosion (AFEX) treatment parameters for enzymatic hydrolysis of corn stover, *Bioresour. Technol.*, 96, 2014, 2005.
40. Wyman, C.E. et al., Comparative sugar recovery data from laboratory scale application of leading pretreatment technologies to corn stover, *Bioresour. Technol.*, 96, 2026, 2005.
41. Sassner, P., Galbe, M., and Zacchi, G., Steam pretreatment of *Salix* with and without SO_2 impregnation for production of bioethanol, *Appl. Biochem. Biotechnol.*, 121–124, 1101, 2005.
42. Teymouri, F. et al., Ammonia fiber explosion treatment of corn stover, *Appl. Biochem. Biotechnol.*, 113–116, 951, 2004.
43. Heinze, T., Schwikal, K., and Barthel, S., Ionic liquids as reaction medium in cellulose functionalization, *Macromol. Biosci.*, 5, 520, 2005.
44. Dadi, A.P., Varanasi, S., and Schall, C.A., Enhancement of cellulose saccharification kinetics using an ionic liquid pretreatment step, *Biotechnol. Bioeng.*, 95, 904, 2006.
45. Zhu, S. et al., Dissolution of cellulose with ionic liquids and its application: a mini-review. *Green Chem.*, 8, 325, 2006.
46. Liyang, L. and Hongzhang, C., Enzymatic hydrolysis of cellulose materials treated with ionic liquid (BMIM), *Chin. Sci. Bull.*, 51, 2432, 2006.
47. Akin, D.E. et al., Alterations in structure, chemistry, and biodegradability of grass lignocellulose treated with the white rot fungi *Ceriporiopsis subvermispora* and *Cyathus stercoreus*, *Appl. Environ. Microbiol.*, 61, 1591, 1995.
48. Taniguchi, M. et al., Evaluation of pretreatment with *Pleurotus ostreatus* for enzymatic hydrolysis of rice straw, *J. Biosci. Bioeng.*, 100, 637, 2005.
49. Cen, P. and Xia, L., Production of cellulase by solid-state fermentation, *Adv. Biochem. Eng. Biotechnol.*, 65, 69, 1999.
50. Lee, Y.Y., Iyer, P., and Torget, R.W., Dilute-acid hydrolysis of lignocellulosic biomass, *Adv. Biochem. Eng. Biotechnol.*, 65, 93, 1999.

51. Hamelinck, C., van Hooijdonck, G., and Faaij, A.P.C., *Prospects for Ethanol from Lignocellulosic Biomass: Techno-economic Performance as Development Progresses*, Report NWS-E-2003-55, Copernicus Institute, Utrecht University, Utrecht, 2003.

52. Graf, A. and Kohler, T., *Oregon Cellulose-Ethanol Study*, Oregon Office of Efficiency, Salem, OR, 2000.

53. Weil, J.R. et al., Removal of fermentation inhibitors formed during pretreatment of biomass by polymeric adsorbents, *Ind. Eng. Chem. Res.*, 41, 6132, 2002.

54. Brink, D.L., Merriman, M.M., and Gullekson, E.E., *Ethanol Fuel, Organic Chemicals, Single-Cell Proteins: a New Forest Products Industry*, Gen. Tech. Rep. PSW-100, Pacific Southwest Forest and Range Experiment Station, Forest Service, U.S. Department of Agriculture, 1987.

55. Mosier, N.S., Ladisch, C.M., and Ladisch, M.R., Characterization of acid catalytic domains for cellulose hydrolysis and glucose degradation, *Biotechnol. Bioeng.*, 79, 610, 2002.

56. Taherzadeh, M.J., Niklasson, C., and Lidén, G., Conversion of dilute-acid hydrolyzates of spruce and birch to ethanol by fed-batch fermentation, *Bioresour. Technol.*, 69, 59, 1999.

57. Taherzadeh, M.J. et al., Conversion of furfural in aerobic and anaerobic batch fermentation of glucose by *Saccharomyces cerevisiae*, *J. Biosci. Bioeng.*, 87, 169, 1999.

58. Nichols, N.N. et al., Culture containing biomass acid hydrolysate and *Coniochaeta lignaria* fungus, U.S. Patent 7,067,303, June 27, 2006.

59. Mosier, N.S. et al., Reaction kinetics, molecular action, and mechanisms of cellulolytic proteins, *Adv. Biochem. Eng. Biotechnol.*, 65, 23, 1999.

60. Lynd, L.R. et al., Microbial cellulose utilization: fundamentals and biotechnology, *Micro. Mol. Biol. Rev.*, 66, 506, 2002.

61. Takashima, S. et al., Molecular cloning and expression of the novel fungal β-galactosidase genes from *Humicola grisea* and *Trichoderma reesei*, *J. Biochem.*, 125, 728, 1999.

62. Nogawa, M. et al., L-Sorbose induces cellulase gene transcription in the cellulolytic fungus *Trichoderma reesei*, *Curr. Genet.*, 38, 329, 2001.

63. Kleywegt, G.J. et al., The crystal structure of the catalytic core domain of the endoglucanase I from *Trichoderma reesei* at 3.6 Å resolution, and a comparison with related enzymes, *J Mol. Biol.*, 272, 383, 1997.

64. Wilson, D.B. and Irwin, D.C., Genetics and properties of cellulases, *Adv. Biochem. Eng. Biotechnol.*, 65, 1, 1999.

65. Swiss-Prot accessed via the proteomics server of the Swiss Institute of Bioinformatics at www.expasy.org.

66. Garcia-Vallvé, S., Romeu, A., and Palau, J., Horizontal gene transfer of glycosyl transferases of the rumen fungi, *Mol. Biol. Evol.*, 17, 352, 2000.

67. Himmel, M.E. et al., Cellulases: structure, function, and applications, in Wyman, C.E. (Ed.), *Handbook on Bioethanol: Production and Utilization*, Taylor & Francis, London, 1996, chap. 8.

68. Taylor, L.E. et al., Complete cellulase system in the marine bacterium *Saccharophagus degradans* strain 2-40ᵀ, *J. Bacteriol.*, 188, 3849, 2006.

69. Demain, A.L., Newcomb, M., and Wu, J.H.D., Cellulase, clostridia, and ethanol, *Micro. Mol. Biol. Rev.*, 69, 124, 2005.

70. Tolan, J.S. and Foody, B., Cellulase from submerged fermentation, *Adv. Biochem. Eng. Biotechnol.*, 65, 41, 1999.

71. Gilkes, N.R. et al., Domains in microbial β-1,4-glycanases: sequence conservation, function, and enzyme families, *Microbiol. Mol. Biol. Rev.*, 55, 303, 1991.

72. www.afmb.mrs.fr/CAZY.

73. Pires, V.M.R. et al., The crystal structure of the family 6 carbohydrate binding module from *Cellvibrio mixtus* endoglucanase 5A in complex with oligosaccharides reveals two distinct binding sites with different ligand specificities, *J. Biol. Chem.*, 279, 21560, 2004.

74. van Tilbeurgh, H. et al., Limited proteolysis of the cellobiohydrolase I from *Trichoderma ressei*. separation of functional domains, *FEBS Lett.*, 204, 223, 1986.

75. Fukuda, T. et al., Enhancement of cellulase activity by clones selected from the combinatorial library of the cellulose-binding domain by cell surface engineering, *Biotechnol. Prog.*, 22, 933, 2006.

76. Rouvinen, J. et al., Three-dimensional structure of cellobiohydrolase II from *Trichoderma reesei*, *Science*, 249, 380, 1990.

77. Divne, C. et al., The three-dimensional crystal structure of the catalytic core of cellobiohydrolase I from *Trichoderma reesei*, *Science*, 265, 524, 1994.

78. Davies, G. and Henrissat, B., Structures and mechanisms of glycosyl hydrolases, *Structure*, 3, 853, 1995.

79. Srisodsuk, M. et al., Role of the interdomain linker peptide of *Trichoderma reesei* cellobiohydrolase I in its interaction with crystalline cellulose, *J. Biol. Chem.*, 268, 20756, 1993.

80. Receveur, V. et al., Dimension, shape, and conformational flexibility of a two-domain fungal cellulase in solution probed by small angle X-ray scattering, *J. Biol. Chem.*, 277, 40887, 2002.

81. Lamed, R., Setter, E., and Bayer, E.A., Characterization of a cellulose-binding, cellulase-containing complex in *Clostridium thermocellum*, *J. Bacteriol.*, 156, 828, 1983.

82. Bayer, E.A. et al., The cellulosomes: multienzyme machines for degradation of cell wall polysaccharides, *Ann. Rev. Microbiol.*, 58, 521, 2004.

83. Schwartz, W.H., The cellulosome and cellulose degradation by anaerobic bacteria, *Appl. Microbiol. Biotechnol.*, 56, 634, 2001.

84. Klyosov, A.A., Cellulases of the third generation, in Aubert, J.-P., Beguin, P., and Millet, J. (Eds.), *Biochemistry and Genetics of Cellulose Degradation*, Academic Press, London, 97, 1988.

85. Gruno, M. et al., Inhibition of the *Trichoderma reesei* cellulases by cellobiose is strongly dependent on the nature of the substrate, *Biotechnol. Bioeng.*, 86, 503, 2004.

86. Gong, C.-S., Ladisch, M.R., and Tsao, G.T., Cellobiase from *Trichoderma viride*: purification, properties, kinetics, and mechanism, *Biotechnol. Bioeng.*, 19, 959, 1977.

87. Wang, L. et al., Changes in the structural properties and rate of hydrolysis of cotton fibers during extended enzymatic hydrolysis, *Biotechnol. Bioeng.*, 93, 443, 2005.

88. Yang, B., Willies, D.M., and Wyman, C.E., Changes in the enzymatic hydrolysis rate of Avicel cellulose with conversion, *Biotechnol. Bioeng.*, 94, 1122, 2006.

89. Eriksson, T., Karlsson, J., and Tjerneld, F., A model explaining declining rate in hydrolysis of lignocellulose substrates with cellobiohydrolase I (Cel7A) and endoglucanase I (Cel7B) of *Trichoderma reesei*, *Appl. Biochem. Biotechnol.*, 101, 41, 2002.

90. Zhang, Y.-H.P. and Lynd, L.R., A functionally based model for hydrolysis of cellulose by fungal cellulase, *Biotechnol. Bioeng.*, 94, 888, 2006.

91. Mandels, M., Applications of cellulases, *Biochem. Soc. Trans.*, 13, 414, 1985.

92. Pourquié, J. et al., Scale up of cellulase production and utilization, in Aubert, J.-P., Beguin, P., and Millet, J. (Eds.), *Biochemistry and Genetics of Cellulose Degradation*, Academic Press, London, 1988, p. 71.

93. Kadam, K.L., Cellulase production, in Wyman, C.E. (Ed.), *Handbook on Bioethanol: Production and Utilization*, Taylor & Francis, London, 1996, chap. 11.

94. Durand, H. et al., Classical and molecular genetics applied to *Trichoderma reesei* for the selection of improved cellulolytic industrial strains, in Aubert, J.-P., Beguin, P., and Millet, J. (Eds.), *Biochemistry and Genetics of Cellulose Degradation*, Academic Press, London, 1988, 135.

95. Schubert, C., Can biofuels finally take center stage?, *Nat. Biotechnol.*, 24, 777, 2006.

96. Seiboth, B. et al., Role of four major cellulases in triggering cellulase gene expression by cellulose in *Trichoderma reesei*, *J. Bacteriol.*, 179, 5318, 1997.

97. Fowler, T. and Brown, R.D., The *bgl1* gene encoding extracellular β-glucosidase from *Trichoderma reesei* is required for rapid induction of the cellulase complex, *Mol. Microbiol.*, 6, 3225, 1992.

98. Strauss, J. et al., *Cre1*, the carbon catabolite repressor protein from *Trichoderma reesei*, *FEBS Lett.*, 376, 103, 1995.

99. Ilmén, M. et al., Functional analysis of the cellobiohydrolase I promoter of the filamentous fungus *Trichoderma reesei*, *Mol. Gen. Genet.*, 253, 303, 1996.

100. Ilmén, M. et al., Regulation of cellulase gene expression in the filamentous fungus *Trichoderma reesei*, *Appl. Environ. Microbiol.*, 63, 1298, 1997.

101. Saloheimo, A. et al., Isolation of the *ace1* gene encoding a Cys_2-His_2 transcription factor involved in the regulation of activity of the cellulase promoter *cbh1* of *Trichoderma reesei*, *J. Biol. Chem.*, 275, 5817, 2000.

102. Aro, N. et al., ACEII, a novel transcriptional activator involved in regulation of cellulase and xylanase genes of *Trichoderma reesei*, *J. Biol. Chem.*, 276, 24309, 2001.

103. Mandels, M., Parrish, F.W., and Reese, E.T., Sophorose as an inducer of cellulase in *Trichoderma reesei*, *J. Bacteriol.*, 83, 400, 1962.

104. Sternberg, D. and Mandels, G.R., Induction of cellulolytic enzymes in *Trichoderma reesei* by sophorose, *J. Bacteriol.*, 139, 761, 1979.

105. Nisizawa, T. et al., Inductive formation of cellulase by sophorose in *Trichoderma viride*, *J. Biochem.*, 70, 375, 1970.

106. Vaheri, M., Leisola, M., and Kaupinnen, V., Transglycosylation products of cellulase system of *Trichoderma reesei*, *Biotechnol. Lett.*, 1, 41, 1979.

107. Carle-Urioste, J.C. et al., Cellulase induction in *Trichoderma reesei* by cellulose requires its own basal expression, *J. Biol. Chem.*, 272, 10169, 1997.

108. Suto, M. and Tomita, F., Induction and catabolite repression mechanisms of cellulase in fungi, *J. Biosci. Bioeng.*, 92, 305, 2001.

109. El-Gogary, S. et al., Mechanism by which cellulose triggers cellobiohydrolase I gene expression in *Trichoderma reesei*, *Proc. Natl. Acad. Sci., USA*, 86, 6138, 1989.

110. Mantano, Y. et al., Cellulose promotes extracellular assembly of *Clostridium cellulovorans* cellulosomes, *J. Bacteriol.*, 176, 6952, 1994.

111. Mitchinson, C., Improved cellulases for the biorefinery: a review of Genencor's progress in the DOE subcontract for cellulase cost reduction for bioethanol. Stanford GCEP Biomass Energy Workshop, April 2004, accessed at: http:/gcep.Stanford.edu/pdfs/energy_workshops_04_04/biomass_mitchinson.

112. Foreman, P.K. et al., Transcriptional regulation of biomass-degrading enzymes in the filamentous fungus *Trichoderma reesei*, *J. Biol. Chem.*, 278, 31988, 2003.

113. Durand, H., Clanet, M., and Tiraby, G., Genetic improvement of *Trichoderma reesei* for large scale cellulase production, *Enz. Microb. Technol.*, 10, 341, 1988.

114. Saville, B.A. et al., Characterization and performance of immobilized amylase and cellulase, *Appl. Biochem. Biotechnol.*, 113–116, 251, 2004.

115. Yuan, X. et al., Immobilization of cellulase using acrylamide grafted acrylonitrile copolymer membranes, *J. Membrane Sci.*, 55, 101, 1999.

116. Tu, M. et al., Immobilization of β-glucosidase on Eupergit C for lignocellulose hydrolysis, *Biotechnol. Lett.*, 28, 151, 2006.

117. Saloheimo, M. et al., Swollenin, a *Trichoderma reesei* protein with sequence similarity to the plant expansins, exhibits disruption activity on cellulosic materials, *Eur. J. Biochem.*, 269, 4202, 2002.

118. Reese, E.T., Sui, R.G.H., and Levinson, H.S., The biological degradation of soluble cellulose derivatives and its relationship to the mechanism of cellulose hydrolysis, *J Bacteriol.*, 59, 485, 1950.

119. Hildén, L. and Johansson, G., Recent developments on cellulases and carbohydrate-binding modules with cellulose affinity, *Biotechnol. Lett.*, 26, 1663, 2004.

120. Kaar, W.E. and Holtzapple, M.T., Benefits from Tween during enzymic hydrolysis of corn stover, *Biotechnol. Bioeng.*, 59, 419, 1998.
121. Henriksson, G. et al., Studies of cellulose binding by a cellobiose dehydrogenase and comparison with cellobiohydrolase I, *Biochem. J.*, 324, 833, 1997.
122. Berlin, A. et al., Optimization of enzyme complexes for lignocellulose hydrolysis, *Biotechnol. Bioeng.*, 97, 286, 2007.
123. Berlin, A. et al., Evaluation of cellulase preparations for hydrolysis of hardwood substrates, *Appl. Biochem. Biotechnol.*, 129–132, 528, 2006.
124. Kim, E. et al., Factorial optimization of a six-cellulase mixture, *Biotechnol. Bioeng.*, 58, 494, 1998.
125. Rosgaard, L. et al., Efficiency of new fungal cellulase systems in boosting enzymatic degradation of barley straw lignocellulose, *Biotechnol. Prog.*, 22, 493, 2006.
126. Pace, N.R., A molecular view of microbial diversity and the biosphere, *Science*, 276, 734, 1997.
127. Sonenshein, A.L., Hoch, J.A., and Losick, R. (Eds.), Bacillus subtilis *and its Closest Relatives. From Genes to Cells*, ASM Press, Washington, DC, 2002, Appendix 2.
128. Shallom, D. and Shoham, Y., Microbial hemicellulases, *Curr. Opin. Microbiol.*, 6, 219, 2003.
129. Zaide, G. et al., Biochemical characterization and identification of catalytic residues in α-glucuronidase from *Bacillus stearothermophilus* T-6, *Eur. J. Biochem.*, 268, 3006, 2001.
130. Numan, M.T. and Bhosle, N.B., α-L-Arabinofuranosidases: the potential applications in biotechnology, *J. Ind. Microbiol. Biotechnol.*, 33, 247, 2006.
131. Birgisson, H. et al., A new thermostable α-L-arabinofuranosidase from a novel thermophilic bacterium, *Biotechnol. Lett.*, 26, 1347, 2004.
132. Srensen, H.R. et al., A novel GH43 α-L-arabinofuranosidase from *Humicola insolens*: mode of action and synergy with GH51 α-L-arabinofuranosidases on wheat arabinoxylan, *Appl. Microbiol. Biotechnol.*, 73, 850, 2006.
133. Wu, S., Liu, B., and Zhang, X., Characterization of a recombinant thermostable xylanase from deep-sea thermophilic bacterium *Geobacillus* sp. MT-1 in East Pacific, *Appl. Microbiol. Biotechnol.*, 72, 1210, 2006.
134. Hinman, N.D. et al., Xylose fermentation: an economic analysis, *Appl. Biochem. Biotechnol.*, 20–21, 391, 1989.
135. Schell, D.J. et al., A technical and economic analysis of acid-catalyzed steam explosion and dilute sulfuric acid pretreatments using wheat straw or aspen wood chips, *Appl. Biochem. Biotechnol.*, 28–29, 87, 1991.
136. Dien, B.S. et al., Enzymatic saccharification of hot-water pretreated corn fiber for production of monosaccharides, *Enz. Microb. Tech.*, 39, 1137, 2006.
137. Brigham, J.S., Adney, W.S., and Himmel, M.E., Hemicellulases: diversity and applications, in Wyman, C.E. (Ed.), *Handbook on Bioethanol: Production and Utilization*, Taylor & Francis, London, 1996, chap. 7.
138. Kabel, M.A. et al., Standard assays do not predict the efficiency of commercial cellulase preparations towards plant materials, *Biotechnol. Bioeng.*, 93, 56, 2006.
139. Ooshima, H., Burns, D., and Converse, A.O., Adsorption of cellulase from *Trichoderma reesei* on cellulose and lignacious residue in wood pretreated by dilute sulfuric acid with explosive decompression, *Biotechnol. Bioeng.*, 36, 446, 1990.
140. Palonen, H. et al., Adsorption of purified *Trichoderma reesei* CBH I and EG II and their catalytic domains to steam pretreated softwood and isolated lignins, *J. Biotechnol.*, 107, 65, 2004.
141. Tatsumoto, K. et al., Digestion of pretreated aspen substrates: hydrolysis rates and adsorptive loss of cellulase enzymes, *Appl. Biochem. Biotechnol.*, 18, 159, 1988.

142. Kawai, S., Nakagawa, M., and Ohashi, H., Degradation mechanisms of a nonphenolic β-O-4 lignin model dimer by *Trametes versicolor* laccase in the presence of 1-hydroxy-benzotriazole, *Enz. Microb. Technol.*, 30, 482, 2002.
143. Palonen, H. and Viikari, L., Role of oxidative enzymatic treatments on enzymatic hydrolysis of softwood, *Biotechnol. Bioeng.*, 86, 550, 2004.
144. Arora, D.S., Chander, M., and Gill, P.K., Involvement of lignin peroxidase, manganese peroxidase and laccase in the degradation and selective ligninolysis of wheat straw, *Int. Biodeterior. Biodegrad.*, 50, 115, 2002.
145. Yang, B. and Wyman, C.E., BSA treatment to enhance enzymatic hydrolysis of cellulose in lignin containing substrates, *Biotechnol. Bioeng.* 94, 611, 2006 — see also U.S. Patent Application 2006/0088922 ("Lignin Blockers and Uses Thereof," filing date September 19, 2005).
146. Tolan, J.S., Iogen's demonstration process for producing ethanol from cellulosic biomass, in Kamm, B., Gruber, P.R., and Kamm, M. (Eds.), *Biorefineries — Industrial Processes and Products*. Volume 1. *Status Quo and Future Directions*, Wiley-VCH Verlag, Weinheim, 2006, chap. 9.
147. Foody, B., Tolan, J.S., and Bernstein, J.D., Pretreatment process for conversion of cellulose to fuel ethanol, U.S. Patent 5,916,780, issued June 29, 1997.
148. Howard, R.L. et al., Lignocellulose biotechnology: issues of bioconversion and enzyme production, *African J. Biotechnol.*, 2, 603, 2003.
149. Rosales, E., Couto, S.R., and Sanromán, A., New uses of food waste: application to laccase production by *Trametes hisuta*, *Biotechnol. Lett.*, 24, 701, 2002.
150. Pandey, A., Soccol, C.R., and Mitchell, D., New developments in solid state fermentation. I. Bioprocesses and products, *Process Biochem.*, 35, 1153, 2000.
151. Perlack, R.D. et al., *Biomass as a feedstock for a bioenergy and bioproducts industry: the technical feasibility of a billion-ton annual supply*, U.S. Department of Energy, April 2005 — available at http.//www.osti.gov/bridge.
152. Kadam, K.L. et al., Softwood forest thinnings as a biomass source for ethanol production: a feasibility study for California, *Biotechnol. Prog.*, 16, 947, 2000.
153. Wingren, A., Galbe, M., and Zacchi, G., Techno-economic evaluation of producing ethanol from softwood: comparison of SSF and SHF and identification of bottlenecks, *Biotechnol. Prog.*, 19, 1109, 2003.
154. McLaughlin, S.B. et al., High-value renewable energy from prairie grasses, *Environ. Sci. Technol.*, 36, 2122, 2002.
155. Wiesenthal, T. et al., *How Much Bioenergy Can Europe Produce Without Harming the Environment?*, European Environment Agency, Copenhagen, Denmark, Office for Official Publications of the European Communities, Luxembourg, 2006.
156. Gray, K.A., Zhao, L., and Emptage, M., Bioethanol, *Curr. Opin. Chem. Biol.*, 10, 141. 2006.
157. Mulakala, C. and Reilly, P.J., *Hypocrea jecorina* (*Tricoderma reesei*) Cel7A as a molecular machine: a docking study, *Proteins*, 60, 598, 2005.
158. Ragauskas, A.J. et al., The path forward for biofuels and biomaterials, *Science*, 311, 484, 2006.
159. Henke, J.M., Klepper, G., and Schmitz, N., Tax exemption for biofuels in Germany: is bio-ethanol really an option for climate policy?, *Energy*, 30, 2617, 2005.

3 Biotechnology of Bioethanol Production from Lignocellulosic Feedstocks

3.1 TRADITIONAL ETHANOLOGENIC MICROBES

The fundamental challenge in selecting or tailoring a microorganism to produce ethanol from the mixture of sugars resulting from the hydrolysis of lignocellulosic feedstocks is easily articulated: the best ethanol producers are incompetent at utilizing pentose sugars (including those that are major components of hemicelluloses, that is, D-xylose and L-arabinose), whereas species that can efficiently utilize both pentoses and hexoses are less efficient at converting sugars to ethanol, exhibit poor tolerance of high ethanol concentrations, or coproduce high concentrations of metabolites such as acetic, lactic, pyruvic, and succinic acids in amounts to compromise the efficiency of substrate conversion to ethanol.[1–4]

Because bioprospecting microbial species in many natural habitats around the global ecosphere has failed to uncover an ideal ethanologen for fuel ethanol or other industrial uses, considerable ingenuity has been exhibited by molecular geneticists and fermentation specialists in providing at least partial solutions for the two most popular "combinatorial biology" strategies of

- Endowing traditional yeast ethanologens with novel traits, including the ability to utilize pentoses
- "Reforming" bacterial species and nonconventional yeasts to be more efficient at converting both pentoses and hexoses to ethanol

A third option, that is, devising conditions for mixed cultures to function synergistically with mixtures of major carbon substrates, is discussed in chapter 4 (section 4.5).

Adding to the uncertainty is the attitude of the traditionally conservative alcohol fermentation industry toward the introduction of organisms that lack the accepted historic advantages of the yeast *Saccharomyces cerevisiae* in being generally regarded as safe (GRAS) and, by extrapolation, capable of being sold as an ingredient in animal feed once the fermentation process is completed.[5] At various times in the past 40–50 years, thermotolerant yeast strains have been developed to accelerate the fermentation process at elevated temperature.[6] Bioprocesses have been advocated and, to varying degrees, developed, in which polymeric carbohydrate inputs are both hydrolyzed with secreted enzymes and the resulting sugars and oligosaccharides are

taken up and metabolized to ethanol by the cell population, the so-called simultaneous saccharification and fermentation (SSF) strategy.[7] Because the potential advantages of SSF are best understood in the light of the differing fermentation hardware
requirements of multistage and single-stage fermentations, consideration of SSF
technologies is postponed until the next chapter (section 4.5).

3.1.1 Yeasts

The principal wine yeast *S. cerevisiae* is, in addition to its well-known desirable properties (as described above), seemingly the best "platform" choice for lignocellulose-
derived substrates because strains are relatively tolerant of the growth inhibitors found
in the acid hydrolysates of lignocellulosic biomass.[8] Its biotechnological limitations,
on the other hand, derive from its relatively narrow range of fermentable substrates:[9]

- Glucose, fructose, and sucrose are rapidly metabolized, as are galactose
  and mannose (constituents of plant hemicelluloses) and maltose (a disaccharide breakdown product of starch).
- The disaccharides trehalose and isomaltose are slowly utilized, as are
  the trisaccharides raffinose and maltotriose (another breakdown product
  of starch), the pentose sugar ribose, and glucuronic acid (a sugar acid in
  plant hemicelluloses).
- Cellobiose, lactose, xylose, rhamnose, sorbose, and maltotetraose are
  nonutilizable.

   S. cerevisiae is, from the standpoint of classical microbial physiology, best
described as "facultatively fermentative." That is, it can metabolize sugars such as
glucose either entirely to CO_2 and water given an adequate O_2 supply or (under microaerobic conditions) generate large amounts of ethanol; this ability for dual metabolism
is exhibited by a large number of yeast species.[10] In complete anaerobiosis, however,
growth eventually ceases because compounds essential for cell growth (e.g., unsaturated fatty acids and sterols) cannot be synthesized without the involvement of O_2.[11] At
moderate temperature, defined isolates of baker's, brewer's, or wine yeasts can accumulate ethanol as the main fermentation product from glucose, sucrose, galactose,
and molasses (table 3.1).[12–15] Molasses is a particularly relevant sugar source for industrial use; the principal carbohydrate in molasses, from either sugarcane or sugarbeet,
is sucrose, but there are also variable (sometimes large) amounts of glucose, fructose,
as well as trisaccharides and tetrasaccharides based on glucose and fructose.[16]
   Outside this range of sugars provided naturally in grape and cane sugar juices, wine
yeasts such as *S. cerevisiae* and *S. bayanus* can only utilize a limited range of monosaccharides and disaccharides for ethanol production. Across the many known facultatively
fermentative yeasts, the ability to efficiently use sugars other than glucose is highly variable (table 3.2).[17] A seriously problematic limitation for ethanol production is that many
possibly suitable yeasts can only respire disaccharides, that is, can grow on the sugars
under aerobic conditions but cannot produce ethanol under any degree of anaerobiosis;
this is the so-called Kluyver effect.[18] In a survey of 215 glucose-fermenting yeast species, 96 exhibited the Kluyver effect with at least one disaccharide — and two of these
disaccharides are of great importance for ethanol production, maltose (a degradation

TABLE 3.1

Ethanol Production by Yeasts on Different Carbon Sources

Yeast	Carbon source	Temperature (°C)	Fermentation time (hr)	Maximum ethanol (g/l)	Reference
Saccharomyces cerevisiae	Glucose, 200 g/l	30	94	91.8	12
Saccharomyces cerevisiae	Sucrose, 220 g/l	28	96	96.7	13
Saccharomyces cerevisiae	Galactose, 20-150 g/l	30	60	40.0	14
Saccharomyces cerevisiae	Molasses, 1.6-5.0 g/l	30	24	18.4	15
Saccharomyces pastorianus	Glucose, 50 g/l	30	30	21.7	12
Saccharomyces bayanus	Glucose, 50 g/l	30	60	23.0	12
Kluyveromyces fragilis	Glucose, 120 g/l	30	192	49.0	12
Kluyveromyces marxianus	Glucose, 50 g/l	30	40	24.2	12
Candida utilis	Glucose, 50 g/l	30	80	22.7	12

product of starch) and cellobiose (a degradation product of cellulose).[17] This effect is only one of four important O_2-related metabolic phenomena, the others being*

- The Pasteur effect, that is, the inhibition of sugar consumption rate by O_2[19]
- The Crabtree effect, that is, the occurrence (or continuance) of ethanol formation in the presence of O_2 at high growth rate or when an excess of sugar is provided[20]
- The Custers effect, that is, the inhibition of fermentation by the absence of O_2 — found only in a small number of yeast species capable of fermenting glucose to ethanol under fully aerobic conditions[21]

For efficient ethanol producers, the probable optimum combination of phenotypes is

1. Pasteur-positive, (that is, with efficient use of glucose and other readily utilizable sugars for growth when O_2 levels are relatively high)
2. Crabtree-positive, for high rates of ethanol production when supplied with abundant fermentable sugar from as soon as possible in the fermentation
3. Custers-negative, that is, insensitive to fluctuating, sometimes very low, O_2 levels
4. Kluyver-negative, for the widest range of fermentable substrates

---

* There is nothing archaic about this area of microbial physiology, one of the effects being emphasized in the title of a 2002 patent (Production of Lactate Using Crabtree Negative Organisms in Varying Culture Conditions, U.S. Patent 6,485,947, November 26, 2002) awarded to Cargill Dow Polymers, LLC, Minnetonka, MN.

TABLE 3.2
Fermentation of Galactose, Five Disaccharides, and Two Trisaccharides by Yeasts

Yeast	Galactose	Maltose	Sucrose	Trehalose	Melibiose	Lactose	Cellobiose	Melezitose	Raffinose
Ambrosiozyma monospora	-	-	-		-	-	-	-	-
Candida chilensis	K	K	K	K	-	K	K	K	-
Candida salmanticensis	+	+	+	+	+	-	+	+	+
Candida silvicultrix	+	-	+	-	+	-	+	K	+
Candida shehatae	+	+	-	+	-	K	K	K	-
Kluyveromyces marxianus	+	+	+	+	-	+	K	K	K
Pachysolen tannophilus	K	-	-	-	-	-	K	-	-
Pichia hampshirensis	-	K	K	K	-	-	K	K	-
Pichia stipitis	+	+	+	+	-	K	K	K	-
Pichia subpelliculosa	K	+	+	K	+	-	K	K	+
Saccharomyces bayanus	+	+	+	+	+	-	-	K	+
Saccharomyces cerevisiae	+	+	+	K	+	-	-	+	+
Saccharomyces kluyveri	+	K	+	K	+	-	K	-	+
Saccharomyces pastorianus	+	+	+	K	+	-	-	-	+
Schizosaccharomyces pombe	-	+	+	-	+	-	-	-	+
Zygosaccharomyces fermentati	+	+	+	+	+	-	K	+	+

K: exhibits aerobic respiratory growth but no fermentation (Kluyver effect); -: may include delayed use (after 7 days)

Source: Data from Barnett et al.[17]

Not all of these effects can be demonstrated with common wine yeasts (or only under special environmental or laboratory conditions), but they are all of relevance when considering the use of novel (or nonconventional) yeasts or when adapting the growth and fermentative capacities of yeast ethanologens to unstable fermentation conditions (e.g., low O_2 supply, intermittent sugar inflow) for optimum ethanol production rates. Nevertheless, *S. cerevisiae*, with many other yeast species, faces the serious metabolic challenges posed by the use of mixtures of monosaccharides, disaccharides, and oligosaccharides as carbon sources (figure 3.1). Potential biochemical "bottlenecks" arise from the conflicting demands of growth, cell division, the synthesis of cellular constituents in a relatively fixed set of ratios, and the requirement to balance redox cofactors with an inconsistent supply of both sugar substrates and O_2:

- Oligosaccharide hydrolysis
- Disaccharide hydrolysis and uptake
- Hexose transport into the cells
- Conversion of hexoses to glucose 6-phosphate
- The glycolytic pathway for glucose 6-phosphate catabolism
- Pyruvate dehydrogenase (PDH) and alcohol dehydrogenase (ADH)
- The tricarboxylic (Krebs) cycle for aerobic respiration and the provision of precursors for cell growth
- Mitochondrial respiration

In the concrete circumstances represented by an individual yeast species in a particular nutrient medium and growing under known physical conditions, specific combinations of these parameters may prove crucial for limiting growth and fermentative ability. For example, during the >60 years since its discovery, various factors have been hypothesized to influence the Kluyver effect, but a straightforward product inhibition by ethanol could be the root cause. In aerobic cultures, ethanol, suppresses the utilizability of those disaccharides that cannot be fermented, the rate of their catabolism being "tuned" to the yeast culture's respiratory capacity.[22] The physiological basis for this preference is that Kluyver-positive yeasts lack high-capacity transporter systems for some sugars to support the high substrate transport into cells necessary for fermentative growth, whereas energy-efficient respiratory growth simply does not require a high rate of sugar uptake.[23,24]

The function of O_2 in limiting fermentative capacity is complex; in excess, it blocks fermentation in many yeasts, but a limited O_2 supply enhances fermentation in other species.[10,20] Detailed metabolic analyses have shown that the basic pathways of carbon metabolism in ethanologenic yeasts are highly flexible on a quantitative basis of expression, with major shifts in how pathways function to direct the "traffic flow" of glucose-derived metabolites into growth and oxidative or fermentative sugar catabolism.[25,26] The Crabtree effect, that is, alcoholic fermentation despite aerobic conditions, can be viewed as the existing biochemical networks adapting to consume as much of the readily available sugar (a high-value carbon source for microorganisms) as possible — and always with the possibility of being able to reuse the accumulated ethanol as a carbon source when the carbohydrate supply eventually becomes depleted.[22,27,28]

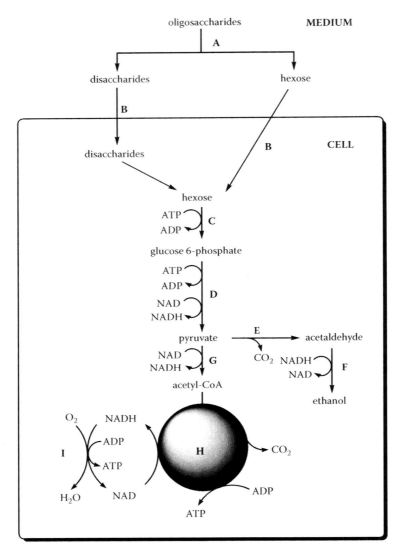

FIGURE 3.1 Biochemical outline of the uptake and metabolism of oligosaccharides and hexoses by yeasts. Indicated steps: A, glycosidases; B, sugar transport and uptake; C, entry into glucose-phosphate pool; D, glycolysis; E, pyruvate decarboxylase; F, alcohol dehydrogenase; G, pyruvate dehydrogenase; H, tricarboxylic acid cycle (mitochondrial); I, electron transport (mitochondrial).

Even when glucose fermentation occurs under anaerobic or microaerobic conditions, the fermentation of xylose (and other sugars) may still require O_2. For example, when xylose metabolism commences by its reduction to xylitol (catalyzed by NADPH-dependent xylose reductase), the subsequent step is carried out under the control of an NAD-dependent xylitol dehydrogenase, thus resulting in a disturbed redox balance of reduced and oxidized cofactors if O_2 is not present, because NADH cannot then be reoxidized, and fermentation soon ceases (figure 3.2).[29,30]

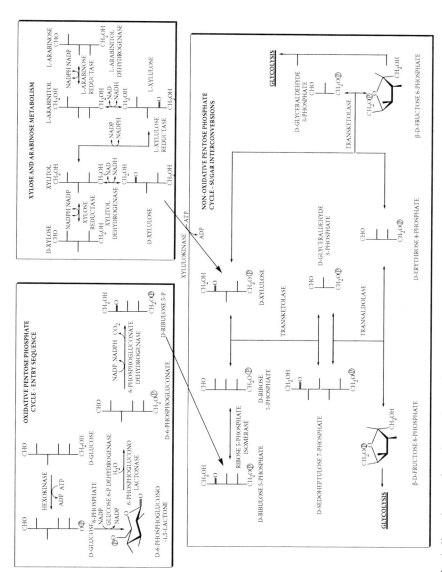

FIGURE 3.2 Metabolic pathways of D-xylose and L-arabinose utilization by bacteria and yeasts: interconnections with oxidative and nonoxidative pentose phosphate pathways (for clarity, sugar structures are drawn without hydroxyl groups and both H atoms and C–H bonds on the sugar backbones).

It is precisely this biochemical complexity in yeasts that makes accurate control of an ethanol fermentation difficult and that has attracted researchers to fermentative bacteria where metabolic regulation is more straightforward and where the full benefits of advances in biochemical engineering hardware and software can be more readily exploited. The remarkably high growth rate attainable by *S. cerevisiae* at very low levels of dissolved O_2 and its efficient transformation of glucose to ethanol that made it originally so attractive for alcohol production maintain, however, its A-list status in the rankings of biologically useful organisms (figure 3.3).[31] Whole genome sequencing has shown that the highly desirable evolution of the modern *S. cerevisiae* yeast ethanologen has occurred over more than 150 million years, resulting in a Crabtree-positive species that can readily generate respiratory-deficient, high alcohol-producing "petite" cells immune to the Pasteur effect in a readily acquired and efficient fermentative lifestyle.[32,33]

3.1.2 BACTERIA

Bacteria are traditionally unwelcome to wine producers and merchants because they are causative spoiling agents; for fuel ethanol production, they are frequent contaminants in nonsterile mashes where they produce lactic and acetic acids, which, in high concentration, inhibit growth and ethanol production by yeasts.[34,35] In a pilot plant constructed and operated to demonstrate ethanol production from corn fiber-derived sugars, for example, *Lactobacilli* were contaminants that could utilize arabinose, accumulating acids that impaired the performance of the ethanologenic yeast; the unwanted bacteria could be controlled with expensive antibiotics, but this experience shows the importance of constructing ethanologens to consume all the major carbon

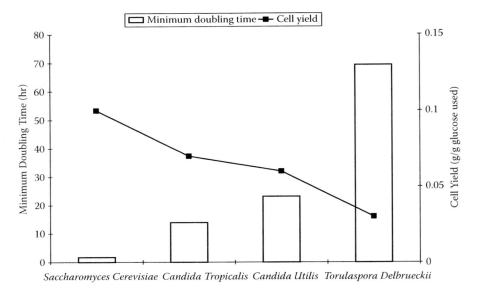

FIGURE 3.3   Growth of yeasts in anaerobic batch cultures after growth previously under O_2 limitation. (Data from Visser et al.[31])

sources in lignocellulosic substrates so as to maximize the competitive advantage of being the dominant microbial life form at the outset of the fementation.[36]

Bacteria are much less widely known as ethanol producers than are yeasts but *Escherichia, Klebsiella, Erwinia*, and *Zymomonas* species have all received serious and detailed consideration for industrial use and have all been the hosts for recombinant DNA technologies within the last 25 years (table 3.3).[37-43] With time, and perhaps partly as a result of the renewed interest in their fermentative capabilities, some bacteria considered to be strictly aerobic have been reassessed; for example, the common and much-studied soil bacterium *Bacillus subtilis* changed profoundly in its acknowledged ability to live anaerobically between the 1993 and 2002 editions of the American Society for Microbiology's monograph on the species and its relatives; *B. subtilis* can indeed ferment glucose to ethanol, 2,3-butanediol, and lactic acid, and its sequenced genome contains two ADH genes.[44] The ability of bacteria to grow at much higher temperatures than is possible with most yeast ethanologens led to proposals early in the history of the application of modern technology to fuel ethanol production that being able to run high-yielding alcohol fermentations at 70°C or above (to accelerate the process and reduce the economic cost of ethanol recovery) could have far-reaching industrial implications.[45,46]

Bacteria can mostly accept pentose sugars and a variety of other carbon substrates as inputs for ethanol production (table 3.3). Unusually, *Zymomonas mobilis* can only use glucose, fructose, and sucrose but can be easily engineered to utilize pentoses by gene transfer from other organisms.[47] This lack of pentose use by the

TABLE 3.3
Bacterial Species as Candidate Fuel Ethanol Producers

Species	Strain type	Carbon source	Ethanol productivity (g/g sugar used)	Reference
Erwinia chrysanthemi	PDC transconjugant	Xylose	0.45	37
Erwinia chrysanthemi	PDC transconjugant	Arabinose	0.33	37
Klebsiella planticola	PDC transconjugant	Xylose	0.40	38
Zymomonas mobilis	Patented laboratory strain	Amylase-digested starch	0.46	39
Klebsiella oxytoca	*Z. mobilis pdc* and *adhB* genes	Xylose	0.42	40
Klebsiella oxytoca	*Z. mobilis pdc* and *adhB* genes	Arabinose	0.34	40
Klebsiella oxytoca	*Z. mobilis pdc* and *adhB* genes	Glucose	0.37	40
Bacillus stearothermophilus	Lactate dehydrogenase mutant	Sucrose	0.30	41
Escherichia coli	*Z. mobilis pdc* and *adhB* genes	Corn fiber acid hydrolysate	0.41	43

wild-type organism probably restricted its early commercialization because otherwise Z. mobilis has extremely desirable features as an ethanologen:

- It is a GRAS organism.
- It accumulates ethanol in high concentration as the major fermentation
  product with a 5–10% higher ethanol yield per unit of glucose used and
  with a 2.5-fold higher specific productivity than S. cerevisiae.[48]
- The major pathway for glucose is the Entner-Doudoroff pathway (figure 3.4);
  the inferior bioenergetics of this pathway in comparison with glycolysis means
  that more glucose is channeled to ethanol production than to growth, and the
  enzymes required comprise up to 50% of the total cellular protein.[48]
- No Pasteur effect on glucose consumption rate is detectable, although interactions between energy and growth are important.[49]

Escherichia coli and other bacteria are, as discussed in chapter 2 (section 2.2), prone
to incompletely metabolizing glucose and accumulating large amounts of carboxylic
acids, notably acetic acid; with some authors, this has been included under the
heading of the "Crabtree effect."[50,51] For *E. coli* as a vehicle for the production of
recombinant proteins, acetate accumulation is an acknowledged inhibitory factor;
in ethanol production, it is simply a metabolic waste of glucose carbon. Other than
this (avoidable) diversion of resources, enteric and other simple bacteria are easily
genetically manipulated, grow well in both complex and defined media, can use a
wide variety of nitrogen sources for growth, and have been the subjects of decades
of experience and expertise for industrial-scale fermentations — *Z. mobilis* also was
developed for ethanol production more than 20 years ago, including its pilot-scale
use in a high-productivity continuous process using hollow fiber membranes for cell
retention and recycling.[52]

3.2   METABOLIC ENGINEERING OF NOVEL ETHANOLOGENS

3.2.1   INCREASED PENTOSE UTILIZATION BY ETHANOLOGENIC YEASTS BY GENETIC MANIPULATION WITH YEAST GENES FOR XYLOSE METABOLISM VIA XYLITOL

It has been known for many years that *S. cerevisiae* cells can take up xylose from
nutrient media; the transport system is one shared by at least 25 sugars (both natural
and synthetic), and only major alterations to the pyranose ring structure of hexoses
(e.g., in 2-deoxyglucose, a compound that lacks one of the hydroxyl groups of glucose) reduce the affinity of the transport system for a carbohydrate.[53] Moreover, both
D-xylose and L-arabinose can be reduced by *S. cerevisiae*, the products being the
sugar alcohols, xylitol, and L-arabinitol, respectively; three separate genes encode
enzymes with overlapping selectivities for xylose and arabinose as substrates.[54]

Indeed, the wild-type *S. cerevisiae* genome does contain genes for both xylose
reductase and xylitol dehydrogenase, thus being able to isomerize xylulose from
xylose, and the resulting xylulose (after phosphorylation catalyzed by a specific
xylulokinase) can enter the pentose phosphate pathway (figure 3.2).[55,56] Overexpressing the endogenous yeast genes for xylose catabolism renders the organism
capable of growth on xylose in the presence of glucose as cosubstrate under aerobic

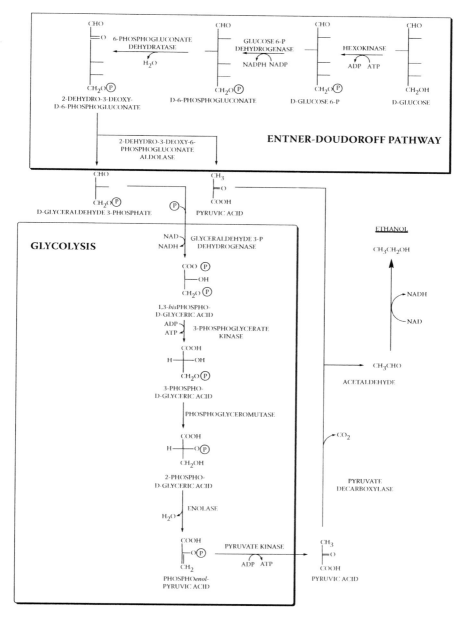

FIGURE 3.4   The Entner-Doudoroff pathway of glucose catabolism in *Z. mobilis* and the Embden-Meyerhof-Parnas pathway of glycolysis in most bacteria and yeasts in providing pyruvic acid as a substrate for the homoethanol pathway.

conditions, although no ethanol is formed; *S. cerevisiae* may, therefore, have evolved originally to utilize xylose and other pentoses, but this has been muted, possibly because its natural ecological niche altered or the organism changed its range of favored environments.[55]

Ethanol production from xylose is a rare phenomenon; of 200 species of yeasts tested under controlled conditions in the laboratory, only six accumulated ethanol to more than 1 g/l (0.1% by volume): *Pichia stipitis, P. segobiensis, Candida shehatae, C. tenuis, Brettanomyces naardenensis*, and *Pachysolen tannophilus*.[56] Even rarer is the ability among yeasts to hydrolyze xylans, only *P. stipitis* and *C. shehatae** having xylanase activity; *P. stipitis* could, moreover, convert xylan into ethanol at 60% of the theoretical yield as computed from the xylose content of the polymer.[58] Three naturally xylose-fermenting yeasts have been used as donors for genes encoding enzymes of xylose utilization for transfer to *S. cerevisiae*: *P. stipitis, C. shehatae*, and *C. parapsilosis*.[59–61] These organisms all metabolize xylose by the enzymes of the same low-activity pathway known in *S. cerevisiae* (figure 3.2), and the relevant enzymes appear to include arabinose as a possible substrate — at least, when the enzymes are assayed in the laboratory.

P. stipitis has been the most widely used donor, probably because it shows relatively little accumulation of xylitol when growing on and fermenting xylose, thus wasting less sugar as xylitol.[62] This advantageous property of the yeast does not appear to reside in the enzymes for xylose catabolism but in the occurrence of an alternative respiration pathway (a cyanide-insensitive route widely distributed among yeasts of industrial importance); inhibiting this alternative pathway renders *P. stipitis* quite capable of accumulating the sugar alcohols xylitol, arabinitol, and ribitol.[63] Respiration in *P. stipitis* is repressed by neither high concentrations of fermentable sugars nor by O_2 limitation (i.e., the yeast is Crabtree-negative), and as an ethanologen, *P. stipitis* suffers from the reduction of fermentative ability by aerobic conditions.[64]

Transferring genetic information from *P. stipitis* in intact nuclei to *S. cerevisiae* produced karyoductants, that is, diploid cells where nuclei from one species have been introduced into protoplasts of another, the two nuclei subsequently fusing, with the ability to grow on both xylose and arabinose; the hybrid organism was, however, inferior to *P. stipitis* in ethanol production and secreted far more xylitol to the medium than did the donor; its ethanol tolerance was, on the other hand, almost exactly midway between the tolerance ranges of *S. cerevisiae* and *P. stipitis*.[65] For direct genetic manipulation of *S. cerevisiae*, however, the most favored strategy (starting in the early 1990s) has been to insert the two genes (*xyl1* and *xyl2*) from *P. stipitis* coding for xylose reductase (XR) and xylitol dehydrogenase (XDH), respectively.[66] Differing ratios of expression of the two foreign genes resulted in smaller or higher amounts of xylitol, glycerol, and acetic acid, and the optimum XR:XDH ratio of 0.06:1 can yield no xylitol, less glycerol and acetic acid, and more ethanol than with other engineered *S. cerevisiae* strains.[67]

The first patented *Saccharomyces* strain to coferment xylose and glucose to ethanol (not *S. cerevisiae* but a fusion between *S. diastaticus* and *S. ovarum* able to produce ethanol at 40°C) was constructed by the Laboratory of Renewable

---

* So similar are *P. stipitis* and *C. shehatae* that they were at one time described as anamorphs of each other but the application of modern nucleic acid analytical techniques established that they are distinct biological species; from their ribosomal RNA sequences, *C. shehatae* has only recently (in biological time) diverged from the *Pichia* group, but both are well seperated from *S. cerevisiae*.[57]

Resources Engineering at Purdue University, with four specific traits tailored for industrial use:[68–70]

1. To effectively direct carbon flow from xylose to ethanol production rather than to xylitol and other by-products
2. To effectively coferment mixtures of glucose and xylose
3. To easily convert industrial strains of *S. cerevisiae* to coferment xylose and glucose using plasmids with readily identifiable antibiotic resistance markers controlling gene expression under the direction of promoters of *S. cerevisiae* glycolytic genes
4. To support rapid bioprocesses with growth on nutritionally rich media

In addition to XR and XDH, the yeast's own xylulose-phosphorylating xylulokinase (XK, figure 3.2) was also overexpressed via high-copy-number yeast–*E. coli* shuttle plasmids.[68] This extra gene manipulation was crucial because both earlier and contemporary attempts to transform *S. cerevisiae* with only genes for XR and XDH produced transformants with slow xylose utilization and poor ethanol production.[68] The synthesis of the xylose-metabolizing enzymes not only did not require the presence of xylose, but glucose was incapable of repressing their formation. It was known that *S. cerevisiae* could consume xylulose anaerobically but only at less than 5% of the rate of glucose utilization; XK activity was very low in unengineered cells, and it was reasoned that providing much higher levels of the enzyme was necessary to metabolize xylose via xylulose because the *P. stipitis* XDH catalyzed a reversible reaction between xylitol and xylulose, with the equilibrium heavily on the side of xylitol.[70–72] Such strains were quickly shown to ferment corn fiber sugars to ethanol and later utilized by the Iogen Corporation in their demonstration process for producing ethanol from wheat straw.[73,74]

The vital importance of increased XK activity in tandem with the XR/XDH pathway for xylose consumption in yeast was demonstrated in *S. cerevisiae*: not only was xylose consumption increased but xylose as the sole carbon source could be converted to ethanol under both aerobic and anaerobic conditions, although ethanol production was at its most efficient in microaerobiosis (2% O_2).[75] Large increases in the intracellular concentrations of the xylose-derived metabolites (xylulose 5-phosphate and ribulose 5-phosphate) were demonstrated in the XK-overexpressing strains, but a major drawback was that xylitol formation greatly exceeded ethanol production when O_2 levels in the fermentation decreased.

In the intervening years (and subsequently), considerable efforts have been dedicated to achieving higher ethanol productivity with the triple XR/XDH/XK constructs. Apart from continuing attempts to more fully understand the metabolism of xylose by unconventional or little-studied yeast species, two main centers of attention have been evident:

1. Strategies for harmonizing the different cofactor requirements in the pathway, that is, NAPDH-dependent (or NAPDH-preferring) XR and NAD-requiring XDH, and thus reducing xylitol formation
2. Overexpressing a wider array of other pentose-metabolizing enzymes to maximize the rate of xylose use or (broadening the metabolic scope) increasing kinetic factors in the central pathways of carbohydrate metabolism

Because the early attempts to overexpress *P. stipitis* genes for XR and XDH in *S. cerevisiae* often resulted in high rates of xylitol formation, if NADPH formed in the oxidative pentose phosphate pathway (figure 3.2) equilibrated with intracellular NAD to form NADH, the reduced availability of NAD could restrict the rate of the XDH reaction in the direction of xylitol oxidation to xylulose; adding external oxidants capable of being reduced by NADH improved ethanol formation and reduced xylitol formation. Two of these were furfural and 5-hydroxyfurfural, known to be sugar degradation products present in lignocellulose acid hydrolysates — see chapter 2, section 2.3.4.[78] The adventitious removal of toxic impurities by these reactions probably explained why xylitol accumulation was very low when lignocellulose acid hydrolysates were used as carbon sources for XR- and XDH-transformed *S. cervisiae*.[79] This line of reasoning does not accord entirely with the high activities of XR measurable in vitro with NADH (63% of the rate with NADPH), but site-specific mutagenesis on the cloned *P. stipitis* XR gene could alter the activity with NADH to 90% of that with NADPH as cofactor and concomitantly greatly reduce xylitol accumulation although with only a marginally increased xylose utilization rate.[69] Further optimization of the xylitol pathway for xylose assimilating was, therefore, entirely possible. Simply coalescing the XR and XDH enzymes into a single fusion protein, with the two active units separated by short peptide linkers, and expressing the chimeric gene in *S. cerevisiae* resulted in the formation of a bifunctional enzyme; the total activities of XR and XDH were similar to the activities when monomeric enzymes were produced, but the molar yield of xylitol from xylose was reduced, the ethanol yield was higher, and the formation of glycerol was lower, suggesting that the artificially evolved enzyme complex was more selective for NADH in its XR domain as a consequence of the two active sites generating and utilizing NADH being near each other.[80]

There have also been four direct methodologies tested for altering the preference of XR to use the NADPH cofactor:

1. A mutated gene for a *P. stipitis* XR with a lower affinity for NADPH replaced the wild-type XR gene and increased the yield of ethanol on xylose while decreasing the xylitol yield but also increasing the acetate and glycerol yields in batch fermentation.[81]

2. The ammonia-assimilating enzyme glutamate dehydrogenase in *S. cerevisiae* (and other yeasts) can be either NADPH or NADH specific, setting an artificial transhydrogenase cycle by simultaneously expressing genes for both forms of the enzyme improved xylose utilization rates and ethanol productivity.[82]

3. Deleting the gene for the NADPH-specific glutamate dehydrogenase aimed to increase the intracellular NADH concentration and the competition between NADH and NADPH for XR but greatly reduced growth rate, ethanol yield, and xylitol yield on a mixture of glucose and xylose; overexpressing the gene for the NADH-specific enzyme in the absence of the NADPH-requiring form, however, restored much of the loss in specific growth area and increased both xylose consumption rate when glucose had been exhausted and the ethanol yield while maintaining a low xylitol yield.[83]

4. Redox (NADPH) regeneration for the XR reaction was approached from a different angle by expressing in a xylose-utilizing *S. cerevisiae* strain the

gene for an NADP-dependent D-glyceraldehyde 3-phosphate dehydrogenase, an enzyme providing precursors for ethanol from either glucose or xylose; the resulting strain fermented xylose to ethanol at a faster rate and with a higher yield.[84]

Two recent discoveries offer novel biochemical and molecular opportunities: first, the selectivity of *P. stipitis* XDH has been changed from NAD to NADP by multiple-site-directed mutagenesis of the gene, thereby harmonizing the redox balance with XR; second, an NADH-preferring XR has been demonstrated in the yeast *Candida parapsilopsis* as a source for a new round of genetic and metabolic engineering.[61,85]

Beyond the initial conversions of xylose and xylitol, pentose metabolism becomes relatively uniform across kingdoms and genera. Most microbial species — and plants, animals, and mammals (including *Homo sapiens*) — can interconvert some pentose structures via the nonoxidative pentose phosphate pathway (figure 3.2). These reactions are readily reversible, but extended and reorganized, the pathway can function to fully oxidize glucose via the glucose 6-phosphate dehydrogenase and 6-phosphogluconate dehydrogenase reactions (both forming NADPH), although the pathway is far more important for the provision of essential biosynthetic intermediates for nucleic acids, amino acids, and cell wall polymers; the reactions can even (when required) run "backward" to generate pentose sugars from triose intermediates of glycolysis.[86] Increasing the rate of entry of xylulose into the pentose phosphate pathway by overexpressing endogenous XK activity has been shown to be effective for increasing xylose metabolism to ethanol (while reducing xylitol formation) with XR/XDH transformants of *S. cerevisiae*.[87] In contrast, disrupting the oxidative pentose phosphate pathway genes for either glucose 6-phosphate or 6-phosphoglu-conate oxidation increased ethanol production and decreased xylitol accumulation from xylose by greatly reducing (or eliminating) the main supply route for NADPH; this genetic change also increased the formation of the side products acetic acid and glycerol, and a further deleterious result was a marked decrease in the xylose consumption rate — again, a predictable consequence of low NADPH inside the cells as a coenzyme in the XR reaction.[88] Deleting the gene for glucose 6-phosphate dehydrogenase in addition to introducing one for NADP-dependent D-glyceraldehyde 3-phosphate dehydrogenase was an effective means of converting a strain ferment-ing xylose mostly to xylitol and CO_2 to an ethanologenic phenotype.[84]

The first report of overexpression of selected enzymes of the main nonoxidative pentose phosphate pathway (transketolase and transaldolase) in *S. cerevisiae* harbor-ing the *P. stipitis* genes for XR and XDH concluded that transaldolase levels found naturally in the yeast were insufficient for efficient metabolism of xylose via the pathway: although xylose could support growth, no ethanol could be produced, and a reduced O_2 supply merely impaired growth and increased xylitol accumulation.[89] In the most ambitious exercise in metabolic engineering of pentose metabolism by *S. cerevisiae* reported to date, high activities of XR and XDH were combined with overexpression of endogenous XK and four enzymes of the nonoxidative pentose phosphate pathway (transketolase, transaldolase, ribulose 5-phosphate epimerase, and ribose 5-phosphate ketoisomerase) and deletion of the endogenous, nonspecific NADPH-dependent aldose reductase (AR), catalyzing the formation of xylitol from xylose.[90] In comparison with a strain with lower XR and XDH activities and no other genetic modification other than XK overexpression, fermentation performance on a mixture of glucose (20 g/l) and xylose (50 g/l) was improved, with higher ethanol production, much lower xylitol formation, and faster utilization of xylose; deleting the nonspecific AR had no effect when XR and XDH activities were high, but glycerol accumulation was higher (figure 3.5).

To devise strains more suitable for use with industrially relevant mixtures of carbohydrates, the ability of strains to use oligosaccharides remaining undegraded to free hexose and pentose sugars in hydrolysates of cellulose and hemicelluloses is essential. Research groups in South Africa and Japan have explored combinations of heterologous xylanases and β-xylosidases:

- A fusion protein consisting of the *xynB* β-xylosidase gene from *Bacillus pumilus* and the *S. cerevisiae* Mfa1 signal peptide (to ensure the correct posttranslational processing) and the *XYN2* β-xylanase gene from *Hypocrea jecorina* were separately coexpressed in *S. cerevisiae* under the control of the glucose-derepressible *ADH2* ADH promoter and terminator; coproduction of these xylan-degrading enzymes hydrolyzed birch wood xylan, but no free xylose resulted, probably because of the low affinity of the β-xylosidase for its xylobiose disaccharide substrate.[91]
- A similar fusion strategy with the *xlnD* β-xylosidase gene from *Aspergillus niger* and the *XYN2* β-xylanase gene from *H. jecorina* enabled the yeast to hydrolyze birch wood xylan to free xylose.[92]

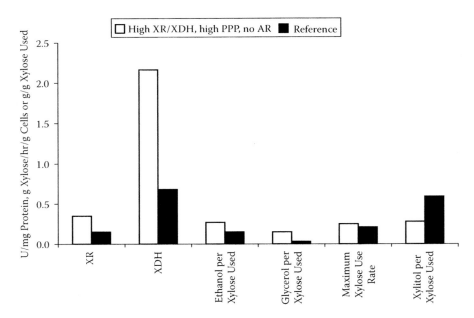

FIGURE 3.5 Effects of increased xylose reductase and xylitol dehydrogenase activities on xylose utilization by *S. cerevisiae*. (Data from Karhumaa et al., 2007.[101])

- A xylan-utilizing *S. cerevisiae* was constructed using cell surface engineering based on β-agglutinin (a cell surface glycoprotein involved in cell-cell interactions) to display xylanase II from *Hypocrea jecorina* and a β-xylosidase from *Aspergillus oryzae*; with *P. stipitis* XR and XDH and overexpressed endogenous XK, the strain could generate ethanol directly from birch wood xylan with a conversion efficiency of 0.3 g/g carbohydrate used.[93]

High XR, XDH, and XK activities combined with the expression of a gene from *Aspergillus acleatus* for displaying β-glucosidase on the cell surface enabled *S. cerevisiae* to utilize xylose- and cellulose-derived oligosaccharides from an acid hydrolysate of wood chips, accumulating 30 g/l of ethanol from a total of 73 g/l of hexose and pentose sugars in 36 hr.[94]

3.2.2 INCREASED PENTOSE UTILIZATION BY ETHANOLOGENIC YEASTS BY GENETIC MANIPULATION WITH GENES FOR XYLOSE ISOMERIZATION

Historically, the earliest attempts to engineer xylose metabolic capabilities into *S. cerevisiae* involved the single gene for xylose isomerase (XI, catalyzing the interconversion of xylose and xylulose) from bacteria (*E. coli* and *B. subtilis*), but these failed because the heterologous proteins produced in the yeast cells were enzymically inactive.[4,69] A greater degree of success was achieved using the XI gene (*xylA*) from the bacterium *Thermus thermophilus*; the transformants could exhibit ethanol formation in O_2-limited xylose fermentations.[95]

A crucial breakthrough, however, was made in 2003, when the gene encoding XI in the fungus *Piromyces* sp. strain E2 (capable of anaerobic growth on xylose) was recognized as part of the known "bacterial" pathway for xylose catabolism and, for the first time, was revealed to be functional in a eukaryote.[96] The same research group at the Delft Technical University, The Netherlands, soon demonstrated that *xylA* gene expressed in *S. cerevisiae* gave high XI activity but could not by itself induce ethanol production with xylose as the carbon source.[97] The additional genetic manipulations required for the construction of ethanologenic strains on xylose were the overexpression of XK, transketolase, transaldolase, ribulose 5-phosphate epimerase, and ribulose 5-phosphate isomerase and the deletion of nonspecific AR, followed by selection of "spontaneous" mutants in xylose-limited continuous cultures and anaerobic cultivation in automated sequencing-batch reactors on glucose-xylose media.[98–100] The outcome was a strain with negligible accumulation of xylitol (or xylulose) and a specific ethanol production three- to fivefold higher than previously publicized strains.[4] Mixtures of glucose and xylose were sequentially but completely consumed by anaerobic cultures of the engineered strain in anaerobic batch culture, with glucose still being preferred as the carbon source.[99]

A side-by-side comparison of XR/XDH- and XI-based xylose utilizations in two isogenic strains of *S. cerevisiae* with genetic modifications to improve xylose metabolism (overexpressed XK and nonoxidative pentose phosphate pathway enzymes and deleted AR) arrived, however, at widely different conclusions for the separately optimal parameters of ethanol production:[101]

- In chemically defined medium, the XI-containing variant showed the highest ethanol yield (i.e., conversion efficiency) from xylose.
- The XR/XDH transformant had the higher rate of xylose consumption, specific ethanol production, and final ethanol concentration, despite accumulating xylitol.
- In a lignocellulose hydrolysate, neither transformant accumulated xylitol, but both were severely affected by toxic impurities in the industrially relevant medium, producing little or no ethanol, xylitol, or glycerol and consuming little or no xylose, glucose, or mannose.

The bacterial XI gene from *T. thermophilus* was revisited in 2005 when this pathway for xylose utilization was expressed in *S. cerevisiae* along with overexpressed XK and nonoxidative pentose phosphate pathway genes and deleted AR; the engineered strain, despite its low measured XI activity, exhibited for the first time aerobic growth on xylose as sole carbon source and anaerobic ethanol production at 30°C.[102]

3.2.3 ENGINEERING ARABINOSE UTILIZATION BY ETHANOLOGENIC YEASTS

Xylose reductase, the first step in the pathway of xylose catabolism in most yeast species, functions as an enzyme equally well with L-arabinose as with D-xylose, with a slightly higher affinity (lower K_m) and a higher maximal rate (V_{max}) for L-arabinose.[59] Polyol dehydrogenases, active on xylitol, on the other hand, find either L- or D-arabinose to be a poor substrate.[103] Although the XDH activity from *P. stipitis* was kinetically investigated in 1989, little is known about its functional physiology; the catalyzed reaction is reversible but activity is unlikely to be regulated by the NAD/NADH balance inside the cell.[104] This yeast also contains a second XDH, quite distinct from the well characterized *xyl2* gene product, but its role is presently undefined in either xylose catabolism or ethanol production.[105] The NAD-specific XR from *S. cerevisiae* itself is even less well characterized, although the enzyme activity is induced by xylose with the wild-type organism.[106]

An outline of known enzyme-catalyzed metabolic relationship for pentitols and pentoses is given in figure 3.6; some of these pathways are of increasing contemporary interest because either they or their engineered variants could lead to the synthesis by whole cells (or in biotransformations with isolated enzymes) of "unnatural" or rare sugars useful for the elaboration of antibiotic or antiviral drugs — this is discussed later in chapter 8 when the Green Chemistry of the biorefinery concept for processing agricultural residues is discussed in depth.

Progress in defining the actual pathways operating in known ethanologenic yeasts was rapid after the year 2000:

- Analysis of mutations in *P. stipitis* revealed that the catabolism of both D-xylose and L-arabinose proceeded via xylitol.[107]
- An NAD-dependent L-arabinitol 4-dehydrogenase activity was demonstrated in *H. jecorina* induced by growth on L-arabinose; the enzyme forms L-xylulose and will accept ribitol and xylitol as substrates but not D-arabinitol.[108]

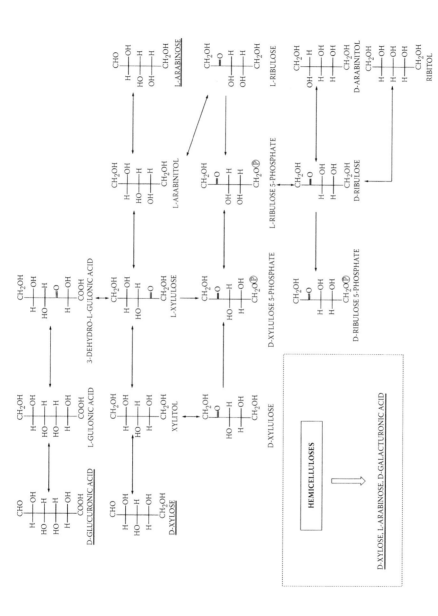

FIGURE 3.6 Metabolic interconversions of pentitols, pentose sugars, and D-glucuronic acid — not all of these enzyme-catalyzed steps may be present in individual bacterial and yeast species.

- A gene encoding an L-xylulose reductase (forming xylitol; NADP-dependent) was then demonstrated in *H. jecorina* and overexpressed in *S. cerevisiae*; the L-arabinose pathway uses as its intermediates L-arabinitol, L-xylulose, xylitol, and (by the action of XDH) D-xylulose; the xylulose reductase exhibited the highest affinity to L-xylulose, but some activity was shown toward D-xylulose, D-fructose, and L-sorbose.[109]
- In *H. jecorina*, deletion of the gene for XDH did not abolish growth because *lad1*-encoded L-arabinitol 4-dehydrogenase compensated for this loss — however, doubly deleting the two dehydrogenase genes abolished the ability to grow on either D-xylose or xylitol.[110]

With this knowledge, expressing the five genes for L-arabinose catabolism in *S. cerevisiae* enabled growth on the pentose and, although at a low rate, ethanol production from L-arabinose under anaerobiosis.[111] In the same year (2003), the genes of the shorter bacterial pathway for L-arabinose catabolism were inserted into *S. cerevisiae*.[112] The bacterial pathway (active in, e.g., *B. subtilis* and *E. coli*) proceeds via L-ribulose, L-ribulose 5-phosphate, and D-xylulose 5-phosphate (figure 3.6), using the enzymes L-arabinose isomerase, L-ribulokinase, and L-ribulose 5-phosphate epimerase. The coexpression of an arabinose-transporting yeast galactose permease allowed the selection on L-arabinose-containing media of an L-arabinose-utilizing yeast transformant capable of accumulating ethanol at 60% of the theoretical maximum yield from L-arabinose under O_2-limiting conditions.[112]

3.2.4 COMPARISON OF INDUSTRIAL AND LABORATORY YEAST STRAINS FOR ETHANOL PRODUCTION

Most of the freely available information regarding ethanologenic yeasts has been derived from "laboratory" strains constructed by research groups in academia; some of these strains (or variants thereof) have certainly been applied to industrial-scale fermentations for bioethanol production, but published data mostly refer to strains either grown in chemically defined media or under laboratory conditions (e.g., continuous culture) or with strains that can — even under the best available test conditions — accumulate very little ethanol in comparison with modern industrial strains used in potable alcohol or fuel alcohol production (figure 3.7). In addition, strains constructed with plasmids may not have been tested in nonselective media, and plasmid survival in fermentations is generally speculative although genetic manipulations are routine for constructing "self-selective" plasmid-harboring strains where a chromosomal gene in the host is deleted or disrupted and the auxotrophic requirement is supplied as a gene contained on the plasmid.[113]

Nevertheless, benchmarking studies comparing "laboratory" and "industrial" strains constructed for pentose utilization have appeared, the industrial examples including genetically manipulated polyploid strains typical of *S. cerevisiae* "working" strains from major brewers or wineries; accounts of engineering such strains for xylose utilization began to be published after 2002.[114,115] A comparison of four laboratory and five industrial strains surveyed both genetically manipulated and genetically undefined but selected xylose consumers (table 3.4).[116] The industrial strains were inferior to the laboratory strains for the yields of both ethanol and xylitol from

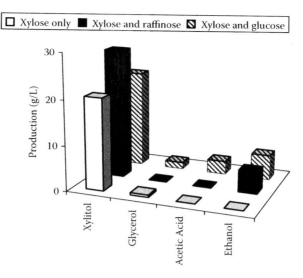

FIGURE 3.7 Product formation by a xylose-utilizing laboratory strain of *S. cerevisiae.* (Data from van Zyl et al.[113])

xylose in minimal media. Resistance to toxic impurities present in acid hydrolysates of the softwood Norway spruce (*Picea abies*) was higher with genetically trans-formed industrial strains, but a classically improved industrial strain was no more hardy than the laboratory strains (table 3.4); similarly, while an industrial strain evolved by genetic manipulation and then random mutagenesis had the fastest rate of xylose use, a laboratory strain could accumulate the highest ethanol concentration on minimal medium with 50 g/l of each glucose and xylose. None of the strains had an ideal set of properties for ethanologenesis in xylose-containing media; long-term chemostat cultivation of one industrial strain in microaerobic conditions on xylose as the sole carbon source definitely improved xylose uptake, but neither ethanol nor xylitol yield. Three of the industrial strains could grow in the presence of 10% solu-tions of undetoxified lignocellulose hydrolysate; the most resistant strain grew best (at 4.3 g/l as compared with 3.7–3.8 g/l) but had a marginally low ethanol production (16.8 g/l as compared with 16.9 g/l), perhaps because more of the carbon substrate was used for growth in the absence of any chemical limitation.

The industrial-background strain TMB 3400 (table 3.4) had no obvious meta-bolic advantage in anaerobic batch fermentations with xylose-based media when compared with two laboratory strains, one catabolizing xylose by the XR/XDH/XK pathway, the other by the fungal XI/XK pathway (figure 3.8).[101] In the presence of undetoxified lignocellulose hydrolysate, however, only the industrial strain could grow adequately and exhibit good ethanol formation (figure 3.8).

Industrial and laboratory strains engineered for xylose consumption fail to metabolize L-arabinose beyond L-arabitinol.[116] With laboratory and industrial strains endowed with recombinant xylose (fungal) and arabinose (bacterial) pathways tested in media containing glucose, xylose, and arabinose, the industrial strain accumulated higher concentrations of ethanol, had a higher conversion efficiency of ethanol per

TABLE 3.4
Laboratory and Industrial Strains of *Saccharomyces cerevisiae* for Bioethanol Production

Strain	Genetic description	Xylitol yield (g/g xylose consumed)	Ethanol yield (g/g xylose consumed)	Maximum hydrolysate[a] (% v/v)
		Laboratory		
TMB3001	XR/XDH/XK overexpressing	0.30	0.33	10
TMB EP	Evolved population from TMB3001	0.31	0.30	
C1	Clone isolated from TMB EP	0.24	0.32	10
C5	Clone isolated from TMB EP	0.28	0.34	10
		Industrial		
F12	XR/XDH/XK overexpressing, polyploid	0.40	0.26	45
A4	XR/XDH/XK overexpressing, polyploid	0.41	0.24	
BH42	Strain selected for improved xylose catabolism	0.36	0.28	10
TMB3399	XR/XDH/XK overexpressing, polypoloid	0.39	0.23	15
TMB3400	Mutagenized and selected from TMB3399	0.41	0.24	15

Source: Data from Sonderegger et al., 2004.[116]

[a] Dilute acid hydrolysate of Norway spruce wood

unit of total pentose utilized, and also converted less xylose to xylitol and less arabinose to arabinitol — although 68% of the L-arabinose consumed was still converted only as far as the polyol.[117]

Interactions between hexose and pentose sugars in the fermentations of lignocellulose-derived substrates has often been considered a serious drawback for ethanol production; this is usually phrased as a type of "carbon catabolite repression" by the more readily utilizable hexose carbon source(s), and complex phenotypes can be generated for examination in continuous cultures.[118] In batch cultivation, xylose supports slower growth and much delayed entry into ethanol formation in comparison with glucose.[119] An ideal ethanologen would co-utilize multiple carbon sources, funneling them all into the central pathways of carbohydrate metabolism — ultimately to

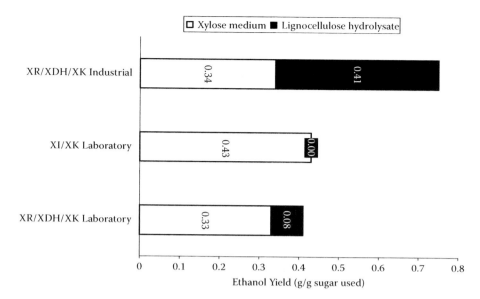

FIGURE 3.8 Ethanol yields from xylose and total sugar of two laboratory and one industrial strain of recombinant *S. cerevisiae*. (Data from Karhumaa et al.[101])

pyruvic acid and thence to acetic acid, acetaldehyde, and ethanol (figures 3.2 and 3.4). No publicly disclosed strain meets these requirements. An emerging major challenge is to achieve the rapid transition from proof-of-concept experiments in synthetic media, using single substrates and in the absence of toxic inhibitors, to demonstrations that constructed strains can efficiently convert complex industrial substrates to ethanol.[120] In addition to the key criteria of high productivity and tolerance of toxic impurities, process water economy has been emphasized.[121]

Integration of genes for pentose metabolism is becoming increasingly routine for *S. cerevisiae* strains intended for industrial use; different constructs have quantitatively variable performance indicators (ethanol production rate, xylose consumption rate, etc.), and this suggests that multiple copies of the heterologous genes must be further optimized as gene dosages may differ for the individual genes packaged into the host strain.[122] Such strains can be further improved by a less rigorously defined methodology using "evolutionary engineering," that is, selection of strains with incremental advantages for xylose consumption and ethanol productivity, some of which advantages can be ascribed to increases in measurable enzyme activities for the xylose pathway or the pentose phosphate pathway and with interesting (but not fully interpretable) changes in the pool sizes of the intracellular pathway intermediates.[123,124] Efficient utilization of xylose appears to require complex global changes in gene expression, and a reexamination of "natural" *S. cerevisiae* has revealed that classical selection and strain improvement programs can develop yeast cell lines with much shorter doubling times on xylose as the sole carbon source as well as increased XR and XDH activities in a completely nonrecombinant approach.[125] This could easily be applied to rationally improve yeast strains with desirable properties that can be isolated in the heavily selective but artificial environment of an industrial fermentation plant — a practice deliberately pursued for centuries in breweries and wineries but equally applicable to facilities for the fermentation of spent sulfite liquor from the pulp and paper industry.[126]

Defining the capabilities of both industrial and laboratory strains to adapt to the stresses posed by toxic inhibitors in lignocellulose hydrolysates remains a focus of intense activity.[127–130] Expression of a laccase (from the white rot fungus, *Trametes versicolor*) offers some promise as a novel means of polymerizing (and precipitating) reactive phenolic aldehydes derived from the hydrolytic breakdown of lignins; *S. cerevisiae* expressing the laccase could utilize sugars and accumulate ethanol in a medium containing a spruce wood acid hydrolysate at greatly increased rates in comparison with the parental strain.[131] *S. cerevisiae* also contains the gene for phenylacrylic acid decarboxylase, an enzyme catalyzing the degradation of ferulic acid and other phenolic acids; a transformant overexpressing the gene for the decarboxylase utilized glucose grew up to 25% faster, utilized mannose up to 45% faster, and accumulated ethanol up to 29% more rapidly than did a control transformant not overexpressing the gene.[132]

3.2.5   IMPROVED ETHANOL PRODUCTION BY NATURALLY PENTOSE-UTILIZING YEASTS

Serious development of yeasts other than *S. cerevisiae* has been muted; this is partly because of the ease of genetic transformation of *S. cerevisiae* and closely related strains with bacterial yeast shuttle vectors. Among nonconventional yeasts, *C. shehatae* has, however, properties highly desirable for bioethanol production from lignocellulosic substrates:[133]

- Ethanol production is more efficient from a mixture of glucose and xylose than from either sugar alone.
- Ethanol production can be demonstrated at elevated temperatures (up to 45°C).
- Ethanol formation from xylose is not affected by wide variation in the xylose concentration in the medium.
- Ethanol can be produced from rice straw hemicellulose hydrolysates.

With a straightforward liquid hot water pretreatment of alfalfa fibers, *C. shehatae* could produce ethanol at a concentration of 9.6 g/l in a batch fermentation with a conversion efficiency of 0.47 g/g sugar consumed; hemicellulose utilization was, however, poor because of the presence of inhibitors.[134] The methylotrophic yeast *Hansenula polymorpha* can, on the other hand, ferment xylose as well as glucose and cellobiose; this species is thermotolerant, actively fermenting sugars at up to 45°C and with a higher ethanol tolerance than *P. stipitis* (although less than *S. cerevisiae*); a vitamin B$_2$ (riboflavin)-deficient mutant exhibited increased ethanol productivity from both glucose and xylose under suboptimal riboflavin supply and the consequent growth restriction.[135]

P. stipitis, the host organism for genes of a xylose metabolism pathway successfully expressed in *S. cerevisiae*, has been developed as an ethanologen by a research group at the University of Wisconsin, Madison, since the early 1990s.[136] Part of this work was the development of a genetic system for *P. stipitis*, which was used to endow the yeast with the ability to grow and produce ethanol anaerobically.

P. stipitis is Crabtree-negative and is poorly productive for ethanol. *S. cerevisiae* derives its ability to function anaerobically from the presence of a unique enzyme, dihydroorotate dehydrogenase (DHOdehase), converting dihydroorotic acid to orotic acid in the pyrimidine biosynthetic pathway for nucleic acids; in *S. cerevisiae*, DHOdehase is a cytosolic enzyme catalyzing the reduction of fumaric acid to succinic acid,* and the enzyme may constitute half of a bifunctional protein with a fumarate reductase or be physically associated with the latter enzyme inside the cell.[137] Expression of the *S. cerevisiae* gene for DHOdehase (*ScURA1*) in *P. stipitis* enabled rapid anaerobic growth in a chemically defined medium with glucose as sole carbon source when essential lipids were supplied; 32 g/l of ethanol was produced from 78 g/l glucose in a batch fermentation.[138]

In mixtures of hexoses and pentoses, xylose metabolism by *P. stipitis* is repressed, whereas glucose, mannose, and galactose are all used preferentially, and this may limit the potential of the yeast for the fermentation of lignocellulosic hydrolysates; neither cellobiose nor L-arabinose inhibits induction of the xylose catabolic pathway by D-xylose.[139] Ethanol production from xylose is also inhibited by the $CaSO_4$ formed by the neutralization of sulfuric acid hydrolysates of lignocellulosic materials with $Ca(OH)_2$, whereas Na_2SO_4 (from NaOH) had no effect on either xylose consumption or ethanol production and $(NH_4)_2SO_4$ (from NH_4OH) reduced growth but enhanced the xylose utilization rate, the rate of ethanol production, and the final ethanol concentration.[140] *P. stipitis* has been shown to produce ethanol on an array of lignocellulosic substrates: sugarcane bagasse, red oak acid hydrolysate, wheat straw, hardwood hemicellulose hydrolysate, and corn cob fractions.[141–145]

Strains of the thermotolerant *Kluyveromyces* yeasts are well known to modern biotechnology as vehicles for enzyme and heterologous protein secretion.[146] *K. marxianus* is one of the extraordinary biodiversity of microbial flora known to be present in fermentations for the spirit *cachaça* in Brazil (chapter 1, section 1.2); more than 700 different yeast species were identified in one distillery during a season, although *S. cerevisiae* was the usual major ethanologen except in a small number of cases where *Rhodotorula glutinis* and *Candida maltosa* predominated.[147] Strains of *K. marxianus* isolated from sugar mills could ferment glucose and cane sugar at temperatures up to 47°C and to ethanol concentrations of 60 g/l, although long fermentation times (24–30 hr) and low cell viability were operational drawbacks.[148] In another study, all eight strains that were screened for D-xylose use were found to be active and one *K. marxianus* strain was capable of forming ethanol at 55% of the theoretical maximum yield from xylose.[149] A medium based on sugarcane molasses was fermented to a final ethanol level of 74 g/l at 45°C, but osmotic stress was evident at high concentrations of either molasses or mixtures of sucrose and molasses.[150] Brazilian work has shown that *K. marxianus* is strictly Crabtree-negative, requiring (at least in laboratory chemostat experiments) the O_2 supply to be shut down for ethanol to be formed; a high tendency to divert sugars via the oxidative pentose phosphate pathway may be the major obstacle to this yeast as an ethanologen, but metabolic engineering could be applied to redirect carbon flow for fermentative efficiency.[151]

---

* This fermentative route is not available to a strictly aerobic yeast such as *Schizosaccharomyces pombe*, which contains a mitochondrial DHOdehase and requires a fully functional mitochondrial electron transport chain for DHOdehase activity.[137]

Some other yeast species are considered in the next chapter when microbes naturally capable of hydrolyzing polysaccharides (or engineered to do so) as well as fermenting the resulting sugars are considered.

3.3 ASSEMBLING GENE ARRAYS IN BACTERIA FOR ETHANOL PRODUCTION

3.3.1 METABOLIC ROUTES IN BACTERIA FOR SUGAR METABOLISM AND ETHANOL FORMATION

Two principal routes for glucose catabolism are known to classical biochemistry in ethanologenic bacteria: the Embden-Meyerhof-Parnas (EMP) pathway of glycolysis (also present in yeasts, fungi, plants, and animals) and, in a restricted range of bacteria (but more widely for the catabolism of gluconic acid), the Entner-Doudoroff (ED) pathway (figure 3.4).[4,86,152] The two initial steps of the ED pathway resulting in 6-phosphogluconic acid are those of the oxidative pentose phosphate pathway, but Z. mobilis is unique in operating this sequence of reactions under anaerobic conditions.[153] The EMP and ED pathways converge at pyruvic acid; from pyruvate, a range of fermentative products can be produced, including acids such as lactate and decarboxylated acids such as 2,3-butanediol (see figure 2.3). In S. cerevisiae and other major ethanol-producing yeasts, ethanol formation requires only two reactions from pyruvate: pyruvate decarboxylase (PDC) catalyzes the formation of acetaldehyde and ADH catalyzes the NADH-oxidizing reduction of acetaldehyde to ethanol; in E. coli, in marked contrast, no PDC naturally exists and pyruvate is catabolized by the PDH reaction under aerobic conditions or by the pyruvate formate lyase (PFL) reaction under anaerobiosis. To maintain the redox balance under fermentative conditions, a spectrum of products is generated from glucose in E. coli and many other enteric bacterial species in which ethanol is often a minor component (table 3.5).[154]

TABLE 3.5
Mixed Acid Fermentation Products Accumulated by *Escherichia coli*

Product	Conversion (mole/mole glucose consumed)	Carbon recovery (%)	Redox balance[a]
Lactic acid	1.09	54	0.0
Acetic acid	0.32	11	0.0
Formic acid	0.02	0	1.0
Succinic acid	0.18	12	1.0
Ethanol	0.41	14	–2.0
CO_2	0.54	9	2.0
H_2	0.18		–1.0
Σ		100	1.0

Source: Data from Moat and Foster.[154]

[a] $= \Sigma O$ atoms $- 0.5 \Sigma H$ atoms

3.3.2   GENETIC AND METABOLIC ENGINEERING OF
BACTERIA FOR BIOETHANOL PRODUCTION

3.3.2.1   Recombinant *E. coli*: Lineages and Metabolic Capabilities

Attempts (beginning in the 1980s) to genetically "improve" *E. coli* and other bacteria for efficient ethanol production with recombinant gene technology foundered because they relied on endogenous ADH activities competing with other product pathways.[4] Success soon followed when the research group at the University of Florida (Gainesville) combined the *Z. mobilis* genes for PDC and ADH with that bacterium's homo-ethanol pathway in a single plasmid, the PET operon; in various guises, the PET system has been used to engineer ethanol production in *E. coli* and other bacteria.[155,156] The original PET operon used the β-galactosidase promoter; to construct a vector suitable for directing chromosomal integration by transformation, a PET cartridge was devised to include the PET operon plus chloramphenicol transacetylase for selection, flanked by truncated fragments of the *E. coli* PFL (*pfl*) open reading frame (without promoter) to target homologous recombination; upon integration, PET expression was under the control of the chromosomal *pfl* promoter — although a single-step selection on chloramphenicol was required to identify a suitably highly expressing mutant.[157]

A crucial property of the *Z. mobilis* PDC is its relatively high affinity (K_m) for pyruvate as a substrate: 0.4 mM as compared with 2 mM for PFL, 7 mM for lactate dehydrogenase and 0.4 mM for PDH; in consequence, much reduced amounts of lactic and acetic acids are formed, and the mix of fermentation products is much less toxic to growth and inhibitory to the establishment of dense cell populations.[152] Further mutations have been utilized to delete genes for the biosynthesis of succinic, acetic, and lactic acids to further reduce the waste of sugar carbon to unwanted metabolic acids.[158]

When a number of the well-characterized *E. coli* laboratory strains were evaluated for their overall suitability for PET transformation, strain B exhibited the best hardiness to environmental stresses (ethanol tolerance and plasmid stability in nonselective media) and superior ethanol yield on xylose.[159,160] This strain was isolated in the 1940s and was widely used for microbiological research in the 1960s; more importantly, it lacks all known genes for pathogenicity.[161] A strain B-derived, chromosomally integrated isolate (KO11), also with a disrupted fumarate reductase gene (for succinate formation), emerged as a leading candidate for industrial ethanol production.[162] It shows ethanol production with much-improved selectivity against the mixed-acid range of fermentation products* (figure 3.9, cf. table 3.5).[157] The capabilities of KO11 have been evaluated and confirmed in laboratories in the United States and Scandinavia.[43,163] An encouragingly wide range of carbon substrates have been shown to support ethanol production:

- Pine wood acid hydrolysates[164]
- Sugarcane bagasse and corn stover[165]
- Corn cobs, hulls, and fibers[166,167]
- Dilute acid hydrolysate of rice hulls[168]

---

* Redox balance is achieved because ethanol (oxidation value –2) and CO_2 (oxidation value +2) effectively cancel one another out, lactate and acetate having zero values, thus leaving a small inaccuracy for the trace of succinate (oxidation value +1) formed in the slightly "leaky" mutant.

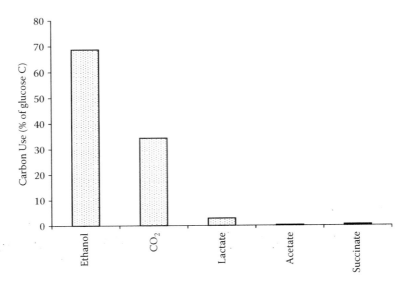

FIGURE 3.9   Fermentative efficiency of recombinant *E. coli* strain KO11 on 10% (v/v) glucose. (Data from Ohta et al.[157] and Dien et al.[162])

- Sweet whey and starch[169,170]
- Galacturonic acid and other components in orange peel hydrolysates[171]
- The trisaccharide raffinose (a component of corn steep liquors and molasses)[172]

To increase its hardiness to ethanol and inhibitors present in acid hydrolysates of lignocellulosic materials, strain KO11 was adapted to progressively higher ethanol concentrations during a period of months, the culture being reselected for resistance to chloramphenicol at regular intervals; the resulting strain (LY01) fermented 140 g/l xylose in 96 hr (compared with 120 hr with LO11).[173] Increased ethanol tolerance was accompanied (fortuitously) by increased resistance to various growth inhibitors, including aromatic alcohols and acids derived from ligninolysis and various aromatic and nonaromatic aldehydes (including 4-hydroxybenzaldehyde, syringaldehyde, vanillin, hydroxymethylfurfural [HMF], and furfural).[174–176] Of this multiplicity of inhibitors, the aromatic alcohols proved to be the least toxic to bacterial growth and metabolism, and *E. coli* strains can be at least as refractory to growth inhibitors as are other microbial ethanologens.

More wide-ranging genetic manipulation of ethanologenic *E. coli* strains has explored features of the molecular functioning of the recombinant cells as measured by quantitative gene expression and the activities of the heterologous gene products.[177–180] A long-recognized problem with high-growth-rate bacterial hosts engineered to contain and express multiple copies of foreign genes is that of "metabolic burden," that is, the diversion of nutrients from biosynthesis and cell replication to supporting the expression and copying of the novel gene complement often results in a reduced growth rate in comparison with the host strain. Chromosomal integration of previously plasmid-borne genes does not avoid this metabolic demand, as became evident when attempts were made to substitute a rich laboratory growth medium with possible cheaper industrial media based on ingredients such as corn steep liquor: uneconomically high concentrations were required to match the productivities observable in laboratory tests, but this could be only partially improved by lavish additions of vitamins, amino acids, and other putative growth-enhancing medium ingredients.[177] More immediately influential was increasing PDC activity by inserting plasmids with stronger promoters into the chromosomally integrated KO11 strain. Because the laboratory trials in the 1990s achieved often quite low cell densities (3 g dry weight of cells per liter), lower by one or two orders of magnitude than those attainable in industrial fermentations, it is highly probable that some ingenuity will be required to develop adequate media for large-scale fermentations while minimizing operating costs.

Genetic manipulation can, however, aid the transition of laboratory strains to commercially relevant media and the physical conditions in high-volume fermentors. For example, growth and productivity by the KO11 strain in suboptimal media can be greatly increased by the addition of simple additional carbon sources such as pyruvic acid and acetaldehyde; this has no practical significance because ethanol production cannot necessarily be a biotransformation from more expensive precursors but, together with other physiological data, implies that the engineered *E. coli* cells struggle to adequately partition carbon flow between the demands for growth (amino acids, etc.) and the requirement to reoxidize NADH and produce ethanol as an end product.[178,179] Expressing in KO11, a *B. subtilis* citrate synthase, whose activity is not affected by intracellular NADH concentrations, improves both growth and ethanol yield by more than 50% in a xylose-containing medium; this novel enzyme in a coliform system may act to achieve a better balance and direct more carbon to 2-oxoglutarate and thence to a family of amino acids required for protein and nucleic acid biosynthesis.[178] Suppressing acetate formation from pyruvate by deleting the endogenous *E. coli* gene (*ackA*) for acetate kinase probably has a similar effect by altering carbon flow around the crucial junction represented by pyruvate (see figure 2.3).[179]

In nutrient-rich media, expressing the *Z. mobilis* homoethanol pathway genes in *E. coli* increases growth rate by up to 50% during the anaerobic fermentation of xylose.[180] Gene array analysis reveals that, of the nearly 4,300 total open reading frames in the genome, only 8% were expressed at a higher level in KO11 in anaerobic xylose fermentations when compared with the B strain parent but that nearly 50% of the 30 genes involved in xylose catabolism to pyruvate were expressed at higher levels in the recombinant (figure 3.10). Calculations from bioenergetics show that xylose is a much poorer source of biochemical energy (generating only 33% of the ATP yield per molecule oxidized), and a physiological basis for the changes in growth rate can be deduced from the genomics data in the greatly elevated expressions of the genes encoding the initial two enzymes of xylose catabolism, although some of the other changes in the pentose phosphate and glycolytic pathways may be important for intracellular fluxes.[180]

A further broadening of the substrate range of the KO11 strain was effected by expressing genes (from *Klebsiella oxytoca*) encoding an uptake mechanism for cellobiose, the disaccharide product of cellulose digestion; an operon was introduced on a plasmid into KO11 containing the two genes for the phosphoenolpyruvate-dependent phosphotransferase transporter for cellobiose (generating phosphorylated cellobiose)

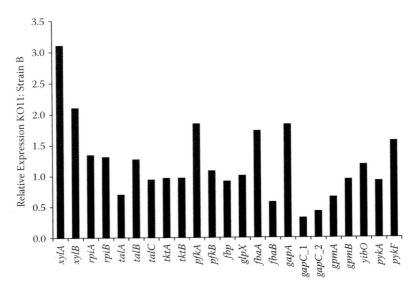

FIGURE 3.10 Gene expression pertinent to xylose metabolism by recombinant *E. coli* strain KO11. (Data from Tao et al.[180])

and phospho-β-glucosidase (for hydrolyzing the cellobiose phosphate intracellularly).[181] The *K. oxytoca* genes proved to be poorly expressed in the *E. coli* host, but spontaneous mutants with elevated specific activities for cellobiose metabolism were isolated and shown to have mutations in the plasmid that eliminated the engineered *casAB* promoter and operator regions; such mutants rapidly fermented cellobiose to ethanol, with an ethanol yield of more than 90% of the theoretical maximum and (with the addition of a commercial cellulase) fermented mixed-waste office paper to ethanol.

A second and distinct major lineage of recombinant *E. coli* was initiated at the National Center for Agricultural Utilization Research, U.S. Department of Agriculture, Peoria, Illinois. The starting point was the incomplete stability of the KO11 strain: phenotypic instability was reported in repeated batch or continuous cultivation, resulting in declining ethanol but increasing lactic acid production.[182,183] Moreover, the results were different when glucose or xylose provided the carbon supply for continuous culture: on glucose alone, KO11 appeared to be stable, but ethanol productivity declined after five days, and the antibiotic (chloramphenicol) selective marker began to be lost after 30 days.[184,185] Novel ethanologenic strains were created by expressing the PET operon on a plasmid in *E. coli* FMJ39, a strain with deleted genes for lactate dehydrogenase and PFL and, in consequence, incapable of fermentative growth on glucose.[186] The introduced homoethanol pathway genes complemented the mutations and positively selected for plasmid maintenance to enable active growth by fermentation pathways, that is, self-selection under the pressure of fermenting a carbon source.[187] The plasmid was accurately maintained by serial culture and transfer under anaerobic conditions with either glucose or xylose as the carbon source and with no selective antibiotic present but quickly disappeared during growth in the aerobic conditions in which the parental strain grew normally.

One of the resulting strains was further adapted for growth on xylose (FBR3); this construct can ferment a 10% (w/v) concentration of glucose, xylose, arabinose, or a mixture of all three sugars at 35°C during a period of 70–80 hr and producing up to 46.6 g/l of ethanol and at up to 91% of the theoretical maximum yield.[188] The strains are also able to ferment hydrolysates prepared from corn hulls and germ meal within 60 hr and at a yield of 0.51 g ethanol/g sugar consumed.[189,190] Variants of the strains have been constructed that are relatively deficient in glucose uptake because they carry a mutation in the phosphoenolpyruvate-glucose phosphotransferase system. In organisms with this transport mechanism, the presence of glucose represses the uptake of other sugars; obviating this induces the cells to utilize xylose and arabinose simultaneously with glucose rather than sequentially after the glucose supply begins to be exhausted and to make ethanol production more rapidly — although the overall productivity (carbon conversion efficiency) is minimally affected.[191] The USDA strains will be considered further in section 3.3.2.5 when performance data for the rival bacterial ethanologens are compared.

Molecular evolution of an efficient ethanol pathway in *E. coli* without resorting to heterologous gene expression was finally accomplished in 2006.[192] Work published over a decade earlier from the University of Sheffield had shown that in the PDH complex, the route of pyruvate oxidation in *E. coli* under aerobic conditions was encoded by a four-gene operon (*pdhR*, *aceE*, *aceF*, and *lpd*), with pyruvate (or a derivative of pyruvate) acting as the inducing agent.[193] A mutant of *E. coli* strain K12 was isolated with the essential genetic mutation occurring in the PDH operon; the phenotype was a novel pathway endowing the capacity to ferment glucose or xylose to ethanol with a yield of 82% under anaerobic conditions, combining a PDC-type enzyme activity with the endogenous ADH activity.[192] This may aid the introduction of bacterial ethanol production in regions (including some member states of the European Union) where genetically manipulated organisms are viewed with popular (or populist) concern.

Commercial take-up of recombinant *E. coli* has proved very slow, but in 2007, demonstration facilities using *E. coli* strains licensed from the University of Florida (table 3.6) have been opened or constructed in the United States and Japan (section 4.8) — in fact, the Japanese site is the world's first commercial plant to produce ethanol from wood by a bacterial fermentation.

3.3.2.2   Engineering *Z. mobilis* for Xylose and Arabinose Metabolism

This organism has been known first as *Termobacterium mobile* and subsequently as *Pseudomonas linderi* since 1912; although it was first known in Europe as a spoiling agent in cider, its function in the making of potable beverages such as palm wines is well established in Africa, Central* and South America, the Middle East, South Asia, and the Pacific islands and can ferment the sugar sap of the *Agave* cactus to yield *pulque*.[194] The species was described as "undoubtedly one of the most unique bacterium [sic] within the microbial world."[195] Its unusual biochemistry has already

---

* Tropical counterparts of yogurts and fermented milk drinks prepared using *Z. mobilis* were known to the Aztecs for their therapeutic properties.[194]

TABLE 3.6
U.S. Patents Awarded for Yeast and Bacterial Ethanol Producer Strains Capable of Utilizing Lignocellulosic Substrates

Date	Title	Assignee	Patent Number
January 11, 1983	Direct fermentation of D-xylose to ethanol by a xylose-fermenting yeast mutant	Purdue Research Foundation, West Lafayette, IN	4368268
September 18, 1984	Process for manufacturing alcohol by fermentation	Kyowa Hakko Kogyo Co. Ltd.	4472501
December 25, 1984	Production of ethanol by yeast using xylulose	Purdue Research Foundation, West Lafayette, IN	4490468
April 16, 1985	Direct fermentation of D-xylose to ethanol by a xylose-fermenting yeast mutant	Purdue Research Foundation, West Lafayette, IN	4511656
May 5, 1987	Process for enhanced fermentation of xylose to ethanol	United States	4663284
June 20, 1989	Process for producing ethanol from plant biomass using the fungus *Paecilomyces* sp.	United States	4840903
March 19, 1991	Ethanol production by *Escherichia coli* strains co-expressing *Zymomonas* PDC and ADH genes	University of Florida, Gainesville, FL	5000000
December 13, 1994	Combined enzyme mediated fermentation of cellulous [sic] and xylose to ethanol …	United States	5372939
May 2, 1995	Xylose isomerase gene of *Thermus aquaticus*	Nihon Shokuhin Kako Co. Ltd., Tokyo, Japan	5411886
June 13, 1995	Ethanol production by recombinant hosts	University of Florida, Gainesville, FL	5424202
October 3, 1995	Process for producing alcohol	CSIR, New Delhi, India	5455163
January 9, 1996	Ethanol production in Gram-positive microbes	University of Florida, Gainesville, FL	5482846
May 7, 1996	Recombinant *Zymomonas* for pentose fermentation	Midwest Research Institute, Kansas City, MO	5514583
January 27, 1998	Pentose fermentation by recombinant *Zymomonas*	Midwest Research Institute, Kansas City, MO	5712133
March 10, 1998	Recombinant *Zymomonas* for pentose fermentation	Midwest Research Institute, Kansas City, MO	5726053
August 4, 1998	Recombinant yeasts for effective fermentation of glucose and xylose	Purdue Research Foundation, West Lafayette, IN	5789210
December 1, 1998	Single *Zymomonas mobilis* strain for xylose and arabinose fermentation	Midwest Research Institute, Kansas City, MO	5843760

Date	Title	Assignee	Patent No.
February 2, 1999	Xylose utilization by recombinant yeasts	Xyrofin Oy, Helsinki, Finland	5866382
June 29, 1999	Ethanol production in Gram-positive microbes	University of Florida, Gainesville, FL	5916787
August 22, 2000	Recombinant cells that highly express chromosomally integrated heterologous genes	University of Florida, Gainesville, FL	6107093
August 15, 2000	Recombinant organisms capable of fermenting cellobiose	University of Florida Research Foundation, Inc., Gainesville, FL	6102690
August 28, 2001	Stabilization of PET operon plasmids and ethanol production in bacterial strains ...	United States	6280986
October 23, 2001	Genetically modified cyanobacteria for the production of ethanol ...	Enol Energy, Inc., Toronto, Canada	6306639
May 21, 2002	SHAM-insensitive terminal-oxidase gene from xylose-fermenting yeast	Wisconsin Alumni Research Foundation, Madison, WI	6391599
December 24, 2002	Pentose fermentation of normally toxic lignocellulose prehydrolysate with strain of *Pichia stipitis* ...	Midwest Research Institute, Kansas City, MO	6498029
May 20, 2003	Recombinant *Zymomonas mobilis* with improved xylose ultilization	Midwest Research Institute, Kansas City, MO	6566107
June 24, 2003	Production of ethanol from xylose	Xyrofin Oy, Helsinki, Finland	6582944
February 1, 2005	Ethanol production in recombinant hosts	University of Florida Research Foundation, Inc., Gainesville, FL	6849434
March 2, 2004	Genetically modified cyanobacteria for the production of ethanol ...	Enol Energy, Inc., Toronto, Canada	6699696
July 4, 2006	High-speed, consecutive batch or continuous, low-effluent process ...	Bio-Process Innovation, Inc., West Lafayette, IN	7070967
August 15, 2006	Transformed microorganisms with improved properties	Valtion Teknillin Tutkimuskeskus, Espoo, Finland	7091014

been described (section 3.3.1), but multiple curious features of its metabolism made it a promising target for industrial process development:

- Lacking an oxidative electron transport chain, the species is energetically grossly incompetent, that is, it can capture very little of the potential bioenergy in glucose — in other words, it is nearly ideal from the ethanol fermentation standpoint.
- What little energy production is achieved can be uncoupled from growth by an intracellular wastage (ATPase).
- It shows no Pasteur effect, seemingly oblivious to O_2 levels regarding glucose metabolism — but acetate, acetaldehyde, and acetoin are accumulated with increasing oxygenation.

During the 1970s, biotechnological interest in Z. *mobilis* became intense.* A patent for its use in ethanol production from sucrose and fructose was granted in mid-1989 (table 3.6). In the same year, researchers at the University of Queensland, Brisbane, Australia, demonstrated the ability of Z. *mobilis* to ferment industrial substrates such as potato mash and wheat starch to ethanol, with 95–98% conversion efficiencies at ethanol concentrations up to 13.5% (v/v).[196] The Australian process for producing ethanol from starch was scaled up to more than 13,000 l.[39]

The capability to utilize pentose sugars for ethanol production — with lignocellulosic substrates as the goal — was engineered into a strain recognized in 1981 as a superior ethanologen; strain CP4 (originally isolated from fermenting sugarcane juice) exhibited the most rapid rate of ethanol formation from glucose, achieved the highest concentration (>80 g/l from 200 g/l glucose), could ferment both glucose and sucrose at temperatures up to 42°C, and formed less polymeric fructose (levan) from sucrose than the other good ethanol producers. On transfer to high-glucose medium, CP4 had the shortest lag time before growth commenced and one of the shortest doubling times of the strains tested.[197] Researchers at the University of Sydney then undertook a series of studies of the microbial physiology and biochemistry of the organism and upscaling fermentations from the laboratory:

- Pilot-scale (500 l)-evaluated mutant strains selected for increased ethanol tolerance while improving ethanol production from sucrose and molasses became targets for strain development.[198]
- To reduce malodorous H_2S evolution by candidate strains, cysteine auxotrophs were isolated from studies of sulfur-containing amino acids.[198]
- Technical and engineering developments greatly increased the productivity of selected Z. *mobilis* strains (as discussed in the context of bioprocess technologies in chapter 4).[198]
- Direct genetic manipulation was explored to broaden the substrate range.[199]
- High-resolution ^{31}P nuclear magnetic resonance (NMR) of intracellular phosphate esters in cells fermenting glucose to ethanol showed that kinetic limitations could be deduced early in the ED pathway (figure 3.4), in the conversion of glucose 6-phosphate to 6-phosphonogluconate, and in the glycolytic pathway (phosphoglyceromutase), defining targets for rational genetic intervention.[200]

---

* By 1993, a review on Z. *mobilis* could already reference 362 publications.

Strains of *Z. mobilis* were first engineered to catabolize xylose at the National Renewable Resources Laboratory, Golden, Colorado. Four genes for xylose utilization by *E. coli* were introduced into *Z. mobilis* strain CP4 and expressed: xylose isomerase (*xylA*), xylulokinase (*xylB*), transketolase (*tktA*), and transaldolase (*talB*) on a plasmid under the control of strong constitutive promoters from *Z. mobilis*.[201,202] The transformant CP4 (pZB5) could grow on xylose as the carbon source with an ethanol yield of 86% of the theoretical maximum; crucially, xylose and glucose could be taken up by the cells simultaneously using a permease because no active (energy-expending), selective transport system for glucose exists in *Z. mobilis*; the transport "facilitator" for glucose is highly specific, and only mannose and (weakly) galactose, xylose, sucrose, and fructose appear to be taken up by this mechanism.[203] Using a plasmid containing five genes from *E. coli*, *araA* (encoding L-arabinose isomerase), *araB* (L-ribulose kinase), *araD* (L-ribulose 5-phosphate-4-epimerase), plus *tkta* and *talB*, a strain (ATCC39767[pZB206]) was engineered to ferment L-arabinose and produce ethanol with a very high yield (96%) but at a slow rate, ascribed to the low affinity of the permease uptake mechanism for L-arabinose.[204] A third NERL strain was ATCC39767 (identified as a good candidate for lignocellulose conversion based on the evidence of its growth in yellow poplar wood acid hydrolysates) transformed with a plasmid introducing genes for xylose metabolism and subsequently adapted for improved growth in the presence of hydrolysate inhibitors by serial subculture in progressively higher concentrations of the wood hydrolyate.[202,205]

A strain cofermenting glucose, xylose, and arabinose was constructed by chromosomal integration of the genes; this strain (AX101, derived from ATCC39576) was genetically stable, fermented glucose and xylose much more rapidly than it did arabinose, but produced ethanol at a high efficiency (0.46 g/g sugar consumed) and with only minor accumulations of xylitol, lactic acid, and acetic acid.[206–208] The major practical drawback for the AX101 strain is its sensitivity to acetic acid (formed in lignocellulosic hydrolysates by the breakdown of acetylated sugars); this sensitivity was demonstrated in trials of the strain with an agricultural waste (oat hulls) substrate pretreated by the two-stage acid process developed by the Iogen Corporation in Canada, although the bacterial ethanologen outperformed a yeast in both volumetric productivity and glucose to ethanol conversion.[209]

The University of Sydney researchers have also transformed their best candidate ethanologen with the NERL pZB5 plasmid to introduce xylose utilization; strain ZM4(pZB5) produced 62 g/l of ethanol from a medium of 65 g/l of both glucose and xylose, but its ethanol tolerance was lower than that of the *Z. mobilis* wild type.[210] The recombinant *Z. mobilis* shares the energy limitation on xylose observed with *E. coli*.[180] NMR examination of strain ZM4(pZB5) growing on glucose-xylose mixtures demonstrated low levels of nucleotide phosphate sugars inside the cells when xylose was mainly supporting metabolism; because these intracellular components are biosynthetic precursors for cell replication, the energy limitation has a clear biochemical mechanism for growth restriction.[211] In addition to the metabolic burden imposed by the plasmids, the production of unwanted by-products (xylitol, acetate, lactate, acetoin, and dihydroxyacetone) and the formation of xylitol phosphate as a possible inhibitor of enzyme-catalyzed processes may all contribute to the poorer fermentation performance on xylose as a carbon source. Further NMR investigations showed that acetic acid at growth-inhibitory concentrations decreased nucleotide phosphate sugars inside the cells and caused acidification of the cytoplasm, both complex bio-chemical factors difficult to remedy by genetic manipulation.[212] Taking one step back, a mutant of the ZM4 strain with greater tolerance to acetate was isolated by classical selection procedures; electroporating the pZB5 plasmid into this Ac[R] strain resulted in a xylose-fermenting strain with demonstrably improved resistance to sodium acetate at a concentration of 12 g/l.[213] Overexpressing a heterologous xylulokinase gene under the control of a native Z. mobilis promoter did not, however, increase growth or xylose metabolism on a xylose-containing medium, indicating that constraints on xylose uti-lization reside elsewhere in the catabolic pathway or in xylose uptake.[214] The xylulo-kinase-catalyzed step was more convincingly rate-limiting for xylose utilization with a Z. mobilis strain constructed at the Forschungszentrum Jülich (Germany) with K. pneumoniae XI and XK, as well as E. coli transketolase and transaldolase genes; overexpression of XK was deduced to be necessary and sufficient to generate strains capable of fermenting xylose to ethanol at up to 93% of the theoretical yield.[215]

The potential impact of the acetate inhibition of Z. mobilis is so severe with com-mercial process that investigations of the effect have continued to explore new molecu-lar targets for its abatement. With starting acetate concentrations in the range 0–8 g/l in fermentations of glucose and xylose mixtures, high acetate slowed the increase in intracellular ATP (as a measure of bioenergetic "health").[216] Expressing a gene from E. coli encoding a 24-amino acid proton-buffering peptide protects Z. mobilis strain CP4 from both low pH (<3.0) and acetic acid; optimization of this strategy may be success-ful with high-productivity strains for lignocellulose hydrolysate fermentation.[217]

3.3.2.3 Development of *Klebsiella* Strains for Bioethanol

The bacterium K. oxytoca was isolated from paper and pulp mills and grows around other sources of wood; in addition to growing on hexoses and pentoses, it can uti-lize cellobiose and cellotriose but does not secrete endoglucanase.[162] A University of Florida research group transformed strain M5A1 with the xylose-directing PET operon; unlike experience with E. coli, lower plasmid copy numbers gave higher ethanol productivity than with higher plasmid copy number.[218] A PET transformant could produce ethanol at up to 98% of the theoretical yield and was highly suitable for lignocellulose substrates because it utilized xylose twice as fast as glucose — and twice as fast as did E. coli strain KO11.

Stabilizing the PET operon was accomplished by chromosomal integration at the site of the PFL (*pfl* gene); screening for mutants hyperresistant to the selectable chloramphenicol marker resulted in the P2 strain with improved fermentation kinetics capable of producing 44–45 g/l of ethanol from glucose or cellobiose (100 g/l) within 48 hr.[219] Strain P2 has been demonstrated to generate ethanol from the cellulosic and lignocellulosic materials sugarcane bagasse, corn fiber, and sugarbeet pulp.[167,220,221]

As a candidate industrial strain for bioethanol production, P2 can utilize a wide range of low-molecular-weight substrates, including the disaccharides sucrose, malt-ose, cellobiose, and xylobiose, the trisaccharides raffinose, cellotriose, and xylotriose, and the tetrasaccharide stachyose.[172,181,219,222] This relatively nonspecific diet has led to the cloning and expression of a two-gene K. oxytoca operon for xylodextrin utilization in E. coli strain KO11; the gene product of *xynB* is a xylosidase (which also has weak arabinosidase activity), whereas that of the adjacent gene in the K. oxytoca genome

($xynT$) is a membrane protein previously found in Na^+/melibiose symporters* and related proteins functioning in transmembrane export and import.[223] The enhanced recombinant $E. coli$ could metabolize xylodextrins containing up to six xylose residues; unexpectedly, xylodextrin utilization was more rapid than by the donor $K. oxytoca$.

3.3.2.4   Other Bacterial Species

Interest in $Erwinia$ bacteria for ethanol production dates back at least to the late 1950s; in 1971, the explanation for the unusually high ethanol production by $Erwinia$ species was identified as a PDC/ADH pathway, decarboxylating pyruvate to acetaldehyde followed by reduction to ethanol, akin to that in $Z. mobilis$; ethanol is the major fermentative product, accompanied by smaller amounts of lactic acid.[224] Soft-rot bacteria secrete hydrolases and lyases to solubilize lignocellulosic polymers, and the PET operon was used to transform $E. carotovora$ and $E. chrysanthemi$ to produce ethanol from cellobiose, glucose, and xylose; both strains fermented cellobiose at twice the rate shown by cellobiose-utilizing yeasts.[42] The genetically engineered $E. chrysanthemi$ could ferment sugars present in beet pulp but was inferior to $E. coli$ strain KO11 in ethanol production, generating more acetate and succinate in mixed-acid patterns of metabolism.[221]

$Lactococcus lactis$ is another GRAS organism; its use in the industrial production of lactic acid is supplemented by its synthesis of the bacteriocin nisin, the only such product approved for food preservation.[225] When a PDC-encoding gene from $Zymobacter palmae$ was inserted into $L. lactis$ via a shuttle vector, the enzyme was functionally expressed, but, although a larger amount of acetaldehyde was detected, a slightly higher conversion of glucose to ethanol was measured (although glucose was used more slowly), and less lactic acid was accumulated, no increased ethanol production could be achieved, presumably because of insufficient endogenous ADH activity.[226] The same group at USDA's National Center for Agricultural Utilization Research examined $L. plantarum$ as an ethanologen for genetic improvement; strain TF103, with two genes for lactate dehydrogenase deleted, was transformed with a PDC gene from the Gram-positive bacterium $Sarcinia ventriculi$ to redirect carbon flow toward ethanol production, but only slightly more ethanol was produced (at up to 6 g/l).[227] Other attempts to metabolically engineer lactic acid bacteria have been similarly unsuccessful (although more ethanol is produced than by the parental strains and the conversion of glucose to ethanol is increased by nearly 2.5-fold), the bacteria remaining eponymously and predominantly lactic acid producers; although $Z. mobilis$ pdc and adh genes in PET operons are transcribed, the enzyme activities can be very low when compared with $E. coli$ transformants.[228,229]

$Zb. palmae$ was isolated on the Japanese island of Okinawa from palm sap by scientists from the Kirin Brewery Company, Yokohama, Japan. A facultative anaerobe, the bacterium can ferment glucose, fructose, galactose, mannose, sucrose, maltose, melibiose, raffinose, mannitol, and sorbitol, converting maltose efficiently to ethanol with only a trace of fermentative acids.[230] Its metabolic characteristics indicate potential as an ethanologen; broadening its substrate range to include xylose followed previous

---

* Symport is the simultaneous transport of a substrate and a cation (Na^+, H^+, etc.) in the same direction, while antiport is the exchange of two like-charged compounds (for example, Na^+ and H^+) via a common carrier; 40% of substrate uptake into bacterial cells requires one or the other of these two types of ion-driven transport.[154]

work with *Z. mobilis*, expressing *E. coli* genes for xylose isomerase, xylulokinase, transaldolase, and transketolase.[231] The recombinant *Zb. palmae* completely cofermented a mixture of 40 g/l each of glucose and xylose simultaneously within eight hours at 95% of the theoretical yield. Introducing a *Ruminococcus albus* gene for β-glucosidase transformed *Zb. palmae* to cellobiose utilization; the heterologous enzyme was more than 50% present on the cell surface or inside the periplasm, and the recombinant could transform 2% cellobiose to ethanol at 95% of the theoretical yield.[232] The PDC enzyme of the organism is, as discussed briefly above, an interesting target for heterologous expression in ethanologenic bacteria; it has the highest specific activity and lowest affinity for its substrate pyruvate of any bacterial PDC, and it has been expressed in *E. coli* to approximately 33% of the soluble protein. Codon usage for the gene is quite similar to that for *E. coli* genes, implying a facile recombinant expression.[233]

Cyanobacteria (blue-green algae) have generally lost their fermentative capabilities, now colonizing marine, brackish, and freshwater habitats where photosynthetic metabolism predominates; of 37 strains in a German culture collection, only five accumulated fermentation products in darkness and under anaerobic conditions, and acids (glycolic, lactic, formate, and oxalate) were the major products.[234] Nevertheless, expression of *Z. mobilis pdc* and *adh* genes under the control of the promoter from the operon for the CO_2-fixing ribulose 1,5-*bis*-phosphate carboxylase in a *Synechococcus* strain synthesized ethanol phototrophically from CO_2 with an ethanol:acetaldehyde molar ratio higher than 75:1.[235] Because cyanobacteria have simple growth nutrient requirements and use light, CO_2, and inorganic elements efficiently, they represent a system for longer-term development for the bioconversion of solar energy (and CO_2) by genetic transformation, strain and process evolution, and metabolic modeling. The U.S. Department of Energy (DOE) is funding (since 2006) DNA sequencing studies of six photosynthetic bacteria at Washington University in St. Louis, Missouri, and the DOE's own sequencing facility at Walnut Creek, California, using a biodiversity of organisms from rice paddies and deep ocean sources to maximize biochemical and metabolic potential.

3.3.2.5 Thermophilic Species and Cellulosome Bioproduction Technologies

By 1983, in experimental laboratory programs, selected *Bacillus* strains had achieved ethanol formation to 20 g/l (from 50 g/l sucrose as the carbon substrate) at 60°C, with ethanol as the major fermentation product; acetic and formic acids remained serious by-products, however, and evidence from laboratory studies suggested that ethanol accumulation followed (and depended on) the formation of those growth-inhibiting acids.[46] The ability to run ethanol fermentations at 70–80°C with thermophilic microbes remains both a fascination and a conscious attempt to accelerate bioprocesses, despite the low ethanol tolerance and poor hexose-converting abilities of anaerobic thermophilic bacteria. In 2004, exploratory work at the Technical University of Denmark tested isolates from novel sources (hot springs, paper pulp mills, and brewery wastewater), using three main criteria for suitable organisms:[236]

1. The ability to ferment D-xylose to ethanol
2. High viability and ethanol productivity with pretreated wheat straw
3. Tolerance to high sugar concentrations

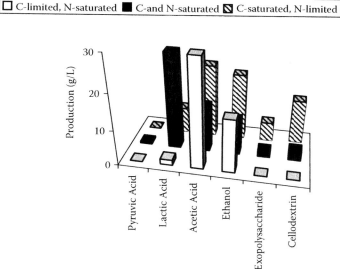

□ C-limited, N-saturated ■ C-and N-saturated ▨ C-saturated, N-limited

FIGURE 3.11   Cellobiose utilization by *Clostridium cellulolyticum* in continuous culture at low dilution rate, 0.023–0.035/hr. (Data from Desvaux.[237])

Five good (but unidentified) strains were identified by this screening program, all from hot springs in Iceland,* the best isolate could grow in xylose solutions of up to 60 g/l.

Thermophilic and mesophilic clostridia also have their advocates, especially with reference to the direct fermentation of cellulosic polymers by the cellulosome multienzyme complexes, as discussed in chapter 2, section 2.4.1). Bypassing the cellulosome is possible if cellulose degradation products (rather than polymeric celluloses) are used as carbon sources — this equates to using bacteria with cellulase-treated materials, including agricultural residues and paper recyclates. Laboratory studies with *C. cellulolyticum* tested cellobiose in this fashion but with chemostat culture so as to more closely control growth rates and metabolism.[237] The results demonstrated that a more efficient partitioning of carbon flow to ethanol was possible than with cellulose as the substrate but that the fermentation remained complex, with acids being the major products (figure 3.11). Nevertheless, clostridia are open to metabolic engineering to reduce the waste of carbohydrates as acids and polymeric products or as vehicles for "consolidated bioprocessing" where cellulase production, cellulose hydrolysis, and fermentation all occur in one step — this is covered in chapter 4 (section 4.5).

3.3.3   CANDIDATE BACTERIAL STRAINS FOR COMMERCIAL ETHANOL PRODUCTION IN 2007

Definitive comparisons of recombinant ethanologenic bacteria in tightly controlled, side-by-side comparisons have not been made public. From data compiled in 2003 from various sources with *E. coli*, *K. oxytoca*, and *Z. mobilis* strains fermenting

---

* Geysers from Yellowstone National Park also yielded one promising isolate.

mixtures of glucose, xylose, and arabinose, conflicting trends are evident.[162] For maximum ethanol concentration, the ranking order was

$$Z.\ mobilis\ AX101 \gg K.\ oxytoca\ P2 = E.\ coli\ FBR5,$$

but for ethanol yield (percentage of maximum possible conversion), *E. coli* was superior:

$$E.\ coli\ FBR5 > K.\ oxytoca\ P2 = Z.\ mobilis\ AX101,$$

and the rankings of ethanol production rate (grams per liter per hour) were again different:

$$E.\ coli\ FBR5 \gg Z.\ mobilis\ AX101 \gg K.\ oxytoca\ P2.$$

All three strains can utilize arabinose, xylose, and glucose, but *Z. mobilis* AX101 cannot utilize the hemicellulose component hexose sugars galactose or mannose. In tests of *E. coli*, *K. oxytoca*, and *Erwinia chrysanthemi* strains, only *E. coli* KO11 was able to convert enzyme-degraded polygalacturonic acid (a pectin polymer) to ethanol.[221]

Enteric bacteria (including *E. coli* and *Klebsiella* sp.) have the additional hurdle to overcome of being perceived as potentially injurious to health, and *K. oxytoca* has been implicated in cases of infectious, hospital-acquired, and antibiotic-associated diseases.[238–240] *K. oxytoca* is well known as a producer of a broad-spectrum β-lactamase, an enzyme capable of inactivating penicillins and other β-lactam antibiotics.[241] Immunosuppression of patients under medical supervision or as a result of pathogenesis has led to the identification of infections by hitherto unknown yeast species or by those not considered previously to be pathogenic, including *Kluyveromyces marxianus*, five species of *Candida*, and three species of *Pichia*.[17] Biosafety issues and assessments will, therefore, be important where planning (zoning) permissions are required to construct bacterial bioethanol facilities.

Because commercially relevant biomass plants for lignocellulosic ethanol only began operating in 2004 (chapter 2, section 2.7), future planned sites in divers parts of the world will inevitably make choices of producing organism that will enormously influence the continued development and selection of candidate strains.[242] Different substrates and/or producing regions may arrive at different choices for optimized ethanol producer, especially if local enzyme producers influence the choice, if licensing agreements cannot be made on the basis of exclusivity, or if national interests encourage (or dictate) seamless transfer of technologies from laboratories to commercial facilities. More than a decade ago, a publication from the National Renewable Energy Laboratory ranked *Z. mobilis* ahead of (in descending order of suitability): recombinant *Saccharomyces*, homofermentative *Lactobacillus*, heterofermentative *Lactobacillus*, recombinant *E. coli*, xylose-assimilating yeasts, and clostridia.[202] Their list of essential traits included

- High conversion yield
- High ethanol tolerance
- Tolerance to hydrolysates
- No O_2 requirement (i.e., a facultative anaerobe)

- Low fermentation pH (to discourage contaminants)
- High fermentation selectivity
- Broad substrate utilization range
- GRAS status

The secondary list of 19 "desirable" traits included being Crabtree-positive (see above, section 3.1.1), high growth rates, tolerance to high salts, high shear, and elevated temperature. No commercialization of the *Z. mobilis* biocatalyst is yet established, but if bacterial ethanologens remain serious candidates for commercial bioethanol production, clear evidence for this should appear in the next decade as increasing numbers of scale-up bioethanol facilities are constructed for a variety of biomass feedstocks.

The increasing list of patents issued to companies and institutions for ethanologens testifies to the endeavors in this field of research and development (table 3.6). More problematic is that the field has been gene-led rather than genome- or (more usefully still) metabolome-led, that is, with full cognizance and appreciation of the flexibility and surprises implicit in the biochemical pathway matrices. The "stockpiling" of useful strains, vectors, and genetic manipulation techniques has built a process platform for the commercialization of ethanol production from lignocellulose biomass. The scientific community can find grounds for optimism in the new insights in metabolic engineering described in this chapter; what is less clear is the precise timescale — 10, 15, or more years — required to translate strain potential into industrial-scale production.[243]

3.4 EXTRAPOLATING TRENDS FOR RESEARCH WITH YEASTS AND BACTERIA FOR BIOETHANOL PRODUCTION

3.4.1 "TRADITIONAL" MICROBIAL ETHANOLOGENS

Given the present indeterminacy regarding "best" choices for producing organism and biomass substrate, it is timely to summarize recent advances in the metabolic and genetic sciences relevant to yeast and bacterial ethanologens, and from them to project forward key areas for their application to industrial ethanol production after 2007. Four themes can readily be identified:

1. A better understanding of stress responses
2. Relating transport phenomena to sugar utilization and ethanol productivity
3. Defining the roles of particular genes in the metabolism of ethanologens
4. Applying whole-genome methodologies and in silico simulations to optimize the balance of carbon flows along synergistic or competing pathways

Yeast ethanologens show dose-dependent inhibition by furfural and HMF and are generally more sensitive to furfural; a *S. cerevisiae* strain converted HMF to 2,5-*bis*-HMF, a previously postulated alcohol formed from the parent aldehyde, and this suggests that an enzymic pathway for detoxifying these components of lignocellulosic hydrolysates should be attainable by gene cloning and/or mutation to improve the activities and substrate preferences of the enzymes.[244] That such a reduction reaction as that to form 2,5-*bis*-HMF might utilize NADPH as cofactor was shown by screening a gene disruption library from *S. cerevisiae*: one type of mutation was in *zwf1*, the gene encoding glucose 6-phosphate dehydrogenase, the NADPH-generating entry point to the oxidative pentose pathway; overexpression of *zwf1* allowed growth in furfural concentrations normally toxic, whereas similar effects with non-redox-associated gene products could be explained as a result of inhibiting the overall activity of the pathway.[245] The *ADH6* gene of *S. cerevisiae* encodes a reductase enzyme that reduces HMF and NADPH supports a specific activity of the enzyme some 150-fold higher than with NADH; yeast strains overexpressing this gene had a higher in vivo conversion rate of HMF in both aerobic and anaerobic cultures.[246]

A key trait for biofuels research is to enable growth in the presence of very high concentrations of both glucose and ethanol, the former in the batch medium (discussed in chapter 4, section 4.3), the latter accumulating as the fermentation proceeds. In yeasts, analysis of the effects of ethanol concentration and temperature point to nonspecific actions on the cell membranes as leading to the variety of cellular responses, including reduced fermentation rate and viability.[247] Adapting cells to ethanol causes changes in the cellular membrane composition, in particular the degree of unsaturation of the fatty acids and the amount of sterols present, both indicative of alterations in the fluidity characteristics of the lipid-based membranes.[248] The low tolerance of *Pichia stipitis* to ethanol can be correlated with a disruption of protein transport across the cell membrane, an energy-requiring process involving a membrane-localized ATPase; the system is activated by glucose, but ethanol uncouples ATP hydrolysis from protein transport, thus wasting energy for no purpose, and structural alteration of the ATPase to affect its sensitivity to ethanol could be a quick-win option for yeast ethanologens in general.[249]

In *Z. mobilis*, there is evidence for a glucose-sensing system to confer (or increase) the cells' tolerance to the osmotic stress imposed by high sugar concentrations; the exact functions of the products of the four genes simultaneously transcribed in a glucose-regulated operon are unknown, but manipulating this genetic locus would aid adaptation to high concentrations (>100 g/l) of glucose in industrial media.[250] The announcement of the complete sequencing of the genome of *Z. mobilis* by scientists in Seoul, South Korea, in 2005 provided rationalizations for features of the idiosyncratic carbohydrate metabolism in the absence of three key genes for the EMP/tricarboxylic acid cycle paradigm, that is, 6-phosphofructokinase, 2-oxoglutarate dehydrogenase, and malate dehydrogenase.[251] The nonoxidative pentose phosphate pathway is mostly missing, although the genes are present for the biosynthesis of phosphorylated ribose and thence histidine and nucleotides for both DNA and RNA (figure 3.12). The 2-Mbp circular chromosome encodes for 1,998 predicted functional genes; of these, nearly 20% showed no similarities to known genes, suggesting a high probability of coding sequences of industrial significance for ethanologens. In comparison to a strain of *Z. mobilis* with lower tolerance to ethanol and rates of ethanol production, glucose uptake, and specific growth rate, strain ZM4 contains 54 additional genes, including four transport proteins and two oxidoreductases, all potentially mediating the higher ethanol productivity of the strain. Moreover, two genes coding for capsular carbohydrate synthesis may be involved in generating an altered morphology more resistant to osmotic stress. Intriguingly, 25 of the "new" genes showed similarities to bacteriophage genes, indicating a horizontal transfer of genetic material via phages.[251]

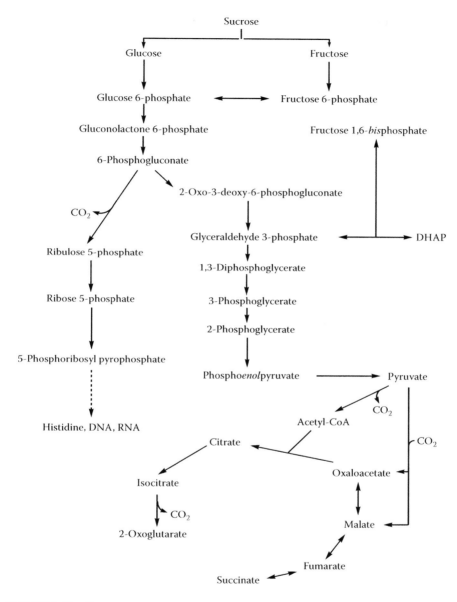

FIGURE 3.12   Fragmentary carbohydrate metabolism and interconversions in *Z. mobilis* as confirmed by whole-genome sequencing. (From Seo et al.[251])

Analysis of mutants of the yeast *Yarrowia lipolytica* hypersensitive to acetic acid and ethanol identified a single gene that could direct multiple phenotypic changes, including colony morphology, the morphology of intracellular membranes, and the appearance of vacuoles and mitochondria, and induce early death on glucose even without the presence of acetate or ethanol; this *GPR1* gene may be a target for manipulation and mutation to increase solute tolerance.[252] A more direct approach in *S. cerevisiae* involved multiple mutagenesis of a transcription factor and selection of dominant mutations that conferred increased ethanol tolerance and more efficient glucose conversion to ethanol: 20% for growth in the presence of an initial glucose concentration of 100 g/l, 69% for space-time yield (grams of ethanol per liter per hour), 41% for specific productivity, and 15% for ethanol yield from glucose.[253] It is likely that the expression of an ensemble of hundreds of genes was increased by the mutations; of the 12 most overexpressed genes, deletions mutations abolished the increased glucose tolerance in all but one gene, and selective overexpression of each of the top three up-regulated genes could replicate the tolerance of the best mutant. Because this work was, by necessity, carried out on a laboratory haploid strain, it is possible that similar changes could have occurred gradually during many generations in industrial strains. The effect on gene expression was, however, greatly diminished at 120 g/l of glucose.

Comparable analyses of gene expression and function are now being introduced into commercial potable alcohol research. A common driver between breweries and industrial bioethanol plants is (or will be) the use of concentrated, osmotically stressful fermentation media. In breweries, the capability to use high-gravity brewing in which the wort is more concentrated than in traditional practice saves energy, time, and space; the ethanol-containing spent wort can be diluted either with lower alcohol-content wort or with water. Variants of brewer's yeast (*S. pastorianus*) were obtained subjecting UV-mutagenized cells to consecutive rounds of fermentation in very-high-gravity wort; variants showing faster fermentation times and/or more complete sugar utilization were identified for investigations of gene expression with microarray technologies.[254] Of the 13 genes either overexpressed or underexpressed in comparison with a control strain, a hexokinase gene was potentially significant as signaling a reduced carbon catabolite repression of maltose and maltotriose uptake; more speculatively, two amplified genes in amino acid biosynthesis pathways could point to increased requirements for amino acid production for growth in the highly concentrated medium. Although deletion mutants are usually difficult to isolate with polyploid industrial strains, the dose-variable amplification of target genes is more feasible.

Strains of *S. cerevisiae* genetically engineered for xylose utilization depend, unlike naturally xylose-consuming yeasts, on xylose uptake via the endogenous hexose transporters; with a high XR activity, xylose uptake by this ad hoc arrangement limits the pentose utilization rate at low xylose concentrations, suggesting that expression of a yeast xylose/pentose transporter in *S. cerevisiae* would be beneficial for the kinetics and extent of xylose consumption from hemicellulose hydrolysates.[255] A high-affinity xylose transporter is present in the cell membranes of the efficient xylose fermenting yeast *Candida succiphila* when grown on xylose and in *Kluyveromyces marxianus* when grown on glucose under fully aerobic conditions.[256]

In ethanologenic bacteria, there are also grounds for postulating that radically changing the mechanism of xylose uptake would improve both anaerobic growth and xylose utilization. Mutants of *E. coli* lacking PFL or acetate kinase activity fail to grow anaerobically on xylose, that is, the cells are energy-limited without the means to generate ATP from acetyl phosphate; in contrast, the mutants can grow anaerobically on arabinose — and the difference resides in the sugar transport mechanisms: energy-expending active transport for xylose but an energy-conserving arabinose.[257] Reconfiguring xylose uptake is, therefore, an attractive option for removing energetic constraints on xylose metabolism. In yeast also, devising means of conserving energy during xylose metabolism would reduce the bioenergetic problems posed by xylose catabolism — gene expression profiles during xylose consumption resemble those of the starvation response, and xylose appears to induce metabolic behavioral responses that mix features of those elicited by genuinely fermentable and respirative carbon sources.[258]

Looking beyond purely monomeric sugar substrates, the soil bacterium *Geobacillus stearothermophilus* is highly efficient at degrading hemicelluloses, possessing a 30-gene array for this purpose, organized into nine transcriptional units; in the presence of a xylan, endo-1,4-β-xylanase is secreted, and the resulting xylooligosaccharides enter the cell by a specialized transport system, now known to be a three-gene unit encoding an ATP-binding cassette transport system, whose transcription is repressed by xylose but activated by a response regulator.[259] This regulation allows the cell to rapidly amplify the expression of the transport system when substrates are available. Inside the cell, the oligosaccharides are hydrolyzed to monomeric sugars by the actions of a family of xylanases, β-xylosidase, xylan acetylesterases, arabinofuranosidases, and glucuronidases. Such a complete biochemical toolkit for hemicellulose degradation should be considered for heterologous expression in bacterial bioethanol producers.

Control of the gene encoding the major PDC in wild-type *S. cerevisiae* was shown by analysis of gene transcription to be the result of ethanol repression rather than glucose induction; an ethanol-repression *ERA* sequence was identified in the yeast DNA; the Era protein binds to the PDC1 promoter sequence under all growth conditions, and an autoregulatory (A) factor stimulates PDC1 transcription but also binds to the pdc1 enzyme — the more enzyme is formed, the more A factor is bound, thus making the transcription of the FDC1 gene self-regulating.[260] This may be a useful target for strain improvement; industrial strains may, however, have already been "blindly" selected at this locus.

Focusing on single-gene traits, a research group at Tianjin University (China) has recently demonstrated three approaches to reduce glycerol, acetate, and pyruvate accumulation during ethanol production by *S. cerevisiae*:

1. Simply deleting the gene encoding glycerol formation from glycolysis, that is, glycerol 3-phosphate dehydrogenase reduced the growth rate compared with the wild type but reduced acetate and pyruvate accumulation. Simultaneously overexpressing the *GLT1* gene for glutamate synthase (catalyzing the reductive formation of glutamate from glutamine and 2-oxo-glutarate) restored the growth characteristics while reducing glycerol accumulation and increasing ethanol formation.[261]
2. Deleting the gene (*FSP1*) encoding the transporter protein for glycerol also improved ethanol production while decreasing the accumulation of glycerol, acetate, and pyruvate — presumably increasing the intracellular concentration of glycerol (by blocking its export) accelerates its metabolism by a reversal of the glycerol kinase and glycerol 3-phosphate dehydrogenase reactions.[262]
3. Improved production of ethanol was also achieved by deleting *FPS1* and overexpressing *GLT1*, accompanied by reduced accumulation of glycerol, acetate, and pyruvate.[263]

A more subtle metabolic conundrum may limit ethanol accumulation in Z. *mobilis*: although the cells are internally "unstructured" prokaryotes, lacking membrane-limited intracellular organelles, different redox microenvironments may occur in which two forms of ADH (rather than one) could simultaneously catalyze ethanol synthesis and oxidation, setting up a futile cycle; this has only been demonstrated under aerobic growth conditions, but the functions of the two ADH enzymes (one Zn-dependent, the other Fe-dependent) inside putative redox microcompartments within the cell warrant further investigation.[264]

S. *cerevisiae*, along with E. *coli*, B. *subtilis*, and a few other organisms, formed the initial "Rosetta Stone" for genome research, providing a well-characterized group of genes for comparative sequence analysis; coding sequences of other organisms lacking detailed knowledge of their biochemistry could be assigned protein functionalities based on their degree of homology to known genes. Genome-scale metabolic networks can now be reconstructed and their properties examined in computer simulations of biotechnological processes. Such models can certainly generate good fits to experimental data for the rate of growth, glucose utilization, and (under microaerobiosis) ethanol formation but have limited predictive value.[265] One limitation is that the biosynthetic genes comprise too conservative a functional array, lacking the sensing mechanisms and signaling pathways that sense and respond to nutritional status and other environmental factors.[266] These elements function to establish developmental and morphological programs in the wild-type ecology and have unpredictable effects on growth patterns in the fluctuating nutritional topologies of large imperfectly mixed industrial fermentors. If computer-aided design for metabolism is to be feasible for industrially relevant microbes, added sophistications must be incorporated — equally, however, radical restructuring is necessary to accommodate the gross differences in metabolic patterns between bacteria such as E. *coli* and, for example, the Entner-Doudoroff pathway organisms or when P. *stipitis* metabolism is compared in detail with the far better characterized pathway dynamics in S. *cerevisiae*.[267,268] Comparing metabolic profiles in 14 yeast species revealed many important similarities with S. *cerevisiae*, but *Pichia augusta* may possess the ability to radically rebalance the growth demand for NADPH from the pentose phosphate pathway (figure 3.2); this suggests a transhydrogenase activity that could be used to rebalance NADH/NADPH redox cofactors for xylose metabolism (section 3.2.1).[269]

In the immediate and short-term future, therefore, applied genomics can successfully define the impacts of more focused changes in quantitative gene expression. Developing a kinetic model of the pentose phosphate and Entner-Doudoroff pathways in Z. *mobilis*, and comparing the predictions with experimental results, showed that the activities of xylose isomerase and transaldolase were of the greatest significance for ethanol production but that overexpressing the enzymic link between the two pathways (phosphoglucose isomerase) was unnecessary.[270] Such systematic analysis can help not only in maximizing fermentation efficiencies but also in minimizing the degree of expression of heterologous enzymes to reduce "metabolic burdens" in recombinant strains. Although in silico predictions failed to account for continued xylitol accumulation by a recombinant, xylose-utilizing S. *cerevisiae* strain on xylose (suggesting continued problems in balancing cofactor requirement in reductase/dehydrogenase reactions and/or still unidentified limitations to later catabolic steps in xylose consumption), computer simulations did accurately highlight glycerol formation as a key target for process improvement: expressing a nonphosphorylating NADPH-dependent glyceraldehyde 3-phosphate dehydrogenase increased carbon flow to pyruvate (rather than being diverted to glycerol formation), reduced glycerol accumulation, and improved ethanol production by up to 25%.[271,272]

3.4.2 "Designer" Cells and Synthetic Organisms

After more than two decades of intensive molecular genetic research, *S. cerevisiae* remains the ethanologen of choice. Although industrial strains are genetically largely undefined, it is easily demonstrated that laboratory strains can function perfectly adequately for ethanol production, even for the complex process of manufacturing potable spirits with defined requirements for sensory parameters in the finished product such as volatiles, higher alcohols, and glycerol content.[273] Nevertheless, it is more likely that applying the knowledge gained to the genomic improvement of hardy industrial *Saccharomyces* (or other Crabtree-positive yeast) strains with proven track records of fermenting very concentrated media to high volumetric yields of ethanol would generate highly suitable biocatalysts for the demanding tasks of growing and metabolizing sugars and oligosaccharides in lignocellulosic hydrolysates.

Even better would be the importing of a methodology developed in the world of industrial biotechnology, that is, "genome breeding," as outlined by Kyowa Hakko Kogyo,* Tokyo, Japan. By comparing the whole genome sequence from the wild-type *Corynebacterium glutamicum*, the major producing organism for L-lysine and L-glutamate, with gene sequence from an evolved highly productive strain, it was possible to identify multiple changes; with these data, transforming a "clean" wild type to a hyperproducer of lysine was accomplished with only three specific and known mutations in the biosynthetic genes.[274] This minimal mutation strain had distinct productivity advantages over the industrial strains that had been developed by chance mutation over decades — a reflection of how many unwanted changes may (and did) occur over prolonged periods of random mutagenesis and selection in the twentieth century.[275] Repeating such an exercise with any of the multitude of commercial alcohol-producing *Saccharomyces* yeasts would rapidly identify specific genomic traits for robustness and high productivity on which to construct pentose-utilizing and other capabilities for bioethanol processes.

A radically different option has been outlined in a patent application at the end of May 2007.[276] Work undertaken at the J. Craig Venter Institute, Rockville, Maryland, defined a minimal set of 381 protein-encoding genes from *Mycoplasma genitalium*, including pathways for carbohydrate metabolism, nucleotide biosynthesis, phospholipid biosynthesis, and a cellular set of uptake mechanisms for nutrients, that would suffice to generate a free living organism in a nutritionally rich culture medium. Adding in genes for pathways of ethanol and/or hydrogen formation would result in a biofuels producer with maximum biochemical and biotechnological simplicity. The timeline required for practical application and demonstration of such a synthetic

---

* The assignee of an early U.S. patent for a microbe (a *Kluyveromyces* yeast) capable of producing alcohol from xylose and cellobiose (table 3.6).

organism is presently unclear, although the research is funded by the DOE, under the genome projects of the Department's Office of Science, with the target of developing a novel recombinant cyanobacterial system for hydrogen production from water and a cellulosome system for the production of ethanol and/or butanol in suitable clostridial cells.[277] The drawbacks to such an approach are that it would be highly dependent on the correct balance of supplied nutrients to the organism (with its limited capabilities), requiring highly precise nutrient feeding mechanisms and a likely protracted optimization of the pathways to rival rates of product formation already attainable with older patented or freely available ethanologens.

REFERENCES

1. Wyman, C.E., Ethanol from lignocellulosic biomass: technology, economics, and opportunities, *Bioresour. Technol.*, 50, 3, 1994.
2. Lynd, L.R., Overview and evaluation of fuel ethanol production from cellulosic biomass: technology, economics, the environment, and policy, *Ann. Rev. Energy Environ.*, 21, 403, 1996.
3. Chandrakant, P. and Bisaria, V.S., Simultaneous bioconversion of cellulose and hemicellulose to ethanol, *Crit. Rev. Biotechnol.*, 18, 295, 1998.
4. Aristidou, A.A., Application of metabolic engineering to the conversion of renewable resources to fuels and fine chemicals: current advances and future prospects, in El-Mansi, E.M.T., Bryce, C.A., Demain, A.L., and Allman, A.R. (Eds.), *Fermentation Microbiology and Biotechnology*, 2nd ed., CRC Press, Boca Raton, 2007, chap. 9.
5. Lin, Y. and Tanaka, S., Ethanol fermentation from biomass resources: current state and prospects, *Appl. Microbiol. Biotechnol.*, 69, 627, 2006.
6. Szczodrak, J. and Targonski, Z., Selection of thermotolerant yeast strains for simultaneous saccharification and fermentation of cellulose, *Biotechnol. Bioeng.*, 31, 300, 1988.
7. Roychoudhury, P.K., Ghose, T.K., and Ghosh, P., Operational strategies in vacuum-coupled SSF for conversion of lignocellulose to ethanol, *Enz. Microb. Technol.*, 14, 581, 1992.
8. Hahn-Hägerdal, B. et al., Metabolic engineering of *Saccharomyces cerevisiae* for xylose utilization, *Adv. Biochem. Eng./Biotechnol.*, 73, 53, 2001.
9. Yoon, S.H., Mukerjee, R., and Robyt, J.F., Specificity of yeast (*Saccharomyces cerevisiae*) in removing carbohydrates by fermentation, *Carbohydr. Res.*, 338, 1127, 2003.
10. van Dijken, J.P. et al., Alcoholic fermentation by non-fermenting yeasts, *Yeast*, 2, 123, 1986.
11. Lagunas, R., Misconceptions about the energy metabolism of *Saccharomyces cerevisiae*, *Yeast*, 2, 221, 1986.
12. Vallet, C. et al., Natural abundance isotopic fractionation in the fermentation reaction: influence of the nature of the yeast, *Bioorg. Chem.*, 24, 319, 1996.
13. Çaylak, B. and Vardar Sukan, F. Comparison of different production processes for bioethanol, *Turk. J. Chem.*, 22, 351, 1998.
14. da Cruz, S.H., Batistote, M., and Ernandes, J.R., Effect of sugar catabolite repression in correlation with the structural complexity of the nitrogen source on yeast growth and fermentation, *J. Inst. Brew.*, 109, 349, 2003.
15. Ergun, M. and Mutlu, S.F., Application of a statistical technique to the production of ethanol from sugar beet molasses by *Saccharomyces cerevisiae*, *Bioresour. Technol.*, 73, 251, 2000.
16. Cejka, A., Preparation of media, in *Biotechnology*, volume 2, Brauer, H. (Ed.), VCH Verlagsgesellschaft, Weinheim, 1985, chap. 26.
17. Barnett, J.A., Payne, R.W., and Yarrow, D., *Yeasts: Characteristics and Identification*, 3rd ed., Cambridge University Press, Cambridge, 2000.

18. Sims, A.P. and Barnett, J.A., The requirement of oxygen for the utilization of maltose, cellobiose, and D-galactose by certain anaerobically fermenting yeasts (Kluyver effect), *J. Gen. Microbiol.*, 106, 277, 1976.

19. Lloyd, D. and James, C.J., The Pasteur effect in yeasts: mass spectrometric monitoring of oxygen uptake, and carbon dioxide and ethanol production, *FEMS Microbiol. Lett.*, 42, 27, 1987.

20. van Dijken, J.P. and Scheffers, W.A., Redox balances in the metabolism of sugars by yeasts, *FEMS Microbiol. Rev.*, 32, 199, 1986.

21. Wijsman, M.R. et al., Inhibition of fermentation and growth in batch cultures of the yeast *Brettanomyces intermedius* upon a shift from aerobic to anaerobic conditions (Custers effect), *Antonie van Leeuwenhoek*, 50, 183, 1984.

22. Weusthuis, R.A. et al., Is the Kluyver effect caused by product inhibition?, *Microbiology*, 140, 1723, 1994.

23. Malluta, E.F., Decker, P., and Stambuk, B.U., The Kluyver effect for trehalose in *Saccharomyces cerevisiae*, *J. Basic Microbiol.*, 40, 199, 2000.

24. Fukuhara, H., The Kluyver effect revisited, *FEMS Yeast Res.*, 3, 327, 2003.

25. Fredlund, E. et al., Oxygen- and glucose-dependent regulation of central carbon metabolism in *Pichia anomola*, *Appl. Environ. Microbiol.*, 70, 5905, 2004.

26. Frick, O. and Wittman, C., Characterization of the metabolic shift between oxidative and fermentative growth in *Saccharomyces cerevisiae* by comparative ^{13}C flux analysis, *Microb. Cell Fact.*, 4: 30, 2005.

27. Sierkstra, L.N. et al., Regulation of glycolytic enzymes and the Crabtree effect in galactose-limited continuous cultures of *Saccharomyces cerevisiae*, *Yeast*, 9, 787, 1993.

28. Kiers, J. et al., Regulation of alcoholic fermentation in batch and chemostat cultures of *Kluyveromyces lactis* CBS 2359, *Yeast*, 14, 459, 1998.

29. Bruinenberg, P.M. et al., The role of the redox balance in the anaerobic fermentation of xylose by yeasts, *Eur. J. Appl. Microbiol. Biotechnol.*, 18, 287, 1983.

30. Bruinenberg, P.M. et al., NADH-linked aldose reductase: the key to alcoholic fermentation of xylose by yeasts, *Appl. Microbiol. Biotechnol.*, 19, 256, 1984.

31. Visser, W. et al., Oxygen requirements of yeasts, *Appl. Environ. Microbiol.*, 56, 3785, 1990.

32. Hutter, A. and Oliver, S.G., Ethanol production using nuclear petite yeast mutants, *Appl. Microbiol. Biotechnol.*, 49, 511, 1998.

33. Merico, A. et al., Fermentative lifestyle in yeasts belonging to the *Saccharomyces* complex, *FEBS J.*, 274, 976, 2007.

34. Narendranath, N.V., Thomas, K.C., and Ingledew, W.M., Acetic acid and lactic acid inhibition of growth of *Saccharomyces cerevisiae* by different mechanisms, *J. Am. Soc. Brew. Chem.*, 59, 187, 2001.

35. Graves, T. et al., Interaction effects of lactic acid and acetic acid at different temperatures on ethanol production by *Saccharomyces cerevisiae* in corn mash, *Appl. Microbiol. Biotechnol.*, 73, 1190, 2007.

36. Schell, D.J. et al., Contaminant occurrence, identification and control in a pilot-scale corn fiber to ethanol conversion process, *Bioresour. Technol.*, 98, 2942, 2007.

37. Tolan, J.S. and Finn, R.K., Fermentation of D-xylose and L-arabinose to ethanol by *Erwinia chrysanthemi*, *Appl. Environ. Microbiol.*, 53, 2033, 1987.

38. Tolan, J.S. and Finn, R.K., Fermentation of D-xylose to ethanol by genetically modified *Klebsiella planticola*, *Appl. Environ. Microbiol.*, 53, 2039, 1987.

39. Millichip, R.J. and Doelle, H.W., Large-scale ethanol production from milo *(sorghum)* using *Zymomonas mobilis*, *Proc. Biochem.*, 24, 141, 1989.

40. Bothast, R.J. et al., Fermentation of L-arabinose, D-xylose and D-glucose by ethanologenic recombinant *Klebsiella oxytoca* strain P2, *Biotechnol. Lett.*, 16, 401, 1994.

41. San Martin, R. et al., Development of a synthetic medium for continuous anaerobic growth and ethanol production with a lactate dehydrogenase mutant of *Bacillus stearothermophilus*, *J. Gen. Microbiol.*, 138, 987, 1992.

42. Beall, D.S. and Ingram, L.O., Genetic engineering of soft-rot bacteria for ethanol production from lignocellulose, *J. Ind. Microbiol.*, 11, 151, 1993.

43. Dien, B.S. et al., Conversion of corn milling fibrous coproducts into ethanol by recombinant *Escherichia coli* strains KO11 and SL40, *World J. Microbiol. Biotech.*, 13, 619, 1997.

44. Nakano, M.M. and Zuber, P., Anaerobiosis, in Sonensheim, A.L., Hoch, J.A., and Losick, R. (Eds.), Bacillus subtilis *and Its Closest Relatives: From Genes to Cells*, ASM Press, Washington, DC, 2002, chap. 28.

45. Zeikus, J.G. et al., Thermophilic ethanol fermentation, *Basic Life Sci.*, 18, 441, 1981.

46. Hartley, B.S. and Payton, M.A., Industrial prospects for thermophiles and thermophilic enzymes, *Biochem. Soc. Symp.*, 48, 133, 1983.

47. Feldmann, S.D., Sahm, H., and Sprenger, G.A., Pentose metabolism in *Zymomonas mobilis* wild-type and recombinant strains, *Appl. Microbiol. Biotechnol.*, 38, 354, 1992.

48. Sprenger, G.A., Carbohydrate metabolism in *Zymomonas mobilis*: a catabolic pathway with some scenic routes, *FEMS Microbiol. Lett.*, 145, 301, 1996.

49. Zikmanis, P., Kruce, R., and Auzina, L., An elevation of the molar growth yield of *Zymomonas mobilis* during aerobic exponential growth, *Arch. Microbiol.*, 167, 167, 1997.

50. Gschaedler, A. et al., Effects of pulse addition of carbon sources on continuous cultivation of *Escherichia coli* containing recombinant *E. coli gapA* gene, *Biotechnol. Bioeng.*, 63, 712, 1999.

51. Aristidou, A.A., San, K.Y., and Bennett, G.N., Improvement of biomass yield and recombinant gene expression in *Escherichia coli* by using fructose as the primary carbon source, *Biotechnol. Prog.*, 15, 140, 1999.

52. Skotnicki, M.L. et al., High-productivity alcohol fermentations using *Zymomonas mobilis*, *Biochem. Soc. Symp.*, 48, 53, 1983.

53. Cirillo, V.P., Relationship between sugar structure and competition for the sugar transport system in bakers' yeast, *J. Bacteriol.*, 95, 603, 1968.

54. Träff, K.L., Jönsson, L.J., and Hahn-Hägerdal, B., Putative xylose and arabinose reductases in *Saccharomyces cerevisiae*, *Yeast*, 19, 1233, 2002.

55. Toivari, M.H. et al., Endogenous xylose pathway in *Saccharomyces cerevisiae*, *Appl. Environ. Microbiol.*, 70, 3681, 2004.

56. Toivola, A. et al., Alcoholic fermentation of D-xylose by yeasts, *Appl. Environ. Microbiol.*, 47, 1221, 1984.

57. Kurtzman, C.P., *Candida shehatae* — genetic diversity and phylogenetic relationships with other xylose-fermenting yeasts, *Antonie van Leeuwenhoek*, 57, 215, 1990.

58. Lee, H. et al., Utilization of xylan by yeasts and its conversion to ethanol by *Pichia stipitis* strains, *Appl. Environ. Microbiol.*, 52, 320, 1986.

59. Verduyn, C. et al., Properties of the NAD(P)H-dependent xylose reductase from the xylose-fermenting yeast *Pichia stipitis*, *Biochem. J.*, 226, 669, 1985.

60. Ho, N.W.Y. et al., Purification, characterization, and amino acid terminal sequence of xylose reductase from *Candida shehatae*, *Enz. Microb. Technol.*, 12, 33, 1990.

61. Lee, J.K., Bong-Seong, K., and Sang-Yong, K., Cloning and characterization of the *xyl1* gene, encoding an NADH-preferring xylose reductase from *Candida parapsilosis*, and its functional expression in *Candida tropicalis*, *Appl. Environ. Microbiol.*, 69, 6179, 2003.

62. Skoog, K. and Hahn-Hägerdal, B., Effect of oxygenation on xylose fermentation by *Pichia stipitis*, *Appl. Environ. Microbiol.*, 56, 3389, 1990.

63. Jeppsson, H., Alexander, N.J., and Hahn-Hägerdal, B., Existence of cyanide-insensitive respiration in the yeast *Pichia stipitis* and its possible influence on product formation during xylose utilization, *Appl. Environ. Microbiol.*, 61, 2596, 1995.

64. Passoth, V., Zimmermann, M., and Klinner, U., Peculiarities of the regulation of fermentation and respiration in the Crabtree-negative, xylose-fermenting yeast *Pichia stipitis*, *Appl. Biochem. Biotechnol.*, 57–58, 201, 1996.
65. Kordowska-Wiater, M., and Targonski, Z., Application of *Saccharomyces cerevisiae* and *Pichia stipitis* karyoductants to the production of ethanol from xylose, *Acta Microbiol. Pol.*, 50, 291, 2001.
66. Kötter, P. et al., Isolation and characterization of the *Pichia stipitis* xylitol dehydrogenase gene, *XYL2*, and construction of a xylose utilizing *Saccharomyces cerevisiae* transformant, *Curr. Genet.*, 18, 493, 1990.
67. Walfridsson, M. et al., Expression of different levels of enzymes from the *Pichia stipitis XYL1* and *XYL2* genes in *Saccharomyces cerevisiae* and its effects on product formation during xylose utilization, *Appl. Microbiol. Biotechnol.*, 48, 218, 1997.
68. Ho, N.W.Y., Chen, Z., and Brainard, A.P., Genetically engineered *Saccharomyces* yeast capable of effective cofermentation of glucose and xylose, *Appl. Environ. Microbiol.*, 64, 1852, 1998.
69. Ho, N.W.Y. et al., Successful design and development of genetically engineered *Saccharomyces* yeast capable of effective cofermentation of glucose and xylose from cellulosic biomass to fuel ethanol, *Adv. Biochem. Eng./Biotechnol.*, 65, 163, 1999.
70. Ho, N.W.Y. and Tsao, G.T., Recombinant yeasts for effective fermentation of glucose and xylose, U.S. Patent 5,789,210, August 4, 1998.
71. Yu, S., Jeppsson, H., and Hahn-Hägerdal, B., Xylulose fermentation by *Saccharomyces cerevisiae* and xylose-fermenting yeast strains, *Appl. Microbiol. Biotechnol.*, 44, 314, 1995.
72. Deng, X.X. and Ho, N.W.Y., Xylulokinase activity in various yeasts including *Saccharomyces cerevisiae* containing the cloned xylulokinase gene, *Appl. Biochem. Biotechnol.*, 24–25, 193, 1990.
73. Tolan, J.S., Iogen's demonstration process for producing ethanol from cellulosic biomass, in Kamm, B., Gruber, P.R., and Kamm, M. (Eds.)., *Biorefineries — Industrial Processes and Products.* Volume 1. *Status Quo and Future Directions,* Wiley-VCH Verlag, Weinheim, 2006, chap. 9.
74. Moniruzzaman, M. et al., Fermentation of corn fibre sugars by an engineered xylose utilizing *Saccharomyces cerevisiae* strain, *World. J. Microbiol. Biotechnol.*, 13, 341, 1997.
75. Toivari, M.H. et al., Conversion of xylose to ethanol by recombinant *Saccharomyces cerevisiae*: importance of xylulokinase (*XKS1*) and oxygen availability, *Metab. Eng.*, 3, 236, 2001.
76. Kötter, P. and Ciriacy, M., Xylose fermentation by *Saccharomyces cerevisiae*, *Appl. Microbiol. Biotechnol.*, 38, 776, 1993.
77. Tantirungkij, M. et al., Construction of xylose-assimilating *Saccharomyces cerevisiae*, *J. Ferment. Bioeng.*, 75, 83, 1993.
78. Wahlbom, C.F. and Hahn-Hägerdal, B., Furfural, 5-hydroxymethyl-furfural, and acetoin act as external electron acceptors during anaerobic fermentation of xylose in recombinant *Saccharomyces cerevisiae*, *Biotechnol. Bioeng.*, 78, 172, 2002.
79. Öhgren, K. et al., Simultaneous saccharification and co-fermentation of glucose and xylose in steam-pretreated corn stover at high fiber content with *Saccharomyces cerevisiae* TMB3400, *J. Biotechnol.*, 126, 488, 2006.
80. Anderlund, M. et al., Expression of bifunctional enzymes with xylose reductase and xylitol dehydrogenase activity in *Saccharomyces cerevisiae* alters product formation during xylose fermentation, *Metab. Eng.*, 3, 226, 2001.
81. Jeppsson, M. et al., The expression of a *Pichia stipitis* xylose reductase mutant with higher K_m for NADPH increases ethanol production from xylose in recombinant *Saccharomyces cerevisiae*, *Biotechnol. Bioeng.*, 93, 665, 2006.
82. Aristidou, A. et al., Redox balance in fermenting yeast, *Monograph — European Brewing Convention*, 28, 161, 2000.

83. Roca, C., Nielsen, J., and Olsson, L., Metabolic engineering of ammonium assimilation in xylose-fermenting *Saccharomyces cerevisiae* improves ethanol production, *Appl. Environ. Microbiol.*, 69, 4732, 2003.

84. Verho, R. et al., Engineering redox cofactor regeneration for improved pentose fermentation in *Saccharomyces cerevisiae*, *Appl. Environ. Microbiol.*, 69, 5892, 2003.

85. Watanabe, S., Kodaki, T., and Makino, K., Complete reversal of coenzyme specificity of xylitol dehydrogenase and increase of thermostability by the introduction of structural zinc, *J. Biol. Chem.*, 280, 10340, 2005.

86. Michal, G. (Ed.), *Biochemical Pathways. An Atlas of Biochemistry and Molecular Biology*, John Wiley & Sons, New York, 1999, chap. 3.

87. Johansson, B. et al., Xylulokinase overexpression in two strains of *Saccharomyces cerevisiae* also expressing xylose reductase and xylose dehydrogenase and its effect on fermentation of xylose and lignocellulosic hydrolysate, *Appl. Environ. Microbiol.*, 67, 4249, 2001.

88. Jeppsson, B. et al., Reduced oxidative pentose phosphate pathway flux in recombinant xylose-utilizing *Saccharomyces cerevisiae* strains improves the ethanol yield from xylose, *Appl. Environ. Microbiol.*, 68, 1604, 2002.

89. Walfridsson, M. et al., Xylose-metabolizing *Saccharomyces cerevisiae* strains overexpressing the *TKL1* and *TKL2* genes encoding the pentose phosphate pathway enzymes transketolase and transaldolase, *Appl. Environ. Microbiol.*, 61, 4184, 1995.

90. Karhumaa, K. et al., High activity of xylose reductase and xylitol dehydrogenase improves xylose fermentation by recombinant *Saccharomyces cerevisiae*, *Appl. Microbiol. Biotechnol.*, 73, 1039, 2007.

91. La Grange, D.C. et al., Coexpression of the *Bacillus pumilus* β-xylosidase (*xynB*) gene with the *Trichoderma reesei* β-xylanase 2 (*xyn2*) gene in the yeast *Saccharomyces cerevisiae*, *Appl. Microbiol. Biotechnol.*, 54, 195, 2000.

92. La Grange, D.C. et al., Degradation of xylan to D-xylose by recombinant *Saccharomyces cerevisiae* coexpressing the *Aspergillus niger* β-xylosidase (*xlnD*) and the *Trichoderma reesei* xylanase II (*xyn2*) genes, *Appl. Environ. Microbiol.*, 67, 5512, 2001.

93. Katahira, S. et al., Construction of a xylan-fermenting yeast strain through codisplay of xylanolytic enzymes on the surface of xylose-utilizing *Saccharomyces cerevisiae* cells, *Appl. Environ. Microbiol.*, 70, 5407, 2004.

94. Katahira, S. et al., Ethanol fermentation from lignocellulosic hydrolysate by a recombinant xylose- and cellooligosaccharide-assimilating yeast strain, *Appl. Microbiol. Biotechnol.*, 72, 1136, 2006.

95. Walfridsson, M. et al., Ethanolic fermentation of xylose with *Saccharomyces cerevisiae* harboring the *Thermus thermophilus xylA* gene, which expresses an active xylose (glucose) isomerase, *Appl. Environ. Microbiol.*, 62, 4648, 1996.

96. Xarhangi, H.R. et al., Xylose metabolism in the anaerobic fungus *Piromyces sp.* strain E2 follows the bacterial pathway, *Arch. Microbiol.*, 180, 134, 2003.

97. Kuyper, M. et al., High-level functional expression of a fungal xylose isomerase: the key to efficient ethanolic fermentation of xylose by *Saccharomyces cerevisiae?*, *FEMS Yeast Res.*, 4, 69, 2003.

98. Kuyper, M. et al., Minimal metabolic engineering of *Saccharomyces cerevisiae* for efficient anaerobic xylose fermentation: a proof of principle, *FEMS Yeast Res.*, 4, 655, 2004.

99. Kuyper, M. et al., Metabolic engineering of a xylose-isomerase-expressing *Saccharomyces cerevisiae* strain for rapid anaerobic xylose fermentation, *FEMS Yeast Res.*, 5, 399, 2005.

100. Kuyper, M. et al., Evolutionary engineering of mixed-sugar utilization by a xylose-fermenting *Saccharomyces cerevisiae* strain, *FEMS Yeast Res.*, 5, 925, 2005.

101. Karhumaa, K. et al., Comparison of the xylose reductase-xylitol dehydrogenase and the xylose isomerase pathways for xylose fermentation by recombinant *Saccharomyces cerevisiae*, *Microb. Cell Fact.*, 6: 5, 2007.
102. Karhumaa, K., Hahn-Hägerdal, B., and Gorwa-Grauslund, M.-F., Investigation of limiting metabolic steps in the utilization of xylose by recombinant *Saccharomyces cerevisiae* using metabolic engineering, *Yeast*, 22, 359, 2005.
103. McCracken, L.D. and Gong, C.-S., D-Xylose metabolism by mutant strains of *Candida* sp., *Adv. Biochem. Eng./Biotechnol.*, 27, 33, 1983.
104. Rizzi, M. et al., A kinetic study of the NAD^+-xylitol-dehydrogenase from the yeast *Pichia stipitis*, *J. Ferm. Bioeng.*, 67, 25, 1989.
105. Persson, B. et al., Dual relationships of xylitol and alcohol dehydrogenases in families of two protein types, *FEBS Lett.*, 324, 9, 1993.
106. Richard, P., Toivari, M.H., and Penttilä, M., Evidence that the gene *YLR070c* of *Saccharomyces cerevisiae* encodes a xylitol dehydrogenase, *FEBS Lett.*, 457, 135, 1999.
107. Shi, N.Q. et al., Characterization and complementation of a *Pichia stipitis* mutant unable to grow on D-xylose or L-arabinose, *Appl. Biochem. Biotechnol.*, 84–86, 201, 2000.
108. Richard, P. et al., Cloning and expression of a fungal L-arabinitol 4-dehydrogenase gene, *J. Biol. Chem.*, 276, 40631, 2001.
109. Richard, P. et al., The missing link in the fungal L-arabinose catabolic pathway, identification of the L-xylulose reductase gene, *Biochemistry*, 41, 6432, 2002.
110. Seiboth, B. et al., D-xylose metabolism in *Hypocrea jecorina*: loss of the xylitol dehydrogenase step can be partially compensated for the *lad1*-encoded L-arabinitol-4-dehydrogenase, *Eukaryot. Cell*, 2, 867, 2003.
111. Richard, P. et al., Production of ethanol from L-arabinose by *Saccharomyces cerevisiae* containing a fungal L-arabinose pathway, *FEMS Yeast Res.*, 3, 185, 2003.
112. Becker, J. and Boles, E., A modified *Saccharomyces cerevisiae* strain that consumes L-arabinose and produces ethanol, *Appl. Environ. Microbiol.*, 69, 4144, 2003.
113. van Zyl, W.H. et al., Xylose utilisation by recombinant strains of *Saccharomyces cerevisiae* on different carbon sources, *Appl. Microbiol. Biotechnol.*, 52, 829, 1999.
114. Zaldivar, J. et al., Fermentation performance and intracellular metabolite patterns in laboratory and industrial xylose-fermenting *Saccharomyces cerevisiae*, *Appl. Microbiol. Biotechnol.*, 59, 436, 2002.
115. Wahlbom, C.F. et al., Generation of the improved recombinant xylose-utilizing *Saccharomyces cerevisiae* TMB 3400 by random mutagenesis and physiological comparison with *Pichia stipitis* CBS 6054, *FEMS Yeast Res.*, 3, 319, 2003.
116. Sonderegger, M. et al., Fermentation performance of engineered and evolved xylose-fermenting *Saccharomyces cerevisiae* strains, *Biotechnol. Bioeng.*, 87, 90, 2004.
117. Karhumaa, K. et al., Co-utilization of L-arabinose and D-xylose by laboratory and industrial *Saccharomyces cerevisiae* strains, *Microb. Cell Fact.*, 5,18, 2006.
118. Roca, C., Haack, M.B., and Olsson, L., Engineering of carbon catabolite repression in recombinant xylose fermenting *Saccharomyces cerevisiae* strains, *Appl. Microbiol. Biotechnol.*, 63, 578, 2004.
119. Govindaswamy, S. and Vane, L.M., Kinetics of growth and ethanol production on different carbon substrates using genetically engineered xylose-fermenting yeast, *Bioresour. Technol.*, 98, 677, 2007.
120. van Maris, A.J. et al., Alcoholic fermentation of carbon sources in biomass hydrolysates by *Saccharomyces cerevisiae*: current status, *Antonie van Leeuwenhoek*, 90, 391, 2006.
121. Hahn-Hägerdal, B. et al., Towards industrial pentose-fermenting yeast strains, *Appl. Microbiol. Biotechnol.*, 74, 937, 2007.

122. Wang, Y. et al., Establishment of a xylose metabolic pathway in an industrial strain of *Saccharomyces cerevisiae, Biotechnol. Lett.*, 26, 885, 2004.
123. Sonderegger, M. and Sauer, U., Evolutionary engineering of *Saccharomyces cerevisiae* for anaerobic growth on xylose, *Appl. Environ. Microbiol.*, 69, 1990, 2003.
124. Pitkanen, J.P. et al., Xylose chemostat isolates of *Saccharomyces cerevisiae* show altered metabolite and enzyme levels compared with xylose, glucose, and ethanol metabolism of the original strain, *Appl. Microbiol. Biotechnol.*, 67, 827, 2005.
125. Attfield, P.V. and Bell, P.J., Use of population genetics to derive nonrecombinant *Saccharomyces cerevisiae* strains that grow using xylose as a sole carbon source, *FEMS Yeast Res.*, 6, 862, 2006.
126. Lindén, T., Peetre, J., and Hahn-Hägerdal, B. et al., Isolation and characterization of acetic acid-tolerant galactose-fermenting strains of *Saccharomyces cerevisiae* from a spent sulfite liquor fermentation plant, *Appl. Environ. Microbiol.*, 58, 1661, 1992.
127. Martín, C. and Jönsson, L. J., Comparison of the resistance of industrial and laboratory strains of *Saccharomyces cerevisiae* and *Zygosaccharomyces* to lignocellulose-derived fermentation inhibitors, *Enz. Microb. Technol.*, 32, 386, 2003.
128. Brandber, T., Franzen, C.J. and Gustafsson, L., The fermentation performance of nine strains of *Saccharomyces cerevisiae* in batch and fed-batch cultures in dilute-acid wood hydrolysate, *J. Biosci. Bioeng.*, 98, 122, 2004.
129. Garay-Arropyo, A. et al., Response to different environmental stress conditions of industrial and laboratory *Saccharomyces cerevisiae* strains, *Appl. Microbiol. Biotechnol.*, 63, 734, 2004.
130. Hahn-Hägerdal, B. et al., Role of cultivation media in the development of yeast strains for large scale industrial use, *Microb. Cell Fact.*, 4: 31, 2005.
131. Larsson, S., Cassland, P., and Jönsson, L.J., Development of a *Saccharomyces cerevisiae* strain with enhanced resistance to phenolic fermentation inhibitors in lignocellulose hydrolysates by heterologous expression of laccase, *Appl. Environ. Microbiol.*, 67, 1163, 2001.
132. Larsson, S., Nilvebrant, N.O., and Jönsson, L.J., Effect of overexpression of *Saccharomyces cerevisiae* Pad1p on the resistance to phenylacrylic acids and lignocellulose hydrolysates under aerobic and oxygen-limited conditions, *Appl. Microbiol. Biotechnol.*, 57, 167, 2001.
133. Abbi, M., Kuhad, R.C., and Singh, A., Fermentation of xylose and rice straw hydrolysate to ethanol by *Candida shehatae* NCL-3501, *J. Ind. Microbiol.*, 17, 20, 1996.
134. Sreenath, H.K. et al., Ethanol production from alfalfa fiber fractions by saccharification and fermentation, *Proc. Biochem.*, 36, 1199, 2001.
135. Ryabova, O.B., Chmil, O.M., and Sibirny, A.A., Xylose and cellobiose fermentation to ethanol by the thermotolerant methylotrophic yeast *Hansenula polymorpha, FEMS Yeast Res.*, 4, 157, 2003.
136. Jeffries, T.W. and Shi, N.-Q., Genetic engineering for improved xylose fermentation by yeasts, *Adv. Biochem. Eng./Biotechnol.*, 65, 117, 1999.
137. Nagy, M., Lacroute, F., and Thomas, D., Divergent evolution of pyrimidine biosynthesis between anaerobic and aerobic yeasts, *Proc. Natl. Acad. Sci., USA*, 89, 8966, 1992.
138. Shi, N.-Q. and Jeffries, T.W., Anaerobic growth and improved fermentation of *Pichia stipitis* bearing a URA1 gene from *Saccharomyces cerevisiae, Appl. Microbiol Biotechnol.*, 50, 339, 1998.
139. Bicho, P.A. et al., Induction of D-xylose reductase and xylitol dehydrogenase activities in *Pachysolen tannophilus* and *Pichia stipitis* on mixed sugars, *Appl. Environ. Microbiol.*, 54, 50, 1988.
140. Agbogbo, F.K. and Wenger, K.S., Effect of pretreatment chemicals on xylose fermentation by *Pichia stipitis, Biotechnol. Lett.*, 28, 2065, 2006.

141. van Zyl, C., Prior, B.A., and du Preez, J.C., Production of ethanol from sugarcane bagasse hemicellulose hydrolysate by *Pichia stipitis*, *Appl. Biochem. Biotechnol.*, 17, 357, 1988.

142. Tran, A.V. and Chambers, R.P., Ethanol fermentation of red oak acid prehydrolysate by the yeast *Pichia stipitis* CBS 5776, *Enz. Microb. Technol.*, 8, 439, 1986.

143. Nigam, J.N., Ethanol production from wheat straw hemicellulose hydrolysate by *Pichia stipitis*, *J. Biotechnol.*, 87, 17, 2001.

144. Nigam, J.N., Development of xylose-fermenting yeast *Pichia stipitis* for ethanol production through adaptation on hardwood hemicellulose acid prehydrolysate, *J. Appl. Microbiol.*, 90, 208, 2001.

145. Eken-Saraçoğlu, N. and Arslan, Y., Comparison of different pretreatments in ethanol fermentation using corn cob hemicellulosic hydrolysate with *Pichia stipitis* and *Candida shehatae*, *Biotechnol. Lett.*, 22, 855, 2000.

146. van den Berg, J.A. et al., *Kluyveromyces* as a host for heterologous gene expression and secretion of prochymosin, *Bio/Technology*, 8, 135, 1990.

147. Schwan, R.F. et al., Microbiology and physiology of *cachaça* (*aguardente*) fermentations, *Antonie van Leeuwenhoek*, 79, 89, 2001.

148. Anderson, P.J., McNeil, K., and Watson, K., High-efficiency carbohydrate fermentation to ethanol at temperatures above 40°C by *Kluyveromyces marxianus* var. *marxianus* isolated from sugar mills, *Appl. Microbiol. Biotechnol.*, 51, 1314, 1986.

149. Margaritis, A. and Bajpal, P., Direct fermentation of D-xylose to ethanol by *Kluyveromyces* strains, *Appl. Environ. Microbiol.*, 44, 1039, 1982.

150. Gough, S. et al., Fermentation of molasses using a thermotolerant yeast, *Kluyveromyces marxianus* IMB3: simplex optimization of media supplements, *Appl. Microbiol. Biotechnol.*, 6, 187, 1996.

151. Bellaver, L.H. et al., Ethanol formation and enzyme activities around glucose-6-phosphate in *Kluyveromyces marxianus* CBS 6556 exposed to glucose or lactose excess, *FEMS Yeast Res.*, 4, 691, 2004.

152. Ingram, L.O. et al., Enteric bacteria catalysts for fuel ethanol production, *Biotechnol. Lett.*, 15, 855, 1999.

153. Buchholz, S.E., Dooley, M.M., and Eveleigh, D.E., *Zymomonas* — an alcoholic enigma, *Trends Biotechnol.*, 5, 199, 1987.

154. Moat, A.G. and Foster, J.W., *Microbial Physiology*, 3rd ed., Wiley-Liss, New York, 1995, chap. 7.

155. Ingram, L.O. et al., Genetic engineering of ethanol production in *Escherichia coli*, *Appl. Environ. Microbiol.*, 53, 2420, 1987.

156. Ingram, L.O. and Conway, T., Expression of different levels of ethanologenic enzymes of *Zymomonas mobilis* in recombinant strains of *Escherichia coli*, *Appl. Environ. Microbiol.*, 54, 397, 1988.

157. Ohta, K. et al., Genetic improvement of *Escherichia coli* for ethanol production: chromosomal integration of *Zymomonas mobilis* genes encoding pyruvate decarboxylase and alcohol dehydrogenase II, *Appl. Environ. Microbiol.*, 57, 893, 1991.

158. Ingram, L.O. et al., Ethanol production in *Escherichia coli* strains coexpressing *Zymomonas* PDC and ADH genes, U.S. Patent 5,000,000, March 19, 1991.

159. Alterhum, F. and Ingram, L.O., Efficient ethanol production from glucose, lactose, and xylose by recombinant *Escherichia coli*, *Appl. Environ. Microbiol.*, 55, 1943, 1989.

160. Ohta, K., Alterhum, F., and Ingram, L.O., Effects of environmental conditions on xylose fermentation by recombinant *Escherichia coli*, *Appl. Environ. Microbiol.*, 56, 463, 1990.

161. Kuhnert, P. et al., Detection system for *Escherichia coli*-specific virulence genes: absence of virulence determinants in B and C strains, *Appl. Environ. Microbiol.*, 63, 703, 1997.

162. Dien, B.S., Cotta, M.A., and Jeffries, T.W., Bacteria engineered for fuel ethanol production: current status, *Appl. Microbiol. Biotechnol.*, 63, 258, 2003.

163. Hahn-Hägerdal, B. et al., An interlaboratory comparison of the performance of ethanol-producing microorganisms in a xylose-rich acid hydrolysate, *Appl. Microbiol. Biotechnol.*, 41, 62, 1994.

164. Barbosa, M.F. et al., Efficient fermentation of *Pinus* sp. acid hydrolysates by an ethanologenic strain of *Escherichia coli*, *Appl. Environ. Microbiol.*, 58, 1382, 1992.

165. Ashgari, A. et al., Ethanol production from hemicellulose hydrolysates of agricultural residues using genetically engineered *Escherichia coli* strain KO11, *J. Ind. Microbiol.*, 16, 42, 1996.

166. Beall, D.S. et al., Conversion of hydrolysates of corn cobs and hulls into ethanol by recombinant *Escherichia coli* strain B containing genes for ethanol production, *Biotechnol. Lett.*, 14, 857, 1992.

167. Moniruzzaman, M. et al., Ethanol production from AFEX pretreated corn fiber by recombinant bacteria, *Biotechnol. Lett.*, 18, 955, 1996.

168. Moniruzzaman, M. and Ingram, L.O., Ethanol production from dilute acid hydrolysate of rice hulls using genetically engineered *Escherichia coli*, *Biotechnol. Lett.*, 20, 943, 1998.

169. Guimaraes, W.V., Dudley, G.L., and Ingram, L.O., Fermentation of sweet whey by ethanologenic *Escherichia coli*, *Biotechnol. Bioeng.*, 40, 41, 1992.

170. Guimaraes, W.Y. et al., Ethanol production from starch by recombinant *Escherichia coli* containing integrated genes for ethanol production and plasmid genes for saccharification, *Biotechnol. Lett.*, 14, 415, 1992.

171. Grohmann, K. et al., Fermentation of galacturonic acid and other sugars in orange peel hydrolysates by an ethanologenic strain of *Escherichia coli*, *Biotechnol. Lett.*, 16, 281, 1994.

172. Moniruzzaman, M. et al., Extracellular melibiose and fructose are intermediates in raffinose catabolism during fermentation to ethanol by engineered enteric bacteria, *J. Bacteriol.*, 179, 1880, 1997.

173. Yomano, L.P., York, S. W, and Ingram, L.O., Isolation and characterization of ethanol-tolerant mutants of *Escherichia coli* KO11 for fuel ethanol production, *J. Ind. Microbiol. Biotechnol.*, 20, 132, 1998.

174. Zaldivar, J., Martinez, A., and Ingram, L.O., Effect of aldehydes on the growth and fermentation of ethanologenic *Escherichia coli*, *Biotechnol. Bioeng.*, 65, 24, 1999.

175. Zaldivar, J., Martinez, A., and Ingram, L.O., Effect of alcohol compounds found in hemicellulose hydrolysate on the growth and fermentation of ethanologenic *Escherichia coli*, *Biotechnol. Bioeng.*, 68, 524, 2000.

176. Zaldivar, J. and Ingram, L.O., Effect of organic acids on the growth and fermentation of ethanologenic *Escherichia coli* LY01, *Biotechnol. Bioeng.*, 66, 203, 1999.

177. Martinez, A. et al., Biosynthetic burden and plasmid burden limit expression of chromosomally integrated heterologous genes (*pdc, adhB*) in *Escherichia coli*, *Biotechnol. Prog.*, 15, 891, 1999.

178. Underwood, S.A. et al., Flux through citrate synthase limits the growth of ethanologenic *Escherichia coli* KO11 during xylose fermentation, *Appl. Environ. Microbiol.*, 68, 1071, 2002.

179. Underwood, S.A. et al., Genetic changes to optimize carbon partitioning between ethanol and biosynthesis in ethanologenic *Escherichia coli*, *Appl. Environ. Microbiol.*, 68, 6263, 2002.

180. Tao, H. et al., Engineering a homo-ethanol pathway in *Escherichia coli*: increased glycolytic flux and levels of expression of glycolytic genes during xylose fermentation, *J. Bacteriol.*, 183, 2979, 2001.

181. Moniruzzaman, M. et al., Isolation and molecular characterization of high-performance cellobiose-fermenting spontaneous mutants of ethanologenic *Escherichia coli* KO11 containing the *Klebsiella oxytoca casAB* operon, *Appl. Environ. Microbiol.*, 63, 4633, 1997.

182. Lawford, H.G. and Rousseau, J.D., Loss of ethanologenicity in *Escherichia coli* B recombinant pLO1297 and KO11 during growth in the absence of antibiotics, *Biotechnol. Lett.*, 17, 751, 1995.

183. Lawford, H.G. and Rousseau, J.D., Factors contributing to the loss of ethanologenicity of *Escherichia coli* B recombinant pLO1297 and KO11, *Appl. Biochem. Biotechnol.*, 57–58, 293, 1996.

184. Dumsday, G.J. et al., Continuous ethanol production by *Escherichia coli* B recombinant KO11 in continuous stirred tank and fluidized bed fermenters, *Aust. Biotechnol.*, 7, 300, 1997.

185. Dumsday, G.J. et al., Comparative stability of ethanol production by *Escherichia coli* KO11 in batch and chemostat culture, *J. Ind. Microbiol. Biotechnol.*, 23, 701, 1999.

186. Mat-Jan, F., Alam, K.Y., and Clark, D.P., Mutants of *Escherichia coli* deficient in the fermentative lactate dehydrogenase *J. Bacteriol.*, 171, 342, 1989.

187. Hespell, R.B. et al., Stabilization of *pet* operon plasmids and ethanol production in *Escherichia coli* strains lacking lactate dehydrogenase and pyruvate-formate-lyase activities, *Appl. Environ. Microbiol.*, 62, 4594, 1996.

188. Dien, B.S. et al., Fermentation of hexose and pentose sugars using a novel ethanologenic *Escherichia coli* strain, *Enz. Microb. Technol.*, 23, 366, 1998.

189. Dien, B.S., Iten, L.B., and Bothast, R.J., Conversion of corn fiber to ethanol by recombinant *Escherichia coli* strain FBR3, *J. Ind. Microbiol. Biotechnol.*, 22, 575, 1999.

190. Dien, B.S. et al., Development of new ethanologenic *Escherichia coli* strains for fermentations of lignocellulosic biomass, *Appl. Biochem. Biotechnol.*, 84–86, 181, 2000.

191. Nichols, N.N., Dien, B.S., and Bothast, R.J., Use of carbon catabolite repression mutants for fermentation of sugar mixtures to ethanol, *Appl. Microbiol. Biotechnol.*, 56, 120, 2001.

192. Kim, Y., Ingram, L.O., and Shanmugam, K.T., Construction of an *Escherichia coli* K-12 mutant for homo-ethanol fermentation of glucose or xylose without foreign genes, *Appl. Environ. Microbiol.*, 73, 1766, 2007.

193. Quail, M.A., Haydon, D.J., and Guest, J.R., The *pdhR-aceEF-lpd* operon of *Escherichia coli* expresses the pyruvate dehydrogenase complex, *Mol. Microbiol.*, 12, 95, 1994.

194. Swings, J. and De Ley, J., The biology of *Zymomonas*, *Bacteriol. Rev.* 41, 1, 1977.

195. Doelle, H.W. et al., *Zymomonas mobilis* — science and industrial application, *Crit. Rev. Biotechnol.*, 13, 57, 1993.

196. Richards, L., and Doelle, H.W., Fermentation of potato mash, potato mash/maltrin mixtures and wheat starch using *Zymomonas mobilis*, *World J. Microbiol. Biotechnol.*, 5, 307, 1989.

197. Skotnicki, M.L. et al., Comparison of ethanol production by different *Zymomonas* strains, *Appl. Environ. Microbiol.*, 41, 889, 1981.

198. Skotnicki, M.L. et al., High-productivity alcohol fermentations using *Zymomonas mobilis*, *Biochem. Soc. Symp.*, 48, 53, 1983.

199. Rogers, P.L., Goodman, A.E., and Heyes, R.H., *Zymomonas* ethanol fermentations, *Microbiol. Sci.*, 1, 133, 1984.

200. Barrow, K.D. et al., ^{31}P nuclear magnetic resonance studies of the fermentation of glucose to ethanol by *Zymomonas mobilis*, *J. Biol. Chem.*, 259, 5711, 1984.

201. Zhang, M. et al., Metabolic engineering of a pentose metabolism pathway in ethanologenic *Zymomonas mobilis*, *Science*, 267, 240, 1995.

202. Zhang, M. et al., Promising ethanologens for xylose fermentation. Scientific note, *Appl. Biochem. Biotechnol.*, 51–52, 527, 1995.
203. Parker, C. et al., Characterization of the *Zymomonas mobilis* glucose facilitator gene-product (*glf*) in recombinant *Escherichia coli*: examination of transport mechanism, kinetics and the role of glucokinase in glucose transport, *Mol. Microbiol.*, 15, 795, 1995.
204. Deanda, K. et al., Development of an arabinose-fermenting *Zymomonas mobilis* strain by metabolic pathway engineering, *Appl. Environ. Microbiol.*, 62, 4465, 1996.
205. Lawford, H.G. et al., Fermentation performance characteristics of a prehydrolysate-adapted xylose-fermenting recombinant *Zymomonas* in batch and continuous fermentations, *Appl. Biochem. Biotechnol.*, 77–79, 191, 1999.
206. Gao, Q. et al., Characterization of heterologous and native enzyme activity profiles in metabolically engineered *Zymomonas mobilis* strains during batch fermentation of glucose and xylose mixtures, *Appl. Biochem. Biotechnol.*, 98–100, 341, 2002.
207. Lawford, H.G. and Rousseau, J.D., Performance testing of *Zymomonas mobilis* metabolically engineered for cofermentation of glucose, xylose, and arabinose, *Appl. Biochem. Biotechnol.*, 98–100, 429, 2002.
208. Mohagheghi, A. et al., Cofermentation of glucose, xylose, and arabinose by genomic DNA-integrated xylose/arabinose fermenting strain of *Zymomonas mobilis* AX101, *Appl. Biochem. Biotechnol.*, 98–100, 885, 2002.
209. Lawford, H.G. and Rousseau, J.D., Cellulosic fuel ethanol: alternative fermentation process designs with wild-type and recombinant *Zymomonas mobilis*, *Appl. Biochem. Biotechnol.*, 105–108, 457, 2003.
210. Joachimsthal, E.L. and Rogers, P.L., Characterization of a high-productivity recombinant strain of *Zymomonas mobilis* for ethanol production from glucose/xylose mixtures, *Appl. Biochem. Biotechnol.*, 84–86, 343, 2000.
211. Kim, I.S., Barrow, K.D., and Rogers, P.L., Kinetic and nuclear magnetic resonance studies of xylose metabolism by recombinant *Zymomonas mobilis* ZM4(pZB5), *Appl. Environ. Microbiol.*, 66, 186, 2000.
212. Kim, I.S., Barrow, K.D., and Rogers, P.L., Nuclear magnetic resonance studies of acetic acid inhibition of rec *Zymomonas mobilis* ZM4 (pZB5), *Appl. Biochem. Biotechnol.*, 84–86, 357, 2000.
213. Jeon, Y.J. et al., Kinetic analysis of ethanol production by an acetate-resistant strain of *Zymomonas mobilis*, *Biotechnol. Lett.*, 24, 819, 2002.
214. Jeon, Y.J., Svenson, C.J., and Rogers, P.L., Over-expression of xylulokinase in a xylose-metabolizing recombinant strain of *Zymomonas mobilis*, *FEMS Microbiol. Lett.*, 244, 85, 2005.
215. De Graaf, A.A. et al., Metabolic state of *Zymomonas mobilis* in glucose-, fructose-, and xylose-fed continuous cultures as analysed by ^{13}C- and ^{31}P-NMR spectroscopy, *Arch. Microbiol.*, 171, 371, 1999.
216. Saez-Miranda, J.C. et al., Measurement and analysis of intracellular ATP levels in metabolically engineered *Zymomonas mobilis* fermenting glucose and xylose mixtures, *Biotechnol. Prog.*, 22, 359, 2006.
217. Baumler, D.J. et al., Enhancement of acid tolerance in *Zymomonas mobilis* by a proton-buffering peptide, *Appl. Biochem. Biotechnol.*, 134, 15, 2006.
218. Ohta, K. et al., Metabolic engineering of *Klebsiella oxytoca* M5A1 for ethanol production from xylose and glucose, *Appl. Environ. Microbiol.*, 57, 2810, 1991.
219. Wood, B.E. and Ingram, L.O., Ethanol production from cellobiose, amorphous cellulose, and crystalline cellulose by recombinant *Klebsiella oxytoca* containing chromosomally integrated *Zymomonas mobilis* genes for ethanol production and thermostable cellulase genes from *Clostridium thermocellum*, *Appl. Environ. Microbiol.*, 58, 2130, 1992.

220. Doran, J.B., Aldrich, H.C., and Ingram, L.O., Saccharification and fermentation of sugarcane bagasse by *Klebsiella oxytoca* P2 containing chromosomally integrated genes encoding the *Zymomonas mobilis* ethanol pathway, *Biotechnol. Bioeng.*, 44, 240, 1994.

221. Doran, J.B. et al., Fermentations of pectin-rich biomass with recombinant bacteria to produce fuel ethanol, *Appl. Biochem. Biotechnol.*, 84–86, 141, 2000.

222. Burchhardt, G. and Ingram, L.O., Conversion of xylan to ethanol by ethanologenic strains of *Escherichia coli* and *Klebsiella oxytoca*, *Appl. Environ. Microbiol.*, 58, 1128, 1992.

223. Qian, Y. et al., Cloning, characterization, and functional expression of the *Klebsiella oxytoca* xylodextrin utilization operon (*xynTB*) in *Escherichia coli*, *Appl. Environ. Microbiol.*, 69, 5957, 2003.

224. Haq, A. and Dawes, E.A., Pyruvic acid metabolism and ethanol formation in *Erwinia amylovora*, *J. Gen. Microbiol.*, 68, 295, 1971.

225. Breukink, E. and de Kruijff, B., The lantibiotic nisin, a special case or not? *Biochim. Biophys. Acta*, 1462, 233, 1999.

226. Liu, S., Dien, B.S., and Cotta, M.A., Functional expression of bacterial *Zymobacter palmae* pyruvate decarboxylase in *Lactococcus lactis*, *Curr. Microbiol.*, 50, 324, 2005.

227. Liu, S. et al., Metabolic engineering of a *Lactococcus plantarum* double *ldh* knockout strain for enhanced ethanol production, *J. Ind. Microbiol. Biotechnol.*, 33, 1, 2006.

228. Gold, R.S. et al., Cloning and expression of the *Zymomonas mobilis* "production of ethanol" genes in *Lactobacillus casei*, *Curr. Microbiol.*, 33, 256, 1996.

229. Nichols, N.N., Dien, B.S., and Bothast, R.J., Engineering lactic acid bacteria with pyruvate decarboxylase and alcohol dehydrogenase genes for ethanol production from *Zymomonas mobilis*, *J. Ind. Microbiol. Biotechnol.*, 30, 315, 2003.

230. Okamoto, T. et al., *Zymobacter palmae gen. nov. sp. nov.*, a new ethanol-fermenting peritrichous bacterium isolated from palm sap, *Arch Microbiol.*, 160, 333, 1993.

231. Yanase, H. et al., Genetic engineering of *Zymobacter palmae* for ethanol production from xylose, *Appl. Environ. Microbiol.*, 73, 1766, 2007.

232. Yanase, H. et al., Ethanol production from cellobiose by *Zymobacter palmae* carrying the *Ruminococcus albus* β-glucosidase gene, *J. Biotechnol.*, 118, 35, 2005.

233. Raj, K.C. et al., Cloning and characterization of the *Zymobacter palmae* pyruvate decarboxylase gene (*pdc*) and comparison to bacterial homologues, *Appl. Environ. Microbiol.*, 68, 2869, 2002.

234. Heyer, H. and Krumbein, W.E., Excretion of fermentation products in dark and anaerobically incubated cyanobacteria, *Arch. Microbiol.*, 155, 284, 1991.

235. Deng, M.-D., and Coleman, J.R., Ethanol synthesis by genetic engineering in cyanobacteria, *Appl. Environ. Microbiol.*, 65, 523, 1999.

236. Sommer, P., Georgieva, T., and Ahring, B.K., Potential for using thermophilic anaerobic bacteria for bioethanol production from hemicellulose, *Biochem. Soc. Trans.*, 32, 283, 2004.

237. Desvaux, M., Unravelling carbon metabolism in anaerobic cellulolytic bacteria, *Biotechnol. Prog.*, 22, 1229, 2006.

238. Chen, J., Cachay, E.R., and Hunt, G. C, *Klebsiella oxytoca*: a rare cause of infectious colitis. First North American case report, *Gastrointest. Endosc.*, 60, 142, 2004.

239. Sardan, Y.C. et al., A cluster of nosocomial *Klebsiella oxytoca* bloodstream infections in a university hospital, *Infect. Control Hosp. Epidemiol.*, 25, 878, 2004.

240. Hogenauer, C. et al., *Klebsiella oxytoca* as a causative organism of antibiotic-associated hemorrhagic colitis, *N. Engl. J. Med.*, 355, 2418, 2006.

241. Wu, S.W., Liang, Y.H., and Su, X.D., Expression, purification and crystallization of an extended-spectrum β-lactamase from *Klebsiella oxytoca*, *Acta Crystallogr. D Biol. Crystallogr.*, 60, 326, 2004.

242. Yinbo, Q. et al., Studies on cellulosic ethanol production for sustainable supply of liquid fuel in China, *Biotechnol. J.*, 1, 1235, 2006.
243. Stephanopolos, G., Challenges in engineering microbes for biofuels production, *Science*, 315, 801, 2007.
244. Liu, Z.L. et al., Adaptive responses of yeast to furfural and 5-hydroxymethylfurfural and new chemical evidence for HMF conversion to 2,5-*bis*hydroxymethylfuran, *J. Ind. Microbial. Biotechnol.*, 31, 345, 2004.
245. Gorsich, S.W. et al., Tolerance to furfural-induced stress is associated with pentose phosphate pathway genes *ZWF1*, *GND1*, *RPE1*, and *TKL1* in *Saccharomyces cerevisiae*, *Appl. Microbiol. Biotechnol.*, 71, 339, 2006.
246. Petersson, A. et al., A 5-hydroxymethyl furfural reducing enzyme encoded by the *Saccharomyces cerevisiae ADH6* gene conveys HMF tolerance, *Yeast*, 23, 455, 2006.
247. Leão, C. and van Uden, N., Effects of ethanol and other alkanols on the kinetics and the activation parameters of thermal death in *Saccharomyces cerevisiae*, *Biotechnol. Bioeng.*, 24, 1581, 1982.
248. Alexandre, H., Rousseau, I., and Charpentier, C., Ethanol adaptation mechanisms in *Saccharomyces cerevisiae*, *Biotechnol. Appl. Biochem.*, 20, 173, 1994.
249. Meyrial, V. et al., Relationship between effect of ethanol on protein flux across plasma membrane and ethanol tolerance, in *Pichia stipitis*, *Anaerobe*, 3, 423, 1997.
250. Christogianni, A. et al., Transcriptional analysis of a gene cluster involved in glucose tolerance in *Zymomonas mobilis*: evidence for an osmoregulated promoter, *J. Bacteriol.*, 187, 5179, 2005.
251. Seo, J.-S. et al., The genome sequence of the ethanologenic bacterium *Zymomonas mobilis* ZM4, *Nat. Biotechnol.*, 23, 63, 2005.
252. Tzschoppe, K. et al., *Trans*-dominant mutations in the *GPR1* gene cause high sensitivity to acetic acid and ethanol in the yeast *Yarrowia lipolytica*, *Yeast*, 15, 1645, 1999.
253. Alper, H. et al., Engineering yeast transcription machinery for improved ethanol tolerance and productivity, *Science*, 314, 1565, 2006.
254. Blieck, L. et al., Isolation and characterization of brewer's yeast variants with improved fermentation performance under high-gravity conditions, *Appl. Microbiol. Biotechnol.*, 73, 815, 2007.
255. Gardonyi, M. et al., Control of xylose consumption by xylose transport in recombinant *Saccharomyces cerevisiae*, *Biotechnol. Bioeng.*, 82, 818, 2003.
256. Stambuk, B.U. et al., D-Xylose transport by *Candida succiphila* and *Kluyveromyces marxianus*, *Appl. Biochem. Biotechnol.*, 105–108, 255, 2003.
257. Hasona, A. et al., Pyruvate formate lyase and acetate kinase are essential for anaerobic growth of *Escherichia coli* on xylose, *J. Bacteriol.*, 186, 7593, 2004.
258. Salusjarvi, L. et al., Transcription analysis of recombinant *Saccharomyces cerevisiae* reveals novel responses to xylose, *Appl. Biochem. Biotechnol.*, 128, 237, 2006.
259. Shulami, S. et al., A two-component system regulates the expression of an ABC transporter for xylo-oligosaccharides in *Geobacillus stearothermophilus*, *Appl. Microbiol. Biotechnol.*, 73, 874, 2007.
260. Liesen, T., Hollenberg, C.P., and Heinisch, J.J., ERA, a novel *cis*-acting element required for autoregulation and ethanol repression of *PDC1* transcription in *Saccharomyces cerevisiae*, *Mol. Microbiol.*, 21, 621, 1996.
261. Kong, Q.X. et al., Overexpressing *GLT1* in *gpd1Δ* mutant to improve the production of ethanol in *Saccharomyces cerevisiae*, *Appl. Microbiol. Biotechnol.*, 73, 1382, 2007.
262. Zhang, A. et al., Effect of *FPS1* deletion on the fermentative properties of *Saccharomyces cerevisiae*, *Lett. Appl. Microbiol.*, 44, 212, 2007.
263. Kong, Q.X. et al., Improved production of ethanol by deleting *FPS1* and over-expressing *GLT1* in *Saccharomyces cerevisiae*, *Biotechnol. Lett.*, 28, 2033, 2006.

264. Kalnenieks, U. et al., Respiratory behaviour of a *Zymomonas mobilis adhB::kan(r)* mutant supports the hypothesis of two alcohol dehydrogenase isoenzymes catalysing opposite reactions, *FEBS Lett.*, 580, 5084, 2006.
265. Duarte, N., Pallson, B.O., and Fu, P., Integrated analysis of metabolic phenotypes in *Saccharomyces cerevisiae*, *BMC Genomics*, 5: 63, 2004.
266. Gagiano, M., Bauer, F.F., and Pretorius, I.S., The sensing of nutritional status and the relationship to filamentous growth in *Saccharomyces cerevisiae*, *FEMS Yeast Res.*, 2, 433, 2002.
267. Fuhrer, T., Fischer, E., and Sauer, U., Experimental identification and quantification of glucose metabolism in seven bacterial species, *J. Bacteriol.*, 187, 1581, 2005.
268. Fiaux, J. et al., Metabolic-flux profiling of the yeasts *Saccharomyces cerevisiae* and *Pichia stipitis*, *Eukaryot. Cell*, 2, 170, 2003.
269. Blank, L.M., Lehmbeck, F., and Sauer, U., Metabolic-flux and network analysis in fourteen hemiascomycetous yeasts, *FEMS Yeast Res.* 5, 545, 2005.
270. Altintas, M.M. et al., Kinetic modeling to optimize pentose fermentation in *Zymomonas mobilis*, *Biotechnol. Bioeng.*, 94, 273, 2006.
271. Jin, Y.S. and Jeffries, T.W., Stoichiometric network constraints on xylose metabolism by recombinant *Saccharomyces cerevisiae*, *Metab. Eng.*, 6, 229, 2004.
272. Bro, C. et al., *In silico* aided metabolic engineering of *Saccharomyces cerevisiae* for improved bioethanol production, *Metab. Eng.*, 8, 102, 2006.
273. Schehl, B. et al., A laboratory yeast strain suitable for spirit production, *Yeast*, 21, 1375, 2004.
274. Ikeda, M. and Nakagawa, S., The *Corynebacterium glutamicum* genome: features and impacts on biotechnological process, *Appl. Microbiol. Biotechnol.*, 62, 99, 2003.
275. Hunter, I.S., Microbial synthesis of secondary metabolites and strain improvement, in El-Mansi, E.M.T., Bryce, C.A., Demain, A.L., and Allman, A.R. (Eds.), *Fermentation Microbiology and Biotechnology*, CRC Press, Boca Raton, 2007, chap. 5.
276. Glass, J.I. et al., Minimal bacterial genome, U.S. Patent Application 2007/0122826, October 12, 2006.
277. *Genomics: GTL Program Projects*, available at http://genomicsgtl.energy.gov/pubs. shtml.

4 Biochemical Engineering and Bioprocess Management for Fuel Ethanol

4.1 THE IOGEN CORPORATION PROCESS AS A TEMPLATE AND PARADIGM

The demonstration process operated since 2004 is outlined in figure 4.1. In many of its features, the Iogen process is relatively conservative:

- Wheat straw as a substrate — a high-availability feedstock with a low lignin content in comparison with tree wood materials (figure 4.2)[1,2]
- A dilute acid and heat pretreatment of the biomass — the levels of acid are sufficiently low that recovery of the acid is not needed and corrosion problems are avoided
- Separate cellulose hydrolysis and fermentation with a single sugar substrate product stream (hexoses plus pentoses) for fermentation
- Cellulase breakdown of cellulose — Iogen is an enzyme producer
- A *Saccharomyces* yeast ethanologen — relatively ethanol-tolerant and engineered for xylose consumption as well as offering a low incidence of contamination, the ability to recycle the cells, and the option for selling on the spent cells for agricultural use[1]

In the first description of the process (written in and before July 1999), agricultural residues such as wheat straw, grasses, and energy crops (aspen, etc.) were equally "possible" or a "possibility."[3] By the next appearance of the article in 2006* — and as discussed in chapter 2, section 2.6 — cereal straws had become the substrates of choice. Lignin does not form a seriously refractory barrier to cellulase access with wheat straw; this renders organic solvent pretreatment unnecessary. More than 95% of the cellulosic glucose is released by the end of the enzyme digestion step, the remainder being included in the lignin cake that is spray-dried before combustion (figure 4.1).

The Iogen process is viewed as a sequential evolution of the bioethanol paradigm, no more complex than wet mill and dry mill options for corn ethanol production (figures 1.20 and 2.21), substituting acid pretreatment for corn grinding steps, and

---

* Acid prehydrolysis times had been described as "less than 1 minute" in 1999 but a more flexible regime had been instituted after the facility became operational (figure 4.1).

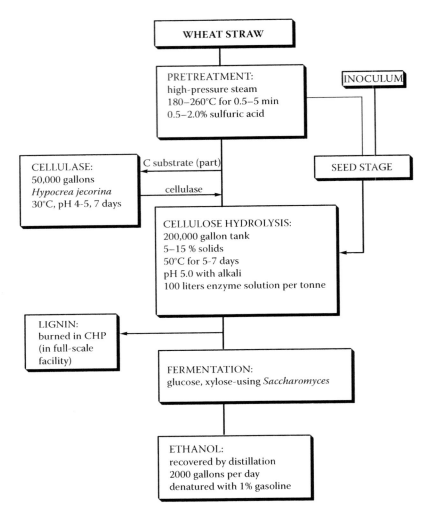

FIGURE 4.1 Outline of Iogen's demonstration process for bioethanol production. (Data from Tolan.[1])

adding on-site cellulase generation, the latter mostly as a strategy to avoid the costs of preservatives and stabilizer but possibly also to use a small proportion of the hydrolyzed cellulose as a feedstock for the enzyme fermentation itself. Salient features of the technology were present in Canadian initiatives from the 1970s and 1980s. The Bio-hol process, financially supported by the Ontario Ministry of Energy and Energy, Mines, and Resources, Canada, opted for *Zymomonas mobilis* as the ethanologen and had established acid hydrolysis pretreatments for wheat straw, soy stalks, corn stover, canola stalks, pine wood, and poplar wood.[4] For both *Z. mobilis* and *S. cerevisiae*, pretreated wheat straw had the distinct advantage of presenting far less of a toxic mixture to the producer organism (figure 4.3); methods for removing growth inhibitors from the biomass acid hydrolysates could reduce the effect by >20-fold.

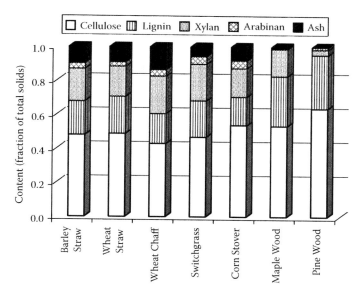

FIGURE 4.2   Bioethanol feedstock compositions. (Data from Foody et al.[2])

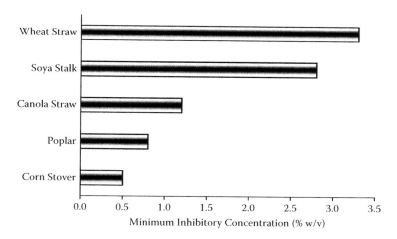

FIGURE 4.3   Growth of *Z. mobilis* on biomass hydrolysates. (Data from Lawford et al.[4])

The Stake Company Ltd. was founded in 1973 to develop and market a process for biomass conversion to sugar streams for both biofuels and animal feeds as well as chemicals derived from lignin and hemicellulose.[5] A continuous feedstock processing system was constructed to handle 4–10 tons of wood chips/hr and licensed to end-users in the United States and France.

Before 2004 (or 1999), moreover, more radical processes were examined in detail — including being upscaled to pilot plant operations — for lignocellulosic ethanol. These proposals included those to avoid the need for cellulase fermentations independent of the main ethanolic fermentation as well as the use of thermophilic bacteria in processes that more closely resembled industrial chemistry than they did the traditional potable alcohol manufacture. Indeed, it is clear that Iogen considered sourcing thermophilic bacteria* and nonconventional yeasts during the 1990s.[3] The achieved reality of the Iogen process will, therefore, be used as a guide to how innovations have successfully translated into practical use — or have failed to do so — reviewing progress over (mostly) the last three decades and offering predictions for new solutions to well-known problems as the bioethanol industry expands geographically as well as in production scale.

4.2   BIOMASS SUBSTRATE PROVISION AND PRETREATMENT

In the calculations inherent in the data for figure 4.2, some interesting conclusions are reached. Although wheat straw has undoubted advantages, other feedstocks outperform wheat straw for some key parameters:

*   Wheat straw has a lower gravimetric ratio of total carbohydrate (cellulose, starch, xylan, arabinan) to lignin than barley straw, corn stover, switchgrass, or even wheat chaff.
*   Wheat straw has a lower cellulose:lignin ratio than all of these nonsoftwood sources (with the exception of switchgrass).
*   Of the seven quoted examples of lignocellulosic feedstocks, wheat chaff and switchgrass have the highest total pentose (xylan and arabinan) contents — quoted as an important quantitative predictor for ethanol yield from cellulose because less cellulase is required (or, conversely, more of the cellulose glucose is made available for fermentation).[1,2]

An important consideration included in Iogen's deliberations on feedstock suitability was the reproducible and predictable supply of wheat straw. The USDA Agricultural Service also itemized sustainable supply as one of their two key factors for biomass feedstocks, the other being cost-effectiveness.[6] Financial models indicate that feedstocks costs are crucial, and any managed reduction of the costs of biomass crop production, harvesting, and the sequential logistics of collection, transportation, and storage before substrate pretreatment will all impact the viability of biofuel facilities — economic aspects of feedstock supply chains are discussed in the next chapter. The Energy Information Administration has constructed a National Energy Modeling System to forecast U.S. energy production, use, and price trends in 25-year predictive segments; the biomass supply schedule includes agricultural residues, dedicated energy crops, forestry sources, and wood waste and mill residues, and wheat straw (together with corn stover, barley straw, rice straw, and sugarcane bagasse) is the specified component of the agricultural residue supply.[7]

A brief survey will, therefore, be made of wheat straw and other leading candidate lignocellulosics, with special emphasis on how different national priorities place

---

* Including *Bacillus* species then being developed by Agrol Ltd. (later, Agrol Biotechnologies Ltd.), a spinout company from the University of London, England.

emphasis on different biomass sources and on what evolving agricultural practices and processing technologies may diversify bioethanol facilities on scales equal to and larger than the Iogen demonstration facility.

4.2.1 WHEAT STRAW — NEW APPROACHES TO COMPLETE SACCHARIFICATION

Although the Iogen process relies on acid pretreatment and cellulase digestion, Danish investigators rank other pretreatment methods (with short residence times, 5–6 minutes) as superior for subsequent wheat straw cellulose digestion by cellulase (24 hours at 50°C):[8]

Steam explosion (215°C) > H_2O_2 (190°C) > water (190°C) > ammonia (195°C) > acid (190°C)

The degradation of cellulose to soluble sugars was enhanced by adding nonionic surfactants and polyethylene glycol during enzymatic hydrolysis; the best results were obtained with a long-chain alcohol ethoxylate in conjunction with steam explosion–pretreated wheat straw, and the additives may either have occupied cellulase binding sites on residual lignin or helped to stabilize the enzyme during the lengthy digestion.[8]

In addition, the attention of Novozymes (one of the world's major enzyme producers) has evidently been attracted post-2004 by wheat straw and the problems of its complete conversion to fermentable sugars:

- Arabinoxylans form an undigested fraction in the "vinasse" (the insoluble fermentation residue) after the end of a wheat-based bioethanol process; a mixture of depolymerizing enzymes from *Hypocrea jecorina* and *Humicola insolens* could solubilize the insoluble material and release arabinose and xylose — although at different rates with different optimal pH values and temperature ranges for the digestion.[9]
- A subsequent study mixed three novel α-L-arabinofuranosidases with an endoxylanase and a β-xylosidase to liberate pentoses from water-soluble and water-insoluble arabinoxylans and vinasse; much lower enzyme activities were required than previously, and this may be a technology for pentose release before wheat straw–substrate fermentations.[10]
- Mixtures of α-L-arabinofuranosidases from *H. insolens*, the white-rot basidomycete *Meripilus giganteus*, and a *Bifidobacterium* species were highly effective in digesting wheat arabinoxylan, the different enzymes acting synergistically on different carbohydrate bonds in the hemicellulose structures.[11]

Arabinans constitute only 3.8% (by weight) of the total carbohydrate (cellulose, starch, xylose, and arabinose) in wheat straw, and a lack of utilization of all the pentose sugars represents a minor inefficiency. Releasing all the xylose (as a substrate for the engineered xylose-utilizing yeast) — xylose constitutes 24% of the total sugars — and completing the depolymerization of cellulose to (insofar as is possible) free glucose are more significant targets for process improvement.

As in Canada, wheat straw has been assessed to be a major lignocellulosic feed-stock in Denmark.* Substrate pretreatment studies from Denmark have, unlike at Iogen, concentrated on wet oxidation, that is, heat, water, and high-pressure O_2 to hydrolyze hemicellulose while leaving much of the lignin and cellulose insoluble; various conditions have been explored, including the combining of thermal hydrolysis, wet oxidation, and steam explosion.[12–15] The major concern with this type of pretreatment method is — with so many biomass substrates — the formation of inhibitors that are toxic to ethanologens and/or reduce ethanol yield.[16–18] These degradation products of lignocellulosic and hemicellulosic polymers include aromatic acids and aldehydes as well as aliphatic carboxylic acids and sugar-derived components; laboratory strains of *Saccharomyces cerevisiae* exhibit differential responses to the growth inhibitors, and cell-free enzyme preparations with cellulase and xylanase activities were severely inhibited by chemically defined mixtures of the known wheat straw inhibitors, with formic acid being by far the most potent inhibitor.[19] The positives to be drawn were, however, that even a laboratory strain could grow in 60% (w/v) of the wheat straw substrate and that a focused removal of one (or a few) inhibitors (in particular, formic acid) may suffice to render the material entirely digestible by engineered yeast strains.

4.2.2 SWITCHGRASS

Switchgrass, a perennial, warm season prairie grass and the leading candidate grass energy crop, could be grown in all rural areas in the continental United States. east of the Rocky Mountains, from North Dakota to Alabama, with the exception of southern Texas, southern Florida, and Maine.[20] Until the mid-1990s, switchgrass was primarily known in scientific agricultural publications as a forage crop for livestock. Shortly after the turn of the millennium, the species had been tested as direct energy source, co-combusted with coal at 7–10% of the energy production levels.[21] Simply burning switchgrass operates at 32% energy efficiency but using pellet grass in space-heating stoves can achieve 85% conversion efficiency. The environmental and agronomic advantages of switchgrass as a direct energy crop are severalfold:[21]

- Like all biomass crops, emissions are low in sulfur and mercury (especially when compared with coal).
- Switchgrass requires modest amounts of fertilizer for optimum growth, much lower application rates than with corn.
- Switchgrass stands are perennial, needing no recurrent soil preparation and so greatly reducing soil erosion and runoff caused by annual tillage.
- An acre of switchgrass could be the energy equivalent of 2–6 tons of coal, the high variability being associated with fertilizer application, climate variation, and others.

In hard economic terms, however, recycling alternative fuels such as municipal solid waste and used tires has been calculated to be preferable to either switchgrass or any

---

* Denmark is entirely above latitude 54°N; the southernmost regions of Canada reach below 50°N; both are too far north to grow biomass crops other than some cereal species and forest tress (and Denmark has a limited land availability for large-scale tree plantations).

form of biomass, independent of the scale of use in mass burn boilers.[22] This analysis is clearly restricted to what can be "acquired" for recycling, and has very different likely outcomes if lower-wastage, non-Western economies and societies were to be similarly analyzed. Using the criterion of bulk burnable material resulting from biomass drying, herbaceous plants have been advocated as the best choice for flexibly harvestable materials intended for power production via steam boilers, this choice being over that of corn stover, tree seedlings such as fast-growing willow, tree trimmings, by-products of lumber production, or switchgrass.[23]

Upland and lowland cultivars of switchgrass differ appreciably in their biomass yield, tolerance to drought, and response to nitrogen fertilizer application; even between both upland and lowland variants, the differences were found to be sufficiently great to merit recommendations for specific types of growth habitat if energy cropping were to be practiced.[24] In the northern prairies, nitrogen fertilizer use results in only variable and inconsistent increases in biomass production; a single annual harvest after the first frost is optimal for polymeric material but with reduced total nitrogen and ash as well as coinciding with low "infestation" by grass weed species; a mixture of switchgrass and big blue stem grass (*Andropogon gerardii* Vitman) has been recommended over dependence on a monoculture approach.[25] Nitrogen application was also found to be of little benefit in a 50-year trial in southern England, where five varieties of switchgrass and one of panic grass (*Panicum amarum* A.S. Hitchin & Chase) were compared; delaying harvest until the dead-stem stage allowed more mineral nutrients to return to the soil.[26]

Like all grasses, switchgrass suffers — as a substrate for bioethanol — from its low polymeric sugar content but elevated contributions of low-molecular-weight material to its dry mass; the lignin component of the insoluble material is reduced when compared with other major lignocellulosic materials (table 1.5). This suboptimal chemistry has spurred attempts to discover means to produce industrially or commercially important biomaterials from switchgrass, in particular, high-value and nutritional antioxidants.[27,28] Soluble phenolics are a potential industrial resource for fine chemicals and are present in the highest concentrations in the top internodes of the grass, whereas lower internodes contained greater amounts of cell wall-linked phenolics such as coumaric and ferulic acids.[28] Steroidal sapogenins, starting points for the synthesis of pharmacologically active compounds, are possible hepatoxins for grazing animals.[29] Grass fibers can also be used as raw material for biocomposites, packagings, and thermoplastics, and switchgrass could be a large-scale substrate for fermentations to biomanufacture biodegradable polyhydroxyalkanoate polymers.[30] Pulps prepared from switchgrass also show promise as reinforcement components in newsprint.[31]

As with the Iogen process for bioethanol, dilute acid hydrolysis has been explored as a pretreatment methodology for switchgrass; in a batch reactor, the optimum conditions were 1.2% sulfuric acid at 180°C for 0.5 minute; subsequent cellulase digestion released 91.4% of the cellulose as glucose and cellobiose.[32] Cellulose and lignin in switchgrass pretreated with dilute acid appeared not to interact when cellulase was added to degrade the insoluble polyglucan, acid-extracted lignin having little or no effect on the rate or extent of cellulose reactivity and saccharification.[33] Mixing switchgrass with aqueous ammonia and heating under pressure at 120°C for 20 minutes aided the subsequent digestion with cellulase and xylanase.[34] Milder conditions were, however, developed for the ammonia fiber explosion technology — 100°C for 5 minutes — and resulted in a 93% solubilization of the polyglucan content of the grass.[35]

Switchgrass has also been included as a test biomass substrate in experimental studies of simultaneous saccharification and fermentation (SSF) (section 4.5).

4.2.3 CORN STOVER

Corn stover is the above-ground plant from which the corn grain has been removed, and the constituent parts are leaves, stalk, tassel, corn cob, and shuck (the husk around the grains when in the intact cob); up to 30% by dry weight of the harvested plant is represented by the collected grain. In one of the earliest technological and economic reviews of corn-based fuel alcohol production, corn stover was included for consideration — but solely as an alternative to coal as a boiler fuel for distillation. In late 1978, the report for the U.S. DOE estimated that corn stover would increase the final cost of fuel ethanol by 4¢/gallon as the use of corn stover as a fuel entailed costs roughly double those of local Illinois coal.[36] The use of corn stover was, therefore, considered to be "justified only if the plant is located in an area where transportation cost would cause a doubling of the coal cost, or environmental considerations would rule against the use of coal; neither of which is very likely." Such arguments left corn stover in the field as an aid against soil erosion for over a decade until the option of lignocellulosic ethanol began to be seriously considered. By 2003, the National Renewable Energy Laboratory, Golden, Colorado, estimated the annual and sustainable production of corn stover as 80–100 million dry tonnes/year, of which 20% might be utilized in the manufacture of "fiber" products and fine chemicals (e.g., furfural); 60–80 million dry tonnes would remain as a substrate for bioethanol production.[37] Five years earlier, an estimate of total corn stover availability had been more than 250 million tonnes, with 30 million being left on the fields for erosion control, leaving 100 million available for biofuels production.[38] With expanding corn acreage and a definite future for corn-based ethanol, a supply of corn stover is ensured — and commercial drivers may direct that starch ethanol and "bioethanol" facilities might be best sited adjacent to one another (see chapter 2, section 2.6).

With corn stover rising up the rankings of biomass substrates for ethanol production in the United States and elsewhere, experimental investigations of pretreatment technologies has proliferated since 2002.[39–50] From this impressive corpus of practical knowledge, some reinforced conclusions are apparent:

- A variety of pretreatment methodologies (see chapter 2, section 2.3.2), using acid or alkali, in batch or percolated modes, can yield material with a high digestibility when cellulase is applied to the insoluble residues (table 4.1).
- With some practical technologies, treatment times can be reduced to a few minutes (table 4.1).
- Particle size is an important parameter for pretreatment kinetics and effectiveness — diffusion of acid within a particle becomes rate-influencing above a critical biomass size and small particle sizes are more readily hydrolyzed with cellulase because of their higher surface area-to-mass ratios — although liquid water at high temperature minimizes this difference and causes ultrastructural changes with the appearance of micrometer-size pores in the material.[40,51]

TABLE 4.1

Pretreatment Methodologies for Corn Stover: Physical Conditions and Cellulose Digestibility

Hydrolysis	Hydrolysis conditions	Hemicellulose solubilization (%)	Enzyme digestion	Glucan conversion (%)	Reference
5% H_2SO_4 or HCl	120°C, 60 min	85	Cellulase	94.7	38
0.2% Na_2CO_3	195°C, 15 min, 12 bar O_2	60	Cellulase	85	40
2% H_2SO_4	190°C, 5 min	—	Cellulase	73[a]	43
Aqueous ammonia	Room temperature, 10–12 min	—	Cellulase	88.5	44
0.5 g $Ca(OH)_2$ per g biomass	55°C, 4 weeks	—	Cellulase	93.2[b]	45
Hot water (pH controlled)	190°C, 15 min	—	Cellulase	90	46
Hot water	Intermittent flowthrough, 200°C	—	Cellulase	90	47
SO_2, steam	200°C, 10 min	—	Cellulase	89	48
0.22–0.98% H_2SO_4	140–200°C	—	Cellulase	92.5[c]	49

[a] Total sugars

[b] Combined post-hydrolysis and pretreatment liquor contents

[c] Combined glucose and xylose recoveries

- Liquid flow-through enhances hemicellulose sugar yields, increases cellulose digestibility to enzyme treatment, and reduces unwanted chemical reactions but with the associated penalties of high water and energy use; some of the benefits of flow-through can be achieved by limited fluid movement and exchange early in the acid digestion process.[42,48,56,57]
- Pretreated corn stovers appear to be much less toxic to ethanologens than other agricultural substrates.[44,46,47]
- Moreover, removal of acetic acid (a degradation product acetylated hemicellulose sugars) has been demonstrated at 25–35°C using activated carbon powder, and a natural fungus has been identified to metabolize furans and actively grow in dilute acid hydrolysates from corn stover.[52,53]
- Predictive mathematical models have been developed for the rheology and delignification of corn stover during and after pretreatment.[54,55]

Detailed studies of individual approaches to pretreatment have resulted in important insights into the fundamental sciences as well as guidelines for their large-scale use with corn stover. With long-duration lime (calcium hydroxide) treatment at moderate temperatures (25–55°C), the enzymic digestibility of the resulting cellulose was highly influenced by both the removal of acetylated hemicellulose residues and delignification, but deacetylation was not seriously influenced by the levels of O_2 or the temperature.[58] Adding a water washing to ammonia-pretreated material removed lignophenolic extractives and enhanced cellulose digestibility.[59] Grinding into smaller particles increased the cellulose digestibility after ammonia fiber explosion, but the chemical compositions of the different particle size classes showed major changes in the contents of xylans and low-molecular-weight compounds (figure 4.4). This could be explained by the various fractions of corn stover being differentially degraded in smaller or larger particles on grinding; for example, the cobs are relatively refractive to size reduction; the smaller particle sizes after AFEX treatment were more cellulase-degradable than were larger particles. Electron microscopic chemical analysis of the surface of the pretreated material provided evidence that lignin-carbohydrate complexes (chapter 2, section 2.3.2) had been disrupted.[59] The high hemicellulose content of corn cobs has been exploited in a development where aqueous ethanol-pretreated material is washed and then hydrolyzed with an endoxylanase; food-grade xylooligosaccharides can easily be purified, and the cellulosic material is readily digestible with cellulase.[60] An additional advantage of corn cobs is that they can be packed at high density, thus reducing the required water inputs and giving a high concentration in the xylan product stream.

The dominance of inorganic acids for acid pretreatment of biomass substrates has only recently been challenged by the use of maleic acid, one of the strongest organic dicarboxylic acids and a potential mimic of the active sites of hydrolase enzymes with two adjacent carboxylic acid residues at their active sites.[61] In comparison with dilute sulfuric acid, maleic acid use resulted in a greatly reduced loss of xylose at high solids loadings (150–200 g dry stover/l), resulting in 95% xylose

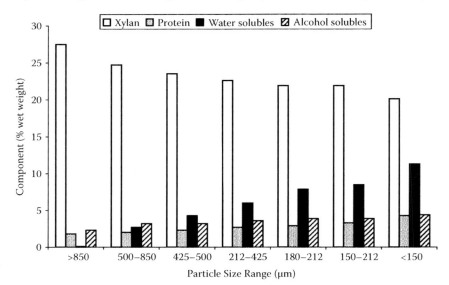

FIGURE 4.4   Size reduction of corn stover and chemical composition of differentially sized particles. (Data from Chundawal et al.[59])

yields, only traces of furfural, and unconditioned hydrolysates that could be used by recombinant yeast for ethanol production; 90% of the maximum glucose release could be achieved by cellulase digestion of the pretreated stover within 160 hours.

Examination of (and experiments with) the cellulase digestion of pretreated corn stover have led to other conclusions for industrial applications:

- Studies of the binding of cellobiohydrolase to pretreated corn stover identified access to the cellulose in cell wall fragments and the crystallinity of the cellulose microfibrils after pretreatment to be crucial.[62]
- Adding small amounts of surfactant-emulsifiers during cellulase digestion of pretreated corn stover also increased the conversion of cellulose, xylan, and total polysaccharide to sugars, by acting to disrupt lignocellulose, stabilize the enzyme, and improve the absorption of the enzyme to the macroscopic substrates.[63]
- With steam-pretreated corn stover, near-theoretical glucose yields could be achieved by combining xylanases with cellulase to degrade residual hemicellulose bound to lignocellulosic components.[64]
- The initial rate of cellulase catalyzed hydrolysis is influenced strongly by the cellulose crystallinity whereas the extent of cellulose digestion is most influenced by the residual lignin.[65] Modern methods of polymer analysis (e.g., diffusive reflectance infrared and fluorescence techniques) used in this work may be adaptable to on-site monitoring of pretreated biomass substrates.
- The formation of glucose from pretreated corn stover catalyzed by cellulase is subject to product inhibition, and the effects of substrate concentration and the amount ("loading") of the enzyme are important in determining kinetic parameters.[66]
- Cellulase and cellobiohydrolase can both be effectively recovered from pretreated and digested corn stover and recycled with consequent cost savings of approximately 15% (50% if a 90% enzyme recovery could be achieved).[67]
- The solid material used for cellulase-catalyzed hydrolysis itself is a source of potential toxic compounds produced during pretreatment but trapped in the bulk solids; activated carbon is (as discussed above) effective in removing acidic inhibitors from the liquid phase resulting from digestion of the reintroduced substrate.[68]

A comparative study of several methods for corn stover pretreatment concluded that alkaline methodologies had the potential to reduce the quantities of cellulase necessary in cellulose digestion but that hemicellulase activities may require supplementation.[69]

4.2.4 SOFTWOODS

The widely quoted assessment of softwoods is that, as a biomass substrate, the lignocellulose is too highly lignified and difficult to process to yield cellulose easily digested by cellulase — in practical terms, excess enzyme is required and imposes unrealistically high costs and protracted digestion times.[1,70] Nevertheless, the massive resources of softwood trees in the Pacific Northwest of the United States, Canada, Scandinavia, northern Europe, and large tracts of Russia maintain softwoods as an attractive potential biomass for fuel alcohol production. Sweden has a particular stake in maximizing the efficiency of ethanol production from softwoods on account of the planned diversion of large amounts of woody biomass to direct heat and power facilities with the phasing out of nuclear generating capacity.[70] Much of the published work on softwood utilization for bioethanol indeed derives from Canadian and Scandinavian universities and research centers.

Although softwoods are low in xylans in comparison with other biomass crops (table 1.5), their content of glucan polymers is high; the requirement for xylose-utilizing ethanologens remains a distinct priority, whereas mannose levels are high and should contribute to the pool of easily utilized hexoses. To make the potential supply of fermentable sugars fully accessible to yeasts and bacteria for fermentation, attention has been focused on steam explosive pretreatments with or without acid catalysts (SO_2 or sulfuric acid) since the 1980s; pretreatment yields a mixed pentose and hexose stream with 50–80% of the total hemicellulose sugars and 10–35% of the total glucose, whereas a subsequent cellulase digestion liberates a further 30–60% of the theoretical total glucose.[70] Because only a fraction of the total glucose may be recoverable by such technologies, a more elaborate design has been explored in which a first stage is run at lower temperature for hemicellulose hydrolysis, whereas a second stage is operated at a higher temperature (with a shorter or longer heating time and with the same, higher, or lower concentration of acid catalyst) to liberate glucose from cellulose.[71] Such a two-stage process results in a sugar stream (before enzyme digestion) higher in glucose and hemicellulose pentoses and hexoses but with much reduced degradation of the hemicellulose sugars and no higher levels of potential growth inhibitors such as acetic acid (figure 4.5).

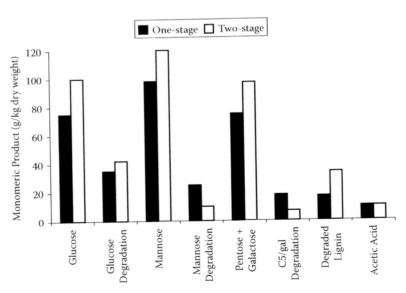

FIGURE 4.5   Sugar stream from single- and dual-stage steam/acid pretreatments of spruce wood chips. (Data from Wingren et al.[71])

Two-stage pretreatment suffers from requiring more elaborate hardware and a more complex process management; in addition, attempting to dewater sulfuric acid-impregnated wood chips before steaming decreases the hemicellulose sugar yield from the first step and the glucose yield after the second, higher-temperature stage.[72] Such pressing alters the wood structure and porosity, causing uneven heat and mass transfer during steaming. Partial air drying appears to be a more suitable substrate for the dual-stage acid-catalyzed steam pretreatment.

Not only is the sugar monomeric sugar yield higher with the two-stage hydrolysis process, the cellulosic material remaining requires only 50% of the cellulase for subsequent digestion.[73] This is an important consideration because steam-pretreated softwood exhibits no evident saturation with added cellulase even at extremely high enzyme loadings that still cannot ensure quantitative conversion of cellulose to soluble sugars: with steam-pretreated softwood material, even lavish amounts of cellulase can only liberate 85% of the glucan polymer at nonviably low ratios of solid material to total digestion volume, and although high temperatures (up to 52°C) and high agitation rates are helpful, enzyme inactivation is accelerated by faster mixing speeds.[74] The residual lignin left after steam pretreatment probably restricts cellulase attack and degradation by forming a physical barrier restricting access and by binding the enzyme nonproductively; extraction with cold dilute NaOH reduces the lignin content and greatly increases the cellulose to glucose conversion, the alkali possibly removing a fraction of the lignin with a high affinity for the cellulase protein.[75]

In addition to the engineering issues, the two-stage technology has two serious economic limitations:

1. Hemicellulose sugar recovery is aided after the first step by washing the slurry with water but the amount of water used is a significant cost factor for ethanol production; to balance high sugar recovery and low water usage, a continuous countercurrent screw extractor was developed by the National Renewable Energy Laboratory that could accept low liquid to insoluble solids ratios.[76]
2. The lignin recovered after steam explosion has a low product value on account of its unsuitable physicochemical properties; organic solvent extraction of lignin produces a higher value coproduct.[77]

As with other biomass substrates, heat pretreatment at extremes of pH generates inhibitors of microbial ethanol production and, before this, cellulase enzyme action.[78,79] A number of detoxification methods have been proposed, but simply adjusting the pH to 10 to precipitate low-molecular-weight sugar and lignin degradation products is effective.[80] Six species of yeasts — including *S. cerevisiae*, *Candida shehatae*, and species found in forest underbrush in the western United States — were tested for adaptation to softwood (Douglas fir) pretreated with dilute acid, and isolates were selected and gradually "hardened" to hydrolysate toxicity for improved ethanol production.[81]

Aqueous ethanol pretreatment of softwoods (the lignol process) has been strongly advocated on account of its ability to yield highly digestible cellulose as well as lignin, hemicellulose, and furfural product streams — the extraction is operated at acid pH at 185–198°C, and some sugar degradation does inevitably occur.[82] After process optimization, a set of conditions (180°C, 60 minutes of treatment time, 1.25% v/v sulfuric acid, and 60% v/v ethanol) yielded[83]

- A solids fraction containing 88% of the cellulose present in the untreated wood chips
- Glucose and oligosaccharides equivalent to 85% of the cellulosic glucose was released by cellulase treatment (48 hours) of the treated lignocellulose
- Approximately 50% of the total xylose recovered from the solubilized fraction
- More than 70% of the lignin solubilized in a form potentially suitable for industrial use in the manufacture of adhesives and biodegradable polymers

4.2.5 SUGARCANE BAGASSE

As with corn stover, bagasse is the residue from sugarcane juice extraction and, as such, is an obligatory waste product. Development programs for bioethanol production from bagasse started in Brazil in the 1990s, and Dedini S/A Indústrias de Base (www.dedini.com.br) now operates a pilot plant facility with the capacity of producing 5,000 l/day of ethanol from bagasse in São Paolo state; up to 109 l of hydrated alcohol can be produced per tonne of wet bagasse, and this could be increased to 180 l/tonne with full utilization of hemicellulose sugars.

The commercial Brazilian process uses organic-solvent-treated bagasse. Other reports from Cuba, Denmark, Sweden, Japan, Austria, Brazil, and the United States describe alternative processes, all with some merits:

- Steam explosion — impregnation with SO_2 before steam explosion gives high yields of pentose sugars with no additional formation of toxic inhibitors when compared with the absence of any acid catalyst[84–86]
- Liquid hot water pretreatment — probably the cheapest method (requiring no catalyst or chemical) and, when operated at less than 230°C, is effective at solubilizing hemicellulose and lignin while leaving cellulose as an insoluble residue for further processing[87,88]
- Peracetic acid — alkaline pre-pretreatment followed by the use of peracetic acid gives synergistic enhancements of cellulose digestibility[89,90]
- Ammonia-water mixtures — vacuum-dried material from the alkaline treatment could be used for enzymatic digestion of cellulose without washing or other chemical procedure[34]
- Dilute acid — this has not yet been fully tested for ethanol production, but as a method for preparing xylose as a substrate for xylitol production, it is capable of yielding hydrolysates with high concentrations of free xylose[91,92]
- Wet oxidation — alkaline wet oxidation at 195°C for 15 minutes produces solid material that is 70% cellulase; approximately 93% of the hemicelluloses and 50% of the lignin is solubilized, and the cellulose can be enzymically processed to glucose with 75% efficiency[93]

Three different species of yeasts have been demonstrated to ferment pentoses and/or hexoses from chemical hydrolysates of sugarcane bagasse: *S. cerevisiae*, *C. shehatae*, *Pichia stipitis*, and *Pachysolen tannophilus*.[94–97] Acid hydrolysates are best detoxified by ion exchange materials or activated charcoal; laccase and high pH precipitation methods are less effective.[95] The most recent of the reports attempted lime treatment to neutralize the acid but found that the novel technique of electrodialysis (migration of ions through membranes under a direct electric field) removed the sulfuric acid and also the acetic acid generated during acid hydrolysis of hemicelluloses so effectively that the reutilization of the sulfuric acid could be contemplated.[*,97] Recombinant xylose-utilizing yeast has been desensitized to hydrolysates containing increasing concentrations of phenolic compounds, furfuraldehydes, and carboxylic acids without loss of the xylose-consuming capacity and while retaining the ability to form ethanol rather than xylitol.[96] Acetic acid and furfural at concentrations similar to those measured in sugarcane bagasse hydrolysates adversely affect both "laboratory" and "industrial" strains (see chapter 3, section 3.2.4) of *S. cerevisiae*.[98] One highly practical solution is that the predominantly pentose-containing hydrolysates from bagasse pretreatments can also be used to dilute the sugarcane juice-based medium for sugar ethanol fermentation while maintaining an equivalent sugar concentration and utilizing a pentose-consuming *P. stipitis* to coferment the sugar mixture.[99]

Sugarcane is, however, not entirely without its industrial biohazards. Bagassosis, caused by airborne cells (or fragments) of *Thermoactinomyces sacchari*, was once very prevalent in workforces exposed to bagasse dust. In the United States and also in Japan (where outbreaks occurred in sugar refineries and lacquerware factories), the disease is thought to have been mostly eradicated during the 1970s by improved product handling and safety practices.[100,101]

4.2.6 OTHER LARGE-SCALE AGRICULTURAL AND FORESTRY BIOMASS FEEDSTOCKS

In addition to isolating grains for processing, cereal-milling plants also generate fiber-rich fractions as a coproduct stream. In the wet milling of corn, the fiber fraction has traditionally been added into a feed product (figure 1.20). The University of Illinois developed a modified dry milling procedure to recover fiber fractions before fermentation: this quick fiber contained 65% by weight of total carbohydrate and 32% by weight of glucans, and dilute acid pretreatment was used before fermentation of the substrate to ethanol by either *Escherichia coli or S. cervisiae*.[102] Destarched, cellulose-rich, and arabinoxylan-rich fractions of the corn fiber support the growth of strains of *Hypocrea jecorina* and their secretion of hydrolases for plant polysaccharides; these enzymes act synergistically with commercial cellulases on corn fiber hydrolysate and represent a valuable source of on-site enzymes for corn fiber product utilization.[103] Similarly, large quantities of wheat bran are produced worldwide as a coproduct of wheat milling; residual starch in the bran material can be hydrolyzed to glucose and oligoglucans by amylolytic enzymes, and acid hydrolysis pretreatment followed by cellulase treatment gives a sugar (pentose and hexose) yield of 80% of the theoretical: 135, 228, and 167 g/kg of starch-free bran for arabinose, xylose, and glucose, respectively.[104]

Rice husks are approximately 36% by weight cellulose and 12% hemicellulose; as such, this agricultural by-product could be a major low-cost feedstock for ethanol production; 60% of the total sugars could be released by acid hydrolysis and treatment

---

* This Chinese work is particularly relevant for improving the economics of bioethanol production because sugarcane bagasse represents the most abundant lignocellulosic agricultural material in southern China.[97]

with a mixture of enzymes (β-glucosidase, xylanase, and esterase) with no formation of furfuraldehyde sugar degradation products.[105] Recombinant *E. coli* could ferment the released sugars to ethanol; high-pH treatment of the hydrolysate reduced the time required for maximal production of ethanol substantially, from 64 to 39 hours. In a study from India, rice straw was pretreated with and without exogenous acid, and the released hemicellulose sugars fermented by a strain of *C. shehatae*; ethanol production was also demonstrated by yeast cells immobilized in calcium alginate beads — an example of an advanced fermentation technology discussed in more detail in the next section.[106]

Fast-growing willow trees are a major focus of research interest as a bioenergy crop in Scandinavia; high sugar recoveries were achieved from lignocellulosic material by steaming sulfuric acid-impregnated material for a brief period (4–8 minutes) at 200°C, and then digesting the cellulose enzymically, liberating glucose with 92% efficiency and xylose with 86% efficiency. The pretreated substrate could also be used for SSF with a *S. cerevisiae* strain.[107]

Many "exotic" plant materials have been included in surveys of potential biomass and bioenergy sources; example of these are considered in chapter 5, section 5.5.2, when sustainability issues are covered at the interfaces among agronomy, the cultivation of bioenergy crops, land use, and food production. As a lignocellulosic, straw from the grass species *Paja brava*, a Bolivian high-plains resident species, can be considered here. Steamed, acid-impregnated material gave hemicellulose fractions at 190°C that could be fermented by three pentose-utilizing yeasts, *P. stipitis*, *C. shehatae*, and *Pachysolen tannophilus*, while a higher temperature (230°C) was necessary for cellulose hydrolysis.[108] Much more widely available worldwide is the mixed solid waste of lumber, paper, tree pruning, and others; this is a highly digestible resource for cellulase, the sugars being readily fermented by *S. cerevisiae* and the residual solids potentially usable for combustion in heat and power generation.[109]

4.3   FERMENTATION MEDIA AND THE "VERY HIGH GRAVITY" CONCEPT

After acid hydrolysis (or some other pretreatment) and cellulase digestion, the product of the process is a mixed carbon source for fermentation by an ethanologenic microbe. Few details have been made public by Iogen about their development of nutritional balances, nitrogen sources, or media recipes for the production stage fermentation.[1] This is not surprising because, for most industrial processes, medium optimization is a category of "trade secret," unless patenting and publication priorities deem otherwise. Few industrial fermentations (for products such as enzymes, antibiotics, acids, and vitamins) rival the conversion efficiency obtained in ethanol production; one notable exception — about which a vast literature is available — is that of citric acid manufacture using yeasts and fungi.[110] Many of the main features of citric acid fermentations have echoes in ethanol processes, in particular the use of suboptimal media for growth to generate "biological factories" of cell populations supplied with very high concentrations of glucose that cannot be used for further growth or the accumulation of complex products but can be readily fluxed to the simple intermediate of glucose catabolism — the biochemistry of *Aspergillus niger* strains used for the production of citric acid at concentrations higher than 100 g/l is as limited (from the viewpoints of biological ingenuity and bioenergetics) as ethanol production by an organism such as *Z. mobilis* (see figure 3.4).

Nevertheless, interest in media development for ethanol production has been intense for many years in the potable alcohol industry, and some innovations and developments in that industrial field have been successfully translated to that of fuel ethanol production.

4.3.1 Fermentation Media for Bioethanol Production

Formulating cost-effective media for the recombinant microorganisms developed for broad-spectrum pentose and hexose utilization (chapter 3, sections 3.2 and 3.3) commenced in the 1990s. For pentose-utilizing *E. coli*, for example, the benchmark was a nutrient-rich laboratory medium suitable for the generation of high-cell-density cultures.[111] Media were then assessed using the criteria that the final ethanol concentration should be at least 25 g/l, the xylose-to-ethanol conversion efficiency would be high (90%), and a volumetric productivity of 0.52 g/l/hr was to be attained; in a defined minimal salts medium, growth was poor, only 15% of that observed in the laboratory medium; supplementation with vitamins and amino acids improved growth but could only match approximately half of the volumetric productivity. The use of corn steep liquor as a complex nitrogen source was (as predicted from its wide industrial use in fermentations) the best compromise between the provision of a complete nutritional package with plausible cost implications for a large-scale process. As an example of the different class of compromise inherent in the use of lignocellulosic substrates, the requirement to have a carbon source with a high content of monomeric xylose and low hemicellulose polymers implied the formation of high concentrations of acetic acid as a breakdown product of acetylated sugar residues; to minimize the associated growth inhibition, one straightforward strategy was that of operating the fermentation at a relatively high pH (7.0) to reduce the uptake of the weak acid inhibitor.

In a study conducted by the National Center for Agricultural Utilization Research, Peoria, Illinois, some surprising interactions were discovered between nitrogen nutrition and ethanol production by the yeast *P. stipitis*.[112] When the cells had ceased active growth in a chemically defined medium, they were unable to ferment either xylose or glucose to ethanol unless a nitrogen source was also provided. Ethanol production was increased by the amino acids alanine, arginine, aspartic acid, glutamic acid, glycine, histidine, leucine, and tyrosine (although isoleucine was inhibitory); a more practical nitrogen supply for industrial fermentations consisted of a mixture of urea (up to 80% of the nitrogen) and hydrolyzed milk protein supplemented with tryptophan and cysteine (up to 60%); the use of either urea or the protein hydrolysate was less effective than the combination of both. Adding small amounts of minerals, in particular, iron, manganese, magnesium, calcium, and zinc salts as well as amino acids could more than double the final ethanol concentration to 54 g/l.

Returning to recombinant *E. coli*, attempts to define the minimum salts concentration (to avoid stress imposed by osmotically active solutes) resulted in the formulation of a medium with low levels of sodium and other alkali metal ions (4.5 mM) and total salts (4.2 g/l).[113] Although this medium was devised during optimization of lactic acid production, it proved equally effective for ethanol production from xylose. Because many bacteria biosynthesize and accumulate internally high concentrations of osmoprotective solutes when challenged with high exogenous levels of salts, sugars, and others, modulating known osmoprotectants was tested and shown to improve the growth of *E. coli* in the presence of high concentrations of glucose, lactate, sodium lactate, and sodium chloride.[114] The minimum inhibitory concentrations of these solutes was increased by either adding the well-known osmoprotectant betaine, increasing the synthesis of the disaccharide trehalose (a dimer of glucose), or both, and the combination of the two was more effective than either alone. Although the cells' tolerance to ethanol was not enhanced, the use of the combination strategy would be expected to improve growth in the presence of the high sugar concentrations that are becoming ever more frequently encountered in media for ethanol fermentations.

Accurately measuring the potential for ethanol formation represented by a cellulosic biomass substrate for fermentation (or fraction derived from such a material) is complex because any individual fermentable sugar (glucose, xylose, arabinose, galactose, mannose, etc.) may be present in a large array of different chemical forms: monomers, disaccharides, oligosaccharides, even residual polysaccharides. Precise chemical assays may require considerable time and analytical effort. Bioassay of the material using ethanologens in a set medium and under defined, reproducible conditions is preferable and more cost effective — and broadly analogous to the use of shake flask tests to assess potency of new strains and isolates and the suitability of batches of protein and other "complex" nutrients in conventional fermentation laboratories.[115]

4.3.2   HIGHLY CONCENTRATED MEDIA DEVELOPED FOR ALCOHOL FERMENTATIONS

Until the 1980s, the general brewing industry view of yeasts for alcohol production was that most could tolerate only low concentrations (7–8% by volume) of ethanol and, consequently, fermentation media (worts) could be formulated to a maximum of 15-16° (Plato, Brix, or Balling, depending on the industry subsector), equivalent to 15–16% by weight of a sugar solution; the events that radically changed this assessment of yeasts and their physiology were precisely and cogently described by one of the key players:[116]

- When brewers' yeasts were grown and measured in the same way as the more ethanol-tolerant distillers' and sake yeasts, differences in ethanol tolerance were smaller than previously thought.
- "Stuck" fermentations, that is, ones with little or no active growth in supraoptimal sugar concentrations, could easily be rescued by avoiding complete anaerobiosis and supplying additional readily utilizable nitrogen for yeast growth.
- By removing insoluble grain residues (to reduce viscosity), recycling clear mashes to prepare more concentrated media from fresh grain, optimizing yeast nutrition in the wort, and increasing cooling capacity, yeast strains with no previous conditioning and genetic manipulation could produce ethanol up to 23.8% by volume.

Very-high-gravity (VHG) technologies have great technical and economic advantages:

1. Water use is greatly reduced.
2. Plant capacity is increased, and fermentor tank volume is more efficiently utilized.

3. Labor productivity is improved.
4. Fewer contamination outbreaks occur.
5. The energy requirements of distillation are reduced because fermented broth is more concentrated (16–23% v/v ethanol).
6. The spent yeast can be more readily recycled.
7. The grain solids removed prefermentation can be a valuable coproduct.

With an increased volume of the yeast starter culture added to the wort (higher "pitching rate") and a prolonged growth phase fueled by adequate O_2 and free amino nitrogen (amino acids and peptides), high-gravity worts can be fermented to ethanol concentrations more than 16% v/v even at low temperatures (14°C) within a week and with no evidence of any ethanol "toxicity."[117–120]

This is not to say, however, that high ethanol concentrations do not constitute a stress factor. High-alcohol-content worts do still have a tendency to cease fermentation, and high ethanol levels are regarded as one of the four major stresses in commercial brewing, the others being high temperature, infection (contamination, sometimes associated with abnormal pH values), and mycotoxins from grain carrying fungal infections of *Aspergillus, Penicillium, Fusarium, Claviceps*, or *Acremonium* species.[121] To some extent, the individual stress factors can be managed and controlled — for example, in extremis, antimicrobial agents that are destroyed during distillation (so that no carryover occurs to the finished products) can be added even in potable alcohol production. It is when the major stresses combine that unique conditions inside a fermentor can be generated. For a potable alcohol producer, these can be disastrous because there is an essential difference between the products of fuel/industrial ethanol and traditional alcoholic beverages: the latter are operated for consistency in flavor and quality of the product; for the beverage producer, flavor and quality outweigh any other consideration — even distinct economic advantages associated with process change and improvement — because of the market risks, especially if a product is to be matured ("aged") for several years before resale.[122] Industrial ethanol is entirely amenable to changes in production practice, strain, trace volatile composition, and even process "excursions" when the stress factors result in out-of-tolerance conditions. Yeast (*S. cerevisiae*) cells may have the ability to reduce short-term ethanol toxicity by entering a "quiescent" state in their average population cell cycle, extending a phase of growth-unassociated ethanol production in a laboratory process developed to produce 20% ethanol by volume after 45 hours.[123]

From the work on VHG fermentations, the realization was gained that typical media were seriously suboptimally supplied with free amino acids and peptides for the crucial early growth phase in the fermentation; increasing the free amino nitrogen content by more than fourfold still resulted in the exhaustion of the extra nitrogen within 48 hours (figure 4.6). With the correct supplements, brewer's yeast could consume all the fermentable sugars in a concentrated medium (350 g/l) within eight days at 20°C or accumulate 17% (v/v) ethanol within three days.[124] Fresh yeast autolysate was another convenient (and cost-effective) means of nitrogen supplementation with an industrial distillery yeast from central Europe — although, with such a strain, while nitrogen additions improved final ethanol concentration and glucose utilization, none of them increased cell viability in the late stages of the fermentation, ethanol yield from sugar, or the maximum rate of ethanol formation.[125] Commercial

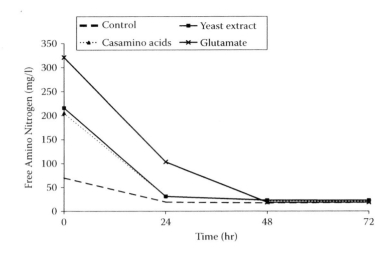

FIGURE 4.6   Utilization of free amino acids and peptides in high-gravity wort for alcohol production. (Data from Thomas and Ingledow.[124])

proteases can liberate free amino acids and peptides from wheat mash, the low-molecular-weight nitrogen sources increasing the maximal growth (cell density) of the yeast cells and reducing the fermentation time in VHG worts from nine to three days — and without (this being an absolute priority) proteolytic degradation of the glucoamylase added previously to the mash to saccharify the wheat starch; rather than adding an extra nitrogen ingredient, a one-time protease digestion could replace medium supplementation.[126] Not all amino acids are beneficial: lysine is severely inhibitory to yeast growth if the mash is deficient in freely assimilated nitrogen, but adding extra nitrogen sources such as yeast extract, urea, or ammonium sulfate abolishes this effect, promotes uptake of lysine, increases cell viability, and accelerates the fermentation.[127]

Partial removal of bran from cereal grains (wheat and wheat-rye hybrids) is an effective means of improving the mash in combination with VHG technology with or without nitrogen supplementation (figure 4.7); in a fuel alcohol plant, this would increase plant efficiency and reduce the energy required for heating the fermentation medium and distilling the ethanol produced from the VHG process.[128] Conversely, adding particulate materials (wheat bran, wheat mash insolubles, soy or horse gram flour, even alumina) improves sugar utilization in VHG media: the mechanism may be to offer some (undefined) degree of osmoprotection.[129,130]

A highly practical goal was in defining optimum conditions for temperature and mash substrate concentration with available yeast strains and fermentation hardware: with a wheat grain-based fermentation, a temperature of 30°C and an initial mash specific gravity of 26% (w/v) gave the best balance of high ethanol productivity, final ethanol concentration, and shortest operating time.[131] The conclusions from such investigations are, however, highly dependent on the yeast strain employed and on the type of beer fermentation being optimized: Brazilian investigators working with

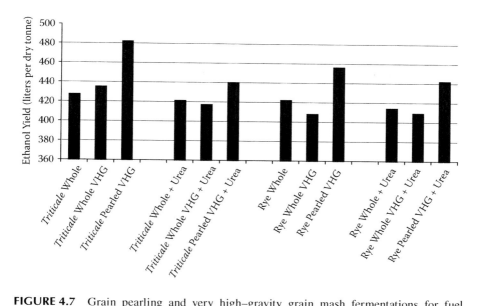

FIGURE 4.7 Grain pearling and very high–gravity grain mash fermentations for fuel ethanol production. (Data from Wang et al.[128])

a lager yeast strain found that a lower sugar concentration (20% w/v) and temperature (15°C) were optimal together with a triple supplementation of the wort with yeast extract (as a peptidic nitrogen source), ergosterol (to aid growth), and the surfactant Tween-80 (possibly, to aid O_2 transfer in the highly concentrated medium).[132]

For VHG fermentations, not only is nitrogen nutrition crucial (i.e., the supply of readily utilizable nitrogen-containing nutrients to support growth) but other medium components require optimization: adding 50 mM of a magnesium salt in tandem with a peptone (to supply preformed nitrogen sources) increased ethanol concentrations from 14.2% to 17% within a 48-hr fermentation.[133] These results were achieved with a medium based on corn flour (commonly used for ethanol production in China), the process resembling that in corn ethanol (chapter 1, section 1.4) with starch digestion to glucose with amylase and glucoamylase enzyme treatments. With a range of nutrients tested (glycine, magnesium, yeast extract, peptone, biotin, and acetaldehyde), cell densities could dramatically differ: the measured ranges were $74–246 \times 10^6$ cells/g mash after 24 hours and $62–392 \times 10^6$ cells/g mash after 48 hours. A cocktail of vitamins added at intervals in the first 28–37 hours of the fermentation was another facile strategy for improving final ethanol concentration, average ethanol production rate, specific growth rate, cell yield, and ethanol yield — and a reduced glycerol accumulation.[134] Small amounts of acetaldehyde have been claimed to reduce the time required to consume high concentrations of glucose (25% w/v) in VHG fermentations; the mechanism is speculative but could involve increasing the intracellular NAD:NADH ratio and accelerating general sugar catabolism by glycolysis (figure 3.1).[135] Side effects of acetaldehyde addition included increased accumulation of the higher alcohols 2,3-butanediol and 2-methylpropanol, exemplifying again how immune fuel ethanol processes are to unwanted "contaminants" and flavor agents so strictly controlled in potable beverage production. Mutants of brewer's yeast capable of faster fermentations, more complete utilization of wort carbohydrates ("attenuation"), and higher viability under VHG conditions are easily selected after UV treatment; some of these variants could also exhibit improved fermentation characteristics at low operating temperature (11°C).[136]

Ethanol diffuses freely across cell membranes, and it seems to be impossible for yeast cells to accumulate ethanol against a concentration gradient.[137] This implies that ethanol simply floods out of the cell during the productive phases of alcoholic fermentations; the pioneering direct measurement of unidirectional rate constants through the lipid membrane of Z. *mobilis* confirmed that ethanol transport does not limit ethanol production and that cytoplasmic ethanol accumulation is highly unlikely to occur during glucose catabolism.[138] Nevertheless, even without such an imbalance between internal and external cellular spaces, product inhibition by ethanol is still regarded as an inhibitor of yeast cell growth, if not of product yield, from carbohydrates.[139] Yeasts used for the production of sake in Japan are well known as able to accumulate ethanol in primary fermentations to more than 15% (v/v), and both Japanese brewing companies and academic centers have pursued the molecular mechanisms for this:

- With the advent of genomics and the complete sequencing of the *S. cerevisiae* genome, whole-genome expression studies of a highly ethanol-tolerant strain showed that ethanol tolerance was heightened in combination with resistance to the stresses imposed by heat, high osmolarity, and oxidative conditions, resulting in the accumulation internally of stress protectant compounds such as glycerol and trehalose and the overexpression of enzymes, including catalase (catalyzing the degradation of highly reactive hydrogen peroxide).[140]
- Inositol synthesis as a precursor of inositol-containing glycerophospholipids in cellular membranes is a second factor in membrane properties altering (or altered by) ethanol tolerance.[141]
- Disrupting the *FAA1* gene encoding a long-chain fatty acid acyl-CoA synthetase and supplying exogenously the long-chain fatty acid palmitic acid were highly effective in stimulating growth of yeast cells in the presence of high ethanol concentrations.[142]
- Ethanol stress provokes the accumulation of the amino acid L-proline, otherwise recognized as a defense mechanism against osmotic stress; disrupting a gene for proline catabolism increased proline accumulation and ethanol tolerance.[143]
- Part of the proline protective effect involves proline accumulation in internal vacuoles — heat shock responses are, however, not changed, and this clearly differentiates cellular and biochemical mechanisms in the various stress reactions.[144]

Multiple sites for how sake yeasts have adapted (and, presumably, can further adapt) to high ethanol concentrations strongly suggest that continued "blind" selection of mutants that are fitter (in an imposed, Darwinian sense) to function despite the stresses of VHG media might be fruitful in the short to medium term.[145] Eventually, however, the need to rationally change multiple sites simultaneously to continue improving the biological properties of yeast ethanologens will require a more proactive use of genomic knowledge.[146] The positive properties of sake yeasts can, however, be easily transmitted to other yeast strains to ferment high-gravity worts.[147] A compromise between "scientific" and traditional methodologies for fuel ethanol production may be to generate fusants between recombinant ethanologens and osmo- and ethanol-tolerant sake strains.

A last footnote for sake brewing (but not for bioethanol production) is that the high ethanol concentrations generated during the fermentation extract the antioxidant protein thioredoxin from the producing cells so that readily detectable levels of the compound persist in the final sake product.[148] In addition to its antioxidant function, thioredoxin is anti-inflammatory for the gastric mucosa and, by cleaving disulfide bonds in proteins, increases protein digestibility, and sake can be considered as a development stage for "functional foods."

4.4 FERMENTOR DESIGN AND NOVEL FERMENTOR TECHNOLOGIES

4.4.1 CONTINUOUS FERMENTATIONS FOR ETHANOL PRODUCTION

With enormous strides made in the development of high-alcohol fermentations prepared using high- and very-high-gravity media, and compatible with low operating temperatures, fast turnaround times, and high conversion efficiency from carbohydrate feedstocks, what has the "traditional" ethanol industry achieved with fermentor hardware and process design and control?

Beginning in the late 1950s in New Zealand, the Dominion Breweries introduced and patented a novel "continuous" brewing process in which part of the fermented beer wort was recycled back to the wort of the start of the fermentation step (figure 4.8).[149] Within two years, a rival continuous process had been announced (and patented) in Canada by Labatt Breweries, and independently conceived research was being published by the Brewing Industry Research Foundation in the United Kingdom.[150] By 1966, the concept had been simplified and reduced to a single-tower structure in which the bulk of the yeast cells were retained (for up to 400 hours) while wort might only reside in the highly anaerobic conditions in most of the tower for only 4 hours before a "beer" product emerged.*[151]

However, despite much initial enthusiasm and the evolution of technologies into a family tree of "open" (where yeast cells emerge at rapid rate from the process), "closed" (where yeast cells are mostly retained), "homogenous" (approximating a standard stirred-tank fermentor), and "heterogeneous" (with gradients of cells, substrates, and products across several vessels) systems devised for the continuous brewing of beer, very few were developed to the production scale and most became defunct. The innovations failed the test of four basic parameters of brewing practice:[150]

1. Commercial brewers offer multiple product lines and flexible production schedules to a marketplace that has become more sophisticated, discerning, and advertisement-influenced.

---

* It is sometimes asserted that brewers disliked the increased emphasis on hygiene and "scientific" precision but offering such precise, reproducible technology was a positive and populat selling point in the 1970s, explicitly stated in advertisements for major brands of nationally sold beers and lagers in the United Kingdom and elsewhere.

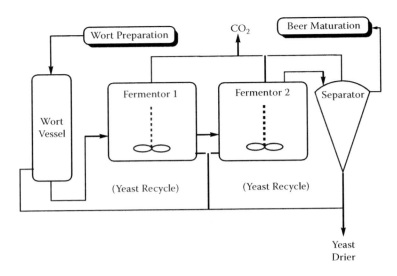

FIGURE 4.8   Continuous ethanol fermentation with yeast recycling for beer manufacture.

2. Continuous systems offer a fixed rate of beer production (and increasing flow rate tends to lead to wash out).
3. Only highly flocculent strains of yeast can be used.
4. Desired flavors and aromas cannot always be generated, whereas undesirable levels of malodorous compounds are easily formed.

Moreover, there were contemporaneous developments in brewing technology that led, for example, to the marked reduction in malting and fermentation time in batch processes, the losses during malting and wort boiling, increased α-acid content and hops (and, consequently, reductions in hop usage), and the elimination of the prolonged "lagering" storage necessary for lager beer production (an increasing popular product line worldwide). Such incremental improvements were easy to retrofit in established (often venerably old) brewing facilities, which demanded mostly tried and trusted hardware for new sites. Other than in New Zealand, continuous brewing systems have proved generally unpopular — and the rise of globally recognized brands has resulted in the backlash of movements for traditional, "real ale," locally produced (in microbreweries) options, continuing the trend of consumer choice against a perceived blandness in the output of the market leaders. Large multinationals, offering a branded product for global markets, remain the most likely to invest in "high-technology" brewing.

In contrast, industrial and fuel ethanol production is immune to such drivers, and by the late 1970s, the cold-shouldering of new technologies by potable alcohol producers had persuaded academic and research groupings of chemical engineers to focus on ethanol as an alternative fuel with a renewed interest in highly engineered solutions to the demands for process intensification, process control, and cost reduction. For example,

• Utilizing the cell recycling principle developed in New Zealand but adding a vacuum-enhanced continuous recovery of ethanol from the fermentation vessel, a high-productivity process was devised that incorporated sparging with O_2 to support active cell growth over a prolonged period and bleed out the fermented broth to withdraw nonvolatile compounds of potential toxicity to the producing cells; for a 95% ethanol product, such technology offered a 50% reduction in production costs over batch fermentation.[152]

- A single-stage gas-lift tower fermentor with a highly flocculent yeast was designed to run with nearly total retention of the cell population and generate a clear liquid effluent; analysis of the vapor-phase ethanol concentration in the headspace gas gave, via computer control, an accurate control of input and output flow rates.[153]

- A more straightforward use of laboratory-type continuous fermentors operating on the fluid overflow principle and not requiring a flocculent yeast achieved a maximal ethanol yield (89% of the theoretical from carbohydrate) at a low dilution rate (0.05/hr), showed 95% utilization of the inflowing sugars up to a dilution rate of 0.15/hr, and only suffered from washout at 0.41/hr; the fermentation was operated with a solubilized mixed substrate of sugars from Jerusalem artichoke, a plant species capable of a very high carbohydrate yield on poor soils with little fertilizer application.[154]

For large-scale fuel ethanol manufacture, however, it is another feature of the continuous process that has proved of widest application, that is, the use of multiple fermentors linked in series with (or without) the option of recycling the fermenting broth, sometimes described as "cascading" (figure 4.9).[122] Several evolutionary variants of this process paradigm have been developed by the Raphael Katzen Associates and Katzen International, Cincinnati, Ohio, from design concepts for corn ethanol dating from the late 1970s.[36] In the former USSR, batteries of 6–12 fermentors were linked in multistage systems for the production of ethanol from miscellaneous raw materials in 70 industrial centers, but, unlike the West, such advanced engineering was applied to other types of fermentation, including beer, champagne, and fruit wines, as well as fodder yeast (as a form of single-cell protein); these were supported by scientists at the All-Union Research Institute of Fermentation Products who developed sophisticated mathematical analyses, unfortunately in a literature almost entirely in Russian.[155]

Multistage fermentations have been merged with VHG media in laboratory systems with 99% of consumption of media containing up to 32% w/v glucose.[156] In such a continuously flowing system, ethanol yield from glucose increased from the first to the last fermentor in the sequence at all glucose concentrations tested (15–32% w/w); ethanol was, therefore, produced more efficiently in the later stages of the inhomogeneous set of fermentations that are set up in quasi-equilibrium within the sequence of the linked fermentation vessels. This implies mathematical modeling to control productivity will be difficult in multistage processes because parameters such as ethanol yield from sugars will be variable and possibly difficult to predict with complex feedstocks. In similar vein, a Chinese prototype with a working volume of 3.3 l achieved a 95% conversion of glucose to ethanol; although oscillations were observed in residual glucose, ethanol concentrations, and cell densities, some success was found in devising models to predict yeast cell lysis and viability loss.[157] In Brazil, attention was focused on the problems of running ethanol fermentations at high ambient (tropical)

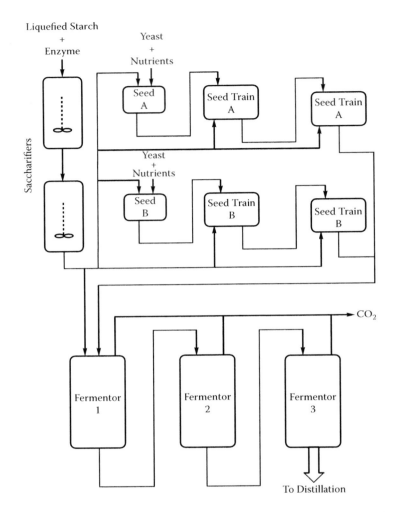

FIGURE 4.9 Cascaded saccharification, yeast propagation, prefermentation, and ethanol fermentation. (After Madson and Monceaux.[122])

temperatures, and a five-fermentor system was devised in which a temperature gradient of 8°C was established: high-biomass cultures were generated at up to 43°C, and a high-viability process with continuous ethanol production was demonstrable at temperatures normally considered supraoptimal for brewer's yeast.[158] A much simpler design, one transposed from brewery work in the United Kingdom, fermented sugarcane juice at sugar concentrations up to 200 g/l in a tower with continuous recycling of a highly flocculent yeast strain; a constant dilution rate and pH (3.3) gave outflow ethanol concentrations of up to 90 g/l with conversion rates as high as 90% of the theoretical maximum.[159] In France, the concept of separate compartments for growth and ethanologenesis was pursued in a two-stage fermentor with efficient recycling of the yeast cells; a steady state could be reached where the residual glucose concentration in the second stage was close to zero.[160]

However described — continuous, cascaded, multistage, and others — sequential fermentations offer a step change in fermentation flexibility that may be of great value for lignocellulosic feedstocks: different vessels could, for example, be operated at higher or lower pH, temperature, and degree of aeration to ferment different sugars and could even accommodate different ethanologenic organisms. Figure 4.10 is a schematic of a process with split pentose and hexose sugar streams with five fermentors in the continuous cascaded sequence, the liquid stream that moves from fermentor to fermentor being in contact with a stripping gas to remove the ethanol

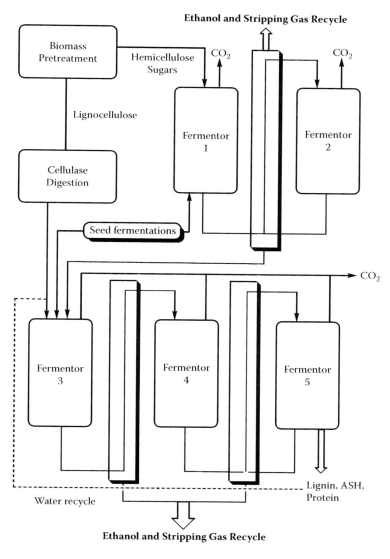

FIGURE 4.10 A cascaded fermentation system for ethanol production from a lignocellulosic biomass with continuous ethanol removal.

(thus avoiding the buildup of inhibitory concentrations); the process is a composite one based on patents granted to Bio-Process Innovation, West Lafayette, Indiana.[161–163] A low-energy solvent absorption and extractive distillation process recovers the ethanol from the stripping gas, and the gas is then reused, but the same arrangement of fermentation vessels can be used with conventional removal of ethanol by distillation from the outflow of the final fermentor. The pentose and hexose fermentations could, moreover, employ the same recombinant organism or separate ethanologens — an example of this was devised by researchers in China who selected *Pichia stipitis* as the organism to ferment pentoses in an airlift loop tower while choosing *S. cerevisiae* for glucose utilization in an overflow tower fermentor for the residual glucose; with an appropriate flow rate, a utilization of 92.7% was claimed from the sugars prepared from sugarcane bagasse.[95]

4.4.2 FED-BATCH FERMENTATIONS

Beyond ethanol production, the modern mainstream fermentation industry manufacturing primary metabolites including vitamins and amino acids, secondary metabolites (including antibiotics), and recombinant proteins almost invariably opts for fed-batch fermentation technologies and has invested much time and expertise in devising feeding strategies for carbon sources (including sugars, oligosaccharides, and amylase-digested starches) to operate at a minimum concentration of free sugars and avoid carbon catabolite repressions. Some large-scale processes are run to vanishingly small concentrations of free glucose, and the feed rate is regulated not by direct measurement but indirectly by effects of transient glucose accumulation on physical parameters such as pH (responding to acid accumulation during glucose overfeeding) or the trends in dissolved O_2 concentration.[164]

Yeast (*S. cerevisiae* in its "baker's yeast" guise) cells are one of the three principal production platforms for recombinant proteins for the biopharmaceuticals market, the others being *E. coli* and mammalian cell cultures: of the 10 biopharmaceutical protein products achieving regulatory approval in the United States or the European Union during 2004, seven were produced in mammalian cell lines, two in *E. coli*, and the tenth in *S. cerevisiae*.[165] In such cell systems, ethanol formation is either to be avoided (as a waste of glucose) or carefully regulated, possibly as a means of feedback control to a complex and variable sugar feed to high cell densities where O_2 supply is critical to maintain anabolic, biosynthetic reactions rather than simple fermentation.[166,167]

For fuel ethanol, in contrast, very high rates of sugar consumption and ethanol production are mandatory for competitive, commercial production processes. Temperature is, as always, an important parameter: under European conditions, 30°C is optimal for growth and 33°C for ethanol production.[168] Rather than aiming at microaerobic conditions, a high-aeration strategy is beneficial for stabilizing a highly viable cell mass capable of high ethanol productivity.[168] Glycerol accumulates as a major coproduct, but this waste of sugar metabolism can be minimized by several options for fermentation management:

- Avoiding high temperature "excursions" in the stage of ethanol production — glycerol formation becomes uncoupled to growth and glycerol

accumulation may (in addition to a role in osmotolerance) offer some degree of temperature protection to ethanol-forming yeast cells[168]
- High-aeration regimes greatly reduce glycerol accumulation[169]
- Maintaining a high respiratory quotient (the ratio of CO_2 produced to O_2 consumed) results in a high ethanol-to-glycerol discrimination ratio when the online data are used to feedback control the inlet sugar feed rate[170]

In addition to cane sugar, fed-batch technology has been used to produce ethanol from sugarcane molasses with an exponentially decreasing feed rate.[171] The detailed mathematical model advanced by the Brazilian authors, incorporating the feed rate profile and two further process variables, is probably too complex for small-scale fermentation sites but well within the capabilities of facilities operating (and managing) large, modern fermentors.

The bacterial ethanologen Z. *mobilis* is most productive both for biomass formation and ethanol production from glucose when feeding avoids the accumulation of high concentrations of glucose; an important finding from this work was that attempts to regulate a constant glucose concentration do not optimize the process because of the complex relationship between specific growth rate and glucose supply.[172] In general, experience in the fermentation routes to producing fine chemicals highlights the importance of accurately monitoring analyte levels inside fermentors to avoid excessive accumulations (or depletions) of key substrates and nutrients and triggering repression mechanism and the appearance of metabolic imbalances, all of which negatively impact on process economics.[173] In early 2007, a major collaboration was announced to provide automated, near real-time online monitoring of commercial fuel ethanol fermentations using proprietary methods to sample high-solids and highly viscous fermentation broths.[174] A spectrum of measurements was included in the design remit, including methodologies for ethanol, sugars, and organic acids. Because fed-batch processes are widely considered to be the favored route to the contemporary yeast cell limit of 23% v/v ethanol production, bioprocess management using more sophisticated tools to ensure a steady and slow release of glucose and other monomeric sugars are a major future milestone for industrial-scale bioethanol production.[121]

4.4.3 IMMOBILIZED YEAST AND BACTERIAL CELL PRODUCTION DESIGNS

Many of the problems encountered in regulating both growth and metabolism in actively growing, high-cell-density systems could, in principle, be avoided by immobilizing the producer cells in or on inert supports and reusing them (if their viability can be maintained) for repetitive cycles of production. This is akin to transforming whole-cell biocatalysis to a more chemically defined form known for more than a century in the fine chemicals manufacturing industry. Cost savings in the power inputs required to agitate large-volume fermentors (100,000–500,000 l), in their cooling and their power-consuming aeration, spurred the fermentation industry at large to consider such technologies, and serious attempts to immobilize productive ethanologens began in the 1970s.

Alginate beads, prepared from the hydrophilic carbohydrate polymer alginic acid (extracted from kelp seaweed), are porous, compatible with an aqueous environment, and could be loosely packed; in such a packed-bed bioreactor, perfused with a nutrient and substrate solution, a high volumetric productivity of ethanol production could be maintained for up to 12 days with a less than 10% loss of productive capacity.[175] With cells of the yeast *Kluyveromyces marxianus* supplied with a Jerusalem artichoke tuber extract, the maximum volumetric productivity was 15 times than for a conventional stirred tank fermentor. With *S. cerevisiae* in a fluidized bed contained in a closed circuit, ethanol production was possible with glucose solutions of up to 40% w/v at a relatively low temperature (18°C).[176]

As with continuous fermentation technologies, potable-alcohol manufacturers expressed interest in and actively developed demonstration facilities for immobilized yeast cells but failed to fully exploit the potential of the innovations — again because the product failed to meet organoleptic and other exacting specifications for the industry.[177] Many different matrices were investigated beyond alginate and other seaweed polymers, including ceramic materials, synthetic polymers, and stainless steel fibers, to adsorb or entrap cells in up to five different reactor geometries:

1. The packed bed, the simplest arrangement and capable of upward or downward flow of the liquid phase
2. The fluidized bed, where mixing of gas, liquid, and solid phases occurs continuously
3. Airlift and bubble column bioreactors
4. Conventional stirred tank reactors
5. Membrane bioreactors where the cells are free but retained by a semipermeable membrane

Beer producers have adopted immobilized cells to selectively remove unwanted aroma compounds from the primary fermentation product, "green" beer, and to both produce and remove aldehydes from alcohol-free and low-alcohol beers using strains and mutants of *S. cerevisiae*.[178] For the main fermentation, however, immobilized cells remain a "promising" option for further development, but the added cost of immobilization techniques are impossible to justify for commercial processes without much stronger increases in volumetric productivity.

Cost-effective options have been explored for fuel ethanol production. In Brazil, improved fermentation practices and efficiencies have taken up the slack offered by low-intensity production system, and further progress requires technological innovations; immobilizing yeast ethanologens on sugarcane stalks is an eminently practical solution to providing an easily sourced and import-substituting support matrix:[179]

• A continuously productive immobilized system with cane sugar molasses as the substrate proved to be stable for up to 60 days at a wide range of dilution rates, 0.05–3.00/hr.
• The sugar utilization efficiency was 75% with an ethanol yield of 86% of the theoretical maximum from the supplied mix of carbohydrates.
• An ethanol outflow concentration of more than 45 g/l was measured at a dilution rate of 0.06/hr, decreasing to 43 g/l at 0.11/hr.

A summary of data from other studies of immobilized cells relevant to fuel ethanol production is given in table 4.2.[180–184] The liquid phase can be recirculated in an immobilized system to maximize the utilization of the fermentable sugars; this is particularly useful to harmonize attaining maximal productivity (grams of ethanol produced per liter per hour) and maximum ethanol yield (as a percentage of the maximum theoretically convertible), with these optima occurring at different rates of flow with bacterial and yeast ethanologens.[180]

Among other recent developments, recombinant xylose-utilizing Z. mobilis cells were immobilized in photo-cross-linked resins prepared from polyethylene or polypropylene glycols and shown to efficiently utilize both glucose and xylose from acid hydrolysates of cedar tree wood, rice straw, newspaper, and bagasse — unfortunately, the pretreatment was unfashionable, that is, a concentrated sulfuric process, and difficult to precisely relate to modern trends in lignocellulosic processes.[185] Alginate-immobilized S. cerevisiae cells have also been used in conjunction with a five-vessel cascade reactor in a continuous alcohol fermentation design.[186] Sweet sorghum is a major potential source of fuel ethanol production in China where laboratory studies have progressed immobilized yeast cells to the stage of a 5-l bioreactor with stalk juice (i.e., equivalent to pressed cane sugar juice) as a carbon substrate.[187]

Insofar as immobilized cells represent a stationary phase or a population of very slowly growing cells, they may exhibit enhanced resistance to growth inhibitors and other impurities in substrate solutions that are more injurious to actively dividing cells. Immobilized cells of various microbial species, for example, are considered to be more resistant to aromatic compounds, antibiotics, and low pH, whereas immobilized S. cerevisiae is more ethanol-tolerant.[178] This opens the novel possibility that immobilized yeast could be used to ferment batches of microbial toxin-containing feedstocks: a trichothecene mycotoxin inhibits protein synthesis and mitochondrial function in S. cerevisiae and at 200 mg/l can cause cell death; conversely, fermentation is not affected and glucose metabolism may be positively redirected toward ethanol production, suggesting that trichothecene-contaminated grain could be "salvaged" for fuel ethanol manufacture.[188] Alginate-encapsulated yeast cells are certainly able to withstand toxic sugar degradation products in acid hydrolysates prepared from softwoods (mainly spruce wood); such immobilized cells could fully ferment the glucose and mannose sugars within 10 hr, whereas free cells could not accomplish this within 24 hours, although the encapsulated cells lost their activity in subsequent batch fermentations.[189]

4.4.4   CONTAMINATION EVENTS AND BUILDUP IN FUEL ETHANOL PLANTS

Contamination was a consideration in the poor take-up of continuous fermentation technologies by potable alcohol producers, especially because holding prepared wort for sometimes lengthy periods without yeast inoculation provided an excellent growth medium for adventitious microbes in the brewery.[177] With the accumulation of operating experience in fuel alcohol facilities, bacterial populations have been identified that not only reduce yield but also can prove difficult to eradicate; some bacteria (including lactic acid producers) form biofilms under laboratory conditions and can colonize many (perhaps every) available surface in complex sequences of linked fermentors and the associated pipe work.[190]

TABLE 4.2
Immobilized and Free Cell Systems for Fuel Ethanol Production: Critical Parameters for Process Efficiency

Ethanologen	Support	Carbon source	Immobilized cells			Free cells			Reference
			Maximum ethanol productivity (g/l/hr)	Maximum ethanol yield (%)	Maximum ethanol outflow (g/l)	Maximum ethanol productivity (g/l/hr)	Maximum ethanol yield (%)	Maximum ethanol outflow (g/l)	
S. cerevisiae	75% polypropylene, 20% soybean hulls, 5% soybean flour	Glucose	499	45		5	24		180
Z. mobilis	75% polypropylene, 20% soybean hulls, 5% zein	Glucose	536	50		5	26		180
S. cerevisiae	Calcium alginate	Sugarbeet molasses	10.2	83	46.2	8.7	71	39.4	181
S. cerevisiae		Seed waste		51					182
E. coli	Membrane cell recycle bioreactor	Xylose	25.2		31.5	1.8		35.2	183
E. coli	Clay brick	Xylose	4.5		21.9	1.8		35.2	183
E. coli	Calcium alginate	Xylose	2.0		37.1	1.8		35.2	183
S. cerevisiae	Calcium alginate	Glucose	2.8	38	13.1	0.3	31.2	9.8	184

When bacterial contaminants reach 10^6–10^7 cells/ml, the economic losses for ethanol production can reach 3% of volumetric capacity; if profitability is marginal, this will have a serious impact, and antibiotic regimes have been devised to pulse controlling agents through continuous processes.[191] This prophylactic approach has been applied to continuous ethanol facilities where the total losses will be greater because continuous operations have begun to dominate the larger (>40 million gallons/year) production plants — an antibiotic such as penicillin G is not metabolized and degraded by *S. cerevisiae*, and its addition rate can be poised against its expected chemical degradation at the low pH of the fermentation broth.* Outside the spectrum of known antibiotics, a useful alternative is the curious (and little known) chemical adjunct between urea and hydrogen peroxide; this bacteriocidal agent can effectively control lactobacilli in wheat mash and provides useful levels of readily assimilable nitrogen and O_2 (by enzyme-catalyzed decomposition of the peroxide) to enhance yeast growth and fermentative capacity.[192]

4.5 SIMULTANEOUS SACCHARIFICATION AND FERMENTATION AND DIRECT MICROBIAL CONVERSION

A biological solution that bypasses the severe practical difficulties posed by growing ethanologens in concentrated solutions of potentially toxic hydrolysates of lignocellulosic materials is to replace physicochemical methods of biomass substrate hydrolysis with enzymic breakdown (cellulase, hemicellulase, etc.) under milder conditions — especially if enzyme-catalyzed hydrolysis can be performed immediately before the uptake and utilization of the released sugars in a combined hydrolysis/fermentation bioprocess. Extrapolating back up the process stream and considering a totally enzyme-based hydrolysis of polysaccharides, an "ideal" ethanol process has been defined to include[193]

* Lignin removal during pretreatment to minimize unwanted solids in the substrate
* Simultaneous conversion of cellulose and hemicellulose to soluble sugars
* Ethanol recovery during the fermentation to high concentrations
* Immobilized cells with enhanced fermentation productivity

An even closer approach to the ideal would use enzymes to degrade lignin sufficiently without resort to extremes of pH to fully expose cellulose and hemicellulose before their degradation to sugars by a battery of cellulases, hemicellulases, and ancillary enzymes (esterases, etc.) in a totally enzymic process with only a minimal biomass pretreatment, that is, size reduction. No such process has been devised, but because pretreatment methods could solubilize much of the hemicellulose, two different approaches were suggested in 1978 and 1988 with either cellulolytic microbes (whole cell catalysis) or the addition of fungal cellulase and hemicellulase to the fermentation medium.[194,195] These two options have become known as direct microbial conversion (DMC) and SSF, respectively.

---

* It remains to be seen if bacterial populations secreting β-lactamase evolve in bioethanol plants, although a wide range of antibiotics could (in principle) to be used to mitigate this resistance.

DMC suffers from the biological problems of low ethanol tolerance by the (usually) clostridial ethanologens and poor ethanol selectivity of the fermentation (see section 3.3.2.5).[196] Commercialization has been slow, few studies progressing beyond the laboratory stage. The phytopathogenic* fungus *Fusarium oxysporium* is the sole nonbacterial wild-type microbe actively considered for DMC; the ability of the organism to ferment xylose as well as hexose sugars to ethanol was recognized in the early 1980s, and several strains can secrete cellulose-degrading enzymes.[197,198] Hemicellulose sugars can also be utilized in acid hydrolysates, although with low conversion efficiencies (0.22 g ethanol per gram of sugar consumed).[199] Extensive metabolic engineering of *F. oxysporium* is, therefore, likely to be required for an efficient ethanologen, and detailed analysis of the intracellular biochemical networks have begun to reveal potential sites for intervention.[200–202]

Metabolic engineering of *S. cerevisiae* to degrade macromolecular cellulose has been actively pursued by research groups in South Africa, the United States, Canada, Sweden, and Japan; fungal genes encoding various components of the cellulase complex have successfully been expressed in ethanologenic *S. cerevisiae*, yielding strains capable of utilizing and fermenting either cellobiose or cellulose.[203–206] Calculations show that, based on the growth kinetics and enzyme secretion by cellulose degraders such as *H. jecorina*, approximately 1% of the total cell protein of a recombinant cellulase-secreting *S. cerevisiae* would be required, perhaps up to 120-fold more than has been achieved to date.[207,208]

In contrast, SSF technologies were installed in North America in the early 1990s in production plants generating between 10 million and 64 million gallons of ethanol/year from starch feedstocks.[122] In addition to starch breakdown and sugar fermentation, the technology can also include the stage of yeast propagation in a cascaded multifermentor design (figure 4.11). Extensive research worldwide has defined some factors for successful process development:

- If yeasts are to act as the ethanologens, thermotolerant strains would perform more in harmony with the elevated temperatures at which cellulases work efficiently.[209,210]
- Bacteria are more readily operated in high-temperature bioprocesses, and recombinant *Klebsiella oxytoca* produced ethanol more rapidly under SSF conditions than did cellobiose-utilizing yeasts; coculturing *K. oxytoca* and *S. pastorianus*, *K. marxianus*, or *Z. mobilis* resulted in increased ethanol production in both isothermal and temperature-profiled SSF to increase the cellulase activity.[211]
- Both *K. oxytoca* and *Erwinia* species have the innate abilities to transport and metabolize cellobiose, thus reducing the need to add exogenous β-glucosidase to the cellulase complex; moreover, chromosomally integrating the *E. chrysanthemi* gene for endoglucanase and expressing the gene at a high level results in high enzyme activities sufficient to hydrolyze cellulose and even produce small amounts of ethanol in the absence of added fungal cellulase.[212,213]

---

* Fusarium wilt is a disease that affects more than a hundred species of plants; the fungus colonizes the xylem vessels, blocking water transport to the leaves.

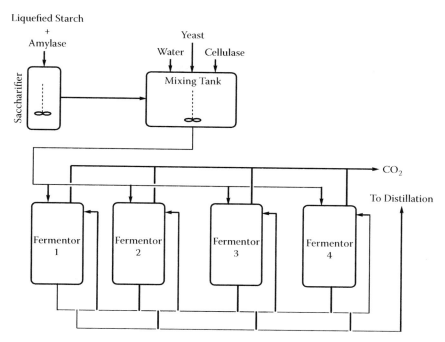

FIGURE 4.11 Simultaneous saccharification, yeast propagation, and fermentation. (After Madson and Monceaux.[122])

- Although high ethanol concentrations strongly inhibit fermentations with recombinant *E. coli* and glucose or xylose as the carbon substrate, SSF with cellulose and added cellulase showed a high ethanol yield, 84% of the theoretical maximum.[214]
- Simulations of the SSF process to identify the effects of varying the operating conditions, pretreatment, and enzyme activity highlight the importance of achieving an efficient cellulose digestion and the urgent need for continued R&D efforts to develop more active cellulase preparations.[215]
- Reducing the quantity of cellulase added to ensure efficient cellulose digestion would also be beneficial for the economics of the SSF concept; adding nonionic surfactants, polyethylene glycol, and a "sacrificial" protein to decrease nonproductive absorption of cellulase to lignin binding sites have also been demonstrated to increase cellulase action so that cellulose digestion efficiency can be maintained at lower enzyme:substrate ratios.[8,216]

The importance of the quantity of cellulase added was underlined by a Swedish study that showed that reducing the enzyme loading by 50% actually increased the production cost of ethanol in SSF by 5% because a less efficient cellulose hydrolysis reduced the ethanol yield.[217] At low enzyme loading, there are considerable advantages by growing the yeast inoculum on the pretreated biomass material (barley straw); the conditioned cells can be used at a reduced concentration (2 g/l, down from 5 g/l), and with an increased solids content in the SSF stage.[218]

The cost of commercially used fungal cellulase has decreased by over an order of magnitude because of the efforts of enzyme manufacturers after 1995.[219] Multiple efforts have been made to increase the specific activity (catalytic efficiency) of cellulases from established and promising novel microbial sources (see section 2.4.1), and recently, the National Center for Agricultural Utilization Research, Peoria, Illinois, has focused on the cellulase and xylanase activities from the anaerobic fungus *Orpinomyces*, developing a robotic sampling and assay system to improve desirable gene mutations for enzymic activity.[220–222] Inserting genes for components of the cellulase complex into efficient recombinant ethanol producers has also continued as part of a strategy to reduce the need to add exogenous enzymes; such cellulases can be secreted at levels that represent significant fractions of the total cell protein and increase ethanol production capabilities.[223–227] This is of particular importance for the accumulation of high concentrations of ethanol because ethanol at more than 65 g/l inhibits the fungal (*H. jecorina*) cellulase commonly used in SSF studies.[228]

SSF has been shown to be superior to independent stages of enzymic hydrolysis and fermentation with sugarcane bagasse, utilizing more of both the cellulose and hemicelluloses.[229] A continued industry-wide commitment to SSF is evident in the numbers of publications on SSF technologies applied to ethanol production with a wide variety of lignocellulosic feedstocks (table 4.3).[32, 230–238] Issues of process economics are discussed in chapter 5. Prominent in the list of lignocellulosic feedstocks in table 4.3 is corn stover, a material that has the unique distinction of having a specific biocatalyst designed for its utilization.[239] This fusion of the biochemical abilities of *Geotrichum candidium* and *Phanerochaete chrysosporium* points toward a long-term option for both SSF and DMC, that of harnessing the proven hypercapabilities of some known microbes to degrade lignocellulose (see section 2.4.1) and converting them to ethanologens by retroengineering into them the ethanol biochemistry of *Z. mobilis* (see section 3.3.2). Before then, attempts will without doubt continue to introduce fungal genes for starch degradative enzymes into candidate industrial ethanologens and explore the possible advantages from combining genetic backgrounds from two microbes into a single hybrid designed for high amylase secretion.[240–242] On a parallel track, commercial use of food wastes such as cheese whey, a lactose-rich effluent stream, has prompted the construction of strains with β-galactosidase to hydrolyze lactose extracellularly and use both the released glucose and galactose simultaneously for ethanol production under anaerobic conditions.[243]

As a final option — and one that mimics the evolution of natural microbial communities in soils, forest leaf litter, water-logged areas, and stagnant pools — cocultivation of a good ethanologen together with an efficient secretor of enzymes to degrade polymeric carbohydrates and/or lignocelluloses is a route avoiding introducing genetically manipulated (GM) organisms and could be adapted to continuous technologies if a close control of relative growth rates and cell viabilities can be achieved. One or more of the microbial partners can be immobilized; table 4.4 includes two examples of this approach together with the cocultivation of different ethanologens to ferment glucose/xylose mixtures and pretreated lignocellulosics.[244–249]

TABLE 4.3

Simultaneous Saccharification and Fermentation Applied to Fuel Ethanol Production from Lignocellulosic Feedstocks

Ethanologen	Lignocellulosic material	County of origin	Year of publication	Reference
S. cerevisiae + *Pachysolen* *tannophilus*	Rice straw	India	1995	230
S. cerevisiae	Hybrid poplar, switchgrass, corn stover	United States	1997	231
P. tannophilus	Timothy grass, alfalfa, reed canary grass, corn stalks, barley straw	Canada	1998	232
Kluyveromyces *marxianus*	Sugarcane leaves, *Antigonum* *leptopus* leaves	India	2001	233
S. cerevisiae	Corn stover	United States	2003	234
S. cerevisiae	Corn stover	Hungary	2004	235
S. cerevisiae	Switchgrass, poplar	Taiwan	2005	236
E. coli (KO11)	Corn stover	United States	2005	32
S. cerevisiae	Corn stover	Sweden	2006	237
S. cerevisiae	Corn stover	United States	2006	238

TABLE 4.4

Cocultivations of Ethanologenic and Ethanologenic Plus Enzyme-Secreting Microbes for DMC/SSF Processes

Ethanologen	Immobilized?	Enzyme secretor	Immobilized?	Reference
S. cerevisiae + Candida *shehatae*	−	*Sclerotum rolfsii*	−	244
S. cerevisiae + Candida *shehatae*	+	None (glucose and xylose mix)	−	245
S. cerevisiae + Pichia *stipitis*	−	None (glucose and xylose mix)	−	246
S. cerevisiae	+	*Aspergillus* *awamori*	−	247
S. cerevisiae + Pachysolen *tannophilus + E. coli*	−	None (softwood hydrolysate)	−	248
S. cerevisiae + Candida *shehatae*	+	None (glucose and xylose mix)	−	249

4.6 DOWNSTREAM PROCESSING AND BY-PRODUCTS

4.6.1 ETHANOL RECOVERY FROM FERMENTED BROTHS

The distillation of ethanol from fermented broths remains the dominant practice in ethanol recovery in large and small ethanol production facilities.[250] Other physical techniques have been designated as having lower energy requirements than simple distillation, and some (vacuum dehydration [distillation], liquid extraction, super-critical fluid extraction) can yield anhydrous ethanol for fuel purposes from a dilute aqueous alcohol feed (figure 4.12).[251] Only water removal by molecular sieving* has, however, been successful on an industrial scale, and all new ethanol plants are built with molecular sieve dehydrators in place.[252]

Nevertheless, the economic costs of dehydration are high, especially when anhydrous ethanol is to be the commercial product (figure 4.12). In the early 1980s, the energy requirements were so high that the practical basis for fuel ethanol production was questioned because the energy required for distillation approximated the total combustion energy of the alcohol product.[253,254] The investment costs of rivals to distillation were, however, so high (up to 8.5 times that of conventional distillation) that little headway was made and attention was focused on improving

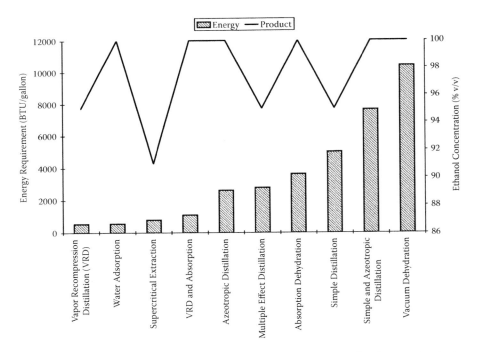

FIGURE 4.12 Energy requirement and ethanol product concentration from technologies developed for separation of ethanol-water mixtures. (Data from Sikyta.[254])

---

* Synthetic zeolite resins are crystalline lattices with pore sizes of 0.3 nm, sufficiently small to allow the penetration of water molecules (0.28 nm in diameter) but exclude ethanol (molecular diameter 0.44 nm).

process efficiencies and energy cycling with the development of low-energy hydrous ethanol distillation plants with 50% lower steam-generating requirements.[252]

The economics of downstream processing are markedly affected by the concentration of ethanol in the fermented broth; for example, the steam required to produce an ethanol from a 10% v/v solution of ethanol is only 58% of that required for a more dilute (5% v/v) starting point, and pushing the ethanol concentration in the fermentation to 15% v/v reduces the required steam to approximately half that required for low conversion broth feeds.[253]

4.6.2 CONTINUOUS ETHANOL RECOVERY FROM FERMENTORS

Partly as another explored route for process cost reduction but also as a means to avoid the accumulation of ethanol concentrations inhibitory to cell growth or toxic to cellular biochemistry, technologies to remove ethanol in situ, that is, during the course of the fermentation, have proved intermittently popular.[254] Seven different modes of separation have been demonstrated in small-scale fermentors:

- A volatile product such as ethanol can be separated from a fermentation broth under vacuum even at a normal operating temperature; a system with partial medium removal and cell recycling was devised to minimize the accumulation of nonvolatile products inhibitory to yeast growth and productivity.[152]
- If the fermentor is operated normally but the culture liquid is circulated through a vacuum chamber, the ethanol formed can be removed on a continuous basis; this arrangement avoids the need to supply O_2 to vessels maintained under vacuum.[255]
- Solvent extraction with a long chain alcohol (*n*-decanol) with immobilized cells of *S. cerevisiae*; up to 409 g/l of glucose (from glucose syrup) could be metabolized at 35°C.[256]
- As with water removal from concentrated ethanol, ethanol can be selectivity adsorbed by different types of resins with hydrophobic surfaces, including cross-linked divinylbenzene polystyrene resins widely used in modern chromatographic separations of alcohols, sugars, and carboxylic acids; such resins work efficiently with ethanol at low ethanol concentrations, and the ethanol can be desorbed with warm dry N_2 gas at 60–80°C.[257,258]
- Hollow-fiber microfiltration is effective for ethanol and other small-molecule products (such as lactic acid) but is slow and difficult to sterilize.[259]
- In membrane pervaporation, the cells are retained by a semipermeable membrane while a partial vacuum is applied to the permeate side; ethanol concentrations could be maintained below 25 g/l for five days while a concentrated ethanol efflux stream of 17% w/v was achieved.[260] Polyvinyl alcohol membranes operate better at elevated temperature, and this suggests that thermophilic ethanologens would be very suitable in a membrane pervaporative process.[261]
- Gas stripping of ethanol can be effected in an air-lift fermentor, a type of vessel originally developed for viscous microbial fermentation broths but also used for some of the more fragile and shear-sensitive mammalian cells in culture; this is another example of a technology that would inevitably

work better with a thermophilic ethanologen and an elevated fermentation temperature.[262] Alternatively (and more economically, with reduced power consumption for gas volume flow), the fermentation broth is circulated through an inert packed column and continuously sparged with a stripping gas (see figure 4.10) — such arrangements can result in highly stable continuous fermentations (for >100 days), with near-theoretical yields of ethanol from concentrated glucose solutions (560 g/l) in corn steep water to provide nutrients.[263,264]

How many (if any) of these advanced downstream technologies become adopted for industrial use will depend heavily on their economics — ethanol stripping is, for example, assessed at providing a significant cost savings for fuel ethanol production from cornstarch.[265] With lignocellulosic substrates being used more widely, especially in developing economies, a much simplified technology can provide surprisingly elegant solutions. Solid-state fermentations* have long been used for fermented foods and sake but can easily be adapted to manufacture (under more stringent conditions and with a reduced labor intensity) many fine chemicals and enzymes.[266] A continuous process has been engineered to process and ferment feedstocks such as fodder beet and sweet sorghum in a horizontal tubular bioreactor, the fermenting material (with a low moisture content) moved along with the aid of a spiral screw.[267,268] Some ethanol volatilization will occur at any temperature above ambient (caused by the fermentation process), but the bulk of the product could be recovered by a gas or air flowing through the container before the ethanol is condensed and transferred to a final dehydration step (as in the gas stripping technology). Although originally devised for farm-scale facilities (by the Alcohol Fuel Research Laboratory, South Dakota State University), this solid-phase bioprocess yielded 87 l of ethanol/tonne of feedstock and was sufficiently productive to allow distillation from 8% v/v outputs. Echoing some of the discussion in chapter 1, the net energy balance (see section 1.6.1) was calculated to be unambiguously positive for fodder beet (2.11 for pasteurized pulped beet fodder, 3.0 for unpasteurized substrate), although much less persuasive for sweet sorghum (1.04 and 1.30, respectively) but was — even in 1984, when world oil prices were unpredictable and high after the price inflation of the 1970s — uncompetitive with then current gasoline prices (figure 1.3).

4.6.3   Solid By-Products from Ethanol Fermentations

The solids remaining at the end of the fermentation (distillers dried grain with solids, or DDGS: see figure 1.21) are a high-protein animal feed — a saleable by-product that has been suggested to be so commercially desirable that reduced ethanol yields could be tolerated to support its increased production, although, in practice, high-sugar residues pose severe practical difficulties to DDGS drying and processing.[122] The rapid rise of ethanol production from cornstarch has, however, demanded some remarketing of this coproduct:[269]

---

* Often (but confusingly for discussions of biofuels) abbreviated to SSF, although arguably with a claim to prior use in this area of biotechnology.

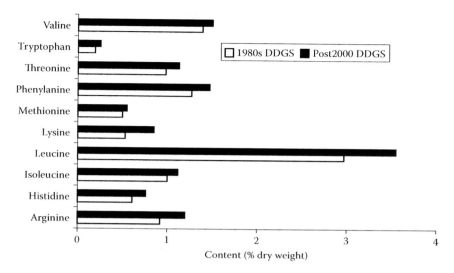

FIGURE 4.13 Essential amino acid content of DDGS: changes in U.S. compositions from 1980s to the present. (Data from Jacques.[269])

- The product is less dark because sugars are more efficiently fermented and less available to react chemically and caramelize in the dried product.
- The essential amino acid contents are higher (figure 4.13).
- Although ruminant animals can certainly benefit from feeding with DDGS, pigs are geographically much closer to ethanol plants in the midwestern United States.

Reducing phosphate content would widen the use of DDGS by addressing animal waste disposal issues, and the development of more efficient methods for removing water in the preparation of the DDGS could greatly reduce processing costs.[270] Adding on a second fermentation (or enzymic biotransformation), a dry-grind processing to generate plant oils and a higher-value animal feed from the DDGS, and separating more useful and saleable fine chemicals from the primary fermentation would increase the total mass of recovered bioproducts to the maximum achievable (figure 4.14).[271] Pricing is crucial because the increased supply of DDGS is likely to significantly reduce its market price, and its alternate use as the feedstock for further ethanol production itself has been worthy of investigation: steam and acid pretreatments can convert the residual starch and fiber into a substrate for yeast-based ethanol production with a yield 73% of the theoretical maximum from the glucans in the initial solids.[272]

A much simpler option is to realize the potential in the fermented solids to provide nutrients and substrates for a new round of yeast (or other ethanologen) growth and ethanol production: such spent media ("spent wash," stillage, or vinasse) can be recycled in the process known as "backsetting," found to be beneficial for yeast growth and a practical means of reducing water usage in a fuel alcohol facility.[273] Backsetting is not without its accepted potential drawbacks, including the accumulation of toxic nonvolatiles in the fermentor, increased mash viscosity, and dead cells causing problems with

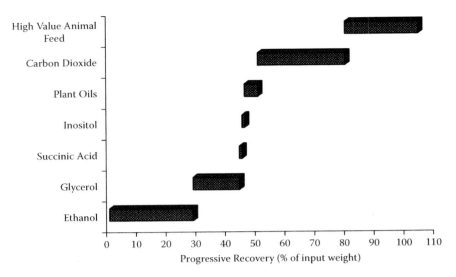

FIGURE 4.14 Projected recovery of product and coproducts from the ethanol fermentation of corn starch. (Data from Dawson.[271])

viability measurements, but as a crude means of adapting the fermentation to a semi-continuous basis, it has its advocates on both environmental and economic grounds. Furthermore, a study by Novozymes demonstrated that mixtures of fungal enzymes could decrease vinasse viscosity and liberate pentose sugars from soluble and insoluble arabinoxylans that would be suitable for fermentation by a suitable pentose-utilizing ethanologen.[274–277] Portuguese work has also shown that such a pentose-rich product stream can be the starting point for the fermentative production of xylitol (a widely used noncalorific sweetener) and arabinitol by the yeast *Debaryomyces hansenii*.[278,279] As a support for the immobilization of yeast cells, brewer's spent grains were a very effective means of supplying "solid-phase" biocatalysts for ethanol production from molasses, and the solids from bioethanol plants could serve a parallel function.[280]

Brazil has, by far, the longest continuous history of devising methods for economically viable disposal for vinasse and the solid waste product (bagasse), especially because neither had at times been considered to be saleable and both could even represent negative value as incurred disposal costs:[281]

- Bagasse combustion in steam turbines generates electricity at 1 MWh/m^3 of produced alcohol.
- Anaerobic digestion of vinasse can produce enough biogas for 0.5 MWh/m^3 of produced alcohol, and both processes have been applied at full scale at distilleries.
- Laboratory studies show that anaerobic digestion would also be beneficial for bagasse, increasing the power output to 2.25 MWh/m^3 of produced alcohol if the nonbiodegradable residue is burned.
- The total potential power generation from biogas and combustion routes would be equivalent of 4% of national power demand.

- Important for minimizing fertilizer use, material dissolved in the digestion wastewater represents 70% of the nutrient demand of sugarcane fields.

4.7 GENETIC MANIPULATION OF PLANTS FOR BIOETHANOL PRODUCTION

Iogen's choice of a wheat straw feedstock was made on practical and commercial grounds from a limited choice of agricultural and other biomass resources in Canada available on a sufficiently large scale to support a bioethanol industry (section 4.1). Wheat, as a monoculture, is inevitably subject to crop losses arising from pathogen infestation. Modern biotechnology and genetic manipulation offer novel solutions to the development of resistance mechanisms as well as yield improvements through increased efficiency of nutrient usage and tolerance to drought, and other seasonal and unpredictable stresses. But the deliberate release of any such genetically modified (GM) species is contentious, highly so in Europe where environmental campaigners are still skeptical that GM technologies offer any advantage over traditional plant breeding and are without the associated risks of monopoly positions adopted by international seed companies, the acquisition of desirable traits by "weed" species, and the horizontal transfer of antibiotic resistance genes to microbes.[282] The positive aspects of plant biotechnology have, in contrast, been succinctly expressed:

> Genetic transformation has offered new opportunities compared with traditional breeding practices since it allows the integration into a host genome of specific sequences leading to a strong reduction of the casualness of gene transfer.[283]

Because large numbers of insertional mutants have been collected in a highly manipulable "model" plant species (*Arabidopsis thaliana*), it has been possible for some years to inactivate any plant gene with a high degree of accuracy and certainty.

4.7.1 ENGINEERING RESISTANCE TRAITS FOR BIOTIC AND ABIOTIC STRESSES

After herbicide resistance in major crop species, the first target area for GM crop development was that of protection against plant pests.[283] Although tolerance of modern herbicides is usually located in the amino acid sequences of a handful of target genes in biosynthetic pathways, plants have multiple inducible mechanisms to fight back against microbial pathogens:[284]

- Permanent ("constitutive") expression of genes encoding chitinases or a ribosomal-inactivating protein confers partial protection against fungal attack.
- Enhancing lignin deposition in response to fungal or bacterial invasion is a possible multigene defense.
- Overexpressing genes encoding biosynthesis of phytoalexin antibiotics have been explored together with the introduction of novel phytoalexin pathways by interspecies gene transfer.
- Specific natural plant disease resistance genes are beginning to be identified and cloned for transfer into susceptible plants.

Wheat is prone to attack by the rust pathogens, *Puccinia graminis* and *P. tritici*; stem and leaf rusts are considered to be major constraints to wheat production world-wide.[285] More immediately alarming from the perspective of Iogen's dependence on wheat biomass is that new and highly infectious variants of the pathogen have been noted in Africa for some years, with newspaper reports in the first quarter of 2007 describing its spread into Asia. A single gene (*Sr2*) has been identified as a broad-spectrum resistance locus for more than 80 years; recently, this gene (or two tightly linked genes) confer resistance and the associated dark pigmentation traits, pseudo-black chaff.[286] Stem rust-susceptible barley has been transformed into a resistant form by an *Agrobacterium* plasmid containing the barley resistance gene, *Rpg1*; a single copy of the gene is sufficient to confer resistance against stem rust.[287]

Of the abiotic stresses that plants experience, drought is a serious limiting factor on growth and productivity even in the Northern Hemisphere, and cereal crops are highly prone to fluctuating yield depending on seasonal rainfall and average tempera-tures in the growing season; to be dependent on a monoculture crop such as wheat (for starch or straw) runs the risk of uncertain prices as well as variable feedstock availability. With the evolution of domesticated cereal species over millennia, genetic diversity has been lost; using the natural genetic diversity of wild species is an invalu-able resource because wild types harbor very broad ranges of tolerance characteris-tics; other exploitable traits include those for salt tolerance (allowing saline water to be the source of irrigation) and, especially if global temperatures increase because of global warming, flowering times and other growth parameters more typical of Medi-terranean regions for transfer to cultivars grown further north.[288] Drought appears to exert physiological effects via oxidative stress signaling pathways, a property shared with freezing, heat, and salinity stresses; protein kinases are often associated with signaling pathways and expression of a protein kinase gene (*NPK1*) from tobacco in maize protects kernel weights when the water supply is reduced.[289]

4.7.2 BIOENGINEERING INCREASED CROP YIELD

How to define *yield* with energy crops? This is both more flexible and less precise than with, for example, crop yield as measured by grain size or weight per plant or per unit area. For a feedstock such as wheat straw, increasing plant biomass accumulation per plant may suffice, and this implies a change in nutrient utilization or absorption from the soil, but this may paradoxically reverse the trend toward "dwarf" crops (i.e., more grain, less stem/stalk) that has been emphasized in the green revolution type of agronomy.[290] In principle, simply achieving higher leaf, stalk, and stem mass per plant is a straightforward target that is not limited by considerations of morphology for crops dedicated to energy supply and/or biofu-els. Mass clonal propagation of commercial trees is now well advanced, and gene transfer technologies have been devised for conifers and hardwoods, that is, forest biotechnology has emerged (probably irreversibly) out of the laboratory and into the global ecosystem.[291]

Traditional breeding and marker-assisted selection can identify genes involved in nutrient use efficiency that can then be used in gene transfer programs to improve features of plant nutrition — for crop plants in intensive agriculture, nitrogen assimilation and recycling within the plant over the stages of plant development are crucial.[292] Although plant biochemistry is increasingly well understood at the molecular level, what is much less clear is how to accurately modulate gene expression (single genes or whole pathways) to achieve harvestable yield increases.[290] As with other "higher" organisms, a greater understanding of how regulatory circuits and networks control metabolism at organ and whole-plant levels as an exercise in systems biology will be necessary before metabolic engineering for yield in crop plants becomes routine.[294]

Nevertheless, successes are now being reported for "gene therapy" with the goal of improving the assimilation of CO_2 into biomass, however this is defined:

- Transgenic rice plants with genes for phospho*enol*pyruvate carboxylase and pyruvate, *ortho*phosphate dikinase from maize, where the two enzymes are key to high photosynthetic carbon fixation under tropical conditions ("C4 metabolism"), increases photosynthetic efficiency and grain yield by up to 35% and has the potential to enhance stress tolerance.[295]
- Once CO_2 is "fixed" by green plants, some of the organic carbon is lost by respiratory pathways shared with microorganisms, and there are several reports that partially disabling the oxidative pathways of glucose metabolism enhances photosynthetic performance and overall growth: for example, in transgenic tomato plants with targeted decreases in the activity of mitochondrial malate dehydrogenase, CO_2 assimilation rates increased by up to 11% and total plant dry matter by 19%.[296]
- Starch synthesis in developing seeds requires ADP-glucose phosphorylase; expressing a mutant maize gene for this enzyme in wheat increases both seed number and total plant biomass, these effects being dependent on increased photosynthetic rates early in seed development.[297]

Much attention has been given to improving the catalytic properties of the primary enzyme of CO_2 fixation, ribulose 1,5-bisphosphate carboxylase (Rubisco), the most abundant single protein on Earth and one with a chronically poor kinetic efficiency for catalysis; although much knowledge has been garnered pertaining to the natural variation in rubisco's catalytic properties from different plant species and in developing the molecular genetics for gene transfer among plants, positive effects on carbon metabolism as a direct result of varying the amounts of the enzyme in leaves have proved very slow to materialize.[298] A more radical approach offers far greater benefits: accepting the inevitable side reaction catalyzed by Rubisco, that is, the formation of phosphoglyceric acid (figure 4.15), transgenic plants were constructed to contain a bacterial pathway to recycle the "lost" carbon entirely inside the chloroplast rather than the route present in plant biochemistry that involved the concerted actions of enzyme in three plant cell organelles (chloroplast, mitochondrion, and peroxisome); transgenic plants grew faster, produced more biomass (in shoot and roots), and had elevated sugar contents.[299]

Is there an upper limit to plant productivity? A temperate zone crop such as wheat is physiologically and genetically capable of much higher productivity and efficiency of converting light and CO_2 into biomass than can be achieved in a "real"

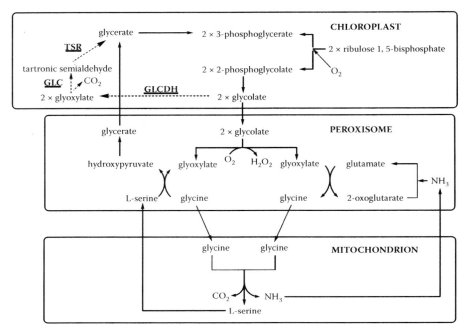

FIGURE 4.15 Intracellular C and N traffic for photorespiration transgenic bacterial glycolate catabolic pathways; the three bacterial gene-encoded enzymes (underlined) are GLCDH, glycolate dehydrogenase; GCL, glyoxylate carboxyligase; TSR, tartronic semialdehyde reductase. (Modified from Kebeish et al.[299])

environment, that is, in hydroponics and with optimal mineral nutrition; again, most studies focus on yield parameters such as grain yield (in mass per unit area) but total leaf mass (a component of straw or stover) will also increase under such ideal conditions.[300] The higher the light intensity, the greater the plant response, but natural environments have measurable total hours of sunshine per year, and the climate imposes average, minimum, and maximum rainfall and temperatures; supplementary lighting is expensive — but, given unlimited renewable energy resources, substantial increases in plant productivity are theoretically possible. From a shorter-term perspective, however, the choice of biomass refuses to be erased from the agenda, and agricultural wastes simply cannot compete with the "best" energy crops. For example, using biofuel yield as the metric, the following ranking can be computed from relative annual yields (liters per hectare):[301]

Corn stover (1.0) < poplar (2.9) < switchgrass (3.2) < elephant grass (*Miscanthus*, 4.4)

Rather than mobilizing molecular resources against the vagaries of climate, concentrating effort on maximizing biomass supply from a portfolio of crops other than those most presently abundant would pay dividends. Overreliance on a few species (vulnerable to pests and climatic variability) would be minimized by expanding the range of such crops.

4.7.3 Optimizing Traits for Energy Crops Intended for Biofuel Production

With a lignocellulosic platform for bioethanol production, one obvious target is (as just discussed) the management of energy crop productivity to maximize the capture of solar energy and atmospheric CO_2; the chemical composition of the biomass is, however, of great practical significance for the industrial bioprocessing of feedstocks:[302,303]

1. Developing crop varieties with reduced lignin contents (especially with softwoods)
2. Crops with increased cellulose and, arguably, hemicellulose contents
3. Plants with the increased capability to degrade cellulose, hemicellulose, and lignin — after harvest (i.e., in a controlled manner capable of minimizing biomass pretreatment)

Of these, modifying lignin content has been the most successful — classical genetics suggests that defining quantitative traits and their genetic loci is relatively easy, and (even better) some of these loci are those for increased cellulose biosynthesis.[304] As collateral, there is the confidence-building conclusion that lignin contents of commercial forest trees have been reduced to improve pulping for the paper industry; the genetic fine-tuning of lignin content, composition, or both is now technically feasible.[305]

Reductions in plant lignin content have been claimed using both single- and multiple-gene modifications (figure 4.16):

- Down-regulating either of the initial two enzymes of lignin biosynthesis, phenylalanine ammonia lyase and cinnamate 4-hydroxylase (C4H), reduces lignin content and impairs vascular integrity in the structural tissues of plants.[305]
- Deletion of the second activity of the bifunctional C4H enzyme, coumarate 3-hydroxylase, results in reduced lignin deposition.[306]
- Later enzymes in the lignin pathway were considered to be less amenable for inhibiting lignification but multiple-gene down-regulation could be effective.[307,308]
- Inactivating O-methyltransferase activity with an aspen gene incorporated into a transmissible plasmid in the antisense orientation reduced lignin formation in *Leucaena leucocephata** by 28%, increased monomeric phenolic levels, and increased the cellulose content by 9% but did not visibly affect the plant phenotype.[309]

Are "lignin-light" plants biologically viable for commercial cultivation? Altered stem lignin biosynthesis in aspen has a large effect on plant growth, reducing total leaf area and resulting in 30% less total carbon per plant; root growth was also

---

* This thornless tree, known by many different names but including false koa, horse tamarind, or jumbie bean, has been included in the list of the 100 worst invaders, forming dense thickets and difficult to eradicate once established. Being "corralled" in energy plantations may be an appropriate use for it — but adjacent farmers and horticulturists may vehemently disagree.

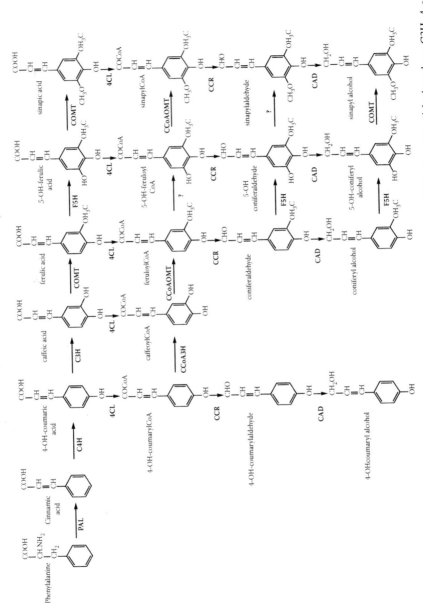

FIGURE 4.16 Outline of biosynthesis of lignin precursors: PAL, phenylalanine ammonia lyase; C4H, cinnamate 4-hydroxylase; C3H, 4-coumarate 3-hydroxylase; COMT, caffeate O-methyltransferase; CCoAOMT, caffeoyl-CoA O-methyltransferase; CCR, cinnamoyl-CoA reductase; CAD, cinnamyl alcohol dehydrogenase; F5H, ferulate 5-hydroxylase. (After Hertzberg et al.[312])

compromised.[310] Vascular impairment can lead to stunted growth.[307] On the other hand, aspen wood in reduced-lignin transgenics was mechanically strong because less lignin was compensated for by increased xylem vessel cellulose.[308] Smaller plants may be grown, as energy crops, in denser plantations; alternatively, plants with reduced stature may be easier to harvest, and various practical compromises between morphology and use can be imagined — this can be seen as analogous to the introduction of dwarfing rootstocks for fruit trees that greatly reduced plant height and canopy spread and facilitated manual and mechanical harvesting.

Also without obvious effects on plant growth and development was the introduction and heterologous expression in rice of the gene from *Acidothermus cellulolyticus* encoding a thermostable endo-1,4-β-glucanase; this protein constituted approximately 5% of the total soluble protein in the plant and was used to hydrolyze cellulose in ammonia fiber explosion-pretreated rice and maize.[311] More ambitiously, enzymes of polysaccharide depolymerization are being actively targeted by plant biotech companies for new generations of crops intended for biofuels. A large number of genes are under strict developmental stage-specific transcriptional regulation for wood formation in species such as hybrid aspen; at least 200 genes are of unknown function, possibly undefined enzymes and transcription factors, but this implies that heterologous glucanases and other enzymes could be produced during plant senescence to provide lignocellulose processing in plants either before the preparation of substrates for conventional ethanol fermentation or in solid-phase bioprocesses (section 4.6.2).[312]

4.7.4 GENETIC ENGINEERING OF DUAL-USE FOOD PLANTS AND DEDICATED ENERGY CROPS

The commercialization of GM technologies for staple food crops such as wheat has faced the obstacles of social and market opposition and resistance.[313] Fortunately (or fortuitously), gene transformation techniques for improving wheat yield have been challenging for plant biotechnology: bread wheat (*Triticum aestivum* L.) has one of the largest and most complex plant polyploid genomes, 80% of which consists of noncoding sequences, deriving from three ancestral genomes; new initiatives to analyze the minority expressed portion of the wheat genome are ongoing, but the structural complexity of the genome has been of enormous value to agronomy over millennia because major chromosomal rearrangements and deletions are well tolerated.[314]

Public perceptions may be more favorable to GM technologies applied to dedicated energy crops, although even here the fear is that of the introduction and spread of unwanted genes into natural populations with consequences that are difficult to accurately predict. Geographical isolation of energy crops (as of GM plants designed to synthesize high-value biopharmaceuticals) is one extreme solution but flies in the face of the limited land availability for nonfood crops. Even if the entire U.S. corn and soybean crops were to be devoted to bioethanol and biodiesel production (with complete elimination of all direct and indirect food uses), only 5.3% of the 2005 gasoline and diesel requirement could be met.[315] Expressed another way, an area of land nearly 20-fold larger than that presently used for corn and soybean cultivation would be required for bioethanol and biodiesel energy crops. Concerns relating to land delineation are highly probable even if a wheat straw/corn stover bioeconomy is used as the main supply of feedstocks for bioethanol production.[316]

Equally inevitable, however, is that GM approaches will be applied to either dual food-energy crops or to dedicated energy crops such as fast-growing willow and switchgrass. The USDA-ARS Western Regional Research Center, Albany, California, has created a gene inventory of nearly 8,000 gene clusters in switchgrass, 79% of which are similar to known protein or nucleotide sequences.[317] A plasmid system has also been developed for the transformation of switchgrass with a herbicide resistance gene.[318] Grasses share coding sequences of many of their genes: the sugarcane genome shows an 81.5% matching frequency with the rice genome and even a 70.5% matching frequency with the genome of the "lower" plant, *Arabidopsis thaliana*.[319] The risk of gene transfer across species and genera in the plant kingdom is, therefore, very real.

A selection of patents relevant to transgenic crops and other plant genetic manipulations for bioethanol production is given in table 4.5.

4.8   A DECADE OF LIGNOCELLULOSIC BIOPROCESS DEVELOPMENT: STAGNATION OR CONSOLIDATION?

Iogen's process in Canada and (insofar as process details can be assessed) from Spanish and other imminent facilities use only a small fraction of the technologies that have been devised for ethanol production from lignocellulosic sources; patents and patent applications covering the last 30 years are exemplified by those listed in table 4.6.

Can any predictions be made about the take-up of innovative process methodologies in the first decade of commercial fuel ethanol production from biomass? Retracing steps across the 10–12 years since the manuscripts were prepared for the 1996 *Handbook on Bioethanol* publication reveals a significant divergence:

- "Ethanol from lignocellulosic biomass (bioethanol) can now be produced at costs competitive with the market price of ethanol from corn."[320]
- "Although bioethanol production is competitive now for blending with gasoline, the goal of the Department of Energy Biofuels Program is to lower the cost of production to $0.18/l, which is competitive with the price of gasoline from petroleum at $25/bbl with no special tax considerations."[321]
- "Despite high capital costs, Charles Wyman of the University of California, Riverside, US, told *Chemistry World* that cellulose technology could be commercialized now if investors would take the risk or government provide more policy and financial assistance. That breakthrough may happen in the next five years."[322]

Despite crude oil prices being considerably higher than $25/bbl since 2002 (figure 1.3), and despite two successive Presidential State of the Union addresses outlining a consistent vision for alternative fuels in the United States, only tax breaks may accelerate cellulosic ethanol production in an uncertain market, or the equivalence between corn and cellulosic ethanol production costs has been broken, or the 1990s may have seen too many overly optimistic forward statements.

Or, the successes of corn and sugar ethanol confused rather than illuminated. Together with biodiesel (a product with a very straightforward technology platform,

TABLE 4.5
Patent Applications and Patents Awarded for Plant Genetic Applications Relevant to Bioethanol Production

Date of filing or award	Title	Applicant/assignee	Patent/application
September 12, 1997	Transgenic plants expressing cellulolytic enzymes	Syngenta AG, Basel, Switzerland	EP 1 574 580 A2
July 23, 1999	Expression of enzymes involved in cellulose modification	Calgene, Davis, CA	WO 00/05381
January 11, 2000	Expression of enzymes involved in cellulose modification	Calgene, Davis, CA	US 6013860
November 16, 2001	Manipulation of the phenolic acid content and digestibility of plant cell walls …	Genencor International, Palo Alto, CA	WO 02/068666 A1
April 11, 2002	Regulation and manipulation of sucrose content in sugarcane	F.C. Botha and J.H. Groenewald	US 2002/0042930 A1
August 27, 2002	Self-processing plants and plant parts	Syngenta AG, Basel, Switzerland	WO 03/018766 A2
November 29, 2005	Genetic engineering of plants through manipulation of lignin biosynthesis	Michigan Technological University, Houghton, MI	US 6969784
January 5, 2006	Manipulation of the phenolic acid content and digestibility of plant cell walls …	Genencor International, Palo Alto, CA	US 2006/0005270 A1
February 2, 2006	Commercial production of polysaccharide degrading enzymes in plants …	Applied Biotechnology Institute, College Station, TX	US 2006/0026715 A1
May 11, 2006	Modification of plant lignin content	ArborGen, LLC	US 2006/0101535 A1
August 15, 2006	Transgenic fiber producing plants with increased expression of sucrose synthetase	Texas Tech University, Lubbock, TX	US 7091400
August 17, 2006	Transgenic plants containing ligninase and cellulase …	M.S. Sticklen et al.	US 2006/0185036 A1

TABLE 4.6
Patent Applications and Patents Awarded for Cellulosic Ethanol Technologies

Date of filing or award	Title	Applicant/assignee	Patent/application
February 22, 1977	Process for making ethanol from cellulosic material using plural ferments	Bio-Industries, Inc., Hialeah, FL	US 4009075
December 9, 1986	Method for the conversion of a P & W substrate to glucose using *Microspora bispora* strain Rutgers P & W	Parsons and Whittemore, Inc. (NY)	US 4628029
October 10, 2000	Ethanol production using a soy hydrolysate-based medium or a yeast autolysate-based medium	University of Florida, Gainesville, FL	US 6130076
December 25, 2001	Ethanol production from lignocellulose	University of Florida, Gainesville, FL	US 6333181
February 27, 2002	Method of processing lignocellulosic feedstock for enhanced xylose and ethanol production	Iogen Bio-Products Corporation, Ontario, Canada	WO 02/070753 A2
July 16, 2002	Method of treating lignocellulosic biomass to produce cellulose	PureVision Technology, Inc., Fort Lupton, CO	US 6419788
March 6, 2003	Organic biomass fractionation process	E.S. Prior	US 2003/0041982 A1
April 29, 2003	Method for processing lignocellulosic material	Forskningscenter Riso, Roskilde, Denmark	US 6555350
June 12, 2003	Methods for cost-effective saccharification of lignocellulosic biomass	E.E. Hood and J.A. Howard	US 2003/0109011 A1
December 9, 2003	Ethanol production with dilute acid hydrolysis using partially dried lignocellulosics	Midwest Research Institute, Kansas City, MO	US 6660506
April 8, 2004	Recombinant hosts suitable for simultaneous saccharification and fermentation	University of Florida Research Foundation, Inc.	US 2004/0067555 A1
March 8, 2004	Methods to enhance the activity of lignocellulose-degrading enzymes	Athenix Corporation, Durham, NC	WO 2004/081185 A2
September 9, 2004	Process to extract phenolic compounds from a residual plant material using a hydrothermal treatment	Centro de Investigaciones Energeticas, Madrid, Spain	US 2004/0176647 A1

Date	Title	Assignee/Inventor	Patent No.
March 31, 2005	Procedure for the production of ethanol from lignocellulosic biomass using a new heat-tolerant yeast	Centro de Investigaciones Energeticas Medioambientales y Tecnologicas, Madrid, Spain	US 2005/0069998 A1
July 28, 2005	Methods for degrading lignocellulosic materials	Novozymes Biotech, Inc., Davis, CA	US 2005/0164355 A1
October 20, 2005	Methods for degrading or converting plant cell wall polysaccharides	Novozymes Biotech, Inc., Davis, CA	US 2005/0233423 A1
January 12, 2006	Methods for glucose production using endoglucanase core protein for improved recovery and reuse of enzyme	D. Wahnon et al.	US 2006/008885 A1
April 11, 2006	Methods and compositions for simultaneous saccharification and fermentation	University of Florida Research Foundation, Inc.	US 7026152
June 13, 2006	Upflow settling reactor for enzymatic hydrolysis of cellulose	B. Foody et al.	US 2006/0154352 A1
April 3, 2007	Pretreatment of bales of feedstock	Iogen Energy Corporation, Ottawa, Canada	US 7198925
June 28, 2007	Methods and processing lignocellulosic feedstock for enhanced xylose and ethanol production	R. Griffin et al.	US 2007/0148751 A1

see chapter 6, section 6.1), conservative technologies proved both profitable and easily scaleable. Biodiesel in particular had the temptation of untapped resources of plant materials (oils) that could, with no time-consuming or costly processing, become a second-generation biofuel after alcohol from sugar and corn.[323] Designs of cellulosic ethanol plants still face the crucial problem of choice, especially that of pretreatment technology (section 4.2) for any of the large-scale feedstocks presently under serious consideration — and there may be simply too many possible choices, all supported by published data sets. In addition, there is the question of timing: the marked acceleration and expansion of fuel ethanol production in the United States started in the late 1990s with corn ethanol and has continued with corn ethanol for at least eight years (figure 1.16). In other words, the actual expansion of the industry could have occurred (and did occur) without exhausting the possible supply of corn, the feedstock with, by far, the easiest substrate preparation route before fermentation that could use the existing corn milling infrastructure.

What now (2008) and in the near future? The almost exclusive focus on corn in North America has certainly caused price inflation that has translated to higher feed costs for beef, pork, and chicken that will (as USDA predicts) result in a declining meat supply.[324] Higher corn prices have also triggered a tortilla inflation crisis in Mexico, and another USDA prognosis is for a 3.5% annual rise in food prices in 2007. These economic changes will only be exacerbated by a continued expansion of corn ethanol outputs without a massive surge in corn grain yield; lignocellulosics break this vicious cycle, adding diversity to the limited range of substrates but inevitably requiring more complex bioprocess technologies — of which the Iogen and Abengoa initiatives represent add-on solutions to existing fuel alcohol production processes. On this analysis, the more advanced process options remain for another five to ten years, and possibly much longer, while investors follow the proven technologies (wheat straw, separate hydrolysis and fermentation, batch or fed-batch fermentation), minimizing production costs by sharing facilities with cereal ethanol production: the first cellulosic ethanol facilities in the United States may be operational in 2010, possibly utilizing sugarcane bagasse feedstock, whereas the same recombinant bacterial technologies will be producing 4 million liters of ethanol from demolition wood waste in Japan by 2008 (www.verenium.com).

The step change beyond mixed cereal/cellulosic ethanol has, however, already been extensively discussed, that of the integrated production of biofuel, power, animal feeds, and chemical coproducts to maximize the number of different saleable commodities to supplement or (if need be, under adverse market conditions) supplant bioethanol as the income stream, that is, "biocommodity engineering."[325] Given the availability of large amounts of biomass sources in probably overlapping and complementary forms (thus necessitating flexibility in the production process to utilize multiple substrates on a seasonal basis), biocommodities will become no more of a process and production challenge than are the different alcoholic outputs from most large breweries or, with a chemical methodology, those from industrial petroleum "cracking" into petrochemicals.

Investors will nevertheless require convincing of the economic viability of bioethanol in the short term, in years where crop yields (corn, wheat, etc.) are better or worse than expected, when farm commodity prices fluctuate, and when international competition for the most widely sought biofuel feedstocks may have become significant and — most importantly — with or without continued "financial assistance" from national governments. The economics of bioethanol will be examined in the next chapter to define the extent of financial subsidies used or that still remain necessary and how high capital costs may inhibit the growth of the nascent industry.

REFERENCES

1. Tolan, J.S., Iogen's demonstration process for producing ethanol from cellulosic biomass, in Kamm, B., Gruber, P.R., and Kamm, M. (Eds.), *Biorefineries — Industrial Processes and Products*. Volume 1. *Status Quo and Future Directions*, Wiley-VCH Verlag, Weinheim, 2006, chap. 9.

2. Foody, B., Tolan, J.S., and Bernstein, J.D., Pretreatment process for conversion of cellulose to fuel ethanol, US Patent 5, 916, 780, June 29, 1999.

3. Tolan, J.S., Alcohol production from cellulosic biomass: the Iogen process, a model system in operation, in Jacques, K.A., Lyons, T.P., and Kelsall, D.R. (Eds.), *The Alcohol Textbook*, 3rd ed, Nottingham University Press, 1999, chap. 9.

4. Lawford, G.R. et al., Scale-up of the Bio-hol process for the conversion of biomass to ethanol, in Moo-Young, M., Hasnain, S., and Lamptey, J. (Eds.), *Biotechnology and Renewable Energy*, Elsevier Applied Science Publishers, London and New York, 1986, p.276.

5. Taylor, J.D., Commercial update on the Stake process, in Moo-Young, M., Hasnain, S., and Lamptey, J. (Eds.), *Biotechnology and Renewable Energy*, Elsevier Applied Science Publishers, London and New York, 1986, p.286.

6. Wright, C.T. et al., Biomechanics of wheat/barley straw and corn stover, *Appl. Biochem. Biotechnol.*, 121–124, 5, 2005.

7. Haq, Z. and Easterley, J.L., Agricultural residue availability in the United States, *Appl. Biochem. Biotechnol.*, 129–132, 3, 2006.

8. Kristensen, J.B. et al., Use of surface active additives in enzymatic hydrolysis of wheat straw lignocellulose, *Enz. Microb. Technol.*, 40, 888, 2007.

9. Sørensen, H.R., Pedersen, S., and Meyer, A.S., Optimization of reaction conditions for enzymatic viscosity reduction and hydrolysis of wheat arabinoxylan in an industrial ethanol fermentation residue, *Biotechnol. Prog.*, 22, 505, 2006.

10. Sørensen, H.R. et al., Enzymatic hydrolysis of wheat arabinoxylan by a recombinant "minimal" enzyme cocktail containing β-xylosidase and novel endo-1,4-β-xylanase and α-L-arabinofuranosidase activities, *Biotechnol. Prog.*, 23, 100, 2007.

11. Sørensen, H.R. et al., A novel GH43 α-L-arabinofuranosidase from *Humicola insolens*: mode of action and synergy with GH51 α-L-arabinofuranosidases on wheat arabinoxylan, *Appl. Microbial. Biotechnol.*, 73, 850, 2006.

12. Ahring, B.K. et al., Pretreatment of wheat straw and conversion of xylose and xylan to ethanol by thermophilic anaerobic bacteria, *Bioresour. Technol.*, 58, 107, 1996.

13. Bjerre, A.B. et al., Pretreatment of wheat straw using combined alkaline wet oxidation and alkaline hydrolysis resulting in convertible cellulose and hemicellulose, *Biotechnol. Bioeng.*, 49, 568, 1996.

14. Schmidt, A.S. and Thomsen, A.B., Optimization of wet oxidation pretreatment of wheat straw, *Bioresour. Technol.*, 64, 139, 1998.

15. Ahring, B.K. et al., Production of ethanol from wet oxidized wheat straw by *Thermoanaerobacter mathranii*, *Bioresour. Technol.*, 68, 3, 1999.

16. Klinke, H.B., Thomsen, A.B., and Ahring, B.K., Potential inhibitors from wet oxidation of wheat straw and their effect on growth and ethanol production by *Thermoanaerobacter mathranii*, *Appl. Microbial. Biotechnol.*, 57, 631, 2001.

17. Klinke, H.B. et al., Characterization of degradation products from alkaline wet oxidation of wheat straw, *Bioresour. Technol.*, 82, 15, 2002.

18. Klinke, H.B. et al., Potential inhibitors from wet oxidation of wheat straw and their effect on ethanol production of *Saccharomyces cerevisiae*: wet oxidation and fermentation by yeast, *Biotechnol. Bioeng.*, 81, 738, 2003.
19. Panagiotou, G. and Olsson, L., Effect of compounds released during pretreatment of wheat straw on microbial growth and enzymatic hydrolysis rates, *Biotechnol. Bioeng.*, 96, 250, 2007.
20. McLoughlin, S.B. et al., High-value renewable energy from prairie grasses, *Environ. Sci. Technol.* 36, 2122, 2002.
21. Tenenbaum, D.A., Switching to switchgrass, *Environ. Health Perspectives*, 110, A18–A19, 2002.
22. Kaylen, K.S., An economic analysis of using alternative fuels in a mass burn boiler, *Bioresour. Technol.*, 96, 1943, 2005.
23. Kamm, J., A new class of plants for a biofuel feedstock energy crop, *Appl. Biochem. Biotechnol.*, 113–116, 55, 2004.
24. Stroup, J.A. et al., Comparison of growth and performance in upland and lowland switchgrass types to water and nitrogen stress, *Bioresour. Technol.*, 86, 65, 2004.
25. Mulkey, V.R., Owens, V.N., and Lee, D.K., Management of warm-season grass mixtures for biomass production in South Dakota, USA, *Bioresour. Technol.*, 99, 609, 2008.
26. Christian, D.G., Riche, A.B., and Yates, N.E., The yield and composition of switchgrass and coastal panic grass grown as a biofuel in southern England, *Bioresour. Technol.*, 83, 115, 2002.
27. Lau, C.S. et al., Extraction of antioxidant compounds from energy crops, *Appl. Biochem. Biotechnol.*, 113–116, 569, 2004.
28. Sarath, G. et al., Internode structure and cell wall composition in maturing tillers of switchgrass (*Panicum virgatum* L.), *Bioresour. Technol.*, 98, 2985, 2007.
29. Lee, S.T. et al., The isolation and identification of steroidal sapogenins in switchgrass, *J. Nat. Toxins*, 10, 273, 2001.
30. Anderson, W.F. et al., Enzyme pretreatment of grass lignocellulose for potential high-value co-products and an improved fermentable substrate, *Appl. Biochem. Biotechnol.*, 121–124, 303, 2005.
31. Law, K.N., Kokta, B.V., and Mao, C.B., Fibre morphology and soda-sulphite pulping of switchgrass, *Bioresour. Technol.*, 77, 1, 2001.
32. Chung, Y.C., Bakalinsky, A., and Penner, M.H., Enzymatic saccharification and fermentation of xylose-optimized dilute acid-treated lignocellulose, *Appl. Biochem. Biotechnol.*, 121–124, 947, 2005.
33. Meunier-Goddik, L. and Penner, M.H., Enzyme-catalyzed saccharification of model celluloses in the presence of lignacious residues, *J. Agric. Food Chem.*, 47, 346, 1999.
34. Kurakake, M. et al., Pretreatment with ammonia water for enzymatic hydrolysis of corn husk, bagasse, and switchgrass, *Appl. Biochem. Biotechnol.*, 90, 251, 2001.
35. Alizadeh, H. et al., Pretreatment of switchgrass by ammonia fiber explosion (AFEX), *Appl. Biochem. Biotechnol.*, 121–124, 1133, 2005.
36. Paul, J.K. (Ed.), Alcohol from corn (50 million gallons/year), in *Large and Small Scale Ethyl Alcohol Processes from Agricultural Raw Materials*, Noyes Data Corporation, Park Ridge, NJ, 1980, part 1.
37. Kadam, K.L. and McMillan, J.D., Availability of corn stover as a sustainable feedstock for bioethanol production, *Bioresour. Technol.*, 88, 17, 2003.
38. Rooney, T., *Lignocellulosic Feedstock Resource Assessment*, Report SR-580-24189, National Renewable Energy Laboratory, Golden, CO, 1998, 123.
39. Varga, E., Szengyel, Z., and Reczey, K., Chemical pretreatments of corn stover for enhancing enzymatic digestibility, *Appl. Biochem. Biotechnol.*, 98–100, 73, 2002.
40. Kim, S.B. and Lee, Y.Y., Diffusion of sulfuric acid within lignocellulosic biomass particles and its impact on dilute-acid treatment, *Bioresour. Technol.*, 83, 165, 2002.

41. Varga, E. et al., Pretreatment of corn stover using wet oxidation to enhance enzymatic digestibility, *Appl. Biochem. Biotechnol.*, 104, 37, 2003.

42. Yang, B. and Wyman, C.E., Effect of xylan and lignin removal by batch and flowthrough pretreatment on the enzymatic digestibility of corn stover cellulose, *Biotechnol. Bioeng.*, 86, 88, 2004.

43. Pimenova, N.V. and Hanley, T.R., Measurement of rheological properties of corn stover suspensions, *Appl. Biochem. Biotechnol.*, 105–108, 383, 2003.

44. Varga, E., Reczey, K., and Zacchi, G., Optimization of steam pretreatment of corn stover to enhance enzymatic digestibility, *Appl. Biochem. Biotechnol.*, 113–116, 509, 2004.

45. Kim, S.B. and Lee, Y.Y., Pretreatment and fractionation of corn stover by ammonia recycle percolation process, *Bioresour. Technol.*, 96, 2007, 2005.

46. Kim, S. and Holtzapple, M.T., Lime pretreatment and enzymatic hydrolysis of corn stover, *Bioresour. Technol.*, 96, 1994, 2005.

47. Mosier, N. et al., Optimization of pH controlled liquid hot water pretreatment of corn stover, *Bioresour. Technol.*, 96, 1986, 2005.

48. Liu, C. and Wyman, C.E., Partial flow of compressed-hot water through corn stover to enhance hemicellulose sugar recovery and enzymatic digestibility of cellulose, *Bioresour. Technol.*, 96, 1978, 2005.

49. Öhgren, K., Galbe, M., and Zacchi, G., Optimization of steam pretreatment of SO_2-impregnated corn stover for fuel ethanol production, *Appl. Biochem. Biotechnol.*, 121–124, 1055, 2005.

50. Lloyd, T.A. and Wyman, C.E., Combined sugar yields for dilute sulfuric acid pretreatment of corn stover followed by enzymatic hydrolysis of the remaining solids, *Bioresour. Technol.*, 96, 1967, 2005.

51. Zeng, M. et al., Microscopic examination of changes of plant cell structure in corn stover due to hot water pretreatment and enzymatic hydrolysis, *Biotechnol. Bioeng.*, 97, 265, 2007.

52. Berson, R.E. et al., Detoxification of actual pretreated corn stover hydrolysate using activated carbon powder, *Appl. Biochem. Biotechnol.*, 121–124, 923, 2005.

53. Nichols, N.N. et al., Bioabatement to remove inhibitors from biomass-derived sugar hydrolysates, *Appl. Biochem. Biotechnol.*, 121–124, 379, 2005.

54. Pimenova, N.V. and Hanley, T.R., Effect of corn stover concentration on rheological characteristics, *Appl. Biochem. Biotechnol.*, 113–116, 347, 2004.

55. Kim, S. and Holtzapple, M.T., Delignification kinetics of corn stover in lime pretreatment, *Bioresour. Technol.*, 97, 778, 2006.

56. Zhu, Y., Lee, Y.Y., and Elander, R.T., Dilute-acid pretreatment of corn stover using a high-solids percolation reactor, *Appl. Biochem. Biotechnol.*, 117, 103, 2004.

57. Zhu, Y., Lee, Y.Y., and Elander, R.T., Optimization of dilute-acid pretreatment of corn stover using a high-solids percolation reactor, *Appl. Biochem. Biotechnol.*, 121–124, 1045, 2005.

58. Kim, S. and Holtzapple, M.T., Effect of structural features on enzyme digestibility of corn stover, *Bioresour. Technol.*, 97, 583, 2006.

59. Chundawal, S.P., Venkatesh, B., and Dale, B.E., Effect of size based separation of milled corn stover on AFEX pretreatment and enzymatic digestibility, *Biotechnol. Bioeng.*, 96, 219, 2007.

60. Zhu, Y. et al., Enzymatic production of xylooligosaccharides from corn stover and corn cobs treated with aqueous ammonia, *Appl. Biochem. Biotechnol.*, 129–132, 586, 2005.

61. Lu, Y. and Mosier, N.S., Biomimetic catalysis for hemicellulose hydrolysis in corn stover, *Biotechnol. Prog.*, 23, 116, 2007.

62. Jeoh, T. et al., Cellulase digestibility of pretreated biomass is limited by cellulose accessibility, *Biotechnol. Bioeng.*, 98, 112, 2007.

63. Kaar, W.E. and Holtzapple, M.T., Benefits from Tween during enzymic hydrolysis of corn stover, *Biotechnol. Bioeng.*, 59, 419, 1998.

64. Öhgren, K. et al., Effect of hemicellulose and lignin removal on enzymatic hydrolysis of steam pretreated corn stover, *Bioresour. Technol.*, 98, 2503, 2007.

65. Laureano-Perez, L. et al., Understanding factors that limit enzymatic hydrolysis of biomass: characterization of pretreated corn stover, *Appl. Biochem. Biotechnol.*, 121–124, 1081, 2005.

66. O'Dwyer, J.P. et al., Enzymatic hydrolysis of lime-pretreated corn stover and investigation of the HCH-1 Model: inhibition pattern, degree of inhibition, validity of simplified HCH-1 Model, *Bioresour. Technol.*, 98, 2969, 2007.

67. Steele, B. et al., Enzyme recovery and recycling following hydrolysis of ammonia fiber explosion-treated corn stover, *Appl. Biochem. Biotechnol.*, 121–124, 901, 2005.

68. Berson, R.E., Young, J.S., and Hanley, T.R., Reintroduced solids increase inhibitor levels in a pretreated corn stover hydrolysate, *Appl. Biochem. Biotechnol.*, 129–132, 612, 2006.

69. Wyman, C.E. et al., Comparative sugar recovery data from laboratory scale application of leading pretreatment technologies to corn stover, *Bioresour. Technol.*, 96, 2026, 2005.

70. Galbe, M. and Zacchi, G., A review of the production of ethanol from softwood, *Appl. Microbiol. Biotechnol.*, 59, 618, 2002.

71. Wingren, A. et al., Process considerations and economic evaluation of two-step steam pretreatment for production of fuel ethanol from softwood, *Biotechnol. Prog.*, 20, 1421, 2004.

72. Kim, K.H., Tucker, M.P., and Nguyen, Q.A., Effects of pressing lignocellulosic biomass on sugar yield in two-stage dilute-acid hydrolysate process, *Biotechnol. Prog.*, 18, 489, 2002.

73. Nguyen, Q.A. et al., Two-stage dilute-acid pretreatment of softwoods, *Appl. Biochem. Biotechnol.*, 84–86, 561, 2002.

74. Tengborg, C., Galbe, M., and Zacchi, G., Influence of enzyme loading and physical parameters on the enzymatic hydrolysis of steam-pretreated softwood, *Biotechnol. Prog.*, 17, 110, 2001.

75. Pan, X. et al., Strategies to enhance the enzymatic hydrolysis of pretreated softwood with high residual lignin content, *Appl. Biochem. Biotechnol.*, 121–124, 1069, 2005.

76. Kim, K.H. et al., Continuous countercurrent extraction of hemicellulose from pretreated wood residues, *Appl. Biochem. Biotechnol.*, 91–93, 253, 2001.

77. Mabee, W.E. et al., Updates on softwood-to-ethanol process development, *Appl. Biochem. Biotechnol.*, 129–132, 55, 2006.

78. Larsson, S. et al., The generation of fermentation inhibitors during dilute acid hydrolysis of softwood, *Enz. Microb. Technol.*, 24, 151, 1999.

79. Tengborg, C., Galbe, M., and Zacchi, G., Reduced inhibition of enzymatic hydrolysis of steam-pretreated softwood, *Enz. Microb. Technol.*, 28, 835, 2001.

80. Larsson, S. et al., Comparison of different methods for detoxification of lignocellulose hydrolyzates of spruce, *Appl. Biochem. Biotechnol.*, 77–79, 91, 1999.

81. Keller, F.A. et al., Yeast adaptation on softwood prehydrolysate, *Appl. Biochem. Biotechnol.*, 70-72, 137, 1998.

82. Pan, X. et al., Biorefining of softwoods using ethanol organosolv pulping: preliminary evaluation of process streams for manufacture of fuel-grade ethanol and co-products, *Biotechnol. Bioeng.*, 90, 473, 2005.

83. Pan, X. et al., Bioconversion of hybrid poplar to ethanol and co-products using an organosolv fractionation process: optimization of process yields, *Biotechnol. Bioeng.*, 94, 851, 2006.

84. Morjanoff, P.J. and Gray, P.P., Optimization of steam explosion as a method for increasing susceptibility of sugarcane bagasse to enzymatic saccharification, *Biotechnol. Bioeng.*, 29, 733, 1987.

85. Kaar, W.E., Gutierrez, C.V., and Kinoshita, C.M., Steam explosion of sugarcane bagasse as a pretreatment for conversion to ethanol, *Biomass Bioenergy*, 14, 277, 1998.
86. Martín, C. et al., Comparison of the fermentability of enzymatic hydrolysates of sugarcane bagasse pretreated by steam explosion using different impregnating agents, *Appl. Biochem. Biotechnol.*, 98–100, 699, 2002.
87. Laser, M. et al., A comparison of liquid hot water and steam pretreatments of sugar cane bagasse for bioconversion to ethanol, *Bioresour. Technol.*, 81, 33, 2002.
88. Sasaki, M., Adschiri, T., and Arai, K., Fractionation of sugarcane bagasse by hydrothermal treatment, *Bioresour. Technol.*, 86, 301, 2003.
89. Teixera, L.C., Linden, J.C., and Schroeder, H.A., Optimizing peracetic acid pretreatment conditions for improved simultaneous saccharification and cofermentation (SSCF) of sugar cane bagasse to ethanol fuel, *Renew. Energy*, 16, 1070, 1999.
90. Teixera, L.C., Linden, J.C., and Schroeder, H.A., Simultaneous saccharification and cofermentation of peracetic acid-pretreated biomass, *Appl. Biochem. Biotechnol.*, 84–86, 111, 2000.
91. Neureiter, M. et al., Dilute-acid hydrolysis of sugarcane bagasse at varying conditions, *Appl. Biochem. Biotechnol.*, 98–100, 49, 2002.
92. Silva, S.S., Matos, Z.R., and Carvalho, W., Effects of sulfuric acid loading and residence time on the composition of sugarcane bagasse hydrolysate and its use as source of xylose for xylitol production, *Biotechnol. Prog.*, 21, 1449, 2005.
93. Martín, C., Klinke, H.B., and Thomsen, A.B. Wet oxidation as a pretreatment method for enhancing the enzymatic convertibility of sugarcane bagasse, *Enz. Microb. Technol.*, 40, 426, 2007.
94. Yang, B. et al., Study on the hydrolyzate of sugarcane bagasse to ethanol by fermentation, *Chin. J. Biotechnol.*, 13, 253, 1997.
95. Chandel, A.K. et al., Detoxification of sugarcane bagasse hydrolysate improves ethanol production by *Candida shehatae* NCIM 3501, *Bioresour. Technol.*, 98, 1947, 2007.
96. Martín, C. et al., Adaptation of a recombinant xylose-utilizing *Saccharomyces cerevisiae* to a sugarcane bagasse hydrolysate with high content of fermentation inhibitors, *Bioresour. Technol.*, 98, 1767, 2007.
97. Cheng, K.-K. et al., Fermentation of pretreated sugarcane bagasse hemicellulose hydrolysate to ethanol by *Pachysolen tannophilus*, *Biotechnol. Lett.*, 29, 1051, 2007.
98. Garay-Arroyo, A. et al., Response to different environmental stress conditions of industrial and laboratory *Saccharomyces cerevisiae* strains, *Appl. Microbiol. Biotechnol.*, 63, 4734, 2004.
99. de Castro, H.F., Oliveira, S.C., and Furlan, S.A., Alternative approach for utilization of pentose stream from sugarcane bagasse by an induced flocculent *Pichia stipitis*, *Appl. Biochem. Biotechnol.*, 105–108, 547, 2003.
100. Lehrer, S.B. et al., Elimination of bagassosis in Louisiana paper manufacturing plant workers, *Clin. Allergy*, 8, 15, 1978.
101. Ueda, A. et al., Recent trends in bagassosis in Japan, *Br. J. Ind. Med.*, 49, 499, 1992.
102. Dien, B.S. et al., Fermentation of "Quick Fiber" produced from a modified corn-milling process into ethanol and recovery of corn fiber oil, *Appl. Biochem. Biotechnol.*, 113–116, 937, 2004.
103. Li, X.L. et al., Profile of enzyme production by *Trichoderma reesei* grown on corn fiber fractions, *Appl. Biochem. Biotechnol.*, 121–124, 321, 2005.
104. Palmarola-Adrados, B. et al., Ethanol production from non-starch carbohydrates of wheat bran, *Bioresour. Technol.*, 96, 843, 2005.
105. Saha, B.C. et al., Dilute acid pretreatment, enzymatic saccharification, and fermentation of rice hulls to ethanol, *Biotechnol. Prog.*, 21, 816, 2005.
106. Abbi, M., Kuhad, R.C., and Singh, A., Fermentation of xylose and rice straw hydrolysate to ethanol by *Candida shehatae* NCL-3501, *J. Ind. Microbiol.*, 17, 20, 1996.

107. Sassner, P. et al., Steam pretreatment of H_2SO_4-impregnated *Salix* for the production of bioethanol, *Bioresour. Technol.*, 99, 137, 2008.
108. Sanchez, G. et al., Dilute-acid hydrolysis for fermentation of the Bolivian straw material *Paja brava*, *Bioresour. Technol.*, 93, 249, 2004.
109. Nguyen, Q.A. et al., Bioconversion of mixed solids waste to ethanol, *Appl. Biochem. Biotechnol.*, 77–79, 455, 1999.
110. Karaffa, L. and Kubicek, C.P., *Aspergillus niger* citric acid accumulation: do we understand this well working black box? *Appl. Microbiol. Biotechnol.*, 61, 189, 2003.
111. Lawford, H.G. and Rousseau, J.D., Studies on nutrient requirements and cost-effective supplements for ethanol production by recombinant *E. coli*, *Appl. Biochem. Biotechnol.*, 57–58, 307, 1996.
112. Slininger, P.J. et al., Nitrogen source and mineral optimization enhance D-xylose conversion to ethanol by the yeast *Pichia stipitis* NRRL Y-7124, *Appl. Microbiol. Biotechnol.*, 72, 1285, 2006.
113. Martinez, A. et al., Low salt medium for lactate and ethanol production by recombinant *E. coli* B, *Biotechnol. Lett.*, 29, 397, 2007.
114. Miller, E.N. and Ingram, L.O., Combined effect of betaine and trehalose on osmotic tolerance of *Escherichia coli* in mineral salts medium, *Biotechnol. Lett.*, 29, 213, 2007.
115. Weimer, P.J. et al., *In vitro* gas production as a surrogate measure of the fermentability of cellulosic biomass to ethanol, *Appl. Microbiol. Biotechnol.*, 67, 52, 2005.
116. Ingledew, W.M., Alcohol production by *Saccharomyces cerevisiae*: a yeast primer, in Jacques, K.A., Lyons, T.P., and Kelsall, D.R. (Eds.), *The Alcohol Textbook,* 3rd ed, Nottingham University Press, 1999, chap. 5.
117. Ingledew, W.M. and Casey, G.P., Rapid production of high concentrations of ethanol using unmodified industrial yeast, in Moo-Young, M., Hasnain, S., and Lamptey, J. (Eds.), *Biotechnology and Renewable Energy,* Elsevier Applied Science Publishers, London and New York, 1986, 246.
118. Casey, G.P. and Ingledew, W.M., High gravity brewing: influence of pitching rate and wort gravity on early yeast viability, *J. Am. Soc. Brew. Chem.*, 41, 148, 1983.
119. Casey, G.P., Magnus, C.A., and Ingledew, W.M., High gravity brewing: nutrient enhanced production of high concentrations of ethanol by brewing yeasts, *Biotechnol. Lett.*, 5, 429, 1983.
120. Casey, G.P., Magnus, C.A., and Ingledew, W.M., High gravity brewing: effects of nutrition on yeast composition, fermentative ability, and alcohol production, *Appl. Microbiol. Biotechnol.*, 48, 639, 1984.
121. Kelsall, D.R. and Lyons, T.P., Management of fermentations in the production of alcohol: moving toward 23% ethanol, in Jacques, K.A., Lyons, T.P., and Kelsall, D.R. (Eds.), *The Alcohol Textbook,* 3rd ed, Nottingham University Press, 1999, chap. 3.
122. Madson, P.W. and Monceaux, D.A., Fuel ethanol production, in Jacques, K.A., Lyons, T.P., and Kelsall, D.R. (Eds.), *The Alcohol Textbook,* 3rd ed, Nottingham University Press, 1999, chap. 17.
123. Cot, M. et al., Physiological behaviour of *Saccharomyces cerevisiae* in aerated fed-batch fermentation for high level production of bioethanol, *FEMS Yeast Res.*, 7, 22, 2006.
124. Thomas, K.C. and Ingledew, W.M., Fuel alcohol production: effects of free amino nitrogen on fermentation of very-high-gravity wheat mashes, *Appl. Environ. Microbiol.*, 56, 2046, 1990.
125. Bafrncová, P. et al., Improvement of very high gravity ethanol fermentation by medium supplementation using *Saccharomyces cerevisiae*, *Biotechnol. Lett.*, 21, 337, 1999.
126. Jones, A.M. and Ingledew, W.M., Fuel alcohol production: assessment of selected commercial proteases for very high gravity wheat mash fermentations, *Enz. Microb. Technol.*, 16, 683, 1994.

127. Thomas, K.C. and Ingledew, W.M., Relationship of low lysine and high arginine concentrations to efficient ethanolic fermentation of wheat mash, *Can. J. Microbiol.*, 38, 626, 1992.

128. Wang, S. et al., Grain pearling and the very high gravity (VHG) fermentation technologies from rye and triticale, *Proc. Biochem.*, 34, 421, 1999.

129. Thomas, K.C., Hynes, S.H., and Ingledew, W.M., Effects of particulate materials and osmoprotectants on very-high-gravity ethanolic fermentation by *Saccharomyces cerevisiae*, *Appl. Environ. Biotechnol.*, 60, 1519, 1994.

130. Reddy, L.V. and Reddy, O.V., Improvement of ethanol production in very high gravity fermentation by horse gram (*Dolichos biflorus*) flour supplementation, *Lett. Appl. Microbiol.*, 41, 440, 2005.

131. Wang, S. et al., Optimization of fermentation temperature and mash specific gravity for fuel ethanol production, *Cereal Chem.*, 76, 82, 1999.

132. Dragone, G. et al., Improvement of the ethanol productivity in a high gravity brewing at pilot plant scale, *Biotechnol. Lett.*, 25, 1171, 2003.

133. Wang, F.Q. et al., Optimization of an ethanol production medium in very high gravity fermentation, *Biotechnol. Lett.*, 29, 233, 2007.

134. Alfenore, S. et al., Improving ethanol production and viability of *Saccharomyces cerevisiae* by a vitamin-feeding strategy during fed-batch process, *Appl. Microbiol. Biotechnol.*, 60, 67, 2002.

135. Barber, A.R., Henningsson, M., and Pamment, N.B., Acceleration of high gravity yeast fermentations by acetaldehyde addition, *Biotechnol. Lett.*, 24, 891, 2002.

136. Blieck, L. et al., Isolation and characterization of brewer's yeast variants with improved fermentation performance under high-gravity conditions, *Appl. Microbiol. Biotechnol.*, 73, 815, 2007.

137. Guijarro, J.M. and Lagunas, R., *Saccharomyces cerevisiae* does not accumulate ethanol against a concentration gradient, *J. Bacteriol.*, 160, 874, 1984.

138. Schoberth, S.M. et al., Ethanol transport in *Zymomonas mobilis* measured by using in vivo nuclear magnetic resonance spin transfer, *J. Bacteriol.*, 178, 1756, 1996.

139. Thatipamala, R., Rohani, S., and Hill, G.A., Effects of high product and substrate inhibitions on the kinetics of biomass and product yields during ethanol batch fermentation, *Biotechnol. Bioeng.*, 40, 289, 2004.

140. Ogawa, Y. et al., Tolerance mechanism of the ethanol-tolerant mutant of sake yeast, *J. Biosci. Bioeng.*, 90, 313, 2000.

141. Furukawa, K. et al., Effect of cellular inositol content on ethanol tolerance of *Saccharomyces cerevisiae* in sake brewing, *J. Biosci. Bioeng.*, 98, 107, 2004.

142. Nozawa, M. et al., A role of *Saccharomyces cerevisiae* fatty acid activation protein 4 in palmitoyl-CoA pool for growth in the presence of ethanol, *J. Biosci. Bioeng.*, 93, 288, 2002.

143. Takagi, H. et al., Effect of L-proline on sake brewing and ethanol stress in *Saccharomyces cerevisiae*, *Appl. Environ. Microbiol.*, 71, 8656, 2005.

144. Matsuura, K. and Takagi, H., Vacuolar functions are involved in stress-protective effect of intracellular proline in *Saccharomyces cerevisiae*, *J. Biosci. Bioeng.*, 100, 538, 2005.

145. Çakar, Z.P. et al., Evolutionary engineering of multiple-stress resistant *Saccharomyces cerevisiae*, *FEMS Yeast Res.*, 5, 569, 2005.

146. van Voorst, F. et al., Genome-wide identification of genes required for growth of *Saccharomyces cerevisiae* under ethanol stress, *Yeast*, 23, 351, 2006.

147. Mukai, N. et al., Beer brewing using a fusant between a sake yeast and a brewer's yeast, *J. Biosci. Bioeng.*, 91, 482, 2001.

148. Inoue, Y. et al., Efficient extraction of thioredoxin from *Saccharomyces cerevisiae* by ethanol, *Appl. Environ. Microbiol.*, 73, 1672, 2007.

149. Campbell, S.L., *The Continuous Brewing of Beer*, available as a download from the New Zealand Institute of Chemistry website, www.nzic.org.nz/ChemProcesses/food.html.
150. Hornsey, I.S., *A History of Beer and Brewing*, The Royal Society of Chemistry, Cambridge, 2003, chap.9.
151. Royston, M.G., Tower fermentation of yeast, *Proc. Biochem.*, 1, 215, 1966.
152. Cysewski, G.R. and Wilke, C.R., Rapid ethanol fermentations using vacuum and cell recycle, *Biotechnol. Bioeng.*, 19, 1125, 1977.
153. Comberbach, D.M., Ghommidh, C., and Bu'Lock, J.D., Kinetic modeling and computer control of continuous alcohol fermentation in the gas-lift tower, in Moo-Young, M., Hasnain, S., and Lamptey, J. (Eds.), *Biotechnology and Renewable Energy*, Elsevier Applied Science Publishers, London and New York, 1986, p. 208.
154. Margaritis, A. and Bajpai, P., Continuous ethanol production from Jerusalem artichoke tubers. I. Use of free cells of *Kluyveromyces marxianus*, *Biotechnol. Bioeng.*, 24, 1473, 1982.
155. Yarovenko, V.L., Theory and practice of continuous cultivation of microorganisms in industrial alcoholic processes, *Adv. Biochem. Eng.*, 9, 1, 1978.
156. Bayrock, D. and Ingledew, W.M., Application of multistage continuous fermentation for production of fuel alcohol by very-high-gravity fermentation technology, *J. Ind. Microbiol. Biotechnol.*, 27, 87, 2001.
157. Bai, F.W. et al., Continuous ethanol production and evaluation of yeast cell lysis and viability loss under very high gravity medium conditions, *J. Biotechnol.*, 10, 287, 2004.
158. Laluce, C. et al., Continuous ethanol production in a nonconventional five-stage system operating with yeast cell recycling at elevated temperatures, *J. Ind. Microbiol. Biotechnol.*, 29, 140, 2002.
159. Paiva, T.C. et al., Continuous alcoholic fermentation process in a tower reactor with recycling of flocculating yeast, *Appl. Biochem. Biotechnol.*, 57–58, 535, 1996.
160. Ben Chaabane, F. et al., Very high ethanol productivity in an innovative continuous two-stage bioreactor with cell recycle, *Bioproc. Biosyst. Eng.*, 29, 49, 2006.
161. Dale, M.C., Wankat, P.C., and Okos, M.R., Immobilized cell reactor-separator with simultaneous product separation and methods for design and use thereof, U.S. Patent 4,665,027, May 12, 1987.
162. Dale, M.C., Method of use of a multi-stage reactor-separator with simultaneous product separation, U.S. Patent 5,141,861, August 25, 1992.
163. Dale, M.C. and Okos, M.R., High speed, consecutive batch or continuous, low effluent process for the production of ethanol from molasses, starches, or sugars, U.S. Patent 7,070,967, July 4, 2006.
164. Åkesson, M. et al., On-line detection of acetate formation in *Escherichia coli* cultures using dissolved oxygen responses to feed transients, *Biotechnol. Bioeng.*, 64, 590, 1999.
165. Walsh, G., Biopharmaceuticals: approvals and approval trends in 2004, *BioPharm. Intl.*, 18 (issue 5, May), 58, 2005.
166. Belo, I., Pinheiro, R., and Mota, M., Fed-batch cultivation of *Saccharomyces cerevisiae* in a hyperbaric bioreactor, *Biotechnol. Prog.*, 19, 665, 2003.
167. Cannizzaro, C., Valentinotti, S., and van Stockar, U., Control of yeast fed-batch process through regulation of extracellular ethanol concentration, *Bioproc. Biosyst. Eng.*, 26, 377, 2004.
168. Aldiguier, A.S. et al., Synergistic temperature and ethanol effect on *Saccharomyces cerevisiae* dynamic behaviour in ethanol bio-fuel production, *Bioproc. Biosyst. Eng.*, 26, 217, 2004.
169. Alfenore, S. et al., Aeration strategy: a need for very high ethanol performance in *Saccharomyces cerevisiae* fed-batch process, *Appl. Microbiol. Biotechnol.*, 63, 537, 2004.

170. Bideaux, C. et al., Minimization of glycerol production during the high-performance fed-batch ethanolic fermentation process in *Saccharomyces cerevisiae*, using a metabolic model as a prediction tool, *Appl. Environ. Microbiol.*, 72, 2134, 2006.
171. Carvalho, J.C. et al., Ethanol production by *Saccharomyces cerevisiae* grown in sugarcane blackstrap molasses through a fed-batch process: optimization by response surface methodology, *Appl. Biochem. Biotechnol.*, 110, 151, 2003.
172. Bravo, S., Mahn, A., and Shene, C., Effect of feeding strategy on *Zymomonas mobilis* CP4 fed-batch fermentations and mathematical modeling of the system, *Appl. Microbiol. Biotechnol.*, 54, 487, 2000.
173. Petersson, A. and Lidén, G., Fed-batch cultivation of *Saccharomyces cerevisiae*, using a metabolic model as a prediction tool, *Appl. Environ. Microbiol.*, 72, 2134, 2006.
174. Novozymes and Groton Biosystems sign an agreement to develop online monitoring for fuel ethanol plants, Groton Biosystems, Inc., Boxborough, MA, February 12, 2007, www.grotonbiosystems.com.
175. Margaritis, A. and Bajpai, P., Continuous ethanol production from Jerusalem artichoke tubers. II. Use of immobilized cells of *Kluyveromyces marxianus*, *Biotechnol. Bioeng.*, 24, 1483, 1982.
176. Moebus, O. and Teuber, M., Production of ethanol by solid particles of *Saccharomyces cerevisiae* in a fluidized bed, *Appl. Microbiol. Biotechnol.*, 15, 194, 1982.
177. Verbelen, P.J. et al., Immobilized yeast cell systems for continuous fermentation applications, *Biotechnol. Lett.*, 28, 1515, 2006.
178. Willaert, R., Cell immobilization and its applications in biotechnology: current trends and future prospects, in El-Mansi, E.M.T., Bryce, C.A., Demain, A.L., and Allman, A.R. (Eds.), *Fermentation Microbiology and Biotechnology*, CRC Press, Boca Raton, 2007, chap. 10.
179. de Vasconcelos, J.N., Lopes, C.E., and de França, F.P., Continuous ethanol production using yeast immobilized on sugar-cane stalks, *Braz. J. Chem. Eng.*, 21, 357, 2004.
180. Kunduru, M.R. and Pometto, A.L., Continuous ethanol production by *Zymomonas mobilis* and *Saccharomyces cerevisiae* in biofilm reactors, *J. Ind. Microbiol.*, 16, 249, 1996.
181. Göksungur, Y. and Zorlu, N., Production of ethanol from beet molasses by Ca-alginate immobilized yeast cells in a packed-bed bioreactor, *Turk. J. Biol.*, 25, 265, 2001.
182. Dey, S., Semi-continuous production of ethanol from agricultural wastes by immobilized coculture in a two-stage bioreactor, *J. Environ. Biol.*, 23, 399, 2002.
183. Qureshi, N. et al., Continuous production of ethanol in high-productivity bioreactors using genetically engineered *Escherichia coli* Fbr5: membrane and fixed cell reactors, *Am. Inst. Chem. Eng.*, paper no. 589g (abstract only), 2005.
184. Najafpour, G., Younesi, H., and Syahidah Ku Ismail, K., Ethanol fermentation in an immobilized cell reactor using *Saccharomyces cerevisiae*, *Bioresour. Technol.*, 92, 251, 2004.
185. Yamada, T., Fatigati, M.A., and Zhang, M., Performance of immobilized *Zymomonas mobilis* 31821 (pZB5) on actual hydrolysates produced by Arkenol technology, *Appl. Biochem. Biotechnol.*, 98–100, 899, 2002.
186. Smogroicová, D., Dömény, Z., and Svitel, J., Modeling of saccharide utilization in primary beer fermentations with yeasts immobilized in calcium alginate, *Appl. Biochem. Biotechnol.*, 94, 147, 2001.
187. Liu, R. and Shen, F., Impacts of main factors on bioethanol fermentation from stalk juice of sweet sorghum by immobilized *Saccharomyces cerevisiae* (CICC 1308), *Bioresour. Technol.*, in press 2008.
188. Koshinsky, H.A., Cosby, R.H., and Khachatourians, G.G., Effects of T-2 toxin on ethanol production by *Saccharomyces cerevisiae*, *Biotechnol. Appl. Biochem.*, 16, 275, 1992.

189. Talebnia, F., Niklasson, C., and Taherzadeh, M.J., Ethanol production from glucose and dilute-acid hydrolyzates by encapsulated *S. cerevisiae*, *Biotechnol. Bioeng.*, 90, 345, 2005.

190. Skinner-Nemec, K.A., Nichols, N.N., and Leathers, T.D., Biofilm formation by bacterial contaminants of fuel ethanol production, *Biotechnol. Lett.*, 29, 379, 2007.

191. Ingledew, W.M., Continuous fermentation in the fuel alcohol industry: how does the technology affect yeast?, in Jacques, K.A., Lyons, T.P. and Kelsall, D.R. (Eds.), *The Alcohol Textbook*, 4th ed, Nottingham University Press, 2003, chap. 11.

192. Narandranath, N.V., Thomas, K.C., and Ingledew, W.M., Urea hydrogen peroxide reduces the numbers of lactobacilli, nourishes yeast, and leaves no residues in the ethanol fermentation, *Appl. Environ. Microbiol.*, 66, 4187, 2000.

193. Padukone, N., Advanced process options for bioethanol production, in Wyman, C.E. (Ed.), *Handbook on Bioethanol: Production and Utilization*, Taylor & Francis, Washington, D.C., 1996, chap. 14.

194. Cooney, C.L. et al., Simultaneous cellulose hydrolysis and ethanol production by a cellulolytic anaerobic bacterium, *Biotechnol. Bioeng. Symp. Ser.*, 8, 103, 1978.

195. Wright, J.D., Ethanol from biomass by enzymatic hydrolysis, *Chem. Eng. Prog.*, August 1998, 62.

196. Philippidis, G.P., Cellulose bioconversion technology, in Wyman, C.E. (Ed.), *Handbook on Bioethanol: Production and Utilization*, Taylor & Francis, Washington, D.C., 1996, chap. 12.

197. Linko, M., Viikari, L., and Suihko, M.L., Hydrolysis of xylan and fermentation of xylose to ethanol, *Biotechnol. Adv.*, 2, 233, 1984.

198. Singh, A. and Kumar, P.K., *Fusarium oxysporium*: status in bioethanol production, *Crit. Rev. Biotechnol.*, 11, 129, 1991.

199. Targonski, Z., Bujak, S., and Baraniak, A., Acid hydrolysis of beech sawdust hemicellulose and ethanol fermentation of hydrolysates by *Fusarium* sp. 27, *Acta Microbiol. Pol.*, 34, 261, 1985.

200. Panagiotou, G. and Christakopoulos, P., NADPH-dependent D-aldose reductases and xylose fermentation in *Fusarium oxysporium*, *J. Biosci. Bioeng.*, 97, 299, 2004.

201. Panagiotou, G. et al., Engineering of the redox imbalance of *Fusarium oxysporium* converting glucose to ethanol, *Metab. Eng.*, 8, 474, 2006.

202. Panagiotou, G. et al., Intracellular metabolite profiling of *Fusarium oxysporium* enables anaerobic growth on xylose, *J. Biotechnol.*, 115, 425, 2005.

203. Fujita, Y. et al., Synergistic saccharification, and direct fermentation to ethanol, of amorphous cellulose by use of an engineered yeast strain codisplaying three types of cellulolytic enzyme, *Appl. Environ. Microbiol.*, 70, 1207, 2004.

204. van Rooyen, R. et al., Construction of cellobiose-growing and fermenting *Saccharomyces cerevisiae*, *J. Biotechnol.*, 120, 284, 2005.

205. McBride, J.E. et al., Utilization of cellobiose by recombinant β-glucosidase-expressing strains of *Saccharomyces cerevisiae*: characterization and evaluation of the sufficiency of expression, *Enz. Microb. Technol.*, 37, 93, 2005.

206. Den Haan, R. et al., Hydrolysis and fermentation of amorphous cellulose by recombinant *Saccharomyces cerevisiae*, *Metab. Eng.*, 9, 87, 2007.

207. Lynd, L.R. et al., Consolidated bioprocessing of cellulosic biomass: an update, *Curr. Opin. Biotechnol.*, 16, 577, 2005.

208. Den Haan, R. et al., Functional expression of cellobiohydrolases in *Saccharomyces cerevisiae* towards one-step conversion of cellulose to ethanol, *Enz. Microb. Technol.*, 40, 1291, 2007.

209. Ballesteros, I. et al., Selection of thermotolerant yeasts for simultaneous saccharification and fermentation (SSF) of cellulose to ethanol, *Appl. Biochem. Biotechnol.*, 28–29, 307, 1991.

210. Ballesteros, I. et al., Optimization of the simultaneous saccharification and fermentation process using thermotolerant yeasts, *Appl. Biochem. Biotechnol.*, 39–40, 201, 1993.

211. Golias, H. et al., Evaluation of a recombinant *Klebsiella oxytoca* strain for ethanol production from cellulose by simultaneous saccharification and fermentation: comparison with native cellobiose-utilizing yeast strains and performance in co-culture with thermotolerant yeasts and *Zymomonas mobilis*, *J. Biotechnol.*, 96, 155, 2002.

212. Ingram, L.O. and Doran, J.B., Conversion of cellulosic materials to ethanol, *FEMS Microbiol. Rev.*, 16, 235, 1995.

213. Zhou, S. and Ingram, L., Engineering endoglucanase-secreting strains of ethanologenic *Klebsiella oxytoca* P2, *J. Ind. Microbiol. Biotechnol.*, 22, 600, 1999.

214. Padukone, N. et al., Characterization of recombinant *E. coli* ATCC 11303 (pLOI 297) in the conversion of cellulose and xylose to ethanol, *Appl. Microbiol. Biotechnol.*, 43, 850, 1995.

215. Philippidis, G.P. and Hatzis, C., Biochemical engineering analysis of critical process factors in the biomass-to-ethanol technology, *Biotechnol. Prog.*, 13, 222, 1997.

216. Yang, B. and Wyman, C.E., BSA treatment to enhance enzymatic hydrolysis of cellulose in lignin containing substrates, *Biotechnol. Bioeng.* 94, 611, 2006.

217. Wingren, A. et al., Effect of reduction in yeast and enzyme concentrations in a simultaneous-saccharification-and-fermentation-based bioethanol process: technical and economic evaluation, *Appl. Biochem. Biotechnol.*, 121–124, 485, 2005.

218. Linde, M., Galbe, M., and Zacchi, G., Simultaneous saccharification and fermentation of steam-pretreated barley straw at low enzyme loadings and low yeast concentration, *Enz. Microb. Technol.*, 40, 1100, 2007.

219. Schubert, C., Can biofuels finally take center stage?, *Nat. Biotechnol.*, 24, 777, 2006.

220. Li, X.L. et al., Properties of a recombinant β-glucosidase from polycentric anaerobic fungus *Orpinomyces* PC-2 and its application for cellulose hydrolysis, *Appl. Biochem. Biotechnol.*, 113–116, 233, 2004.

221. Hughes, S.R. et al., High-throughput screening of cellulase F mutants from multiplexed plasmid sets using an automated plate assay on a functional proteomic robotic workcell, *Proteome Sci.*, 4, 10, 2006.

222. Li, X.L. et al., Expression of an AT-rich xylanase gene from the anaerobic fungus *Orpinomyces* sp. strain PC-2 in and secretion of the heterologous enzyme by *Hypocrea jecorina*, *Appl. Microbiol. Biotechnol.*, 74, 1264, 2007.

223. Murai, T. et al., Assimilation of cellooligosaccharides by a cell-surface engineered yeast expressing β-glucosidase and carboxymethylcellulase from *Aspergillus aculeatus*, *Appl. Environ. Microbiol.*, 64, 4857, 1998.

224. Zhou, S. et al., Enhancement of expression and apparent secretion of *Erwinia chrysanthemi* endoglucanase (encoded by *celZ*) in *Escherichia coli* B, *Appl. Environ. Microbiol.*, 65, 2439, 1999.

225. Cho, K.M. and Yoo, Y.J., Novel SSF process for ethanol production from microcrystalline cellulose using δ-integrated recombinant yeast, *Saccharomyces cerevisiae*, *J. Microbiol. Biotechnol.*, 9, 340, 1999.

226. Zhou, S. et al., Gene integration and expression and extracellular secretion of *Erwinia chrysanthemi* endoglucanase CelY (*celY*) and CelZ (*celZ*) in ethanologenic *Klebsiella oxytoca* P2, *Appl. Environ. Microbiol.*, 67, 6, 2001.

227. Fujita, Y. et al., Direct and efficient production of ethanol from cellulosic material with a yeast strain displaying cellulolytic enzymes, *Appl. Environ. Microbiol.*, 68, 5136, 2002.

228. Wu, Z. and Lee, Y.Y., Inhibition of the enzymatic hydrolysis of cellulose by ethanol, *Biotechnol. Lett.*, 19, 977, 1997.

229. Martín, C. et al., Investigation of cellulose convertibility and ethanolic fermentation of sugarcane bagasse pretreated by wet oxidation and steam explosion, *J. Chem. Technol. Biotechnol.*, 81, 1669, 2006.
230. Chadha, B.S. et al., Hybrid process for ethanol production from rice straw, *Acta Microbiol. Immunol. Hung.*, 42, 53, 1995.
231. Chung, Y.C., Bakalinsky, A., and Penner, M.H., Analysis of biomass cellulose in simultaneous saccharification and fermentation processes, *Appl. Biochem. Biotechnol.*, 66, 249, 1997.
232. Belkacemi, K. et al., Ethanol production from AFEX-treated forages and agricultural residues, *Appl. Biochem. Biotechnol.*, 70-72, 441, 1998.
233. Hari Krishna, S., Janardhan Reddy, T., and Chowdary, G.V., Simultaneous saccharification and fermentation of lignocellulosic wastes to ethanol using a thermotolerant yeast, *Bioresour. Technol.*, 77, 193, 2001.
234. Schell, D.J. et al., Dilute-sulfuric acid pretreatment of corn stover in pilot-scale reactor: investigation of yields, kinetics, and enzymatic digestibilities of solids, *Appl. Biochem. Biotechnol.*, 105–108, 69, 2003.
235. Varga, E. et al., High solid simultaneous saccharification and fermentation of wet oxidized corn stover to ethanol, *Biotechnol. Bioeng.*, 88, 567, 2004.
236. Kim, T.H. and Lee, Y.Y., Pretreatment of corn stover by soaking in aqueous ammonia, *Appl. Biochem. Biotechnol.*, 121–124, 1119, 2005.
237. Öhgren, K. et al., Simultaneous saccharification and co-fermentation of glucose and xylose in steam-pretreated corn stover at high fiber content with *Saccharomyces cerevisiae* TMB3400, *J. Biotechnol.*, 126, 488, 2006.
238. Kim, T.H. et al., Pretreatment of corn stover by low-liquid ammonia percolation process, *Appl. Biochem. Biotechnol.*, 133, 41, 2006.
239. Zhang, Y. et al., Protoplast fusion between *Geotrichum candidium* and *Phanerochaete chrysosporium* to produce fusants for corn stover fermentation, *Biotechnol. Lett.*, 28, 1351, 2006.
240. Skotnicki, M.L. et al., High-productivity alcohol fermentations using *Zymomonas mobilis*, *Biochem. Soc. Symp.*, 48, 53, 1983.
241. Gorkhale, D. and Deobagkar, D., Isolation of intergeneric hybrids between *Bacillus subtilis* and *Zymomonas mobilis* and the production of thermostable amylase by hybrids, *Biotechnol. Appl. Biochem.*, 20, 109, 1994.
242. Ma, Y.-J. et al., Efficient utilization of starch by a recombinant strain of *Saccharomyces cerevisiae* producing glucoamylase and isoamylase, *Biotechnol. Appl. Biochem.*, 31, 55, 2000.
243. Ramakrishnan, S. and Hartley, B.S., Fermentation of lactose by yeast cells secreting recombinant fungal lactase, *Appl. Environ. Microbiol.*, 59, 4230, 1993.
244. Palnitkar, S.S. and Lachke, A.H., Efficient simultaneous saccharification and fermentation of agricultural residues by *Saccharomyces cerevisiae* and *Candida shehatae*, the D-xylose fermenting yeast, *Appl. Biochem. Biotechnol.*, 26, 151, 1990.
245. Lebeau, T., Jouenne, T., and Junter, G.A., Continuous alcoholic fermentation of glucose/xylose mixtures by co-immobilized *Saccharomyces* and *Candida shehatae*, *Appl. Microbiol. Biotechnol.*, 50, 309, 1998.
246. Kordowska-Wister, M. and Targonski, Z., Ethanol fermentation on glucose/xylose mixture by co-cultivation of restricted glucose catabolite repressed mutants of *Pichia stipitis* with respiratory deficient mutants of *Saccharomyces cerevisiae*, *Acta Microbiol. Pol.*, 51, 345, 2002.
247. Farid, M.A., El-Enshasy, H.A., and El-Deen, A.M., Alcohol production from starch by mixed cultures of *Aspergillus awamori* and immobilized *Saccharomyces cerevisiae* at different agitation speeds, *J. Basic Microbiol.*, 42, 162, 2002.

248. Qian, M. et al., Ethanol production from dilute-acid softwood hydrolysate by co-culture, *Appl. Microbiol. Biotechnol.*, 134, 273, 2006.
249. Lebeau, T., Jouenne, T., and Junter, G.A., Long-term incomplete xylose fermentation, after glucose exhaustion, with *Candida shehatae* co-immobilized with *Saccharomyces cerevisiae*, *Microbiol. Res.*, 162, 211, 2007.
250. Madson, P.W., Ethanol distillation: the fundamentals, in Jacques, K.A., Lyons, T.P., and Kelsall, D.R. (Eds.), *The Alcohol Textbook,* 4th ed, Nottingham University Press, 2003, chap. 22.
251. Singh, A. and Mishra, P., *Microbial Pentose Utilization. Current Applications in Biotechnology (Prog. Ind. Microbiol.* 33), 1995, chap. 14.
252. Swain, R.L.B., Development and operation of the molecular sieve: an industry standard, in Jacques, K.A., Lyons, T.P., and Kelsall, D.R. (Eds.), *The Alcohol Textbook,* 4th ed, Nottingham University Press, 2003, chap. 23.
253. Essien, D. and Pyle, D.L., Energy conservation in ethanol production by fermentation, *Proc. Biochem.*, 18 (August, 4), 31, 1983.
254. Sikyta, B., *Techniques in Applied Microbiology (Prog. Ind. Microbiol.* 31), 1995, chap. 12.
255. Ghose, T.K., Roychoudhury, P.K., and Ghosh, P., Simultaneous saccharification and fermentation (SSF) of lignocellulosics to ethanol under vacuum cycling and step feeding, *Biotechnol. Bioeng.*, 26, 377, 1984.
256. Minier, M. and Goma, G., Ethanol production by extractive fermentation, *Biotechnol. Bioeng.*, 24, 1565, 1982.
257. Pitt, W.W., Haag, G.L., and Lee, D.D., Recovery of ethanol from fermentation broths using selective sorption-desorption, *Biotechnol. Bioeng.*, 25, 123, 1983.
258. Einicke, W.D., Gläser, B., and Schöoullner, R., In-situ recovery of ethanol from fermentation broth by hydrophobic adsorbents, *Acta Biotechnol.*, 11, 353, 1991.
259. Nishizawa, Y. et al., Ethanol production by cell recycling with hollow fibers, *J. Ferment. Technol.*, 61, 599, 1983.
260. O'Brien, D.J. et al., Ethanol recovery from corn fiber hydrolysate fermentations by pervaporation, *Bioresour. Technol.*, 92, 15, 2004.
261. Li, G. et al., Time-dependence of pervaporation performance for the separation of ethanol/water mixtures through poly(vinyl alcohol) membrane, *J. Colloid Interface Sci.*, 306, 337, 2007.
262. Dominguez, J.M. et al., Ethanol production from xylose with the yeast *Pichia stipitis* and simultaneous product recovery by gas stripping using a gas-lift fermentor with attached side-arm (GLSA), *Biotechnol. Bioeng.*, 67, 336, 2000.
263. Taylor, F. et al., Continuous fermentation and stripping of ethanol, *Biotechnol. Prog.*, 11, 693, 1995.
264. Taylor, F. et al., Effects of ethanol concentration and stripping temperature on continuous fermentation rate, *Appl. Microbiol. Biotechnol.*, 48, 311, 1997.
265. Taylor, F. et al., Dry-grind process for fuel ethanol by continuous fermentation and stripping, *Biotechnol. Prog.*, 16, 541, 2000.
266. Mazumdar-Shaw, K. and Suryanarayan, S., Commercialization of a novel fermentation concept, *Adv. Biochem. Eng./Biotechnol.*, 85, 29, 2003.
267. Gibbons, W.R., Westby, C.A., and Dobbs, T.L., A continuous, farm-scale, solid-phase fermentation process for fuel ethanol and protein feed production from fodder beets, *Biotechnol. Bioeng.*, 26, 1098, 1984.
268. Gibbons, W.R., Westby, C.A., and Dobbs, T.L., Intermediate-scale, semicontinuous solid-phase fermentation process for fuel ethanol from sweet sorghum, *Appl. Environ. Microbiol.*, 51, 115, 1986.

269. Jacques, K.A., Ethanol production and the modern livestock feed industry: a relationship continuing to grow, in Jacques, K.A., Lyons, T.P., and Kelsall, D.R. (Eds.), *The Alcohol Textbook*, 4th ed, Nottingham University Press, 2003, chap. 27.

270. Rausch, K.D. and Belyea, R.L., The future of coproducts from corn processing, *Appl. Biochem. Biotechnol.*, 128, 47, 2006.

271. Dawson, K.A., Biorefineries: the versatile fermentation plants of the future, in Jacques, K.A., Lyons, T.P. and Kelsall, D.R. (Eds.), *The Alcohol Textbook*, 4th ed, Nottingham University Press, 2003, chap. 28.

272. Tucker, M.P. et al., Conversion of distiller's grain into fuel alcohol and a higher value animal feed by dilute-acid pretreatment, *Appl. Biochem. Biotechnol.*, 113–116, 1139, 2004.

273. Leiper, K.A. et al., The fermentation of beet sugar syrup to produce bioethanol, *J. Inst. Brew.*, 112, 122, 2006.

274. Sørensen, H.R., Meyer, A.S., and Pederson, S., Enzymatic hydrolysis of water-soluble wheat arabinoxylan. I. Synergy between α-L-arabinofuranosidases, endo-1,4-β-xylanases, and β-xylosidase activities, *Biotechnol. Bioeng.*, 81, 726, 2003.

275. Sørensen, H.R. et al., Efficiencies of designed enzyme combinations in releasing arabinose and xylose from wheat arabinoxylan in an industrial ethanol fermentation residue, *Enz. Microb. Technol.*, 36, 773, 2005.

276. Sørensen, H.R., Pederson, S., and Meyer, A.S., Optimization reaction conditions for enzymatic viscosity reduction and hydrolysis of wheat arabinoxylan in an industrial ethanol fermentation residue, *Biotechnol. Prog.*, 22, 505, 2006.

277. Sørensen, H.R., Pederson, S., and Meyer, A.S., Synergistic enzyme mechanisms and effects of sequential enzyme additions on degradation of water insoluble wheat arabinoxylan, *Enz. Microb. Technol.*, 40, 908, 2007.

278. Duarte, L.C. et al., Comparison of two posthydrolysis processes of brewery's spent grain autohydrolysis liquor to produce a pentose-containing culture medium, *Appl. Biochem. Biotechnol.*, 113–116, 1041, 2004.

279. Carvalheiro, F. et al., Supplementation requirements of brewery's spent grain hydrolysate for biomass and xylitol production by *Debaryomyces hansenii* CCMI 941, *J. Ind. Microbiol. Biotechnol.*, 33, 646, 2006.

280. Kopsahelis, N. et al., Comparative study of spent grains and delignified spent grains as yeast supports for alcohol production from molasses, *Bioresour. Technol.*, 98, 1440, 2007.

281. van Haandel, A.C., Integrated energy production and reduction of the environmental impact at alcohol distillery plants, *Water Sci. Technol.*, 52, 49, 2005.

282. Paula, L. and Birrer, F., Including public perspectives in industrial biotechnology and the biobased economy, *J. Agric. Environ. Ethics*, 19, 253, 2006.

283. Lanfranco, L., Engineering crops, a deserving venture, *Rivista di Biologia*, 96, 31, 2003.

284. Lamb, C.J. et al., Emerging strategies for enhancing crop resistance to microbial pathogens, *Biotechnology*, 10, 1436, 1992.

285. Khan, R.R. et al., Molecular mapping of stem and leaf rust resistance in wheat, *Theor. Appl. Genet.*, 111, 846, 2005.

286. Kota, R. et al., Fine genetic mapping fails to dissociate durable stem rust resistance gene *Sr2* from pseudo-black chaff in common wheat (*Triticum aestivum* L.), *Theor. Appl. Genet.*, 112, 492, 2006.

287. Horvath, H. et al., Genetically engineered stem rust resistance in barley using the *Rpg1* gene, *Proc. Natl. Acad. Sci., USA*, 100, 364, 2003.

288. Ellis, R.P. et al., Wild barley: a source of genes for crop improvement in the 21st century?, *J. Exp. Bot.*, 51, 9, 2000.

289. Shou, H., Bordallo, P., and Wang, K., Expression of the *Nicotiana* protein kinase (NPK1) enhanced drought tolerance in transgenic maize, *J. Exp. Bot.*, 55, 1013, 2004.

290. Nagano, H. et al., Genealogy of the "Green Revolution" gene in rice, *Genes Genet. Syst.*, 80, 351, 2005.
291. Merkle, S.A. and Dean, J.F., Forest tree biotechnology, *Curr. Opin. Biotechnol.*, 11, 298, 2000.
292. Good, A.G., Shrawat, A.K., and Muench, D.G., Can less yield more? Is reducing nutrient input into the environment compatible with maintaining crop production?, *Trends Plant Sci.*, 9, 597, 2004.
293. Sinclair, T.R., Purcell, L.C., and Sneller, C.H., Crop transformation and the challenge to increase yield, *Trends Plant Sci.*, 9, 70, 2004.
294. Carrari, F. et al., Engineering central metabolism in crop species: learning the system, *Metab. Eng.*, 5, 191, 2003.
295. Ku, M.S. et al., Introduction of genes encoding C4 photosynthesis enzymes into rice plants: physiological consequences, *Novartis Found. Symp.*, 236, 100, 2001.
296. Nunes-Nesi, A. et al., Enhanced photosynthetic performance and growth as a consequence of decreasing mitochondrial malate dehydrogenase activity in transgenic tomato plants, *Plant Physiol.*, 137, 611, 2005.
297. Smidansky, E.D. et al., Expression of a modified ADP-glucose pyrophosphorylase large subunit in wheat seeds stimulates photosynthesis and carbon metabolism, *Planta*, 225, 965, 2007.
298. Parry, M.A. et al., Manipulation of Rubisco: the amount, activity, function and regulation, *J. Exp. Bot.*, 54, 1321, 2003.
299. Kebeish, R. et al., Chloroplastic photorespiratory bypass increases photosynthesis and biomass production in *Arabidopsis thaliana*, *Nature Biotechnol.*, 25, 593, 2007.
300. Bugbee, B.G. and Salisbury, F.B., Exploring the limits of crop productivity. I. Photosynthetic efficiency of wheat in high irradiance environments, *Plant Physiol.*, 88, 869, 1988.
301. Sanderson, K., A field in ferment, *Nature*, 444, 673, 2006.
302. McLaren, J.S., Crop biotechnology provides an opportunity to develop a sustainable future, *Trends Biotechnol.*, 23, 339, 2005.
303. Sticklen, M., Plant genetic engineering to improve biomass characteristics for biofuels, *Curr. Opin. Biotechnol.*, 17, 315, 2006.
304. Cardinal, A.J., Lee, M., and Moore, K.J., Genetic mapping and analysis of quantitative loci affecting fiber and lignin content in maize, *Theor. Appl. Genet.*, 106, 866, 2003.
305. Baucher, M. et al., Lignin: genetic engineering and impact on pulping, *Crit. Rev. Biochem. Mol. Biol.*, 38, 305, 2003.
306. Anterola, A.M. and Lewis, N.G., Trends in lignin modification: a comparative analysis of the effects of genetic manipulations/mutations on lignification and vascular integrity, *Phytochemistry*, 61, 221, 2002.
307. Abbott, J.C. et al., Simultaneous suppression of multiple genes by single transgenes. Down-regulation of three unrelated lignin biosynthetic genes in tobacco, *Plant. Physiol.*, 128, 844, 2002.
308. Laigeng, L. et al., Combinatorial modification of multiple lignin traits in trees through multigene cotransformation, *Proc. Natl. Acad. Sci., USA*, 100, 4939, 2003.
309. Rastogi, S. and Dwivedi, U.N., Down-regulation of lignin biosynthesis in transgenic *Leucaena leucocephala* harboring O-methyltransferase gene, *Biotechnol. Prog.*, 22, 609, 2006.
310. Hancock, J.E. et al., Plant growth, biomass partitioning and soil carbon formation in response to altered lignin biosynthesis in *Populus tremuloides*, *New Phytol.*, 173, 732, 2007.
311. Oraby, H. et al., Enhanced conversion of plant biomass into glucose using transgenic rice-produced endoglucanase for cellulosic ethanol, *Transgenic Res.*, in press 2007.

312. Hertzberg, M. et al., A transcriptional roadmap to wood formation, *Proc. Natl. Acad. Sci., USA*, 98, 14732, 2001.
313. Bhalla, P.L., Genetic engineering of wheat: current challenges and opportunities, *Trends Biotechnol.*, 24, 305, 2006.
314. Francki, M. and Appels, R., Wheat functional genomics and engineering crop improvement, *Genome Biol.*, 3, 1013.1, 2002.
315. Hill, J. et al., Environmental, economic, and energetic costs and benefits of biodiesel and ethanol biofuels, *Proc. Natl. Acad. Sci., USA*, 103, 11206, 2006.
316. Torney, F. et al., Genetic engineering approaches to improve bioethanol production from maize, *Curr. Opin. Biotechnol.*, 18, 193, 2007.
317. Tobias, C.M. et al., Analysis of expressed sequence tags and the identification of associated short tandem repeats in switchgrass, *Theor. Appl. Genet.*, 111, 956, 2005.
318. Somleva, M.N., Switchgrass (*Panicum virgatum* L.), *Methods Mol. Biol.*, 344, 65, 2006.
319. Vincentz, M. et al., Evaluation of monocot and eudicot divergence using the sugarcane transcriptome, *Plant Physiol.* 134, 951, 2004.
320. Abelson, P.H., Preface, in Wyman, C.E. (Ed.), *Handbook on Bioethanol: Production and Utilization*, Taylor & Francis, Washington, D.C., 1996.
321. Wyman, C.E., Ethanol production from lignocellulosic biomass: overview, in Wyman, C.E. (Ed.), *Handbook on Bioethanol: Production and Utilization*, Taylor & Francis, Washington, D.C., 1996, chap. 1.
322. Van Noorden, R., Against the grain, *Chemistry World*, 4, part 3 (March 2007), p. 13.
323. Anonymous, Canola and soya to the rescue, *The Economist*, 30, May 6, 2006.
324. Anonymous, Dash for green fuel pushes up price of meat in US, *The Times*, April 12, 2007, www.timesonline.co.uk.
325. Lynd, L.R., Wyman, C.E., and Gerngross, T.U., Biocommodity engineering, *Biotechnol. Prog.*, 15, 777, 1999.

5 The Economics of Bioethanol

5.1 BIOETHANOL MARKET FORCES IN 2007

5.1.1 THE IMPACT OF OIL PRICES ON THE "FUTURE" OF BIOFUELS AFTER 1980

"Economists attract ridicule and resentment in equal measures."[1]

The most telling aspect of the above quote is not that it derives from a recent collection of essays originally published in *The Economist*, one of the leading opinion formers in Western liberal economic thought, but that it is the first sentence in the Introduction to that volume. Graphical representations from many economic sources share one common factor, a short time axis. In the world of practical economics, hours, days, weeks, and months dominate the art of telling the near future — for price movements in stock markets, in the profitability of major corporations and their mergers, the collapse of currencies, or surges in commodity prices. Extrapolations are usually linear extrapolations from a small historical database of recent trends. Pundits predict much, but whatever the outcome, their predictions may very soon be forgotten. Economic models may, with hindsight, be wildly optimistic or inaccurate, but by the time "hindsight" is raised as a debating issue, the original set of parameters may have become completely irrelevant.

The history of biofuels since the early 1970s exhibits such cycles of optimism and pessimism, of exaggerated claims or dire prognostications; a series of funding programs have blossomed but — sometimes equally rapidly — faded.[2] The prime mover in that sequence of boom and bust, evangelism and hostility, newspaper headlines and indifference, has invariably been the market price of oil. However undesirable a driver this is in the ongoing discussions on the development of biofuels, from the viewpoint of the biotechnologist in the scientific research community, it can never be ignored.[3] Even with supporting arguments based on energy security and amelioration of greenhouse gas emissions, a high cost of any biofuel relative to that of gasoline, diesel fuel, and heating oil is the main plank in the logic used by skeptics: that however worthy are the goal and vision of biofuels for the future, they simply cannot be afforded and — in a sophisticated twist of the argument — may themselves contribute to the continuing deprivation of energy-poor nations and societies while the energy-rich developed economies impose rationing of fossil fuel use and access to maintain their privileged position. Although lobbyists for the global oil industry clearly have a vested interest in continuously challenging the economic costs of biofuel production, an underlying fear that, whether significant climate change could be lessened by the adoption of biofuels for private transportation (a debatable but quantifiable scenario — see chapter 1, section 1.6.2) and whether "energy security" is simply a novel means of subsidizing inefficient farmers to grow larger and larger harvests of monoculture crops to maintain agricultural incomes and/or employment for a few decades more, only a clear understanding of the financial implications of biofuels can help fix the agenda for rational choices to be made about investing in new technologies across the wide spectrum of rival possible biofuel options in the twenty-first century.

It is, however, undeniable that oil price volatility can shake the confidence of any investor in bioenergy. In the two decades after 1983, the average retail price for gasoline (averaged over all available grades) was 83¢/gallon, but transient peaks and troughs reached 129¢ and 55¢, respectively (figure 5.1). Slumping oil prices almost wiped out the young sugar ethanol-fueled car fleet in Brazil in the 1980s (chapter 1, section 1.2). Since 2003, the conclusion that the era of cheap oil is irreversibly over has been increasingly voiced as the demands of the burgeoning economies of India and China place unavoidable stresses on oil availability and market price.[4] If accurate, that prediction would be the single most important aid and rationale for biofuels as a commercial reality.

5.1.2 Production Price, Taxation, and Incentives in the Market Economy

It is vital at this point to differentiate commercial realities from strategic (or geopolitical) and all other considerations. Although (as discussed in chapter 1) historical, environmental, political, and macroeconomic arguments have all been adduced in support of bioenergy programs, fiscal considerations now play an important role in both encouraging (prompted by political agendas) the take-up of novel alternative fuels and in partitioning the market for first and subsequent generations of rival but not equally readily commercialized biofuels. Indeed, taxation issues were quickly recognized and seized on by proponents of bioethanol, particularly because they were useful to counter the gasoline versus gasohol price differential: for example, in the United States, the indirect costs of regulating air pollution and of military protection for oil supplies from the Middle East are calculable and greatly inflate the nominal price of crude oil but are not (explicitly) passed on to the consumer.[5] This distortion of the transportation fuel market by "hidden" subsidies has also led to economics models in which other indirect factors are included in the cost-benefit analysis:

- Technological developments that improve the national scientific base for employment, patents and overseas licensing, and engineering advances that "spillover" into related fields
- Reduced foreign currency payments and associated "banking" costs — highly important for a developing economy such as Brazil's
- Higher income and sales tax returns from greater rural employment
- Reduced longer-term economic impacts of climate change and air pollution

All these arguments are, to varying degrees, contentious, and skeptics can be found from opposite ends of the economic spectrum, from oil industry analysts to academics who foresee only accelerated land degradation from the industrial agronomy of energy crop cultivation.[6,7]

Taxation as an instrument of social and economic policy has, moreover, obvious limitations if wasteful subsidies or punitive levels of taxation on standard gasoline and diesel products are to be avoided. Consider the following three scenarios:

- Bioethanol production can generate a commercial fuel with pump prices no greater than those of standard gasoline grades at equivalent tax rates, the comparison being valid when average prices during a period of one to five years are calculated, thus avoiding false comparisons at peaks and troughs caused by fluctuations in both agricultural feedstock prices (as an important cost input to biofuel production) and oil price movements if they continue to move inside the wide limits evident since the early 1980s (figure 5.1).

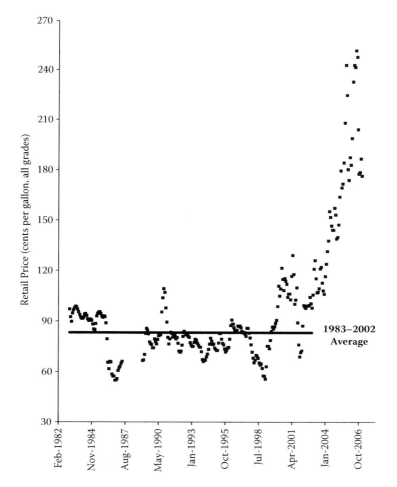

FIGURE 5.1   U.S. gasoline retail prices, 1983–2007: total sales by all sellers, incorporating prices for all available grades. (Data from the U.S. Department of Energy.)

- Bioethanol can be produced commercially at a total (production, distribution, and resale) cost that averaged, during a 5- to 10-year period, 10–50% higher than that of gasoline.
- Bioethanol production can only generate an unsubsidized product with a total cost more than twice that of the refinery gate price of standard gasoline (a price differential quoted for the United States in the late 1990s[8]) — or perhaps, even up to 10 times higher than conventional fuels where, for example, local conditions of climate and biomass availability are consistently much less favorable than for sugarcane production in Brazil or corn in the United States or where only refractory lignocellulosic feedstocks can be accessed with poorly developed bioprocessing technology.

The first (optimistic) case approximates that of Brazilian consumers with flexibly fueled cars after 2000.[9] The second case is the conclusion most often reached in technoeconomic studies, whereas the third scenario is parallel to the emergency or "wartime" case discussed when the energy yields of conventional and alternative fuels were considered in chapter 1 (section 1.6.1): even if biofuels are prohibitively expensive now, technical developments may erode that differential or be obviated if (or when) fossil fuel shortages become acute in the present century (see later, section 5.6). In all three cases, taxation policy can (and will) influence consumer choice and purchasing patterns, whether for short- (tactical) or longer-term (strategic) reasons and when legislation enforces alternative or reconstituted fuels to achieve environmental targets.

A snapshot of data from October 2002 in Brazil, however, reveals the complexity of the interaction between production/distribution costs and imposed taxation on the final at-pump selling price.[10] Although gasohol mixtures, hydrous ethanol, and diesel all had very similar production costs, equivalent to approximately 15¢/l (57¢/gallon) at that time, the final cost to the user was determined by the much higher taxes applied to gasohol (figure 5.2). Brazil exemplifies the extensive use of taxation to determine and direct the perceived prices of gasoline and alternative fuels as a deliberate instrument of national policy. Such management of the fuel economy is likely to be instigated in societies where not only economics but social and environmental considerations are taken into account, but runs the risk of experiencing budgetary shortfalls if the total tax raised is severely reduced when the policy is too successful in achieving its aims — this becomes even worse if private transportation is perceived as being subsidized by other taxation sources (e.g., sales tax, income tax). For all the various interest groups in biofuels development, therefore, the priority is to establish viable production processes with the minimum requirement for tax incentives.

5.2 COST MODELS FOR BIOETHANOL PRODUCTION

Economic considerations have featured in both primary analyses and reviews of the biotechnology of fuel ethanol production published in the last 25 years.[11] Because a lignocellulosic ethanol industry has yet to fully mature, most of those studies have been derived from laboratory or (at best) small pilot-plant data, and estimates for feedstock and capital investment costs have varied greatly, as have assumptions on

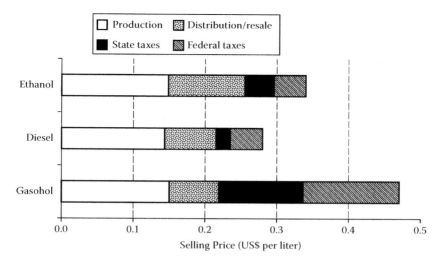

FIGURE 5.2 Production and distribution costs and taxation for motor fuels in Brazil in 2002. (Data from Moreira et al.[10])

the scale of commercial production required to achieve any intended price/cost target for the product. As with estimates of net energy yield and greenhouse gas reductions (chapter 1, sections 1.6.1 and 1.6.2), the conclusions reached are heavily influenced by the extent to which costs can be offset by coproduct generation (as a source of income) and of the complexity of the total production process, not only as a primary cause for increased setup costs but also a potential source of process efficiencies and additional, saleable coproducts.

Few of the influential studies are full business models for bioethanol, in particular avoiding any computations for profitability, often because the main driver has been to establish and substantiate grounds for initial or continued investment by national and/or international funding agencies — and with the implicit assumption that any production process for fuel ethanol outside Brazil suffers by that very comparison because of the lack of such favorable climatic and economic features (in particular, land use, labor cost, and the dovetailing of ethanol production with a fully mature sugarcane industry). Nevertheless, a historical survey of key points in the development of the economic case for bioethanol reveals the convergence toward a set of key parameters that will be crucial for any biofuel candidate in the next 10–50 years.

5.2.1 EARLY BENCHMARKING STUDIES OF CORN AND LIGNOCELLULOSIC ETHANOL IN THE UNITED STATES

During the 1970s, the U.S. Department of Energy (DOE) commissioned four detailed technical and economic reports from consultants on possible production routes for ethanol as a fuel supplement:[12]

- Corn-based manufacturing facilities at scales of 10 million up to 100 million gallons/year

- Wheat straw conversion via enzymic hydrolysis at a 25-million-gallon/year scale of production
- Another intermediate scale process for molasses fermentation to produce 14 million gallons/year
- A farm-based model (25 gallons/hr)

5.2.1.1   Corn-Derived Ethanol in 1978

The assessment of corn-derived ethanol was the most extensive of the reports (60% of the total printed pages in the final collection of papers) and formed a notional blueprint for a facility sited in Illinois with a projected working life of 20 years and operating costs of approximately 95¢/gallon of hydrous ethanol (table 5.1). The final factory gate selling price was computed to be $1.05/gallon (1978 prices) in the base case of the 50-million-gallon/year capacity including the results of a 15% discounted cash flow/interest rate of return analysis; the selling price was a little lower (98¢/gallon) with twice the annual capacity but considerably higher ($1.55/gallon) at only 10 million gallons/year. The quoted comparative price for refinery gasoline was 40¢/gallon; after allowing for the lower energy content of ethanol (70% of that of gasoline — chapter 1, section 1.3), the "real" cost of corn-derived ethanol would have been $1.50/gallon for the 50-million-gallon facility, that is, 3.75-fold higher than gasoline at that time.

Various options were explored in the study to define the sensitivity of the required selling price for ethanol:

1. The DOE required the analyses to define a selling price that would cover not only the annual operating expenses but also yield a return on equity, the base case being a 15% discounted cash flow/interest rate of return; increasing this factor to 20% resulted in a higher selling price ($1.16/gallon for the base-case scenario).
2. Lengthening the depreciation schedule from 10 to 20 years increased the selling price by 2¢/gallon.
3. Increasing the working capital to 20% of the total production cost increased the selling price by 3¢/gallon.
4. A higher investment tax credit (50%) would reduce the selling price by 2¢/gallon.
5. Financing only 80% of the plant investment could reduce the selling price by 10¢/gallon.
6. For every 10% rise in the price of corn, the selling price would increase by 8¢/gallon (after allowing for a triggered rise in the selling price of the solid animal feed coproduct).
7. For every 10% rise in the price of the animal feed coproduct, the selling price would decrease by 4¢/gallon.
8. Replacing local coal by corn stover as the fuel for steam generation would increase the selling price by 4¢/gallon — although a lower total investment (by approximately $1 million) would have been an advantage resulting from the removal of the need for flue gas desulfurization.

TABLE 5.1
Cost Estimates for Ethanol Production from Corn Grain

Manufacturing input	Annual costs ($ million)	Production cost (cents/gallon)
Raw materials		
Corn	44.77	89.5
Yeast	0.32	0.6
Ammonia	0.37	0.7
Coal	2.41	4.8
Other chemicals	0.18	0.4
Utilities		
Cooling water (from plant)	0.00	0.0
Steam (from plant)	0.00	0.0
Electricity	1.65	3.3
Diesel fuel	0.01	0.0
Labor		
Management	0.24	0.5
Supervisors/operators	2.19	4.4
Office and laborers	1.20	2.4
Fixed charges		
Depreciation	5.80	11.6
License fees	0.03	0.1
Maintenance	1.83	3.7
Tax and insurance	0.91	1.8
Miscellaneous		
Freight	2.50	5.0
Sales	1.93	3.9
General/administrative overheads	0.64	1.3
Coproduct credits		
Dark grains	19.18	38.4
Ammonium sulfate	0.41	0.8
Total	47.41	94.8

Source: Data from Paul.[12]

All of these changes are comparatively minor, and other quantified changes to the overall process were likely to have been equally small: ammonium sulfate (a coproduct arising from flue gas desulfurization) was only generated in small amounts, approximately 3 tons/day, and no allowance was made for capturing and selling the CO_2 generated in the fermentation step. No denaturant was included in the final cost breakdown.

Alternative feedstocks were also explored. Milo (grain sorghum) offered a slight reduction in the selling price of ethanol (down to $1.02/gallon) but was considered a small-acreage crop at that time. Both wheat and sweet sorghum were likely to increase the final factory gate selling price to $1.31/gallon and $1.40/gallon, respectively. Although wheat and milo grain could be processed in essentially the same equipment used for corn, sweet sorghum required a higher investment in plant facilities.

5.2.1.2 Wheat Straw-Derived Ethanol in 1978

The first detailed economic costing of a lignocellulosic ethanol process envisaged a process similar in outline to that eventually adopted by Iogen in Canada, where wheat straw was first pretreated with acid before fungal cellulase was used to digest the cellulose for a *Saccharomyces* fermentation of the liberated glucose (no pentose sugars were included as substrates at that date). Results from laboratory studies were extrapolated to a 25-million-gallon/year facility that was designed to be stand-alone and capable of generating the required fungal cellulase on site (table 5.2).

The calculated factory gate plant was much higher than that estimated for corn-derived ethanol, that is, $3.34/gallon — this cost included raw materials, utilities,

TABLE 5.2
Cost Estimates for Ethanol Production from Wheat Straw

Manufacturing input	Production cost (cents/gallon)	Cost (% of total)
	Raw materials	
Wheat straw	38.8	11.6
Cellulose, news print	55.9	16.7
Peptone	79.2	23.7
Other chemicals	20.3	6.1
	Utilities	
Cooling water	1.0	0.3
Process water	2.8	0.8
Steam	30.2	9.0
Electricity	14.2	4.2
	Variable cost	
Operating labor	9.4	2.8
Maintenance labor	8.6	2.6
Control laboratory	1.9	0.6
Maintenance material	8.6	2.6
Operating supplies	0.9	0.3
	Direct cost	
Plant overhead	15.9	4.8
Taxes and insurance	7.7	2.3
Depreciation	38.7	11.6
Total	334.2	100

Source: Data from Paul.[12]

maintenance materials and labor, operating labor and supplies, the facility laboratory, plant overhead, taxes, insurance, and depreciation but made no estimates for general and administrative, sales or research costs, profit, or any by-product credits or disposal charges. Of this total, 61% was attributed to materials, of which the wheat straw feedstock accounted for 11–12%. The other components of the price estimate and various options for reducing the total were assessed:

- Utilities and capital costs were each 14% of the total.
- Labor costs were 11% of the total.
- A high-cost peptone (proteased protein) nitrogen source was included in the fermentation media; adopting a lower-price product could reduce the cost of the product by 9%.
- The conversion of cellulose to soluble sugars was 45% (w/w basis, calculated on the straw weight); increasing this to 60% could effect a 6% reduction in product cost.
- Reducing the enzyme loading (ratio of enzyme to cellulose) by fourfold could reduce the product price to approximately $2/gallon.

No allowance is made in table 5.2 for capital costs or profit; on the basis of 15%/year of fixed capital for general and administrative, sales, and research costs, a 15-year life expectancy of the plant, and a 48% tax rate, the selling price for the 95% ethanol product would have to have been $4.90 or $4.50/gallon for a discounted cash flow return on investment (after taxes) or 15% or 10%, respectively.[12]

An independent estimate for ethanol production from corn stover, funded by the DOE for work carried out at the University of Berkeley, California, was first announced at a symposium in 1978.[13] Scaling up an advanced process option with cell recycling (chapter 4, section 4.4.1) and converting the postdistillation stillage into methane by anaerobic fermentation to generate a combustible source of steam generation, a total production cost of $3.38/gallon was estimated, again with no allowance for plant profitability at a 14-million-gallon/year scale of production.

The main conclusion was, therefore, that the costs of generating fermentable sugars by enzymic hydrolysis dominated production (i.e., not-for-profit) economics; consequently, any major reduction in the cost of generating the cellulase or increase in its specific activity (or stability or ease of recovery for recycling) would be highly effective in lowering the production costs. Although federal and state initiatives in operation by 1979 offered some mitigation of the high costs of biomass-derived ethanol, the sums were small: for example, under the National Energy Act of 1978, alcohol fuels were eligible for DOE entitlements worth 5¢/gallon, and 16 states had reduced or eliminated entirely state gasoline taxes on gasohol mixtures, the largest amount being that in Arkansas, worth 9.5¢/gallon.

5.2.1.3 Fuel Ethanol from Sugarcane Molasses

A 14-million-gallon/year facility for 95% ethanol production from sugarcane molasses was technically the easiest production process to design and cost in the 1970s (table 5.3). To simplify even further, although coproducts (CO_2 and fusel oils) were considered, no economic calculations were made for these; sale of the yeasts grown in

TABLE 5.3
Cost Estimates for Ethanol Production from Molasses

Manufacturing input	Annual cost ($ x 10³)	Cost (% of total)
Molasses[a]	9100	63.2
Other materials	100	0.7
Power	44	0.3
Steam	315	2.2
Water	30	0.2
Labor	770	5.3
Administration	400	2.8
Interest	1178	8.2
Depreciation	1405	9.8
Maintenance	662	4.6
Taxes and insurance	397	2.8
Total	14401	100
Ethanol sales (gallons)	13539	
Yeast sales	942	
Ethanol production cost[b] ($/gallon)	0.994	

Source: Data from Paul.[12]

[a] 82,000 tons per year

[b] After allowing for sales of the yeast coproduct

the fermentations were, however, included to effect a cost reduction of approximately 6.5%. The final calculated manufacturing price of 95% ethanol was $0.995/gallon, "a cost which is today very comparable to producing alcohol from ethylene."[12]

Of the production costs, the raw material molasses was the dominating factor, accounting for 63% of the total — no exact geographical location for the hypothetical facility was given but the quoted molasses price ($50/ton) was the 1978 summer average of molasses delivered to the East Coast and Midwest of the United States.

5.2.1.4   Farm-Scale Ethanol Production

How small can a viable rural ethanol production site be? The final section of the 1978–1980 collection of case studies projected fuel ethanol production from corn on a family-run farm in the Midwest. The cost calculations were very different, however, because there was already a working market for fuel alcohol — and a known price ($1.74/gallon in Iowa, November 1979). Using this figure,* a net operating profit of total revenues could be projected for the first year of operation, ignoring factors such as finished goods and work-in-process inventories that would reduce the actual production cost of ethanol (table 5.4).

---

* "It is conceivable that as the price of gasoline increases to a point greater than the price of ethanol, producers could raise the price of ethanol to equalize the prices of the two liquid fuels."[12]

TABLE 5.4
Cost Estimates for Farm-scale Ethanol Production from Corn

Input	Annual cost ($ x 10³)	Cost (% of total)
Corn[a]	138	51.1
Enzymes	24	8.8
Electricity	3	1.1
Straw	11	4.0
Miscellaneous	18	6.7
Labor	21	7.7
Interest	24	8.8
Depreciation	7	2.5
Sales and marketing	26	9.5
Total	270	100
Ethanol sales (gallons)	132	
Stillage sales	56	
Ethanol production cost[b] ($/gallon)	1.63	
Ethanol sales @ $1.74/gallon ($)	230	
Total income ($)	285	
Gross profit ($)	15	

Source: Data from Paul.[12]

[a] 60,000 bushels per year (3.4 million lb or 1.5 million kg)

[b] After allowing for sales of the stillage coproduct

The operating profit was, however, entirely represented by sales of the fermentation stillage delivered (by truck) to neighbors within 5 miles of the farm. The selling price of such stillage would be much depressed if a large brewery or distiller were located nearby; if no net income could be generated by these means, then the facility would run at a loss. The capture of CO_2 from the fermentation was not considered because the capital cost of the equipment was too high to give a good return on the investment. A localized and small-scale production of fuel ethanol could, therefore, provide all the fuel requirements for running a farm's gasoline-consuming operations and provide a reasonable financial return as a commercial venture — but only if the "agrobusiness" was run as an early example of a biorefinery (chapter 8), producing not only ethanol but a saleable fermentation- and corn-derived coproduct.

Published in 1982, a second survey of technology and economics for farm-scale ethanol production at more than 100 gallons/hr (or, up to 1 million gallons/year) estimated a total annual cost of $1.97/gallon as a breakeven figure.[14] The technical aspects of the process had been investigated in a facility at the South Dakota State University with fermentation vessels of up to 5750 l capacity. The projected price included a $0.41/gallon sales income from the wet grain coproduct, annual amortized capital cost, operating costs, and fixed costs (including insurance, maintenance, and property taxes). Also designed for farm-scale use in South Dakota, a solid-state fermentation process (chapter 4, section 4.6.2) using sweet sorghum as the feedstock and upscaled to produce in principle 83 l of 95% ethanol/hr was predicted to have production costs of $1.80/gallon; the single largest contributor (61%) to production costs was that of the sorghum feedstock.[15]

5.2.2 Corn Ethanol in the 1980s: Rising Industrial Ethanol Prices and the Development of the "Incentive" Culture

A key change in the pricing structure of industrial alcohol in the United States occurred in the decade after 1975: the price of petrochemical ethylene showed an increase of nearly tenfold, and this steep rise in feedstock costs pushed the price of synthetic industrial alcohol from 15¢/l (57¢/gallon) to 53¢/l ($2.01/gallon).[16] Corn prices fell significantly (from $129/ton to $87/ton) between 1984 and early 1988; because the coproduct costs were increasing as a percentage of the corn feedstock cost at that time, the "net corn cost" for ethanol production (the net cost as delivered to the ethanol production plant minus the revenue obtained by selling the coproducts) and the net corn cost per unit volume of ethanol were both halved (figure 5.3).

By 1988, the costs involved in corn-derived ethanol production were entirely competitive with those of synthetic industrial alcohol (table 5.5). The major concern was that unexpectedly high investment costs could place a great strain on the economics of the process if the selling price for ethanol dipped: in general, such costs could be minimized by adding on anhydrous ethanol capacity to an existing beverage alcohol plant or adding an ethanol production process to a starch or corn syrup plant, but expensive grassroots projects could face financial problems. Across the whole range of production facilities (small and large, new or with added capacity,

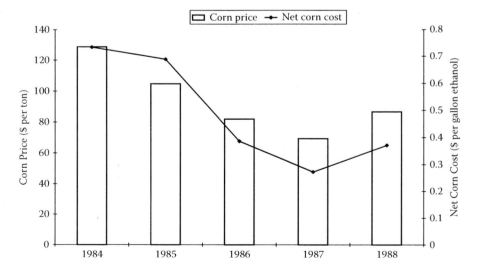

FIGURE 5.3 Corn price movements in the United States in the mid-1980s. (Data from Keim and Venkatasubramanian.[16])

TABLE 5.5

Production Costs for Corn-Derived Ethanol in the United States in 1988

Manufacturing input	Production cost ($ per liter)[a]	Production cost ($ per gallon)[a]	Production cost (% of total)
Direct			
Grain	0.098	0.37	31.4
Steam and electric power	0.040	0.15	12.6
Enzymes	0.010	0.04	3.2
Yeast	0.010	0.04	3.2
Labor	0.010	0.04	3.2
Investment-related	0.145	0.55	46.4
Total	0.313	1.18	

Source: Data from Keim and Venkatasubramanian.[16]

[a]   Average values calculated from the quoted range of values

and with varying investment burdens), a manufacturing price for ethanol could be as low as 18¢/l (68¢/gallon) or as high as 42¢/l ($1.59/gallon).[16] By 1988, the average fuel ethanol selling price had fallen below 30¢/l ($1.14/gallon), an economic movement that would have placed severe pressures on farm-scale production business plans (see section 5.2.1.4). As an incentive to fuel ethanol production and continuing the developments noted above (section 5.2.1.2), federal excise tax concession of 16¢/l (15¢/gallon), discounting by individual states by as much as 21¢/l (61¢/gallon), and direct payments by states to producers amounting to as much as 11¢/l (42¢/gallon), in conjunction with loan guarantees and urban development grants, encouraged the development of production by grassroots initiatives.[16] Industrial-size facilities, built without special incentives, were already reaching capacities higher than 1 billion gallons/year as large corporations began to realize the earning potential of fuel ethanol in what might become a consumer-led and consumer-oriented market.

5.2.3   WESTERN EUROPE IN THE MID-1980S: ASSESSMENTS OF BIOFUELS PROGRAMS MADE AT A TIME OF FALLING REAL OIL PRICES

European commentators and analysts were far less sanguine on the desirability of fuel ethanol as a strategic industry for the future in the mid-1980s. This was a time of steeply falling oil prices, both expressed in real terms and in the actual selling price (figure 1.3). In 1987, two independent assessments of ethanol production from agricultural feedstocks were published in the United Kingdom and Europe.[17–19] Across Europe, the introduction of lead-free fuels heralded an important new market for ethanol and other additives but the competition among these compounds (including MTBE and methanol) was likely to be intense for the estimated 2-million-tonnes/year market by 1998.[17]

The U.K. survey included wheat grain and sugarbeet as possible local sources of carbohydrates; in both cases, raw plant materials dominated the production cost analysis (table 5.6). The monetary value of coproducts were important, although only animal feeds were considered as viable sources of income to offset ethanol production

TABLE 5.6
Cost Estimates for Wheat- and Sugarbeet-Derived Ethanol in the U.K. in 1987

Manufacturing input	Production cost from wheat ($ per liter)[a]	Production cost from wheat ($ per gallon)[a]	Production cost from sugarbeet ($ per liter)[a]	Production cost from sugarbeet ($ per gallon)[a]
Raw materials	0.44	1.68	0.44	1.68
Operating costs	0.09	0.34	0.13	0.50
Capital cost	0.04	0.17	0.10	0.39
Coproduct credit	0.16	0.62	0.10	0.39
Total	0.41	1.57	0.58	2.18

Source: Data from Marrow et al.[17]

[a] Currency exchange values used from January 1987 to convert the original pound sterling data

costs.[17] The market prices for all major agricultural products were determined by the price support policies of the Common Agricultural Policy (CAP), an essential part of the Treaty of Rome (March, 1957), under which the European Economic Community (EEC) was set up and regulated; among its many provisions, the CAP was designed to ensure both a fair standard of living for farmers and reasonable consumer prices, and the CAP operated to guarantee a minimum price for basic agricultural products through intervention prices and protected the community's internal markets against fluctuations in world prices through the establishment of threshold prices. Technical progress was, however, also a goal of the CAP to increase agricultural productivity. The CAP has been controversial inside the EEC and subsequently the European Community and European Union as individual member states have received varying benefits from the policy but it has the advantage of enabling commodity prices to be more predictable, a useful factor when calculating possible trends in feedstock prices for the production of biofuels. In 1987, the likely costs of ethanol from wheat and sugarbeet were greatly in excess of the refinery price of petrol (gasoline), with a cost ratio of 3.2–4.4:1, allowing for the lower energy content of ethanol.[17] The continuing influence of the CAP was moreover highly unlikely to reduce feedstock costs for bioethanol to be price competitive with conventional fuels.

The second volume of the U.K. study gave outline production cost summaries for ethanol derived from wood (no species was specified) using acid and enzymatic hydrolysis for the liberation of glucose from cellulose (table 5.7). Additionally, straw residues were considered from cereals (wheat, barley, and oats), field beans, and oil seed rape (canola) using acid and enzymic hydrolysis; electricity and lignin were modeled as saleable coproducts. No source for ethanol could yield a product with a production cost less than three times that of conventional fuels (figure 5.4). These poor economics resulted in the authors being unable to recommend initiating a large program of work directed toward bioethanol production in the United Kingdom, although continued support of existing research groups was favored to enable the United Kingdom to be able to take advantage of fundamental breakthroughs, especially in lignocellulose conversion. A return to the high oil prices experienced in 1973–4 and 1978–80 was considered unlikely until well into the

TABLE 5.7
Cost Estimates for Wood-Derived Ethanol in the U.K. in 1987

Manufacturing input	Production cost acid hydrolysis ($ per liter)[a]	Production cost acid hydrolysis ($ per gallon)[a]	Production cost enzymic hydrolysis ($ per liter)[a]	Production cost enzymic hydrolysis ($ per gallon)[a]
Raw materials	0.30	1.13	0.27	1.04
Operating costs	0.18	0.67	0.27	1.04
Capital cost	0.14	0.53	0.29	1.08
Coproduct credit	0.14	0.52	0.03	0.11
Total	0.48	1.82	0.81	3.05

Source: Data from Marrow and Coombs.[18]

[a] Currency exchange values used from January 1987 to convert the original pound sterling data

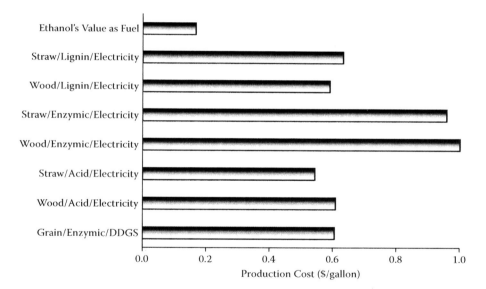

FIGURE 5.4 Production prices in the United Kingdom of ethanol from grain, wood, and straw feedstocks by acid or enzymic hydrolysis and with saleable coproducts. (Data recalculated from Marrow and Coombs.[18])

twenty-first century in any case, because transport costs were inevitably an important element of feedstock costs in the United Kingdom, rising oil prices would tend to increase the total costs for bioethanol production.[18]

The European study presented less detailed economic data but considered a wider range of feedstocks, not all of which were (or are) major agricultural products across the whole of Europe, but which represented potential sources for expanded agricultural production or as dedicated energy crops. For production facilities capable of manufacturing more than 150 million liters (40 million gallons) of ethanol/year, wheat grain was the cheapest source with a production cost of the feedstock equivalent to $0.36/gallon (converting the now obsolete European currency unit values to U.S. dollar at the exchange rate prevailing in early January 1987), followed by corn ($0.62/gallon), sugarbeet ($0.83/gallon), Jerusalem artichoke ($0.87/gallon), potatoes ($1.89/gallon), and wine* ($2.73/gallon).[19] The estimated cost of ethanol production from wheat was 0.49 ECU/l (equivalent to 53¢/l, or $2.01/gallon); in comparison, the Rotterdam refinery price for premium gasoline in late 1986 was approximately 10¢/l. The consultants who assembled the report concluded:

1. Encouragement of a bioethanol program was not in the economic interests of the European Community
2. A large reduction of the feedstock costs would be required to show a net economic benefit from bioethanol production
3. Alternatively, an oil price in the range $30–40/barrel would be required to achieve economic viability for bioethanol

Although the conversion of ethanol to ethylene was technically feasible, a cost analysis of this route indicated that it would be even less viable economically than ethanol production as a fuel additive.

The European study included an analysis of the development of fuel ethanol industries in Brazil and the United States, noting that the bulk of the financial incentives in the U.S. corn ethanol sector benefited the large producers rather than the small operators or the corn farmers; even more disturbing to European decision makers was the conclusion that blenders had benefited disproportionately, enjoying effectively cost-free ethanol as a gasoline additive during 1986, and using the available subsidies to start a price-cutting war between ethanol producers rather than promoting total sales. The prospect of subsidies being essential for establishing and maintaining a bioethanol program in Europe was nevertheless consistent with the European Community and its long-established strategic approach to agricultural development.

5.2.4 Brazilian Sugarcane Ethanol in 1985: After the First Decade of the Proálcool Program to Substitute for Imported Oil

The year 1985 marked the end of the first decade of the national program to use sugarcane-derived ethanol as an import substitute for gasoline. Rival estimates of the cost of Brazilian fuel ethanol varied widely, from $35 to $90/barrel of gasoline replaced. Assessing the economic impact of the various subsidies available to alcohol producers was difficult but indicated a minimum unsubsidized price of $45/barrel on the same gasoline replacement basis.[20] Assuming that gasoline was mostly manufactured from imported petroleum, the overall cost comparison between gasoline and nationally produced ethanol was close to achieving a balance, clearly so if the import surcharge then levied on imported oil was taken into consideration (table 5.8).

---

* The high cost of wine-derived ethanol was particularly unfortunate because wine was very much in surplus in the 1980s (the so-called European wine lake), mostly as a result of overproduction of cheap table wines in France, Italy, and Spain.

TABLE 5.8

Production Costs for Sugar-Derived Ethanol in Brazil by 1985

Cost component	Cost
	Ethanol
Sugarcane	$10–12 per ton
Ethanol yield	65 liters per ton
Distillation	$0.09–0.11 per liter
Production cost	$0.264–0.295 per liter
Replacement ratio 100%	1.2 liter ethanol per liter gasoline
Replacement ratio 20%	1.0 liter ethanol per liter gasoline
Ethanol cost 100%	$50–65 per barrel gasoline replaced
Ethanol cost 20%	$42–47 per barrel gasoline replaced
	Gasoline
Imported petroleum	$29 per barrel
Shipping costs	$2 per barrel
Import surcharge	$6 per barrel
Refining cost	$10 per barrel
Total	$47 per barrel
Total - surcharge	$41 per barrel

Source: Data from Geller.[20]

Anticipating future sentiments expressed for biofuels, justifying full gasoline replacement at then current world oil prices was held to be dependent on the additional benefits of ethanol production, including employment creation, rural development, increased self-reliance, and reduced vulnerability to crises in the world oil market.[20]

5.2.5 ECONOMICS OF U.S. CORN AND BIOMASS ETHANOL ECONOMICS IN THE MID-1990s

In retrospect, it is perhaps surprising that the groundbreaking 1996 monograph on bioethanol contained no estimate of production costs from lignocellulosic substrates but advanced only a brief economic analysis of corn-derived ethanol.[21] U.S. production of corn ethanol had, after the relative doldrums of 1985–1990, begun to surge — a trend only briefly halted in 1995 (see figure 1.16). Wet and dry milling processes had also become competitive as ethanol added a major product to the long-established mix of feed additives, corn oil, and others.

Table 5.9 summarizes average production costs from both wet and dry milling. As in the late 1980s, the mean values masked a wide range of variation: $0.28–0.37/l ($1.06–1.40/gallon) for dry milling and $0.232–0.338/l ($0.88–1.28/gallon) for wet milling. The wet milling process was moreover highly sensitive to the prices of corn and the process coproducts, even to the point of net corn prices becoming negative at times of lower feedstock but higher coproduct prices.[21] Further technical

TABLE 5.9
Production Costs for Corn-Derived Ethanol in the United States by the Mid-1990s

Manufacturing input	Production cost ($ per liter)[a]	Production cost ($ per liter)[a]	Production cost (% of total)
	Dry corn milling		
Net corn costs	0.120	0.45	51.1
Other operating costs	0.105	0.40	44.7
Annualized capital costs	0.010	0.04	4.3
Total	0.235	0.89	
	Wet corn milling		
Net corn costs	0.097	0.37	41.3
Other operating costs	0.103	0.39	43.8
Annualized capital costs	0.085	0.32	36.2
Total	0.285	1.08	

Source: Data from Elander and Putsche.[21]

[a] Average values calculated from the quoted range of values (in 1993 dollars)

advances, including the introduction of corn hybrids with properties tailored for wet milling (e.g., accelerated steeping) and improved fermentor designs allowing cell entrapment were estimated to offer cost reductions of 4–7¢/gallon.

The "missing" analysis of lignocellulosic ethanol in the mid-1990s was (in all probability) supplied by two reviews also published in 1996.[22,23] Two scenarios were considered in the cost modeling: a base case and an advanced technology option (although without a precise date for implementation); the data are summarized in table 5.10. Using plausible technology for the mid-1990s, a production cost of $1.18/ gallon was computed; a fourfold increase in the capacity of the facility, together with innovative bioprocess technologies, was predicted to reduce the production costs to approximately 50¢/gallon.

5.2.6 Lignocellulosic Ethanol in the Mid-1990s: The View from Sweden

Although a few pioneering studies attempted cost estimates of wood-derived ethanol from the 1980s onward, they focused on aspects of the technological processes required rather than making firm conclusions about market prices.[24,25] Swedish studies appear to have been the first to present detailed cost breakdowns for ethanol production from accessible large-scale woody biomass sources.[26,27] The first of the two to be published was highly unusual in that it used recent advances in pentose utilization by recombinant bacteria to model a pentose stream process, that is, using the solubilized sugars from the pretreatment of wood (willow) feedstock — the procedure involved the impregnation of the material with SO_2 and subsequent steaming — and included a detoxification procedure to reduce the levels of inhibitors from the hydrolysate.[26] The fermentation with *Escherichia coli* KO11 (chapter 3, section 3.3.2.1) was assumed to consume 96%

TABLE 5.10

Production Costs for Cellulosic Ethanol in the United States by the Mid-1990s

Manufacturing input	Capital, labor and related (cents per liter)	Energy (cents per liter)	Production cost (cents per liter)	Production cost (% of total)
		Base case[a]		
Feedstock			45.97	39.0
Other raw materials			9.78	8.3
Pretreatment	13.75	6.55	20.30	17.2
Cellulase preparation	1.55	1.67	3.22	2.7
SSF	13.83	3.34	17.17	14.6
Pentose conversion	3.22	0.99	4.21	3.6
Distillation	2.74	5.10	7.84	6.7
Power cycle	28.61	−26.96	1.65	1.4
Other	7.34	0.36	7.70	6.5
Total			117.84	
		Advanced technology[b]		
Feedstock			35.84	71.3
Other raw materials			0.95	1.9
Pretreatment	3.22	5.63	8.85	17.6
Fermentation	1.95	1.00	2.95	5.9
Distillation	1.79	2.83	4.62	9.2
Power cycle	14.06	−22.03	−7.97	−15.8
Other	4.74	0.32	5.06	10.1
Total			50.30	

Source: Data from Lynd.[22]

[a]   658,000 dry tons/year, 60.1 million gallons/year, installed capital $150.3 million

[b]   2,738,000 dry tons/year, 249.9 million gallons/year, installed capital $268.4 million

of the pentose* sugars (and all of the much smaller amount of hexoses) to generate more than 67 million liters (1.8 million gallons) of ethanol/year. The final production cost of 95% aqueous alcohol was equivalent to 48¢/l ($1.82/gallon) after allowance for financial costs and assuming a small net income from CO_2 as a coproduct (table 5.11). An essential parameter was that of the extent to which the cells could be recycled: single-batch use of the cells increased the production cost to 64¢/l ($2.42/gallon). The single largest contributor to the production cost was, however, the financial burden of repaying the investment in the plant, that is, more than 37% of the total annual production cost outlay (table 5.11).

* The authors did not discriminate between xylose and xylooligosaccharides or between xylose and arabinose, and the implicit assumption was that more than 95% of the available pentoses were in a readily metabolizable form, that is, monomeric xylose.

TABLE 5.11

Cost Estimates for Ethanol Production from Pentose Stream from Willow

Manufacturing input	Annual capacity/output (tonne/year)	Annual cost ($ x 10³)	Cost (% of total)
Wood hydrolysate	13877	439	13.1
Ammonia (25%)	370	101	3.0
Phosphoric acid (80%)	179	90	2.7
Magnesium oxide	4.8	1	0.03
Sodium sulfite	252	142	4.2
Calcium oxide	1750	236	7.1
Sulfuric acid (37%)	214	18	0.5
Electricity		80	2.4
Steam	12885	226	6.8
Distillation		391	11.7
Maintenance		73	2.2
Labor		300	9.0
Annual capital costs		1247	37.3
Working capital		8	0.2
Coproduct credit for CO_2	5883	8	-0.2
Total		3344	
Ethanol (m³/year)	6906		
Ethanol production cost ($/liter)	0.48		

Source: Data from von Sivers et al.[26]

A very similar analysis of three different approaches to utilizing the full carbohydrate potential of pine wood, digesting the cellulose component with concentrated acid, dilute acid, and enzymatic methods, calculated a full manufacturing costs for ethanol of between 50¢ and 53¢/l ($1.89–2.01/gallon).[27] The bulk of the production cost (up to 57.5%) was accounted for by the financial costs of installing the hardware for generating fermentable carbohydrates (hexoses as well as pentoses) in more complex total processes with longer cycle times.

5.2.7 SUBSEQUENT ASSESSMENTS OF LIGNOCELLULOSIC ETHANOL IN EUROPE AND THE UNITED STATES

5.2.7.1 Complete Process Cost Models

Acknowledging the great uncertainties in establishing guideline costs for lignocellulosic feedstocks, a Swedish review of the bioethanol industry in Scandinavia and elsewhere in 1999 noted that large-scale processes and improved overall ethanol yield would be highly desirable for future economic production of biofuels.[28] The first trend probably heralded the demise of the farm-scale ethanol plant (section 5.2.1.4), a production model that probably is only relevant to local and private consumption of transportation fuels as the market for fuel ethanol imposes competitive pricing; the second point is a natural conclusion from the vast efforts invested in developing recombinant producing organism and bioprocesses (chapters 3 and 4, respectively).

The Swedish authors (from Lund University) have continued to explore cost models for ethanol from lignocellulosic substrates:

- SSF bioprocesses offer improved economics over standard separate hydrolysis and fermentation because of higher ethanol yields and reduced capital costs; with softwood biomass sources, there are also significant advantages if either process could be operated with higher levels of insoluble material and if recycling the stillage after distillation ("backsetting"), in principle reducing the production cost to $0.42/l, or $1.59/gallon (table 5.12).[29]
- Operating steam pretreatment of softwoods in two steps (at lower and higher temperatures) to maximize the recovery of hemicellulose sugars and cellulosic glucose, respectively, has a higher overall ethanol yield and reduced requirement of enzymes but is more capital-intensive and has a higher energy demand; the net result is no reduction in the production cost of ethanol (table 5.13); further improvement to the process, including a higher insoluble solids content for the second step, might reduce the production cost by 5–6%.[30]

The National Renewable Energy Laboratory, in association with consultant engineers, presented an outline cost model for the industrial-scale production of bioethanol (2000 tonnes/day consumption of feedstock, 52 million gallons of ethanol/year) from a hardwood yellow poplar biomass source.[31] Operating costs were calculated to be approximately 62¢/gallon of ethanol (table 5.14). Discounted cash flow analysis assuming a discount rate of 10% indicated a minimum selling price of $1.44/gallon for a capital investment of $234 million. On the technical level, the key features of the envisaged process were

- Acid pretreatment of the biomass substrate (19% of the installed equipment cost)
- On-site generation of cellulase
- SSF of the pretreated substrate with a *Zymomonas mobilis* capable of utilizing only glucose and xylose
- Wastewater treatment via anaerobic digestion to methane
- Utilization of three available waste fuel streams (methane, residual lignin solids, and a concentrated syrup from evaporation of the stillage) in a fluidized bed combustor, burner, and turbogenerator (33% of the installed equipment cost)

Although the complete array of technology in the model was unproven on a large scale, much of the process could be accurately described as "near term" or "based on the current status of research that is complete or nearly so."[31] The computed minimum selling price for ethanol was 20% higher than that of corn-derived ethanol, with a more than twofold greater investment cost ($4.50/gallon as compared with approximately $2/gallon).

TABLE 5.12

Cost Estimates for Ethanol Production from Softwood Using Different Bioprocess Technologies

Manufacturing input	SSFa base case ($ per liter)	SSFa base case ($ per gallon)	SHFb base case ($ per liter)	SHFb base case ($ per gallon)	SSFa 8% solids ($ per liter)	SSFa 8% solids ($ per gallon)	SHFb 8% solids ($ per liter)	SHFb 8% solids ($ per gallon)
Wood[c]	0.16	0.61	0.19	0.73	0.16	0.61	0.19	0.70
Enzymes	0.08	0.31	0.06	0.21	0.08	0.31	0.06	0.21
Yeast	0.06	0.22	0.00	0.00	0.04	0.14	0.00	0.00
Other operating costs	0.04	0.16	0.05	0.19	0.03	0.13	0.04	0.16
Labor, maintenance, insurance	0.07	0.26	0.01	0.37	0.06	0.24	0.09	0.32
Capital costs	0.16	0.62	0.25	0.93	0.14	0.53	0.20	0.76
Coproduct credits[d]	0.01	0.03	0.01	0.03	0.06	0.22	0.07	0.27
Total	0.57	2.14	0.64	2.41	0.46	1.74	0.50	1.87

Source: Data from Wingren et al.[29]

[a] Simultaneous saccharification and fermentation, 63 billion liters per year (base case)

[b] Separate hydrolysis and fermentation, 55 billon liters per year (base case)

[c] 195,600 tonne raw material per year, operated continuously (8000 hours per year), notionally located in northern Sweden

[d] CO_2 and solid fuel

TABLE 5.13
Cost Estimates for Ethanol Production from Softwood Using Different Pretreatment Options

Manufacturing input	Steam pretreatment one-step[a] ($ per liter)	Steam pretreatment one-step[a] ($ per gallon)	Steam pretreatment two-step[b] ($ per liter)	Steam pretreatment two-step[b] ($ per gallon)
Wood[c]	0.19	0.71	0.18	0.69
Chemicals	0.11	0.43	0.11	0.42
Utilities	0.03	0.01	0.03	0.01
Other operating costs	0.09	0.33	0.09	0.33
Capital costs	0.21	0.78	0.21	0.79
Coproduct credits[d]	0.07	0.27	0.06	0.24
Total	0.55	2.08	0.55	2.09

Source: Data from Wingren et al.[30]

[a] 215°C, residence time 5 min, SO_2 added to 2% of the water content of the wood; 47 billion liters ethanol per year capacity

[b] 190°C, residence time 2 min, then 210°C for 5 min; SO_2 added to 2% of the water content of the wood; 49 billion liters ethanol per year capacity

[c] 200,000 tonnes per year; plant operating time 8000 hours per year

[d] CO_2 and solid fuel

TABLE 5.14
Operating Costs for Yellow Poplar Sawdust-Ethanol in the United States

Input	Production cost (cents per gallon)[a]	Production cost (% of total)
Feedstock[a]	37.0	60.0
Chemicals	8.0	13.0
Nutrients	6.2	10.1
Fossil fuels	0.9	1.5
Water	0.9	1.5
Utility chemicals	1.2	1.9
Solid waste disposal	1.2	1.9
Fixed costs	13.5	21.9
Electricity credit[b]	7.2	−11.7
Total	61.7	

Source: Data from Wooley et al.[31]

[a] Poplar sawdust at $25 per tonne

[b] Excess electricity sold to grid at 4 cents per kWh

TABLE 5.15
Estimated Production Costs for Bioethanol in 2003

Source of ethanol	Production cost (€/GJ)	Production cost ($/liter)[a]	Production cost ($/gallon)
Sugarcane (Brazil)	10–12	0.24–0.29	0.91–1.10
Starch and sugar (U.S. and Europe)	16.2–23	0.39–0.55	1.48–2.08
Lignocellulosic (U.S.)	15–19	0.36–0.46	1.36–1.74
Lignocellulosic (Europe)	34–45	0.82–1.08	3.10–4.09

Source: Data from Hamelinck et al.[32]

[a] Higher heating value of ethanol = 83,961 Btu/gallon (24GJ/liter)

5.2.7.2   Reviews of "Gray" Literature Estimates and Economic Analyses

Outside primary scientific journals, data from a range of sources (including reports prepared for governments and conference proceedings) were compiled on the basis of 2003 costings as a baseline for future cost modeling.[32] Ethanol produced from sugar, starch (grain), and lignocellulosic sources covered production cost estimates from less than $1/gallon to more than $4/gallon (table 5.15). Even with the lower production costs for lignocellulosic ethanol in the United States, taking into account financial outlays and risks ($260 million for a 50-million-gallon annual production plant), an ethanol price of $2.75/gallon would be more realistic.[33]

The International Energy Agency's most recent assessment of sugar- and starch-derived ethanol (2005 reference basis) is that Brazil enjoys the lowest unit costs ($0.20/l, or $0.76/gallon), starch-based ethanol in the United States costs (after production subsidies) an average of around $0.30/l (or $1.14/gallon), and a European cost (including all subsidies) is $0.55/l (or $2.08/gallon).[34] Brazilian production costs for fuel alcohol, close to $100/barrel in 1980, decreased rapidly in the 1980s, and then more slowly, but only a severe shortage of sugarcane or a marked rise in sugar prices would interrupt the downward trend in production costs.[35,36]

With due allowance of the lower fuel value of ethanol, therefore, the historical trend of fuel ethanol production costs versus refinery gate price* of gasoline is showing some degree of convergence (figure 5.5). In particular, the real production costs of both sugar- and corn-derived ethanol have fallen so that the production costs (with all tax incentives in place, where appropriate) now is probably competitive with the production cost of gasoline, as predicted for *biomass* ethanol in 1999.[37] Critics of the corn ethanol program have, however, argued that the price of fuel ethanol is artificially low because total subsidies amount to $0.79/gallon for production costs of $1.21/gallon, that is, some $3 billion are expended in subsidizing the substitution of only 1% of the total oil use in the United States.[38] Although incentives for domestic

---

* Data published by the DOE's Energy Information Administration (www.eia.doe.gov and www.tonto.eia.doe.gov) point to refinery gate prices of gasoline being 70% of the retail price (excluding taxes), although this increases to 90% or more during crude oil price surges; figure 5.5 was constructed from the U.S. conventional gasoline bulk prices after January 1994 and, between 1983 and 1994, the corresponding retail prices × 0.70.

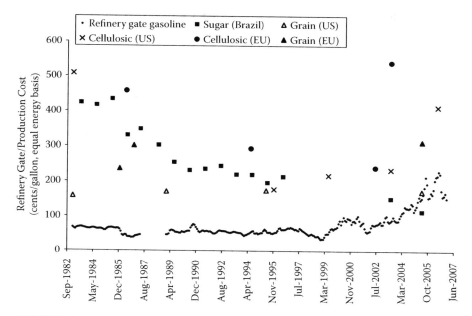

FIGURE 5.5   Refinery gate price of U.S. oil and estimates of bioethanol production costs between 1980 and 2006.

ethanol use in Brazil were discontinued by 1999 (except as part of development policies in the northeast region), a cross-subsidy was created during the 1990s to subsidize ethanol production through taxation on gasoline and diesel; this was operated via a tight government control of the sales prices of gasoline, diesel, and ethanol, and the monopoly represented in the country by PETROBRÁS.[36]

Brazilian sugarcane ethanol has reached the stage of being an importable commodity to the United States, promoting the development of shipping, port handling, and distribution network infrastructures. In Europe, grain alcohol might be cost-competitive — with the much higher tax rates prevailing in Europe, the scope for regulating the end user price is much higher. In contrast, the economics of lignocellulosic ethanol remain problematic, although it is possible that in the United States, at least, production costs may become competitive with gasoline within the next five to ten years unless, that is, crude oil prices decrease significantly again.

5.3   PILOT PLANT AND INDUSTRIAL EXTRAPOLATIONS FOR LIGNOCELLULOSIC ETHANOL

5.3.1   NEAR-FUTURE PROJECTIONS FOR BIOETHANOL PRODUCTION COSTS

The persistently high production costs of lignocellulosic ethanol (particularly in Europe) have catalyzed several attempts to predict trends in 5-year, 10-year, and longer scenarios, with the implicit or explicit rationale that only lignocellulosic biomass is sufficiently abundant to offer a means of substituting a sizeable proportion of the gasoline presently used for transportation.

In 1999, the National Renewable Energy Laboratory published a projection of the economic production costs for lignocellulosic ethanol that, starting from a baseline of $1.44/gallon (in 1997 dollars), computed a decrease to $1.16/gallon with a 12% increase in yield (to 76 gallons/ton of feedstock), with a 12% increase in plant production capacity together with a 12% reduction in new capital costs.[31] A price trajectory envisaged this price deceasing to below $0.80/gallon by 2015 based on developments in cellulase catalytic efficiency and production and in ethanologenic production organisms:

- Improved cellulose-binding domain, active site, and reduced nonspecific binding
- Improved cellulase producers genetically engineered for higher enzyme production
- Genetically engineered crops as feedstocks with high levels of cellulases
- Ethanologens capable of producing ethanol at temperatures higher than 50°C
- Ethanologens capable of direct microbial conversion of cellulose to ethanol

These topics were covered in chapters 2 to 4; to a large extent, they remain research topics, although cellulase production and costs have certainly been improved greatly at the industrial scale of production.

A European study suggested that by 2010 lignocellulosic ethanol production costs could decrease to $0.53/l ($2.00/gallon), $0.31/l ($1.18/gallon) by 2020, and $0.21/l ($0.79/gallon) after 2025.[32] These improvements in process economics were considered to result from the combined effects of higher hydrolysis and fermentation efficiencies, lower specific capital investment, increases of scale, and cheaper biomass feedstock costs, but the prospect of ethanol ever becoming cost-competitive to gasoline was nevertheless considered to be "unlikely" — much of the analysis was, however, undertaken during the early stages of the great surge in oil prices and gasoline production costs after 2002 (figure 5.5). A more recent publication by the same research group (Utrecht, the Netherlands) computed a 2006 production cost for lignocellulosic ethanol of €22/GJ ($2.00/gallon, assuming exchange parity between the currencies) that was anticipated to fall to €11/GJ ($1.00/gallon) by 2030 — the single largest contributor to the production cost in 2006 was capital-related (46%) but this was expected to decrease in both absolute and relative terms so that biomass costs predominated by 2030.[39]

The International Energy Agency's most recent prognosis is for lignocellulosic ethanol (from willow, poplar, and *Miscanthus* biomass sources) to reach production costs below €0.05/l (€0.18/gallon) by 2030 with achieved biomass yield increases per unit land area of between 44% and 100% (table 5.16); biomass-derived ethanol would, on this basis, enjoy lower production costs than other sources in Europe (cereal grain, sugarbeet, etc.).[34] These three biomass options each target different areas:

- Willow already has considerable commercial experience in Sweden, the United Kingdom, and elsewhere and has strong potential in Eastern Europe where growing conditions and economics are favorable.

TABLE 5.16
Estimated Production Costs for Bioethanol from Biofuel Crops in 2005

Biofuel crop	Production cost (€/GJ)	Production cost ($/liter)[a]	Production cost ($/gallon)
Rape (canola)	20	0.48	0.91–1.10
Sugar beet	12	0.29	1.48–2.08
Willow	3–6	0.07–0.14	0.26–0.53
Poplar	3–4	0.07–0.10	0.26–0.38
Miscanthus	3–6	0.07–0.14	0.26–0.53

Source: Data from *World Energy Outlook.*[34]

[a] Higher heating value of ethanol = 83,961 Btu/gallon (24GJ/liter)

- Poplar is already grown for pulp production, with typical rotation cycles of eight to ten years.
- Elephant grass (*Miscanthus*) is a perennial crop suited to warmer climates (where high yields are possible), but its production potential in Europe in general is uncertain.

Transport costs are important for all biomass options, the energy used in transportation being equal to 1–2% of that contained in the biomass for short distances but 6–10% for long-distance shipping; transport, handling, and storage costs were considered to add 10% to ethanol production costs when the feedstocks are shipped by road to a local plant.[34]

5.3.2 SHORT- TO MEDIUM-TERM TECHNICAL PROCESS IMPROVEMENTS WITH THEIR ANTICIPATED ECONOMIC IMPACTS

Each of the exercises in predicting future cost/price trends discussed in section 5.3.1 suffers from a high content of speculative thought because many key technologies are uncosted at industrial scales of production — indeed, many aspects have not progressed beyond the laboratory bench. Some aspects of both upstream (feedstock provision and preparation) and fermentation strategies have, however, progressed to the stage that economic implications can be calculated across a range of lignocellulosic and other processes.

The ultimate starting point for bioethanol production is the collection and delivery of the biomass substrate, and the associated costs. A central assumption for lignocellulosic biomass-derived ethanol is that, not only can a large supply of such material be sourced, but that it also can be provided cheaply. With the much more advanced starch ethanol industry, a far larger body of data is available for analyzing the effect of feedstock cost increase on ethanol production costs; for example, the Agricultural Research Service of USDA has developed a detailed process and cost model for the dry milling of corn that showing that most (>85%) of any increase in the price of corn is transferred to the final product cost and that a relatively small decrease in the starch content (from 59.5% to 55%, w/w) can similarly negatively impact the productivity of an ethanol production unit.[40]

Projecting forward from a 1995 baseline, feedstock costs for a mature biomass ethanol technology were anticipated as being within the price range $34–38.6/dry ton.[22] Switchgrass farming has been estimated to cost $30–36/dry tonne; in comparison with straw or corn stover, the collection of switchgrass is probably less expensive because of the high yield of a denser biomass — nevertheless, delivered costs of the switchgrass to a production plant could be as low as $37/dry tonne of compacted material to $47/tonne of bales.[41] Similarly, corn stover is difficult to handle on account of its low bulk density; chopped corn stover can be compacted into briquettes that can reach a density of 950 g/l, and these more easily transported briquettes are more durable if produced from stover with low water content (5–10%).[42] A more radical option is the pipeline transport of corn stover (e.g., at 20% wet solids concentration); this is cheaper than trucking at more than 1.4 million dry tonnes/year and allows the possibility of conducting a partial saccharification during transport if enzymes are added, thus reducing the need for investment in the fermentation plant and lowering production costs by 7–8¢/gallon.[43]

In general, as bioethanol plant capacity is increased to cut unit production costs, the land area required for collection of sufficient biomass feedstock increases; as biomass supplies are sought from larger and larger distances, the costs of moving the raw material increases, introducing possibly diseconomies into production models. One solution is to introduce more flexibility into the feedstock "diet," taking advantage of whatever surpluses of other biomass material may seasonally occur; for example, a Californian study investigated what biomass supplies could be considered for a 40-million-gallon facility in the San Joaquin Valley: locally grown corn was significantly more expensive than midwestern corn ($1.21/gallon of ethanol versus $0.92/gallon), but surplus raisins and tree fruit and (although much more expensive) grapes and citrus fruit might all be included in a biomass harvest.[44] The proximate example of feedstock diversity is, however, that of cane sugar bagasse: the Dedini Hidrólise Rápida (Rapid Hydrolysis) process uses organic solvent extraction of sugarcane bagasse as a pretreatment method and aims to double the alcohol production per hectare of sugarcane harvested (www.dedini.com.br).

Using paper sludge as a feedstock for ethanol production has been claimed to be profitable as it provides a near ideal substrate for cellulase digestion after a comparatively easy and low-cost pretreatment; even without xylose conversion to ethanol, such a technology may be financially viable at small scales, perhaps as low as 15 tonnes of feedstock processed/day.[45] This is a genuinely low-cost feedstock because, in the absence of any productive use, paper sludge goes into landfills — at a cost to its producer. This has led to several attempts to find viable means of converting such waste into fuel, and in late 2006, New York state contributed $14.8 million to fund the development of a demonstration facility in Rochester (New York) to use paper sludge as well as wood chips, switchgrass, and corn stover as feedstocks, the facility being operated by Mascoma (www.mascoma.com). The pulp and paper industry in Canada is estimated to produce at least 1.3 million tonnes of sludge every year, and at up to 70% cellulose, this raw material is economically competitive with cereal grain as a substrate for ethanol production.[46] Hungarian researchers have also identified paper sludge and other industrial cellulosic wastes as being cost-effective routes to bioethanol.[47]

This illustrates the proposition that biomass ethanol facilities might be designed and constructed to adventitiously utilize a range of biomass substrates as and when they become available. As in the Californian example noted above, investigation of wastes from fresh and processed vegetables defined a sizeable resources of plant material (450,000 tonnes/year) in Spain; easily pretreated with dilute acid, such inputs could be merged with those for starch or lignocellulosic production lines at minimal (or even negative) cost.[48]

Substantial cost savings in cereal-based ethanol production can be achieved by a more integrated agronomic approach: although fermentor stillage could be used as a substitute for mineral fertilizer, total ethanol production was 45% lower if cereals were grown after a previous nitrogen-fixing legume crop; intensifying cereal yields certainly increased crops per unit land area, but ethanol production costs per liter dropped as the ethanol yield per unit land area outweighed the other costs; field trials also suggest that barley may (under German conditions) be economically favorable in comparison with wheat and rye).[49] All these conclusions may be directly applicable to cereal straw as well as cereal grain. Moreover, since one of the strongest candidates for lignocellulosic ethanol production (wheat straw) also has one of the lowest ethanol yields per unit dry mass, the ability to flexibly mix substrates has capacity advantages if feedstock handling and processing regimes can be harmonized (figure 5.6).[50]

Once in the ethanol facility, biomass pretreatment and hydrolysis costs are important contributors to the total production cost burden; for example, with hardwoods and softwoods, enzymes represent 18–23% of the total ethanol production costs; combining enzyme recycling and doubling the enzyme treatment time might

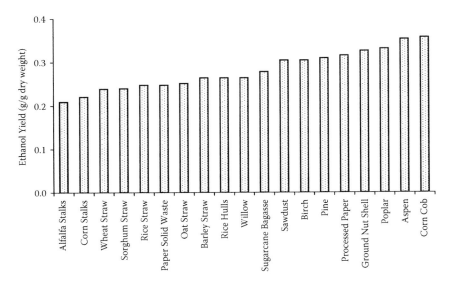

FIGURE 5.6 Maximum ethanol yield from lignocellulosic biomass sources. (Data from Chandel et al.[50])

improve the economic cost by 11%.[51] Bench-scale experiments strongly indicated that production costs could be reduced if advanced engineering designs could be adopted, in particular improving the efficiency of biomass pretreatment with dilute acid in sequential cocurrent and countercurrent stages.[52] Different pretreatment techniques (dilute acid, hot water, ammonia fiber explosion, ammonia recycle percolation, and lime) are all capital-intensive, low-cost reactors being counterbalanced by higher costs associated with catalyst recovery or ethanol recovery; as a result, the five rival pretreatment options exhibited very similar production cost factors.[53] Microbial pretreatments could greatly rescue the costs and energy inputs required by biomass hydrolysis techniques because enzymic digestibility is increased and hydrodynamic properties improved in stirred bioreactors; a full economic analysis remains to be undertaken in industrial ethanol facilities.[54] Continued optimization of chemical pretreatments has resulted in a combined phosphoric acid and organic solvent option that has the great advantage of requiring a relatively low temperature (50°C) and only atmospheric pressure.[55]

Being able to generate higher ethanol concentrations in the fermentation step would improve overall economic performance by reducing the costs of ethanol recovery; one solution is to run fermentations at higher biomass substrate loadings, that is, accomplish biomass liquefaction and saccharification at high solids concentrations; wheat straw could be processed to a paste/liquid in a reactor system designed for high solids content, the material then being successfully fermented by *Saccharomyces cerevisiae* at up to 40% (w/v) dry matter in the biological step.[56]

What is the optimum processing of the various chemical streams (soluble sugars and oligosaccharides, pentoses and hexoses, cellulose, and lignin) resulting from the pretreatment and hydrolysis of lignocellulosic biomass? The utilization of pentose sugars for ethanol production is certainly beneficial for process economics, a conclusion reached as early as 1989 in a joint U.S.–New Zealand study of pine as a source of woody biomass, where ethanol production costs of $0.75/l ($2.83/gallon) were calculated, decreasing by 5% if the pentose stream was used for fermentable sugars.[57] The stillage after distillation is also a source of carbohydrates as well as nutrients for yeast growth; replacement of up to 60% of the fresh water in the fermentation medium was found to be possible in a softwood process, with consequential reductions in production costs of as much as 17%.[58]

Because larger production fermentors are part of the drive toward economies of scale savings in production costs, reformulating media with cheaper ingredients becomes more important. In the fermentation industry at large, devising media to minimize this operating cost parameter has had a long history; as recombinant ethanologens are increasingly engineered (chapter 3), suitable media for large scales of production are mandatory. With *E. coli* KO11, for example, laboratory studies showed that expensive media could be substituted by a soya hydrolysate-containing medium, although the fermentation would proceed at a slightly slower rate; both this and a corn steep liquor-based medium could contribute as little as 6¢/gallon of ethanol produced from biomass hydrolysates.[59,60]

Operating a membrane bioreactor (in a demonstration pilot plant of 7,000-l capacity) showed that the yearly capital costs could be reduced to $0.18–0.13/gallon, with total operating costs for the unit of $0.017–0.034/gallon.[61] Another advanced engineering design included continuous removal of the ethanol product in a gas stream; compared with a conventional batch process, ethanol stripping gave a cost saving of $0.03/gallon with a more concentrated substrate being used, thus resulting in less water to remove downstream.[62] Simply concentrating the fermented broth if the ethanol concentration is low in a conventional process is feasible if reverse osmosis is employed but not if the water is removed by evaporation with its high energy requirement.[63]

Incremental savings in bioethanol production costs are, therefore, entirely possible as processes are evolved; many of the steps involved obviously require higher initial investment when compared with basic batch fermentation hardware, and it is unlikely that radical innovations will be introduced until a firm set of benchmark costings are achieved in semiindustrial and fully production-scale units. Efficient utilization and realization of the sales potential of coproducts remains, on the other hand, an immediate possibility. Coproduct credits have long been an essential feature of estimates of ethanol production (section 5.2); among these, electricity generation has been frequently regarded as readily engineered into both existing and new ethanol production facilities, especially with sugarcane as the feedstock for ethanol production. For example, in Brazil steam turbines powered by combustion of sugarcane bagasse can generate 1 MWh/m^3 (1,000 l) of ethanol, and economic analysis shows that this is viable if the selling price for electricity is more than $30/MWh, the sales price in Brazil in 2005.[64] When electricity credits cannot be realized or where coproducts can be used as substrates for other chemical or biological processes, different criteria come into play — this is discussed in chapter 8 (section 8.2). Pyrolysis of sugarcane wastes to produce "bio-oil" could yield 1.5 tons of saleable products per ton of raw sugar used, but the selling cost of the product will be crucial for establishing a viable coproduction process.[65]

5.3.3 BIOPROCESS ECONOMICS: A CHINESE PERSPECTIVE

China's demand for oil as a transportation fuel is forecast to increase more than tenfold between 1990 and 2030 (from 30 to 396 million tonnes), reaching 50% of that of the United States by that date.[66] Economic analysis has shown that sweet sorghum and its bagasse as well as rice hulls and corn stover have extensive availability in northern China and could represent attractive feedstocks for bioethanol production.[67]

Investigations into gasoline supplementation with endogenously produced ethanol began in 1999, and by 2004, E90 grades were available in eight provinces; a Renewable Energy Law and a National Key R&D Program for cellulosic ethanol were applied to the energy sector during 2005.[68] Other primary factors in China's newly acquired interest in bioethanol include the following:

- Fuel ethanol production was initiated from cereal grain feedstocks, but a government-set selling price of $1.65/gallon required large subsidies because the production costs were more than $1.80/gallon.
- Total biomass production could exceed the International Energy Agency's prediction for transportation fuel needs by 2030 at a low feedstock cost ($22/dry ton).

- Assuming successful implementation of the types of lignocellulosic ethanol technologies on which recent U.S. and European cost models have been based (section 5.2), production cost estimates for Chinese production sites would be in the range of $0.43–0.95/gallon.

Building and operating commercial cellulosic ethanol plants in China thus appears very feasible and would generate exactly the kind of practical experience and knowledge that would induce other nations to invest. The Chinese government has announced the allocation of $5 billion in capital investment in the coming decade for ethanol production capacity with a focus on noncereal feedstocks.[68]

5.4   DELIVERING BIOMASS SUBSTRATES FOR BIOETHANOL PRODUCTION: THE ECONOMICS OF A NEW INDUSTRY

5.4.1   Upstream Factors: Biomass Collection and Delivery

The "billion ton vision" is a program to access a billion tons of dry biomass per year to produce bioethanol (and other biofuels) to replace 30% of U.S. gasoline consumption by 2030.[69] One of the key parameters in a large-scale restructuring of the U.S. national fuel industry is that of supplying biomass raw material at such a high rate and at an economically acceptable cost.

The logistics and transportation costs of such large amounts of low-value, high-volume raw materials have only recently attracted serious consideration. Canadian studies comprise the most detailed considerations of these highly practical questions now that cellulosic ethanol facilities are nearing industrial reality. For an agricultural economy (and climate) like Canada's, wood is highly likely to be a large fraction of the biomass supply, initially from forest harvest residues and "energy plantations" on marginal farmland.[70] For wood chips, larger production plants (up to 38 million dry tons of biomass/year) are more economic than smaller units (2 million tons), and truck delivery is limited to such small units by issues of traffic congestion and community acceptance.[71] Combined road and rail shipping, that is, initial collection by truck followed by trans-shipping to rail, is only economic when the cost per unit distance of the rail sector is less than the trucking-only mode because of the incremental fixed costs: for woody material, the minimum economic rail shipping distance is 125 km (78 miles), whereas for cereal straw, the minimum distance extends to 175 km (109 miles).[72] Existing rail networks impose, however, a serious restriction (that of their location), supplying only sites close to already positioned track; road transport is more versatile. Factoring in additional considerations, including air emissions during transport, definitely favors rail transport.[73] Policy changes and new infrastructure investment appear therefore to be inevitable if the development of bioethanol production is not to be inhibited by objections of cost and pollution.

Focusing on corn stover and wheat straw as raw material inputs, a study of North Carolina concluded that more than 80% of these resources were located in the coastal area; four ethanol plants with feedstocks demands of between 146,000 and 234,000 dry tons/year required collection radiuses between 42 and 67 km (26–42 miles).[74]

The siting of production facilities to minimize transportation costs implies a contradiction with the economies of scale possible with larger production units. This is made more likely if relatively marginal biomass inputs such as municipal solid waste are to be considered.[75] Urban fringes might be close to existing landfill sites and also within short distances of field crop residues, wastes from horticultural industries, and seasonal supplies of tree and plant residues from urban parkland. As discussed previously (section 5.3.2), the ability to design and build smaller-scale bioethanol production units, especially if they can utilize a variable and adventitious supply of feedstocks, would be highly beneficial to match the fragmented nature of the cheapest likely raw materials.

5.4.2 MODELING ETHANOL DISTRIBUTION FROM PRODUCTION TO THE END USER

Constructing a large-scale bioethanol industry also implies a major change in the industrial landscape: whereas oil refineries are predominately coastal, biorefineries would be situated in agricultural areas or (with the development of a mature industry) close to forests and other biomass reserves. Gasoline distribution to retail outlets is without doubt a mature industry — in the United States, shipments of 6.4 billion liters of petroleum and petroleum products are made each day, 66% by pipeline (in 320,000 miles of pipeline), but only 4% by truck and 2% by rail; from the Gulf Coast to New York, shipping costs for gasoline amount to only 0.8 ¢/l.[76] It was the early development of a national distribution system for gasoline that decided the use of this fuel rather than ethanol for the emerging automobile industry before 1920.[77]

To support nationwide consumption of E10, cellulosic ethanol would be 61% of the total, the remainder being corn-derived; assuming that switchgrass will be a major contributor to the feedstock mix, ethanol production would be centered in a wide swathe of states from North Dakota to Georgia, whereas demand would have geographic maxima from west to east (figure 5.7). Ethanol shipping would be predominantly by truck or rail until the industry evolved to take over existing petroleum pipelines or to justify the construction of new ones; linear optimization showed that shipping by truck would entail a cost of $0.13/l ($0.49/gallon), whereas rail transport would entail lower costs, $0.05/l ($0.19/gallon). In contrast, gasoline transportation to retail outlets only incurs costs of $0.003/l ($0.01/gallon).[76] The same study concluded that national solutions, although they would spur innovation and eventually lead to economies of scale, would increase shipping distances and add to total truck movements; an investment of $25 billion would be required for a dedicated ethanol pipeline system, "just to make petroleum pipelines obsolete in the long-term."

Tax incentives and subsidies are, therefore, highly likely to be features of policy making relevant to the adoption of biofuels in OECD economies generally. Funding the ethanol supply chain will be crucial; minimizing shipping costs implies the construction of as many production sites as possible, based on the use of raw materials from multiple geographical areas (forest, agricultural wastes, dedicated energy crops, municipal solid waste, etc.), ideally to match the likely distribution of major urban demand centers for ethanol blends (figure 5.7).

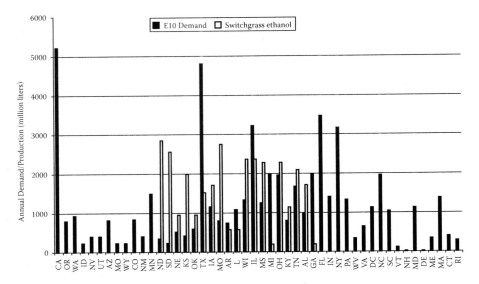

FIGURE 5.7 Hypothetical switchgrass ethanol production and E10 gasoline blend demand across the United States, except Alaska and Hawaii. (Data from Morrow et al.[76])

5.5 SUSTAINABLE DEVELOPMENT AND BIOETHANOL PRODUCTION

5.5.1 DEFINITIONS AND SEMANTICS

Much of the public and scientific debate about biofuels has assumed that the production of bioethanol as the lead biofuel is inherently "sustainable," mostly on the grounds that any agricultural activity is renewable, whereas the extraction of crude oil is necessarily a once-only activity, given the extremely long geological time scale of oil's generation. As a term, "sustainability" suffers from inexactness.[78,79] As Patzek and Pimentel[79] point out, an exact definition can be deduced from thermodynamics and is consequently defined in terms of mathematical and physical properties, that is, a cyclic process is sustainable if (and only if):

1. It is capable of being maintained indefinitely without interruption, weakening, or loss of quality
2. The environment on which this process depends and into which the process expels any "waste" material is itself equally renewable and maintainable

These are very strict criteria and are not exemplified by, for example, an annual replanting of a crop plant such as maize, which depends on outlays of fossil fuel energy (for fertilizers, etc.) and which may seriously deplete the soil or minerals and contribute to soil erosion — although such a system appears to be renewed every year, that is "only" within living memory or the history of agricultural production on Earth, on a geological time scale a minuscule length of time. It is a sad fact of human agricultural activity that periodic crises have accompanied large-scale, organized farming for millennia in (among many other examples) the erosion and salt accumulation that caused the downfall of the Mesopotamian civilization, the overgrazing and poor cultivation practices that have caused the expansion of the Sahara, and the overintensive cultivation of fragile tropical rainforest soils that contributed to the collapse of the Mayan economy and society.[80] Industrial agriculture and intensive farming are relatively recent arrivals, within the last century, and they have brought accelerated land degradation and soil erosion; agrochemical residues in water courses are a global problem, leading to eutrophication of freshwater fishing grounds and threatening the collapse of fragile ecosystems. To environmental lobbyists, therefore, the prospect of energy crop plantations is highly unwelcome if such agronomy requires and is dependent on the heavy use of fertilizers, insecticides, and pesticides, the depletion of soil organic carbon, and (worst of all) the sequestration of limited farmland.

A more analytically useful definition of the renewability of "renewable" energy could focus on a more analyzable time scale, perhaps 160 years (approximately, the life span of the industrial activity that underpins modern society).[79] Perhaps, even more appropriate would be a century, that is, most of the time in which automobiles have been fuelled by gasoline and diesel. For either option, generating a quantitative framework is crucial to any assessment of the practicalities of a biofuels program.

5.5.2 GLOBAL AND LOCAL SUSTAINABLE BIOMASS SOURCES AND PRODUCTION

In 2000 (but using data from 1996), a review, two of whose authors had affiliations to ExxonMobil, concluded that sugarcane grown under Brazilian conditions could generate above-ground and harvestable biomass of 932 GJ/hectare/year; a total of 9.3×10^8 hectares of land could substitute the full global primary energy of fossil fuels (3.2×10^{11} GJ/year), even if only sugarcane stems were used as a fuel with a quantitative extraction of the energy inside the plant material, that is, 343 GJ/hectare.[81] Based on a total area of land used to grow crops worldwide of 1.4×10^9 hectares, the fossil fuel demand could, in principle, be met with 67% of cultivatable land dedicated to sugarcane as an energy crop, or only 24% if all of the harvestable material were used. If, however, the sugarcane stems were to be used for ethanol production, only a third of the useful energy would be converted to the biofuel, and more than twice the global area of land used to grow crops would be needed, that is, ethanol production could not substitute for the world's appetite for fossil fuels.

This is an interesting but disingenuous set of calculations because ethanol had never been advocated as anything other than a convenient energy carrier as a replacement (partial or otherwise) for gasoline in automobiles. From the most recent data published by the International Energy Agency, oil used for transport is expected to continue being approximately 20% of the yearly fossil fuel demand or to slowly increase to more than 50% of the crude oil extraction rate (figure 5.8). To substitute the global demand for oil as a transport fuel, therefore, sugarcane grown for ethanol would need to occupy 40% of the world's arable land (stems only as a feedstock) or 20–25% if the entire harvestable biomass were to be used for ethanol production. The long-perceived potential conflict between land use for food crop production and for bioenergy is highly likely to be a reality, and genetic engineering or other innovative

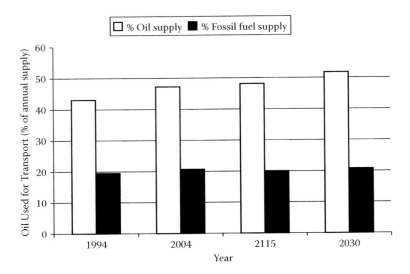

FIGURE 5.8   Oil consumption for transportation in comparison with total oil and fossil fuel usage. (Data from International Energy Agency.[34])

technologies cannot avoid this without massive increases over current bioprocess-limited abilities to transform biomass into ethanol (or any other liquid biofuel).[82] Any conventional or molecular "improvement" of a major crop species must avoid greater dependence on fertilizer and other agrochemical application to achieve greater yields. A particular drawback is the reliance on energy crops with large inputs of nonrenewable energy for the production of biofuels: corn-derived ethanol and, for biodiesel, rape (canola) and soybean seeds (see chapter 6, section 6.1).[83]

Equally, the use of agricultural residues for biofuel production has severe agronomic limitations. Although a worldwide potential residue harvest of 3.8×10^9 tonnes/year has been estimated, equivalent to 7.5 billion barrels of diesel, removing more than 30% of this could greatly increase soil erosion as well as depleting the soil organic carbon content. One solution to this problem is to establish new bioenergy plantations on uncultivated land, perhaps 250 million hectares worldwide.[84] This could be supplemented by the exploitation of weed infestations, for example, water hyacinth in lakes and waterways, naturally occurring sugar-rich forest flowers, or other adventitious resources.[85,86] Of the major candidate energy crops, switchgrass and other prairie grasses appear to offer the best balance among high net energy yields, low nutrient demands, and high soil and water conservation; with its high total root mass, switchgrass can replace soil carbon lost during decades of previous tilling within (perhaps) 20 years.[87] Under U.S. farm conditions, there is mounting evidence that dedicated energy crops would significantly reduce erosion and chemical runoff in comparison with conventional monoculture crops.[88–90] Replacing continuous corn cropping with a corn-wheat rotation and no-till field operations might maximize agricultural residue availability.[91]

Doubts persist, however, that any of these options represent true sustainability, even when consideration is extended to prolific plantations of acacia and eucalyptus trees, both of which rapidly exhaust tropical soils of nutrients and fall far short of even 100 years of high productivity.[79] The nearest approach to true sustainability could be the immediate-locale use of sun-dried wood from well-managed energy plantations, using highly efficient wood-burning stoves; the necessary high rates of biomass "mining" (probably requiring some method of energy-dependent biomass drying) to support predicted rates of growth of gasoline and biofuel consumption appears to be mathematically unsustainable.[79] In any case, intensive energy crop plantations would certainly require a high degree of environmental monitoring and management to avoid biological collapse unless lavish amounts of fertilizers are applied and maximum yields are ensured by equally large supplies of insecticides and pesticides.

A final twist is the impact of global climate change: could global warming (at least for a limited time, perhaps as long as 50–100 years) increase plant growth to such an extent that biomass extraction targets would be more closely met? Even accurately predicting biomass yields from monocultures of fast-growing tree species such as willow in the northeastern United States suffers from methodological weaknesses.[92] The effects of changes in daily minimum and maximum temperatures are complex because they differentially influence crop yield parameters; this is a major area of uncertainty for projecting yield responses to climate change.[93] For three major cereal species (wheat, corn, and barley), and an increase in annual global temperatures since 1980 of approximately 0.4°C, there is evidence for decrease in yield; the magnitude of the effect is small in comparison with the technological yield gains during the same period but suggests that rising temperature might cancel out the expected increase in yield because of increased CO_2 concentrations.[94,95] Using the global environment as an uncontrolled experimental system has its methodological drawbacks, but the results of the one long-term field trial to attempt to isolate the effects of temperature on rice growth indicate that a 15% reduction in yield could result from each 1°C rise in temperature, a much greater effect than predicted by simulation models.[96] The International Panel on Climate Change has, in stark contrast, predicted that CO_2 benefits will exceed temperature-induced yield reductions with a modest rise in temperature. Because this conclusion has received wide media coverage, it is important to examine the detailed conclusions for food, fiber, and forest products:

- Crop productivity is projected to increase slightly at mid to high latitudes for local mean temperature increases of up to 1–3°C depending on the crop and then decrease beyond that in some regions.
- At lower latitudes, especially seasonally dry and tropical regions, crop productivity is projected to decrease for even small local temperature increases (1–2°C), which would increase risk of hunger.
- Globally, the potential for food production is projected to increase with increases in local average temperature over a range of 1–3°C, but above this, it is projected to decrease.
- Adaptations such as altered cultivars and planting times allow low and mid to high cereal yields to be maintained at or above baseline yields for modest warming.
- Increases in the frequency of droughts and floods are projected to affect local production negatively, especially in subsistence sectors at low latitudes.

- Globally, commercial timber productivity rises modestly with climate change in the short to medium term, with large regional variability around the global trend.

Taken together, these International Panel on Climate Change prognoses suggest some short-term improvement in the productivity of a range of plant species, including timber grown as feedstocks for lignocellulosic ethanol, but the predictions become more unreliable as the geographical area narrows and related effects of climate are considered. This highlights the obvious conclusion that geostatistical surveys and controlled experiments should both be pursued vigorously as priority issues for agronomy and plant physiology in the next ten years.

5.5.3 Sustainability of Sugar-Derived Ethanol in Brazil

The case of the mostly widely applauded biofuel scheme to date, that of sugarcane ethanol in Brazil, also has major doubts from the environmental perspective, although the first decade (1976–1985) of the program probably achieved a reasonable soil balance by recycling fermentor stillage as fertilizer, a valuable source of minerals, particularly potassium.[20,36] With the great expansion of the industry subsequently, however, significant pollution problems have emerged. The volume of stillage that can be applied varies from location to location, and in regions with near-surface groundwater, much less stillage can be applied without contaminating the water supply.[36] In the case of the Ipojuca river in northeast Brazil, sugar cultivation and adjacent ethanol production plants use stillage extensively for both fertilization and irrigation, and this has led to water heating, acidification, increased turbidity, O_2 imbalance, and increased coliform bacteria levels.[98] The authors of this joint German-Brazilian study urged that a critical evaluation be made of the present environmental status of the sugar alcohol industry, focusing on developing more environmentally friendly cultivation methods, waste-reducing technologies, and water recycling to protect the region's water resources.

The preservation of surface and groundwater in Brazil in general as a consequence of the sugar alcohol industry's activities and development was ranked "uncertain, but probably possible" (table 5.17).[99] Sugarcane plantations have been found to rank well for soil erosion and runoff criteria in some locations in São Paolo state, although the experimental results date from the 1950s (table 5.18). A much more recent study included in the second, Dutch-Brazilian report showed much poorer results for sugarcane in comparison with other monoculture crops (figure 5.9). Nevertheless, although Brazilian sugarcane alcohol (viewed as an industrial process) makes massive demands on the water supply (21 m³/tonne of cane input), much of this water can (in principle) be recycled; in addition, Brazil enjoys such a large natural supply of freshwater from its eight major water basins (covering an area of 8.5 million km²) that the ratio of water extracted to supply is, on a global basis, exceedingly small: approximately 1%/annum, equivalent to 30-fold less than comparable data for Europe. Local seasonal shortages may, however, occur, and two of the four main sugar production regions have relatively low rainfalls (figure 5.10). Although sugar cultivation has mainly been rain-fed, irrigation is becoming more common.

TABLE 5.17

Selected Sustainability Criteria for Sugar Ethanol Production in Brazil

Criterion	Measurable parameter	Expected compliance
Greenhouse gas emissions	Net reduction 30% by 2007	Probable
Greenhouse gas emissions	Net reduction 50% by 2011	Probable
Competition with food supply	?	Uncertain
Biodiversity	No decline of protected areas in 2007	Very uncertain
	Active protection of local ecosystems by 2011	Very uncertain
Welfare	Compliance with treaties, declarations, etc.	Partial or unknown
Environment		
Waste management	Compliance with existing laws	Uncertain
Use of agrochemicals	Compliance with existing laws	Partial
Use of agrochemicals	Compliance with EU legislation by 2011	Uncertain
Prevention of soil erosion and nutrient depletion	Management plans	Unclear
Preservation of surface and groundwater	Water use and treatment	Probably possible
Airborne emissions	Compliance with EU laws by 2011	Uncertain
Use of GMOs	Compliance with EU laws by 2011	Possible

Source: Modified from Smeets, E. et al.[99]

Even before 1985, expanding sugar cultivation was putting pressure on the area of cultivatable land in Brazil used for food crops — although crops grown for export (including soybeans and coffee, as well as sugar) were more likely than sugarcane to occupy land traditionally used for food crops, so that from 1976 to 1982:[20]

- Land used for basic food crops in Brazil increased from 27.5 to 29.4 million hectares
- Land for export crops increased from 13.4 to 15.6 million hectares
- Land on which sugarcane was grown increased from 2.6 to 3.9 million hectares

The real competition is, therefore, a triple one among food, fuel, and export crops, or, more accurately, between land used for food, fuel, and export crops and land cleared by deforestation, particularly pertinent in Brazil where newly cultivatable land may well have previously been virgin rainforest.

Available data on the social impacts of land clearance are old and official figures are elusive; any overall effects of ethanol production per se on land use are intertwined with those of cash crops and social factors such as income distribution.[99] Given the emotive nature of the subject matter, the debate tends to be acerbic: "As it is currently developing, the Brazilian ethanol industry represents a direct challenge to food sovereignty and agrarian reform. Ethanol production to sustain

TABLE 5.18
Annual Soil Losses by Erosion and Runoff in Experimental Stations in Brazil

Fertile soil, 9.4% slope[a]			Red soil, 8.5% slope[b]		
Crop	Soil loss (tonne/ hectare)	Runoff (mm)	Crop	Soil loss (tonne/ hectare)	Runoff (mm)
Cassava	53	254	Castor beans	56.1	199
Cotton (in rotation)	38	250	Common beans	54.3	180
Soybean (continuous)	35	208	Cotton	51.4	183
Cotton (continuous)	33	228	Cassava	42.6	170
Soybean (in rotation)	26	146	Upland rice	36.6	143
Sugarcane	23	108	Maize (residues incorporated)	30.9	144
Maize (in rotation)	19	151	Peanut	30.6	134
Maize + common beans	14	128	Maize (residues burned)	29.0	131
Maize (continuous)	12	67	Maize + macuna bean (incorporated)	28.2	133
Maize + macuna bean (incorporated)	10	100	Sugarcane	21.0	88
Maize + manure	6.6	97	Maize + lime	19.1	96
Maize + macuna bean (mulched)	3.0	42	Maize + manure	8.9	62
Gordura grass	2.6	46	Jaragua grass	5.5	45

Source: Data from Smeets et al.[99]

[a] Average rainfall = 1,347 mm per year

[b] Average rainfall = 1,286 mm per year

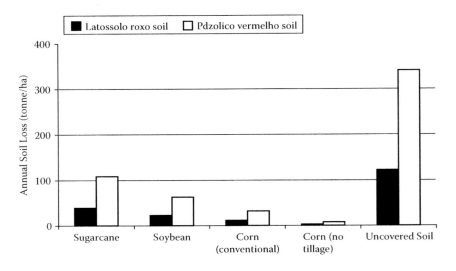

FIGURE 5.9 Soil erosion for two types of Brazilian soils in the 1990s. (Data from Smeets et al.[99])

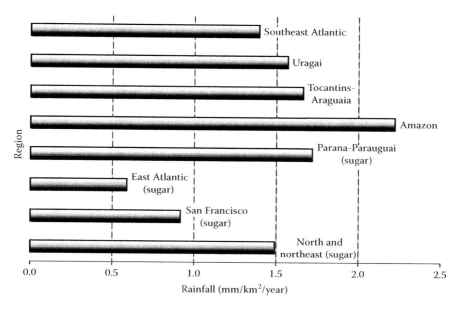

FIGURE 5.10 Annual rainfall in main sugar-producing and other regions of Brazil. (Data from Smeets et al.[99])

the enormous consumption levels of the Global North will not lead the Brazilian countryside out of poverty or help attain food sovereignty for its citizens."[100] On the other hand, to achieve poverty alleviation and the eradication of social exclusion and with support from environmentalists, Brazil proposed the Brazilian Energy Initiative at the 2002 World Summit on Sustainable Development (Johannesburg, South Africa) aiming at the establishment of global targets and timeframes of minimum shares of energy from renewable sources.[101] Headline figures for the global numbers of malnourished people known to international agencies are another datum point with a large uncertainty: from below 1 billion to 3.7 billion.[38,95] As an economist from the Earth Policy Institute was quoted as saying: "The competition for grain between the world's 800 million motorists to maintain their mobility and its two billion poorest people who are simply trying to stay alive is emerging as an epic issue."[102]

Brazil became the global leader in ethanol exports in 2006, exporting 19% (3 billion liters) of its production — 1.7 billion liters of which were imported by the United States — and plans to export 200 billion liters annually by 2025, increasing sugarcane planting to cover 30 million hectares.[100] Sugar for ethanol will increasingly be viewed by nations without a strong industrial base but with suitable climatic conditions for sugarcane growth as a cash crop, in exactly the manner that Brazil regards coffee or soybeans; the example provided by Brazil in creating rural employment at low cost, reducing the economic burden of oil imports, and developing national industrial infrastructure will be one difficult to resist, especially if major sugar producers, including Brazil, India, Cuba, Thailand, South Africa, and Australia, unite to create an expanding alternative fuel market with sugar-derived ethanol.[36] South Africa, for example, has a great and acknowledged need to improve its sugarcane

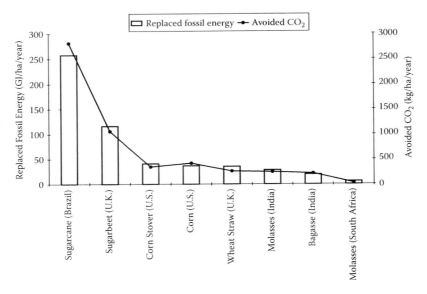

FIGURE 5.11 Agricultural land efficiency in bioethanol production. (Data from von Blottnitz and Curran.[105])

economy, where 97% of its sugarcane growers are small scale, achieving only a quarter of the productivity realized by commercial operators; sugar is produced in a surplus, most of which is exported, but a national plan to encourage biofuels usage is in place, and a first ethanol plant is planned for construction by a South African sugar producer* in neighboring Mozambique.[103]

Academic economists and agronomists are calling (and will continue to call) for an informed debate about land use in the context of increasingly large areas of highly fertile or marginal land being reallocated for energy crops.[104] Although there is good evidence that sugarcane-derived ethanol in Brazil shows the highest agricultural land efficiency in both replacing fossil energy for transportation and avoiding greenhouse gas emissions (figure 5.11), impacts on acidification and human and ecological toxicity and deleterious environmental effects occurring mostly during the growing and processing of biomass are more often ranked as unfavorable than favorable in surveys.[105]

The principal economic drivers toward greater biofuel production in developing economies are, however (and paradoxically), those widely accepted programs to reduce greenhouse gas emissions, increase energy security, and move to a scientifically biobased economy by promoting the use of biofuels (table 5.19). If a new organization of ethanol exporting countries, mostly in the Southern Hemisphere, arises to make up any shortfall in the production of endogenous biofuels in major OECD economies, only a sustained effort to require and enforce agronomically sound and environmentally safe practices on the part of those net importers will provide

---

* South African businesses are planning to open 18 corn-based ethanol production sites within the national borders by 2012.

TABLE 5.19

Support Measures and Targets for Biofuels

Country	Target (% of biofuels in total road fuel consumption)	Target deadline	Production incentives?	Consumption incentives?
		Ethanol		
United States	2.78[a]	2006	✓	✓
Brazil	(40% rise in production)	2010	✓	✓
Japan	(500 million liters)	2010	✗	✗
Canada	3.5	2010	✗	✓
		Biofuels		
European Union	5.75	2010	✓	✓
Sweden	3	2005	✓	✓
France	10	2015	✓	✓
Germany	2	2005	✓	✓
UK	5	2020	✓	✓
India	5	(unspecified)	✓	✓
China	15 (total renewables)	2020	✓	✗
Thailand	2	2010	✓	✓

Source: Modified from *World Energy Outlook.*[66]

[a] 4 billion gallons (2006) rising to 7.5 billion gallons by 2012

a counterbalance to the lure of accelerated growth and development as absolute, uncontested priorities in emerging/developing/transitional economies. The reality, however, is more likely to be close insistence on a reliability of bioethanol supply to Europe, Japan, and the United States, despite fluctuating growing conditions from year to year (as occurred, for example, in the dip in ethanol production in Brazil in 2001–2002): consumers with flexibly fueled vehicles will, under those conditions, turn back to gasoline and diesel, thus inhibiting the progress of biofuels programs.[106] Although short-lived oil gasoline price bubbles are tolerated (in the absence of any real choice), shortfalls in fuel ethanol supply are unlikely to be; considerations such as those of encroachment on to virgin and marginal land in the tropics, deforestation, and loss of soil carbon may be much lower down the agendas of publics and policy-makers inside and outside the OECD.

5.5.4 Impact of Fuel Economy on Ethanol Demand for Gasoline Blends

Improvements in automobile fuel economy would unambiguously improve the chances of an easier and better-managed introduction of biomass-based fuel alcohols: doubling the mileage achieved by gasoline-fueled vehicles in the United States would, for example, reduce the demand for ethanol by 45% or more at ethanol/gasoline blends of 10% or higher (figure 5.12). Mandatory fuel economy standards and voluntary agreements with automobile manufacturers in OECD and other countries aim at varying degrees of improved mileage in passenger cars and light commercial vehicles (table 5.20).

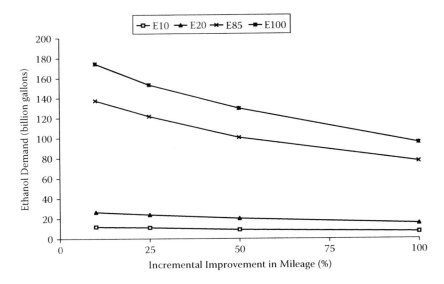

FIGURE 5.12 The impact of fuel economy on projected demand for ethanol in various gasoline blends. (Data from Morrow et al.[76])

TABLE 5.20
Selected Policies on Light-Duty Vehicle Fuel Economy

Country	Target	Target deadline	Policy basis
United States	20.7 mpg to 22.2 mpg	2007	Mandatory
	24 mpg	2011	Mandatory
Japan	23% reduction in fuel consumption (cars)	Progressive	Mandatory
	13% reduction in fuel consumption (light trucks)	Progressive	Mandatory
China	10% reduction in fuel consumption	2005	Mandatory
	20% reduction in fuel consumption	2008	Mandatory
Australia	18% reduction in fuel consumption (cars)	2010	Voluntary
Canada	Increase in corporate average fuel economy in line with U.S. standards	2007–2011	Voluntary

Source: Modified from *World Energy Outlook.*[66]

This (relative) parsimony harmonizes well with three of the principles stated by the National Research Council (NRC) in its report on the future for biobased industrial products:[107]

- Reducing the potential for war or economic disruption due to oil supply interruptions
- Reducing the buildup of atmospheric carbon dioxide

- Shifting our reliance on petroleum products to biobased products that generally have fewer harmful environmental effects

When another principle is added — strengthening rural economies and increasing demand for agricultural commodities — the main issues of the political agenda that has emerged post-2000 in both the United States and OECD economies in general are clear. There is one final argument, however, and one that commenced in the 1950s, that, instead of rendering the question of economic price of biofuels irrelevant, reformulates the question to ask: how will biofuels affect the cost of living and personal disposable income in the twenty-first century?

5.6 SCRAPING THE BARREL: AN EMERGING RELIANCE ON BIOFUELS AND BIOBASED PRODUCTS?

For most of the two decades after 1980, the scientific debate on ethanol and other biofuels revolved around the issue of production cost relative to those of conventional gasoline and diesel. As late as 1999, a lead author of the NRC report was still focusing on the question: is there any real hope that biobased products can compete economically with petroleum-derived products?[108]

Less than a decade on, a radically different perspective on biofuels has been enforced by the dramatic increase in oil prices that have rendered quite irrelevant the doubts expressed in the 1980s and 1990s on the feasibility of ever producing biofuels at an economic cost competitive to that of conventional gasoline. This watershed was evident in a comment made in an article in the August 29, 2006, issue of the *International Herald Tribune*: "As long as crude oil is above $50 a barrel, there is a momentum to biofuels that is unstoppable."[109] Analyses made in the 1980s and 1990s (and earlier) were all individually correct in that they reflected the prevailing economic realities of world oil prices; whatever the production route, bioethanol and other biofuels were only likely to be palatable as mass transportation fuels if heavily subsidized and/or as a result of enthusiastic, deliberate, and sustained governmental action (as in Brazil). In late 2007, the future of biofuels seems even rosier as world oil prices have topped first $80 a barrel, then $90 a barrel, and are predicted by industry analysts (never adverse to risking the "ridicule and resentment" noted at the beginning of this chapter) to even reach $120 a barrel during 2008.

Conversely, it is straightforward to identify the conditions under which the perceived "momentum" in favor of biofuels would falter:

- A severe (or sustained) global economic recession, or
- An unexpected announcement of the discovery of several major untapped oil deposits in politically stable regions of the world and that could be exploited at costs no more than marginally above those presently accepted by the oil industry

Both of these events would act to reduce crude oil prices again toward (if not to) those "enjoyed" for most of the twentieth century (figure 1.3). The outcome would be, however, no more than a postponement of an "event horizon" of far greater historical significance than the year-on-year fluctuations of oil and gasoline retail prices.

The concept of ultimate cumulative production, now usually referred to as estimated ultimate recovery, of global oil was introduced in 1956.[110] Given a finite quantity of oil in the Earth's crust, production could only reach that fixed amount, the oil extraction rate being mathematically fitted by a curve function with a distinct maximum, that is, the "peak oil" theory. By 1956, historical oil production rate maxima were well known: for the Ohio oil field before 1900 and for the Illinois oil field in 1940; projecting forward, the U.S. peak production rate was predicted to occur between 1970 and 1975 and a world peak production rate around the year 2000 — in contrast, world coal production would have a much delayed peak rate (approximately 2150), and total fossil fuel "family" (oil, coal, and natural gas) might last until 2400–2500.*

Although U.S. oil production did, in the event, peak a little before 1975, the global picture has remained unclear, and "oil prophets" (who are equally as capable of producing reactions of deep skepticism and dread as are economists) now predict, in reports and publications appearing between 1997 and 2007, peak oil rates to occur at any date between 2010 and 2120, with a mean value of 2040.[111] The great variability in these rival estimates derives from multiple uncertainties, including those of the extent of future discoverable oil reserves and their timing. The onset of irreversible decline could also be influenced by developments in oil-producing regions, in particular increasing domestic oil consumption might eliminate over time the ability of some countries to export oil to net consumers, reducing the number of net exporters from 35 to between 12 and 28 (another large uncertainty) by 2030 — there is even the "inverse" oil security scenario where Middle Eastern oil exporters attempt to withhold or restrict oil extraction if faced by a concerted attempt on the part of the OECD to reduce dependence on Middle Eastern oil.[112]

To return to a question posed in the preface, when will the oil run out? A different perspective is provided by calculating the quotient of the known oil reserves and the actual consumption rate, accepting that "proven" oil reserves may be over-estimated or underestimated and putting to one side any possible (but unproven) reserves to be substantiated in the future (figure 5.13). For the past 18 years, this estimate has changed little, after increasing markedly between 1979 and 1998 as new discoveries were made; the average time until exhaustion of the supply has been 41.5 years (with a standard deviation of ±1.0 year) since 1998. The dilemma lies in interpreting the detailed trend line: is the mean "life expectancy" of oil reserves now decreasing? After reaching parity in the late 1980s, the rate of discovery has been overtaken by the consumption rate since 2003; if that relative imbalance persists, the original Hubbert prediction will prove to have been accurate (figure 5.14). There certainly is no sign of the time to eventual exhaustion having increased during the past 20 years, individual years of optimism being followed by a succession of years that fit better with a static or decreasing trend (figure 5.13). The Energy Watch Group, a German organization "of independent scientists and experts who investigate sustainable concepts for global energy supply," concluded in October 2007 that

---

* A much less widely quoted prediction by Hubbert concerned nuclear energy, that is, that recoverable uranium in the United States amounted to an energy potential several hundred times that of all the fossil fuels combined and that the world stood on the threshold of an era of far greater energy consumption than that made possible by fossil fuels.

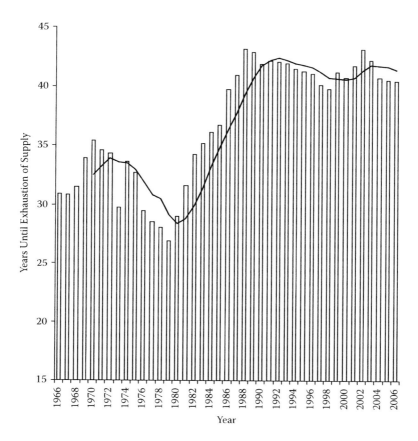

FIGURE 5.13   Crude oil supply longevity as calculated from yearly consumption and estimates of reserves. (Data from *BP Statistical Reviews.*[112a])

so many major production fields were past their peak output that oil supply would decrease rapidly from 81 million barrels/day in 2006 to 58 million barrels/day by 2020 to 39 million barrels/day by 2030, with all regions (apart from Africa) showing reduced production rates by 2020, that is, "peak oil is now."[113] Table 5.21 collects the Energy Watch Group's analysis of the historical sequence of individual nations' peaks of production since 1955 and compares these dates with the dwindling rate of new oil field discoveries after the 1960s. As collateral, the evidence was presented in the German report that big international oil companies, taken in aggregate, have been unable to increase their production in the last decade despite the marked rise in world oil prices. Coincidentally, in October 2007, oil prices exceeded $90/barrel, highlighting the persistence of the new era of high energy prices that bodes ill if diminishing oil supplies seriously exert their inevitable economic effect.

With natural gas supplies, the horizon of exhaustion remains further away, that is, with a mean value of more than 66 years for estimates made after 1988 (figure 5.15). Integrated over the past 19 years, however, the outlook does not offer promise of natural gas supplies having an increased longevity. If anything, the prospect appears

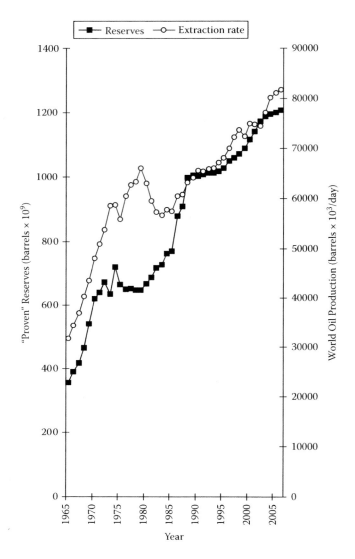

FIGURE 5.14   Crude oil reserves and extraction rate. (Data from *BP Statistical Reviews.*[112a])

to be one now of rapidly dwindling stocks if the trends from 2001 onward prove to be consistent (figure 5.15). As with oil, therefore, the era of discovery of large and accessible reserves may be over.

Unless fuel economy is radically boosted by technological changes and popular take-up of those choices, price pressures on oil products caused by a dwindling or static supply (and an expected increase in demand from expanding Asian economies) will act to maintain high oil and gasoline prices. Modeled scenarios envisaged by the DOE and European energy forums include those with persisting high oil prices*

---

* These have little effect on greenhouse gas emissions because of fuel substitution (oil or gas to coal).[114]

TABLE 5.21

Peak Oil Years for Producing Nations and Trend in New Discoveries

Geographical source	Year of peak	Time period	Average oil discoveries (Gb/year)
Austria	1955	1950–1959	41.2
Germany	1967	1960–1969	55.4
U.S. (Lower 48)	1971	1970-1979	38.8
Canada	1974		
Romania	1976		
Indonesia	1977		
Alaska	1989	1980–1989	20.9
Egypt	1993	1990–1999	15.1
India	1995		
Syria	1995		
Gabon	1997		
Malaysia	1997		
Argentina	1998		
Venezuela	1998		
Colombia	1999		
Ecuador	1999		
U.K.	1999		
Australia	2000	2000–2001	17.0
Oman	2001		
Norway	2001		
Yemen	2001		
Denmark	2004	2004–2005	12.0
Mexico	2004		

Source: Data from Zittel and Schindler.[113]

until 2025.[114] In that context, the supporting case for bioethanol and other biofuels is, therefore, threefold:

1. They are derived from biomass feedstocks that can, with careful management ("husbandry"), be accessible when fossil fuels have become depleted — possibly, within the next four decades.
2. Their use at least partially mitigates greenhouse gas emission, far more so if nonfossil fuel energy sources contribute to their production energy inputs.
3. They can, in crude production cost terms, be competitive with conventional fuels.

Arguments in favor of lignocellulosic ethanol have been almost invariably defensive, guarding against the charge of consumer costs far in excess of what an open market

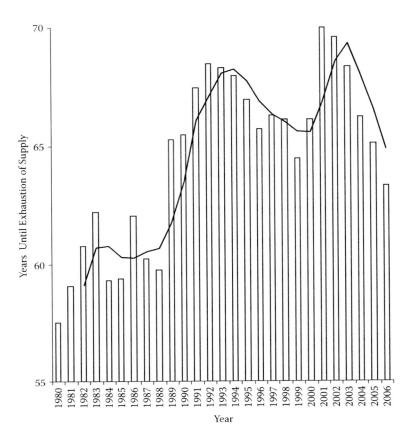

FIGURE 5.15 Natural gas supply longevity as calculated from yearly consumption and estimates of reserves. (Data from *BP Statistical Reviews*.[112a])

can bear, and calling upon a wider range of socioeconomic factors to develop a full accounting of "externality costs."[115] A clearly positive case can, however, now be made without the recourse to such arguments.

Although there are nonbiological routes to substitutes for petroleum products, capable of extending liquid hydrocarbon fuel usage by a factor of up to tenfold, their estimated production costs lie between two- and sevenfold that of conventional gasoline and oil products (figure 5.16).[116] Unavoidably, their production from oil shale, tar sands, natural gas, and coal deposits would add massively to total greenhouse gas emissions. Although it is also possible that technological innovations will enable oil to be extracted with high efficiency from such nonconventional sources as oil shale and tar sands and push the limits of geographical and geological possibilities for neglected or undiscovered deep-ocean oil deposits, these too will be costly.*

---

* The extraction, processing, and use as an automobile fuel of oil shale hydrocarbons would also increase net CO_2 emissions above those of, for example, the use of natural gas.[113]

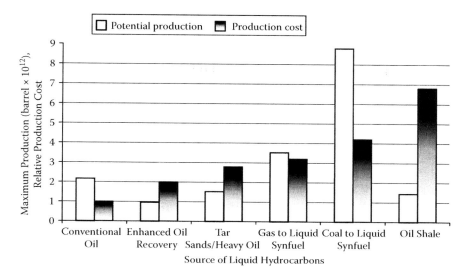

FIGURE 5.16   Potential future global supply of liquid hydrocarbons from fossil resources. (Data from Farrell and Brandt.[116])

Perhaps, the crucial debate in coming decades will be that of allocating public funds in the forms of tax incentives to offset exploration costs (on the one hand, to the oil industry) or R&D costs to a maturing biofuels industry, that is, the crucial policy decisions for investments in new technologies that may be substantially unproven when those choices must be made. The oil industry has, in fact, been highly successful in attracting such subsidies. The last (numerical) word can be left with the U.S. General Accountability Office, whose assistant director for Energy Issues, Natural Resources, and the Environment made a presentation to the "Biomass to Chemicals and Fuels: Science, Technology, and Public Policy" conference at Rice University, Houston, Texas, in September 2006.[117] Tax incentives to the ethanol industry between 1981 and 2005 amounted to only 12% of those for the oil and gas industry between 1968 and 2005 (figure 5.17). These figures underestimate the full sums expended in incentives to the oil and gas industry because they date only from when full records were kept by the U.S. Treasury of revenue losses, not when an incentive was implemented (the Tariff Act of 1913, in the case of the oil and gas industry). Although the magnitudes of some subsidies for conventional fuels are much reduced presently as compared with the situation in the 1970s and 1980s, they still outweigh the sums laid out to support biofuels. An "incentives culture" has, therefore, a long history in shaping and managing energy provision.

The competition is not between fuel ethanol, on the one hand, and substitutes for conventional oil products, on the other, but that between rival technologies for liquid fuels (most of which are biobased) and — a highly strategic issue — using plant biomass as either a source of biofuels or predominantly a source of carbon to replace petrochemical feedstocks in the later twenty-first century. Is the future one of a hydrogen economy for transportation and a cellulose-based supply of

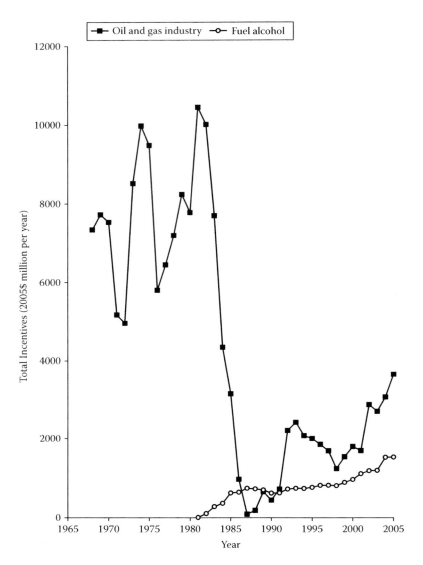

FIGURE 5.17 U.S. tax incentives and subsidies to the oil/gas and fuel alcohol industries. (Data from Agbara.[117])

"green" chemicals? Will nuclear power be the key to providing hydrogen as an energy carrier and will biological processes provide liquid fuels as minor, niche market energy carriers for automobiles? Or is the future bioeconomy (as discussed in chapter 6) a mosaic of different technologies competing to augment nuclear, solar, and other renewable energy sources while gradually replacing dwindling and ever more expensive hydrocarbon deposits as a renewable and (possibly) sustainable bedrock for the chemical industry beyond the twenty-first century?

REFERENCES

1. Cox, S. (Ed.), *Economics. Making Sense of the Modern Economy*, The Economist Newspaper Ltd. in association with Profile Books Ltd., London, 2006.
2. Holden, C., Is bioenergy stalled?, *Science* 227, 1018, 1981.
3. Herrera, S., Bonkers about biofuels, *Nat. Biotechnol.*, 24, 755, 2006.
4. Anonymous, The dragon and the eagle, in Cox, S. (Ed.), *Economics. Making Sense of the Modern Economy*, The Economist Newspaper Ltd. in association with Profile Books Ltd., London, 2006, p. 89.
5. Wheals, A.E. et al., Fuel ethanol after 25 years, *Trends Biotechnol.*, 17, 482, 1999.
6. Pimentel, D., Ethanol fuels: energy security, economics, and the environment, *J. Agric. Environ. Ethics*, 4, 1, 1991.
7. Patzek, T.W. et al., Ethanol from corn: clean, renewable fuel for the future, or drain on our resources and pockets?, *Environ. Dev. Sustain.*, 7, 319, 2005.
8. Zechendorf, B., Sustainable development: how can biotechnology contribute?, *Trends Biotechnol.*, 17, 219, 1999.
9. Martines-Filho, J., Burnquist, H.L., and Vian, C.E.F., Bioenergy and the rise of sugar-cane-based ethanol in Brazil, *Choices*, 2nd Quarter 2006, 21(2), 91.
10. Moreira, J.R., Noguiera, L.A.H., and Parente, V., Biofuels for transport, development, and climate change: lessons from Brazil, in Bradley, R. and Baumert, K.A. (Eds.), *Growing in the Greenhouse: Protecting the Climate by Putting Development First*, World Resources Institute, Washington, D.C., 2005, chap. 3.
11. Hahn-Hägerdal, B. et al., Bio-ethanol: the fuel of tomorrow from the residues of today, *Trends Biotechnol.*, 24, 549, 2006.
12. Paul, J.K. (Ed.), *Large and Small Scale Ethyl Alcohol Processes from Agricultural Raw Materials*, Noyes Data Corporation, Park Ridge, NJ, 1980.
13. Wilke, C.R. et al., Raw materials evaluation and process development studies for conversion of biomass to sugars and ethanol, *Biotechnol. Bioeng.*, 23, 1681, 1981.
14. Westby, C.A. and Gibbons, W.R., Farm-scale production of fuel ethanol and wet grain from corn in a batch process, *Biotechnol. Bioeng.*, 24, 1681, 1982.
15. Gibbons, W.R., Westby, C.A., and Dobbs, T.L., A continuous, farm-scale, solid-phase fermentation process for fuel ethanol and protein feed production from fodder beets, *Biotechnol. Bioeng.*, 26, 1098, 1984.
16. Keim, C.R. and Venkatasubramanian, K., Economics of current biotechnological methods of producing ethanol, *Trends Biotechnol.*, 7, 22, 1989.
17. Marrow, J.E., Coombs, J., and Lees, E.W., *An Assessment of Bio-Ethanol as a Transport Fuel in the UK*, Her Majesty's Stationery Office, London, 1987.
18. Marrow, J.E. and Coombs, J., *An Assessment of Bio-Ethanol as a Transport Fuel in the UK*, Volume 2, Her Majesty's Stationery Office, London, 1987.
19. Commission of the European Communities, *Cost/Benefit Analysis of Production and Use of Bioethanol as a Gasoline Additive in the European Community*, Office for Official Publications of the European Communities, Luxembourg, 1987.
20. Geller, H.S., Ethanol fuel from sugar cane in Brazil, *Ann. Rev. Energy*, 10, 135, 1985.
21. Elander, R.T. and Putsche, V.L., Ethanol from corn: technology and economics, in Wyman, C.E. (Ed.), *Handbook on Bioethanol: Production and Utilization*, Taylor & Francis, Washington, D.C., 1996, chap. 15.
22. Lynd, L.R., Overview and evaluation of fuel ethanol from cellulosic biomass: technology, economics, the environment, and policy, *Ann. Rev. Energy Environ.*, 21, 403, 1996.
23. Lynd, L.R., Elander, R.T., and Wyman, C.E., Likely features and costs of mature biomass ethanol technology, *Appl. Biochem. Biotechnol.*, 57–58, 741, 1996.
24. Wayman, M. and Dzenis, A., Ethanol from wood: economic analysis of an acid hydrolysis process, *Can. J. Chem. Eng.*, 62, 699, 1984.

25. Ladisch, M.R. and Svarckopf, J.A., Ethanol production and the cost of fermentable sugars from biomass, *Bioresour. Technol.*, 36, 83, 1991.
26. von Sivers, M. et al., Cost analysis of ethanol from willow using recombinant *Escherichia coli*, *Biotechnol. Prog.*, 10, 555, 1994.
27. von Sivers, M. and Zacchi, G., A techno-economical comparison of three processes for the production of ethanol from wood, *Bioresour. Technol.*, 51, 43, 1995.
28. von Sivers, M. and Zacchi, G., Ethanol from lignocellulosics: a review of the economy, *Bioresour. Technol.*, 56, 131, 1999.
29. Wingren, A., Galbe, M., and Zacchi, G., Techno-economic evaluation of producing ethanol from softwood: comparison of SSF and SHF and identification of bottlenecks, *Biotechnol. Prog.*, 19, 1109, 2003.
30. Wingren, A. et al., Process considerations and economic evaluation of two-step steam pretreatment for production of fuel ethanol from softwood, *Biotechnol. Prog.*, 20, 1421, 2004.
31. Wooley, R. et al., Process design and costing of bioethanol technology: a tool for determining the status and direction of research and development, *Biotechnol. Prog.*, 15, 794, 1999.
32. Hamelinck, C.N., van Hooijdonk, G., and Faaij, A.P.C., Ethanol from lignocellulosic biomass: techno-economic performance in short-, middle- and long-term, *Biomass Bioenergy*, 28, 384, 2005.
33. Bohlmann, G.M., Process economic considerations for production of ethanol from biomass feedstocks, *Industrial Biotechnol.*, 2, 14, 2006.
34. *World Energy Outlook*, International Energy Agency, Paris, 2006, chap. 14.
35. Goldemberg, J., The evolution of ethanol costs in Brazil, *Energy Policy*, 24, 1127, 1996.
36. Moreira, J.R. and Goldemberg, J., The alcohol program, *Energy Policy*, 27, 229, 1999.
37. Wyman, C.E., Biomass ethanol: technical progress, opportunities, and commercial challenges, *Ann. Rev. Energy Environ.*, 24, 189, 1999.
38. Pimentel, D., Patzek, T., and Cecil, G., Ethanol production: energy, economic, and environmental losses, *Rev. Environ. Contam. Toxicol.*, 189, 25, 2007.
39. Hamelinck, C.N. and Faaij, A.P.C., Outlook for advanced biofuels, *Energy Policy*, 34, 3268, 2006.
40. Kwiatkowski, J.R. et al., Modeling the process and costs of fuel ethanol production by the corn dry-grind process, *Ind. Crops Products*, 23, 288, 2006.
41. Kumar, A. and Sokhansanj, S., Switchgrass (*Panicum virgatum* L.) delivery to a biorefinery using integrated biomass supply analysis and logistics (BSAL) model, *Bioresour. Technol.*, 98, 1033, 2007.
42. Mani, S., Tabil, L.G., and Sokhansanj, S., Specific energy requirement for compacting corn stover, *Bioresour. Technol.*, 97, 1420, 2006.
43. Kumar, A., Cameron, J.B., and Flynn, P.C., Pipeline transport and saccharification of corn stover, *Bioresour. Technol.*, 96, 819, 2005.
44. Burnes, E.I. et al., Economic analysis of ethanol production in California using traditional and innovative feedstock supplies, *Appl. Biochem. Biotechnol.*, 113–116, 95, 2004.
45. Fan, Z. and Lynd, L.R., Conversion of paper sludge to ethanol. II. Process design and economic analysis, *Bioprocess Biosys. Eng.*, 30, 35, 2007.
46. Ag-West Biotech, http://www.agwest.sk.ca/publications/infosource/inf_may98.php.
47. Kádár, Z., Szengyel, Z., and Réczey, K., Simultaneous saccharification and fermentation (SSF) of industrial wastes for the production of ethanol, *Ind. Crops Products*, 20, 103, 2004.
48. del Campo, I. et al., Diluted acid hydrolysis pretreatment of agri-food wastes for bioethanol production, *Ind. Crops Products*, 24, 214, 2006.

49. Rosenberger, A. et al., Costs of bioethanol production from winter cereals: the effect of growing conditions and crop production intensity levels, *Ind. Crops Products*, 15, 91, 2002.

50. Chandel, A.K. et al., Economics and environmental impact of bioethanol production technologies: an appraisal, *Biotechnol. Mol. Biol. Rev.*, 2, 14, 2007.

51. Gregg, D.J., Boussaid, A., and Saddler, J.N., Techno-economic evaluations of a generic wood-to-ethanol process: effect of increased cellulose yields and enzyme recycle, *Bioresour. Technol.*, 63, 7, 1998.

52. Nagle, N., Ibsen, K., and Jennings, E.A., Process economic approach to develop a dilute-acid cellulose hydrolysis process to produce ethanol from biomass, *Appl. Biochem. Biotechnol.*, 77–79, 595, 1999.

53. Eggman, T. and Elander, R.T., Process and economic analysis of pretreatment technologies, *Bioresour. Technol.*, 96, 2019, 2005.

54. Keller, F.A., Hamilton, J.E., and Nguyen, Q.A., Microbial pretreatment of biomass: potential for reducing severity of thermochemical biomass pretreatment, *Appl. Biochem. Biotechnol.*, 105–108, 27, 2003.

55. Zhang, Y.-H.P. et al., Fractionating recalcitrant lignocellulose at modest reaction conditions *Biotechnol. Bioeng.*, 97, 214, 2007.

56. Jorgensen, H. et al., Liquefaction of lignocellulose at high-solids concentrations, *Biotechnol. Bioeng.*, 96, 862, 2007.

57. Manderson, G.J. et al., Price sensitivity of bioethanol produced in New Zealand from *Pinus radiata* wood, *Energy Sources*, 11, 135, 1989.

58. Alkasrawi, M., Galbe, M., and Zacchi, G., Recirculation of process streams in fuel ethanol production from softwood based on simultaneous saccharification and fermentation, *Appl. Biochem. Biotechnol.*, 98–100, 849, 2002.

59. York, S.W. and Ingram, L.O., Soy-based medium for ethanol production by *Escherichia coli*, *J. Ind. Microbiol.*, 16, 374, 1996.

60. Ashgari, A. et al., Ethanol production from hemicellulose hydrolysates of agricultural residues using genetically engineered *Escherichia coli* strain KO11, *J. Ind. Microbiol.*, 16, 42, 1996.

61. Escobar, J.M., Rane, K.D., and Cheryan, M., Ethanol production in a membrane bioreactor: pilot-scale trials in a wet corn mill, *Appl. Biochem. Biotechnol.*, 91–93, 283, 2001.

62. Taylor, F. et al., Dry-grind process for fuel ethanol by continuous fermentation and stripping, *Biotechnol. Prog.*, 16, 541, 2000.

63. Zacchi, G. and Axelsson, A., Economic evaluation of preconcentration in production of ethanol from dilute sugar solutions, *Biotechnol. Bioeng.*, 34, 223, 1989.

64. van Haandel, A.C., Integrated energy production and reduction of the environmental impact at alcohol distillery plants, *Water Sci. Technol.*, 52, 49, 2005.

65. Alonso Pippo, W., Garzone, P., and Cornacchia, G., Agro-industry sugarcane residues disposal: the trends of their conversion into energy carriers in Cuba, *Waste Manag.*, 27, 869, 2007.

66. *World Energy Outlook*, International Energy Agency, Paris, 2006, Annex A.

67. Gnansounou, E., Dauriat, A., and Wyman, C.E., Refining sweet sorghum to ethanol and sugar: economic trade-offs in the context of North China, *Bioresour. Technol.*, 96, 985, 2005.

68. Yang, B. and Lu, Y., The promise of cellulosic ethanol production in China, *J. Chem. Technol. Biotechnol.*, 82, 6, 2007.

69. Perlack, R.D. et al., *Biomass as a Feedstock for a Bioenergy and Bioproducts Industry: the Technical Feasibility of a Billion-Ton Annual Supply*, U.S. Department of Energy, April 2005, available at http.//www.osti.gov/bridge.

70. Mabee, W.E. et al., Canadian biomass reserves for biorefining, *Appl. Biochem. Biotechnol.*, 129–132, 22, 2006.

71. Kumar, A., Cameron, J.B., and Flynn, P.C., Large-scale ethanol fermentation through pipeline delivery of biomass, *Appl. Biochem. Biotechnol.*, 121–124, 47, 2005.

72. Mahmudi, H. and Flynn, P.C., Rail vs truck transport of biomass, *Appl. Biochem. Biotechnol.*, 129–132, 88, 2006.

73. Kumar, A., Sokhansanj, S., and Flynn, P.C., Development of a multicriteria assessment model for ranking biomass feedstock collection and transportation systems, *Appl. Biochem. Biotechnol.*, 129–132, 71, 2006.

74. Shahbazi, A. and Li, Y., Availability of crop residues as sustainable feedstock for bioethanol production in North Carolina, *Appl. Biochem. Biotechnol.*, 129–132, 71, 2006.

75. Li, A., Antizar-Ladislao, B., and Khraisheh, M., Bioconversion of municipal solid waste to glucose for bio-ethanol production, *Bioproc. Biosyst. Eng.*, 30, 189, 2007.

76. Morrow, W.R., Griffin, W.M., and Matthews, H.S., Modeling switchgrass derived cellulosic ethanol distribution in the United States, *Environ. Sci. Technol.*, 40, 2877, 2006.

77. McCarthy, T., The coming wonder? Foresight and early concerns about the automobile, *Environ. History*, 6, 46, 2001.

78. Patzek, T.W., Thermodynamics of the corn-ethanol biofuel cycle, *Crit. Rev. Plant Sci.*, 23, 519, 2004.

79. Patzek, T.W. and Pimentel, D., Thermodynamics of energy production from biomass, *Crit. Rev. Plant Sci.*, 24, 327, 2005.

80. Holdren, J.P. and Ehrlich, P.R., Human population and the global environment, *Am. Sci.*, 62, 282, 1974.

81. Kheshgi, H.S., Prince, R.C., and Marland, G., The potential of biomass fuels in the context of global climate change: focus on transportation fuels, *Ann. Rev. Energy Environ.*, 25, 199, 2000.

82. Torney, F. et al., Genetic engineering approaches to improve bioethanol production from maize, *Curr. Opin. Biotechnol.*, 18, 193, 2007.

83. Dewulf, J., van Langenhove, H., and van de Velde, B., Exergy-based efficiency and renewability assessment of biofuel production, *Environ. Sci. Technol.*, 39, 3878, 2005.

84. Lal, R., World crop residues production and implications of its use as a biofuel, *Environ. Int.*, 31, 575, 2005.

85. Gunnarsson, C.C. and Petersen, C.M., Water hyacinths as a resource in agriculture and energy production: a literature review, *Waste Manag.*, 27, 117, 2007.

86. Swain, M.R. et al., Ethanol fermentation of mahula (*Madhuca latifolia* L.) flowers using free and immobilized yeast (*Saccharomyces cerevisiae*), *Microbiol. Res.*, 162, 93, 2007.

87. Downing, M., McLaughlin, S., and Walsh, M., Energy, economic, and environmental implications of production of grasses as biomass feedstocks, in *Proc. Second Biomass Conference of the Americas: Energy, Environment, Agriculture, and Industry*, National Renewable Energy Laboratory, Golden, CO 1995, p. 288.

88. Graham, R.L., Downing, M., and Walsh, M.E., A framework to assess regional environmental impacts of dedicated energy crop production, *Environ. Manage.*, 20, 475, 1996.

89. Sanderson, M.A. et al., Nutrient movement and removal in a switchgrass biomass-filter strip system treated with dairy manure, *J. Environ. Qual.*, 30, 210, 2001.

90. Nelson, R.G., Ascough, J.C., and Langemeier, M.R., Environmental and economic analysis of switchgrass production for water quality improvement in northeast Kansas, *J. Environ. Manage.*, 79, 336, 2006.

91. Nelson, R.G. et al., Methodology for estimating removable quantities of agricultural residues for bioenergy and bioproduct use, *Appl. Biochem. Biotechnol.*, 113–116, 13, 2004.

92. Arevalo, C.B.M. et al., Development and validation of aboveground biomass estimations for four *Salix* clones in central New York, *Biomass Bioenergy*, 31, 1, 2007.

93. Lobell, D.B. and Ortiz-Monasterio, J.I., Impacts of day versus night temperatures on spring wheat yields. A comparison of empirical and CERES model predictions in three locations, *Agron. J.*, 99, 469, 2007.

94. Lobell, D.B. and Field, C.B., Global scale climate-crop yield relationships and the impacts of recent warming, *Environ. Res. Lett.*, 2, 014002, 2007.

95. Cassman, K.G., Climate change, biofuels, and global food security, *Environ. Res. Lett.*, 2, 011002, 2007.

96. Peng, S. et al., Rice yields decline with higher night temperature from global warming, *Proc. Natl. Acad. Sci., USA*, 101, 9971, 2004.

97. Intergovernmental Panel on Climate Change, *Climate Change 2007:Climate Change Impacts, Adaptation and Vulnerability, Summary for Policymakers*, April 6, 2007 (available at: http://ipcc.ch/SPM6avr07.pdf).

98. Gunkel, G. et al., Sugar cane industry as a source of water pollution — case study on the situation in Ipojuca River, Pernambuco, Brazil, *Water Air Soil Poll.*, 180, 261, 2007.

99. Smeets, E. et al., *Sustainability of Brazilian Bio-Ethanol*, Report NWS-E-2006-110 (ISBN 90-8672-012-9), Senternovem (The Netherlands Agency for Sustainable Development and Innovation), Utrecht, The Netherlands, August 2006.

100. Kenfield, I., *Brazil's Ethanol Plan Breeds Rural Poverty, Environmental Degradation*, International Relations Center, Silver City, NM, March 6, 2007, http://americas.irc-online.org/am/4049.

101. Goldemberg, J., Coelho, S.T., and Lucopn, O., How adequate polices can push renewables, *Energy Policy*, 32, 1141, 2004.

102. Brown, L.R., quoted in McNeely, J.A., The big green fuel lie, *The Independent*, March 5, 2007, http://environment.independent.co.uk/environment/climate_change/article2328821.ece.

103. Morris, G., *Strong Land Use Is Key to Developing South African Biofuels*, Biofuel Review of the Renewable Energy and Energy Efficiency Partnership, April 10, 2007, hhtp://www.reeep.org/index.cfm?articleid=1628.

104. Hill, J., Environmental costs and benefits of transportation biofuel production from food- and lignocellulose-based energy crops. A review, *Agron. Sustain. Dev.*, 27, 1, 2007.

105. von Blottnitz, H. and Curran, M.A., A review of assessments conducted on bio-ethanol as a transportation fuel from a net energy, greenhouse gas, and environmental life-cycle perspective, *J. Cleaner Prod.*, 15, 607, 2007.

106. Wang, M., Learning from the Brazilian biofuel experience, *Environ. Res. Lett.*, 1, 011002, 2007.

107. National Research Council, *Biobased Industrial Products: Research and Commercialization Priorities*, 2000, http://books.nap.edu/catalog/5295.html.

108. Dale, B.E., Biobased industrial products: bioprocess engineering when cost really counts, *Biotechnol. Prog.*, 15, 775, 1999.

109. Leow, C. and Kishan, S., Biofuels' fortunes glisten as price of energy soars, *International Herald Tribune*, 29 August, 2006.

110. Hubbert, M.K., *Nuclear Energy and the Fossil Fuels*, Publication no. 95, Shell Development Company (Exploration and Production Research Division), Houston, TX, June 1956.

111. United States Government Accountability Office, *Crude Oil: Uncertainty About Future Oil Supply Makes It Important to Develop a Strategy for Addressing a Peak and Decline in Oil Production*, GAO-07-283 Washington, D.C., February 2007.

112. Hallock, J.L. et al., Forecasting the limits to the availability and diversity of global conventional oil supply, *Energy*, 29, 1673, 2004.

112a.*BP Statistical Review of World Energy*, British Petroleum, London, 2006.

113. Zittel, W. and Schindler, J., *Crude Oil. The Supply Outlook*, EWG-Series No. 3/2007, October 2007, www.energywatchgroup.org.

114. Vielle, M. and Viguier, L., On the climate change effects of high oil prices, *Energy Policy*, 35, 844, 2007.

115. Claassen, P.A.M. et al., Utilization of biomass for the supply of energy carriers, *Appl. Microbiol. Biotechnol.*, 52, 741, 1999.

116. Farrell, A.E. and Brandt, A.R., Risks of the oil transition, *Environ. Res. Lett.*, 1, 014004, 2006.

117. Agbara, G.M., *Federal Energy Tax Incentives and Subsidies and the Current State of Biomass Fuels: A View from Congressional Oversight*, http://www.rice.edu/energy/events/past/biofuels2006.html.

6 Diversifying the Biofuels Portfolio

Biodiesel, Fischer-Tropsch Diesel, and "Bio-ols"

6.1 BIODIESEL: CHEMISTRY AND PRODUCTION PROCESSES

6.1.1 VEGETABLE OILS AND CHEMICALLY PROCESSED BIOFUELS

Practical interest in the oils extracted from plant seeds as sources of usable transportation fuels has a historical lineage back to Rudolf Diesel and Henry Ford. Minimally refined vegetable oils can be blended with conventional diesel fuels and, if a 10% lower energy content of widely available oils (on a volume basis) is acceptable — with the consequent reduction of maximum fuel energy but without modification of the injection system — diesel fuel extenders are cheap and plentiful.[1] There is much anecdotal evidence for diesel tanks being illegally "topped up" with vegetable oils, reducing fuel costs but risking detection via the unusual aromas emanating from the tail pipe.

Industrial production has, however, focused on transforming vegetable oils into a mixture of fatty acid esters by a process frequently described as being akin to the "cracking" of petroleum, that is, a transesterification of triglycerides with low-molecular-weight alcohols (figure 6.1). This produces a higher-volatility mixture ("biodiesel") with physicochemical properties much more similar to those of conventional diesel fluids (table 6.1). Biodiesel is not, as such, a biotechnological product, being manufactured with any suitable vegetable oil from crops with no history of plant biotechnology (or even from animal fats) by an entirely chemical procedure but commentators include biodiesel in the portfolio of emerging biofuels because of its biological origin as a plant seed oil. In 2005, the estimated world production of biodiesel was 2.91 million tonnes of oil equivalent, of which 87% was manufactured in the European Union (62% in Germany), with only the United States (7.5%) and Brazil (1.7%) as other major producers; this total supply amounted to less than 20% of that of global fuel ethanol production.[2] World biodiesel supply had, on the other hand, increased by threefold between 2000 and 2005, and a marked acceleration in the United States as well as in Europe is predicted by the International Energy Agency up to 2030.

Soybean oil dominates U.S. biodiesel production in existing and planned production facilities designed for single-oil use, but more of the operating or planned sites are capable of handling multiple feedstocks, including animal fats, recycled cooking oils,

285

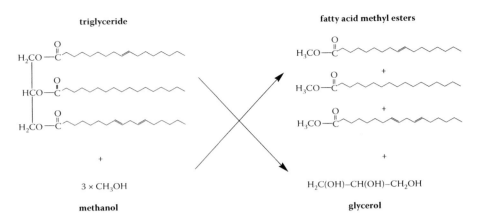

FIGURE 6.1 Transesterification of triglycerides to fatty acid methyl esters (biodiesel).

TABLE 6.1
Canola Seed Oil, Biodiesel, and Diesel

Physical parameter	Diesel	Semirefined oil	Methyl esters (biodiesel)
Onset of volatilization (°C)	70	280	70
End of volatilization (°C)	260	520	250
90% Distillation temperature (°C)	220	500	242
Density, g/ml	0.83–0.85	0.92	0.88
Cetane number	45–54	32–40	48–58
Viscosity, mm²/sec at 20°C	5.5	73	7
Cloud point, °C	1 (seasonal)	−11	−1

Source: Data from Culshaw and Butler[4] and McDonnell et al.[1]

and vegetable oils (figure 6.2).[3] Figures supplied by the National Diesel Board (www. nbb.org) on September 2007 indicate 165 sites then operational in the United States with a total national production capacity of 1.85 billion gallons/year, with a further 80 sites under construction, with a total capacity expansion of 1.37 billion gallons/ year — the range of sizes of the sites is enormous, with (at one extreme) a site capable of producing 100 million gallons, dwarfing the smallest site (annual capacity 50,000 gallons). Almost as striking is the great geographical spread of biodiesel producers, with only 4 of the lower 48 states not being represented by late 2007.*

In the European Union, rape (canola) is the most abundant suitable monoculture crop, with the particular advantage of being readily cultivated in the relatively cold climates of northern Europe.[4] It is, however, the sheer variety of single or mixed sources of oil and fat that could be transformed into biodiesel that has attracted both large-scale and niche-market industrial interest — at one extreme, even used

---

* Hawaii had two sites but Alaska had none.

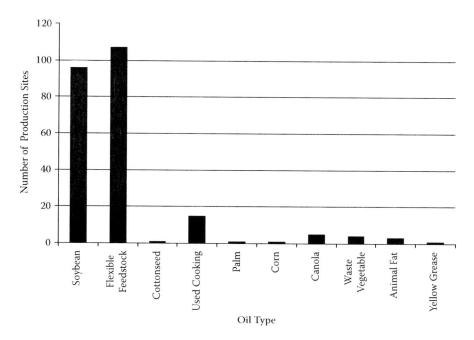

FIGURE 6.2 Breakdown of feedstock utilization by U.S. biodiesel producers. (Data from the National Biodiesel Board, September 2007.)

cooking oil (manufactured initially from corn, sunflower, etc.) can serve as the biological input, a widely publicized example of exemplary social recycling.[5]

6.1.2 BIODIESEL COMPOSITION AND PRODUCTION PROCESSES

Biodiesel is unique among biofuels in not being a single, defined chemical compound but a variable mixture, even from a monoculture crop source. The triglycerides in any plant oil are a mixture of unsaturated and saturated fatty acids esterified to glycerol; fatty materials from land animals have much higher contents of saturated fatty acids (table 6.2).[6] This variability has one far reaching implication: reducing the content of saturated fatty acid methyl esters in biodiesel reduces the cloud point, the temperature below which crystallization becomes sufficiently advanced to plug fuel lines; a diesel suitable for winter use may have a cloud point below −11°C, and "winterization" (treatment at low temperature and removal of solidified material) of biodiesel generates a product with similar improved operability and startup characteristics.[7,8]

The idiosyncratic fatty acid content of canola seed oil, with its preponderance of the very long chain erucic acid (table 6.2), has a quite different significance. Erucic acid has been known since the 1950s to stimulate cholesterol synthesis by animals.[9] The potential adverse health effects (increased risk of circulatory disease) led to legislation on the erucic acid content of edible oils and the development of low-erucic acid cultivars, whereas, by contrast, high-erucic acid oils have a market (estimated to be more than $120 million in 2004) because erucic acid and its derivatives are

TABLE 6.2
Fatty Acid Composition of Plant Oils and Beef Tallow

Material	Saturated						Unsaturated			
	Lauric	Myristic	Palmitic	Stearic	Arachidic	Palmitoleic	Oleic	Linooleic	Liolenic	Other[a]
		(% of total fatty acids)					(% of total fatty acids)			
Corn oil		1.4	10	3.2	5.8	1.5	49.6	34.3		
Canola oil			1				32	15	1	50
Soybean oil	0.2	0.1	9.8	2.4	0.9	0.4	28.9	50.7	6.5	0.1
Sunflower oil			5.6	2.2	0.9		25.1	66.2		
Beef tallow		6.3	27.4	14.1			49.6	2.5		

Source: Data from Lide.[6]

[a] Erucic (canola), C14 monoethenoic (soybean)

feedstocks for the manufacture of slip-promoting agents, surfactants, and other specialized chemicals.[10] High-erucic acid oils would be either desirable or neutral for biodiesel production, but low-erucic cultivars are higher yielding — and, in any case, legal requirements were in place in the European Union by 1992 to geographically separate the two types of "oilseed rape" cultivation to minimize cross-pollination and contamination of agricultural products intended for human consumption.[4]

The majority of the biodiesel producers continue to employ a base-catalyzed reaction with sodium or potassium hydroxide (figure 6.3).[11] This has the economic attractions of low temperatures and pressures in the reaction, high conversion efficiencies in a single step, and no requirement for exotic materials in the construction of the chemical reactor. The liberation of glycerol (sometimes referred to as "glycerine" or "glycerin") in the transesterification reaction generates a potentially saleable coproduct (see section 6.3). The generation of fatty acid methyl esters is the same reaction as that to form volatile derivatives of fatty acids before their analysis by gas liquid chromatographic methods, and the key parameters for optimization are reaction time, temperature, and the molar ratio of oil to alcohol, but choices of

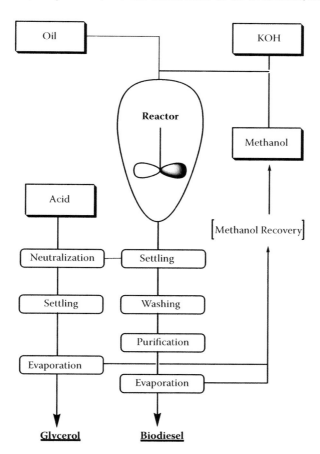

FIGURE 6.3   Schematic of typical biodiesel production process with alkaline catalysis.

the type of catalyst used and the short-chain alcohol coreactant can also be made.[12] Different oil types of plant origin have been the subject of process optimization studies; five recent examples, exemplifying the global nature of R&D activities with biodiesels, are summarized in table 6.3.[13–17] More subtle factors include differential effects on product yield and purity; for example, temperature has a significant positive effect on biodiesel purity but a negative influence on biodiesel yield, and the alcohol:oil molar ratio is only significant for biodiesel purity (with a positive influence).[17] Although the biodiesel yield increased at decreasing catalyst concentration and temperature, the methanol:oil ratio did not affect the material balance.[18]

A variety of novel catalysts have been explored, partly to avoid the use of caustic materials but also to facilitate catalyst recovery and reuse:

- Sulfonated amorphous carbon[19]
- Ion-exchange resins[20]
- Sodium ethoxide[21]
- Solid acid catalysts (e.g., ZnO)[14]

Indeed, the requirement for a catalyst can be eliminated if high temperatures and pressures are used to generate "supercritical" fluid conditions, under which alcohols can either react directly with triglycerides or (in two-stage procedures) with fatty acids liberated from triglycerides.[22–24]

Far greater attention has, however, been paid to developing a biotechnological approach to biodiesel production, employing enzyme catalysts, usually lipases, and employing their catalytic abilities to carry out transesterification (or alcohololysis) rather than straightforward hydrolyses of triglycerides to liberate free fatty acids and glycerol.[25] The principal process advantage of the enzyme-based approach is the ability to use low to moderate temperatures and atmospheric pressure in the reaction vessel while ensuring little or no chemical decomposition (i.e., a high product purity); the main drawback is the much longer incubation times to achieve more than 90% conversion of the triglycerides, that is, up to 120 hours.[26] The barrier to full commercialization is maintaining the (relatively expensive) enzyme active during repeated

TABLE 6.3
Transesterification Optimization for the Production of Biodiesel from Different Oils

Plant oil (country)	Alcohol	Temperature (°C)	Reaction time (hours)	Molar ratio alcohol:oil	Catalyst	Reference
Soybean and castor (Brazil)	Ethyl	70	3	9	NaOH	13
Pongamia pinnata (India)	Methyl	60		10	KOH	14
Waste frying oils (Portugal)	Methyl		1	4.8	NaOH	15
Rapeseed (Korea)	Methyl	60	0.33	10	KOH	16
Sunflower (Spain)	Methyl	25		6	KOH	17

batch use. Rival enzyme products show differing stability, and methanol appears to induce a faster loss of activity than does ethanol.[27] Examples of enzyme-catalyzed processes using oils of plant origin and with prolonged survival of the lipases are summarized in table 6.4.[26–29] Lipase from *Pseudomonas cepacia* was used in an immobilized form within a chemically inert, hydrophobic sol-gel support; under optimal conditions with soybean oil, high methyl and ethyl ester formations were achieved within a 1-hr reaction, and the immobilized lipase was consistently more active than the free enzyme, losing little activity when subjected to repeated uses.[30] Stepwise addition of methanol to delay inactivation of the enzyme is another possible strategy.[31] Reversal of methanol-mediated inactivation of immobilized lipase has been demonstrated with higher alcohols (secondary and tertiary butanols).[32] In the long term, molecular evolution technologies will develop lipases with reduced sensitivity to methanol and increased specific activities; in the short term, whole-cell biocatalysis has obvious potential for industrial application, offering on-site generation of lipase activity in cell lines that could be selected to be robust for oil transesterification.[33–35] As with cellulases (chapter 2, section 2.4.3), investigation of newly discovered microbes or extremophiles may reveal enzymes with properties particularly well suited for industrial use. A lipase-producing bacterium strain screened from soil samples of China, identified as *Pseudomonas fluorescens*, contains a novel psychrophilic lipase (with a temperature optimum of only 20°C); this may represent a highly competitive energy-saving biocatalyst because lipase-mediated biodiesel production is normally carried out at 35–50°C.[36]

Commercially available lipases and lipases identified in a wider spectrum of microbial enzyme producers can efficiently use different low-molecular-weight alcohols as substrates for transesterification. Substituting higher alcohols for methanol can maintain active lipase for much longer periods of continuous batch operation.[37] A research group in Italy has also exploited this lax substrate specificity to produce fatty acid esters from a mixture of linear and branched short-chain alcohols

TABLE 6.4
Enzyme Catalysts for the Transesterification of Oils to Biodiesel

Plant oil	Alcohol	Conditions	Lipase	Enzyme activity	Reference
Soybean	Several	Continuous batch operation at 30°C	Lipozyme TL IM	95% after 10 batches	26
Soybean, sunflower, etc.	Ethyl	Quantitative conversion within 7 hr at 25°C	Novozyme 435	85% after 9 batches	27
Olive	Methyl	Stepwise addition of methanol at 60°C	Novozyme 435	70% after 8 batches	28
Sunflower, etc.	(Ethyl acetate)	Immobilized enzyme at 25°C	Novozyme 435	85% after 12 batches	29

that mimics the residual fusel oil left after ethanol; not only can this utilize a waste product from bioethanol production but the fatty acid esters are potentially important for biodiesels because they improve low-temperature properties.[38]

Crude oils can give poor transesterification rates because of their contents of free fatty acids and other components lost during refining; the free fatty acids (up to 3% of the oils) react with the alkaline catalysts and form saponified products during the transesterification. Crude soybean oil could be converted into methyl esters as well as refined oil if a lipase process was used, although a lengthy incubation period was again required.[39] With an Indonesian seed oil (from *Jatropha curcas*) exhibiting very high free fatty acids (15%), a two-step pretreatment process was devised: the first step was carried out with sulfuric acid as catalyst in a reaction at 50°C and removing the methanol-water layer; the second step was a conventional transesterification using an alkaline catalyst to produce biodiesel at 65°C.[40] Rice bran stored at room temperature can show extensive (>75%) hydrolysis of triglycerides to free fatty acids; the successful processing of the oil fraction also required a two-step methanolysis process (but both steps being acid-catalyzed), resulting in a 98% methyl ester formation in less than 8 hr, and the coproduction of residue with high contents of nutraceuticals such as γ-oryzanol and phytosterols.[41] Supercritical methanol treatment (without any catalyst) at 350°C can generate esters from both triglycerides and free fatty acids, thus giving a simpler process with a higher total yield of biodeisel.[42]

Other innovations in biodiesel production have included the following:

- A six-stage continuous reactor for transesterification of palm oil in Thailand, claimed to produce saleable biodiesel within residence time of six minutes in a laboratory prototype with a production capacity of 17.3 l/hr[43]
- A Romanian bench-scale continuous process for the manufacture of biodiesel from crude vegetable oils under high-power, low-frequency ultrasonic irradiation[44]
- A two-phase membrane reactor developed to produce biodiesel from canola oil and methanol (this combination is immiscible, providing a mass-transfer challenge in the early stages of the transesterification); this Canadian design of reactor is particularly useful in removing unreacted oil from the product, yielding high-purity biodiesel and shifting the reaction equilibrium to the product side[45]
- A novel enzyme-catalyzed biodiesel process was developed to avoid the liberation of glycerol from triglycerides, maximizing the carbon recovery in the product; methyl acetate replaced methanol, and the resulting triacetylglycerol had no negative effect on the fuel properties of the biodiesel[46]

After biodiesel production, the fuel's thermal properties have been improved — in this case, to reduce the onset of volatilization (table 6.1) of soybean-derived biodiesel to below that of conventional diesel — by ozonolysis; the onset freezing temperature of ozonated methyl soyate was reduced from −63°C to −86°C.[47]

The most radical development in biodiesel production has, however, been in Brazil where PETROBRÀS has combined mineral and biological oils in the H-BIO process; few details have been made public, but the essential step is to add a vegetable oil to the straight-run diesel, gasoil, and coker gasoil fractions from the refining process, the total streams then being catalytically hydrogenated.[48] The triglycerides are transformed into linear hydrocarbon chains, similar to the hydrocarbons in the petroleum oil streams; the conversion of triglycerides is high (at least 95%), with a small propane coproduct. Because the process takes advantage of the existing infrastructure of an oil refinery, the potential exists for an orderly transition from conventional diesel to biodiesel blends, with a gradual increase in the "bio" input if (as widely predicted) oil reserves dwindle (chapter 5, section 5.6).

6.1.3 BIODIESEL ECONOMICS

The International Energy Agency's 2006 analysis concludes that biodiesels are not price competitive with conventional diesels if all subsidies to crops and production are excluded; if the biodiesel source is animal fat, however, the derived biodiesel would be competitive at crude oil prices below $50–55/barrel.[2] By 2030, assuming various process improvements and economies of scale, biodiesel from vegetable oils were also predicted to be competitive at crude oil prices below $50–55/barrel; European biodiesel would continue to be more expensive than U.S. biodiesel, with feedstock costs being the largest contributor (figure 6.4). This continues a tradition of cost assessments that commenced in the 1990s. Rapeseed oil-derived biodiesel was estimated to require a total subsidy of between 10% and 186% of the price of conventional diesel in 1992, the variation reflecting the price of the seeds sown to grow the crop; this was equivalent to the cost of biodiesel being between 11% and 286% of the refinery gate cost of conventional diesel — the contemporary price for seeds would have resulted in biodiesel being 243% of the conventional diesel

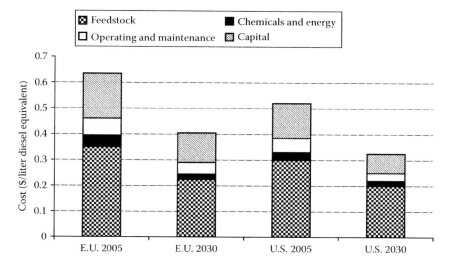

FIGURE 6.4 Production costs of biodiesel in the European Union and United States (including subsidies to crop production) for 2005 and 2030 Reference Scenario. (Data from *World Energy Outlook*.[2])

cost.[4] These calculations required both coproducts (glycerol and rapeseed meal) to be income generators; in addition, the spring-sown crop (although with lower yields per unit of land) had lower production costs.

An evaluation of costs from U.S. soybean and sunflower in 2005 concluded that soybean biodiesel would have production costs 2.8-fold that of conventional diesel (using 2003 price data), whereas sunflower-derived biodiesel was more than five-fold more expensive to produce than petroleum diesel fuel.[49] Sunflower seeds were accepted to have a higher oil content (25.5%) than soybeans (18%) but to also have a lower crop productivity (1500 versus 2700 kg/hectare); in both cases, oil extraction was calculated to be highly energy intensive.

The production costs for biodiesels also depend on the production route. With waste cooking oil as the feedstock, although an alkali-catalyzed process using virgin vegetable oil had the lowest fixed capital cost, an acid-catalyzed process using the waste oil was more economically feasible overall, providing a lower total manufacturing cost, a more attractive after-tax rate of return, and a lower biodiesel breakeven price; in addition, plant capacity was found to be a significant factor affecting the economic viability of biodiesel manufacture.[50,51] The U.S. Department of Agriculture's Agricultural Research Service has developed a computer model to estimate the capital and operating costs of a moderately sized industrial biodiesel production facility (annual production capacity, 10 million gallons):[52]

- Facility construction costs were calculated to be $11.3 million.
- The largest contributors to the equipment cost (accounting for nearly one-third of expenditures) were storage tanks to contain a 25-day capacity of feedstock and product.
- At a value of $0.52/kg for feedstock soybean oil, a biodiesel production cost of $0.53/l ($2.00/gallon) was predicted.
- The single greatest contributor to this value was the cost of the oil feedstock, which accounted for 88% of total estimated production costs. An analysis of the dependence of production costs on the cost of the feedstock indicated a direct linear relationship between the two.
- Process economics included the recovery of coproduct glycerol generated during biodiesel production, and its sale into the commercial glycerol market, which reduced production costs by approximately 6%.

Waste cooking oils, restaurant grease, and animal fats are inexpensive feedstocks; they represent 30% of total U.S. fats and oil production but are currently devoted mostly to industrial uses and animal feed, and because the free fatty acids may represent more than 40% of the material, the production process may be complex.[53] Nevertheless, such unconventional feedstocks may become increasingly important because soybean oil prices reached a peak not seen since 1984: an article posted online in *Biodiesel Magazine* traced the rapid inflation in soybean oil from early 2006.[54] This surge in the price of biodiesel feedstock has occurred despite the stocks of soybean oil being at near-record levels — and in only three individual months between January 2006 and March 2007 did soybean oil use exceed production (figure 6.5). With its main feedstock being increasingly expensive, U.S. biodiesel is

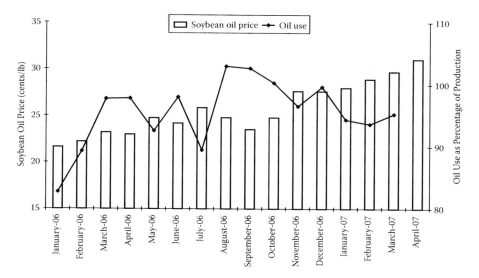

FIGURE 6.5 Soybean oil prices and monthly use, 2006 and 2007. (Data from Kram.[54])

not competitive on price with diesel fuels or heating oil; tax incentives may be necessary to overcome these production price issues.[54]

Confounding these feedstock problems, the great expansion of biodiesel production both in Europe and the United States has caused such a glut of glycerol-containing waste (or coproduct) that, in the absence of glycerol valorization mechanisms in place and on site, disposing of this glycerol is proving an increasingly expensive disposal cost outlay. Because of the enormous potential of this renewable source of a potentially valuable chemical intermediate, however, biodiesel waste glycerol is best considered an example (rather premature) of "biocommodity engineering" and is discussed at length in chapter 8 (section 8.3.3) when the broader topic of replacing petrochemicals by biobased products is considered.

6.1.4 ENERGETICS OF BIODIESEL PRODUCTION AND EFFECTS ON GREENHOUSE GAS EMISSIONS

Inevitably, an essential facet of the public discussions on costs and subsidies of biodiesel production has been that of its potential amelioration of greenhouse gas emissions. If significant, this would augment the case for production and consumption incentives to offset higher production costs than for conventional diesel. At the scientific level, this debate has mirrored that for bioethanol (chapter 1, section 1.6) and has proved equally contentious and acrimonious.

In the early 1990s, net energy balance (NEB) values of up to 3.8:1 were calculated for rapeseed-derived biodiesel, depending on how the coproducts and crop straw were assessed in the calculations (figure 6.6).[4] Unpublished reports and communications quoted in that report were from 1.3 to 2.1 without coproduct credits and from 2 to 3 if thermal credits for the meal and glycerol coproducts were included. Radically different conclusions were reached in a 2005 publication: biodiesel production

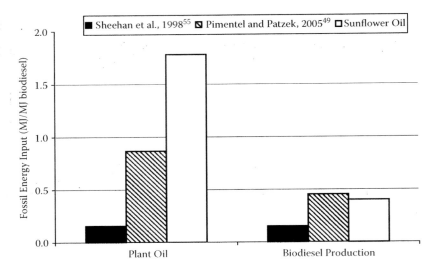

FIGURE 6.6 Estimated fossil energy inputs to biodiesels manufactured from soybean or sunflower oil.

from soybean oil required 27% more fossil energy than the biodiesel energy content, whereas sunflower oil was even less viable (requiring 118% more fossil energy than in the product).[49]

Midway (in time) between these conflicting estimates was a report from the National Renewable Energy Laboratory whose main conclusion was that biodiesel (from soybean oil) yielded 3.2 units of fuel product energy for every unit of fossil energy consumed in its life cycle, whereas conventional diesel yielded only 0.83 unit per unit of fossil fuel consumed, that is, that biodiesel was eminently "renewable."[55] Direct comparison of these conflicting results shows that the disagreements are major both for the stages of soybean cultivation and biodiesel production (figure 6.6). As so often in biofuel energy calculations, part of the discrepancy resides in how the energy content of coproducts is allocated and handled in the equations (chapter 1, section 1.6.1). As the authors of the 2005 study pointed out, if the energy credit of soybean meal is subtracted, then the excess energy required for biodiesel production falls to 2% of the biodiesel energy content.[49] A paper posted on the University of Idaho bioenergy site suggests other factors:[56]

- Lime application was a major agricultural input in the 2005 study but may have been overestimated by a factor of 5, that is, the application rate should have been spread over five years rather than every season; making this change reduces the energy required to only 77% of the biodiesel energy content.
- In the National Renewable Energy Laboratory study, the energy input to produce soybeans and then extract the oil was divided 18:82 in favor of soybean meal, that is, after the weight split of oil and residual material; allocating only 18% of the input energy to soybean oil changes the energy balance dramatically to 5.3 times more biodiesel energy than fossil energy input.

- Even adding in energy requirements for oil transport and transesterification as well as biodiesel transport produces a favorable energy balance of 2.9:1.

The energy balance, in any of these scenarios, is highly dependent on viewing the process as a biorefinery producing coproducts as well as biodiesel. If the energetic (and economic) value of the soybean meal cannot be realized, then the balance will be negative — even using the soybean meal as a "green manure" spread on the soybean fields would only partially offset the major loss of replaced fossil energy in the total process. If biodiesel production from oil seed crops is viewed as an "opportunity" over their alternative uses as, for example, foodstuffs, then the NEB can be recalculated to be favorable.[57] As discussed previously (chapter 1, section 1.6.1), this is a contentious argument, and much media comment in Europe during 2006–7 has pointed to increased areas of arable land being devoted to oilseed rape as demand for the crop as a source of biodiesel increases, thus fulfilling the prediction of a subsidized cash crop.[4] A focus of future attention may be that of realizing an economic return on the greatly increased amounts of seed meal and of finding a viable use for glycerol — refining the glycerol coproduct to a chemically pure form is expensive, and alternative uses of glycerol for small- and medium-scale biodiesel facilities are being explored, for example, its use as an animal feed supplement.[58]

As the number of industrial units producing biodiesel increases, assessments of energy balances should be possible from collected data rather than from calculations and computer simulations. A report on activities in six Brazilian and Colombian biodiesel facilities using palm oil as the agricultural input attempted precisely this.[59] NEBs were in the range of 6.7–10.3, with differences arising because of

- Different rates of fertilizer application
- Different uses of plant residues as fertilizers or as boiler fuel for electricity production
- On-site electricity generation at some sites, whereas others were entirely dependent on purchased electricity
- Differing efficiencies in the generation of coproducts and the recovery of unused palm oil

Taken together as a group, these palm oil biodiesel producers were assessed as being more energy efficient than reference manufacturers in Europe or the United States — the most recent (2006) detailed estimate of biodiesel from soybean oil in the United States arrived at a NEB of 1.93:1, but this was critically dependent on full credits being taken for soybean meal and glycerol coproducts (without them, the balance decreases to only 1.14:1).[60]

The energy balance is an important parameter that defines the extent of the biodiesel's capacity to reduce greenhouse gas emissions, because, in the extreme case, if biodiesel requires more fossil energy in its production than can be usefully recovered in the product, no savings could possibly accrue.[49] With a favorable energy balance for soybean biodiesel, its use could displace 41% of the greenhouse gas emissions relative to conventional diesel.[60] As headline statements, the National

Renewable Energy Laboratory study on biodiesel use for public transport concluded the following:[55]

1. Substituting 100% biodiesel (B100) for petroleum diesel reduced the life cycle consumption of petroleum by 95%, whereas a 20% blend (B20) reduced consumption by 19%.
2. B100 reduced CO_2 emissions by 74.5%, B20 by 15.7%.
3. B100 completely eliminated tailpipe emissions of sulfur oxides and reduced life cycle emissions of CO, sulfur oxides, and total particulate matter by 32%, 35%, and 8%, respectively.
4. Life cycle emissions of NO_x and hydrocarbons were higher (13.4% and 35%, respectively) with B100, but there were small reductions in methane emissions.

Earlier assessments indicated that only 55% of the CO_2 emitted from fossil diesel could be saved if biodiesel were to be used because of the CO_2 emissions inherent in the production of biodiesel and that, other than a marked reduction in sulfur oxides, effects on CO, hydrocarbons, NO_x, and polyaromatic hydrocarbons were inconsistent.[4] As the use of biodiesel has widened globally, the number of publications exploring individual pollutants or groups of greenhouse gas emissions has expanded, especially after 2000 (table 6.5).[61-68] The report of increased mutagenicity in particulate emissions with a biodiesel is unusual as two earlier reports from the same research group in Germany found reduced mutagenicity with rapeseed oil- and soybean-derived biodiesels.[66,69,70] A high sulfur content of the fuel and high engine speeds (rated power) and loads were associated with an increase in mutagenicity of diesel exhaust particles. This is in accord with the desirability of biodiesels because of their very low sulfur contents, zero or barely detectable, as compared with up to 0.6% (by weight) in conventional diesels.[4] There are suggestions that exhaust emissions from biodiesels are less likely to present any risk to human health relative to petroleum diesel emissions, but it has been recommended that the speculative nature of a reduction in health effects based on chemical composition of biodiesel exhaust needs to be followed up with thorough investigations in biological test systems.[71]

TABLE 6.5
Recent Studies on Biodiesels and Their Impact on Aerial Pollutants

Biodiesel source	Pollutants investigated	Reference
Neem oil (Bangladesh)	CO↓, NO_x↑, smoke↓	61
Soybean oil (Turkey)	CO↓, NO_x↑, particulates↓, hydrocarbons↓	62
Rapeseed oil (Korea)	CO↑, NO_x↑, smoke↓, CO_2↑	63
Soybean oil (U.S.)	Particulates↓	64
Waste cooking oil (Spain)	Particulates↓, smoke↓	65
Soybean, rapeseed oil (Germany)	Mutagenicity of particulates↑	66
Palm oil (China)	CO↓, polyaromatics↓, particulates↓, hydrocarbons↓	67
Brassica carinata (Italy)	NO_x↑, particulates↓	68

6.1.5 ISSUES OF ECOTOXICITY AND SUSTAINABILITY WITH EXPANDING BIODIESEL PRODUCTION

Local ecological impacts from biodiesel production units are most likely to be severe in areas of the world with lax environmental policies or enforcement. This is a particularly acute problem for biodiesel production because of the potential to generate large volumes of aqueous wastes with high biological demands. Glycerol is a major waste product unless it is exploited as an income-generating stream, but this is not always economically feasible.[58] Biodiesel wastes containing glycerol can be utilized by a *Klebsiella pneumoniae* strain to produce hydrogen by fermentation as a source of locally generated combustible gas for local heating or on-site use for biomass drying.[72] A second biohydrogen system was based on an *Enterobacter* isolated from methanogenic sludge; glycerol-containing biodiesel wastes were diluted with a synthetic medium to increase the rate of glycerol consumption and the addition of nitrogen sources (yeast extract and tryptone) enhanced the rates of both hydrogen and ethanol formation.[73] In general, however, wastewater streams from biodiesel plants are not considered suitable for microbial remediation because of the high-pH, hexane-extractable oil, low nitrogen concentrations, and the presence of growth inhibitors; an oil-degrading *Rhodotorula mucilaginosa* yeast was found to degrade oil in wastewaters diluted with water to reduce growth inhibition and the content of solid materials, and this was developed into a small-scale treatment method.[74]

Especially in Europe, however, possible environmental damage is more often viewed as an international issue. This can be traced back to the initial period of biodiesel production in the 1980s, when it became evident that land resources inside the European Union were highly likely to limit biodiesel manufacturing capacity with European feedstocks: by 1992, oilseed rape cultivation in the United Kingdom (barely known in the 1970s) was estimated to cover 400,000 hectares of land, enough crop to satisfy no more than a 5% substitution of conventional diesel sales.[4] Twenty-five years later, the United Kingdom was devoting 570,000 hectares to growing oilseed rape but 40% of this was for food use (cooking oil, margarine, etc.); even if all this land and whatever land is "set aside" under Common Agricultural Policy, policies could still only support a 5% (or less) substitution of fossil diesel.[75] Even if all the U.S. soybean crop were to be devoted to biodiesel production, only 6% of U.S. diesel demand could be met.[60] To meet U.S. and EU targets, therefore, importing biodiesel feedstocks is probably unavoidable, and this is (equally probably) dependent on supplies from Africa and Asia, with energy crops grown on recently cleared, deforested land; net importers can limit their imports from countries operating under plans such as the Round Table on Sustainable Palm Oil, an initiative to legitimize the trade in sustainably produced feedstock, but have often proved reluctant to ban imports of unsustainable biofuel sources for fear of breaching World Trade rules.[76]

To increase the pressure on available arable land further, biodiesel from monoculture crops such as soybean and canola support much poorer energy production rates than does corn-based ethanol.[77] The International Energy Agency has predicted that the land requirements for biofuels production will increase from a global figure of 1% in 2004 to 2.5% by 2030 — or, under alterative scenarios, to 3.8% or even 4.2% (58.5 million hectares).[2] The search for high-yielding energy crops suitable

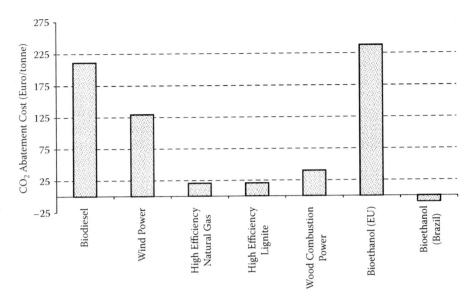

FIGURE 6.7 CO_2 abatement costs of technologies for biofuels or improved power generation. (Data from Frondel and Peters.[79])

for biodiesel production has, therefore, focused on little-known tropical species; in South Africa, for example, the perennial tree *Jatropha moringa* can generate more than three times the biodiesel yield of soybeans per hectare.[78]

Although biodiesel is part of an increasingly well-publicized strategy in OECD countries to combat global warming due to greenhouse gas accumulation, critics have identified environmental problems with domestic monoculture energy crops:[79]

- Soil acidification is caused by SO_2 and NO_x emissions from fertilizers, whereas N_2O emissions also contribute to ozone depletion.
- Fertilizer runoff causes eutrophication, algal blooms, and others.
- Pesticide applications can cause toxic pollution of surface water.
- Biodiesels represent a totally cost-ineffective option for CO_2 abatement under the European CO_2 Emissions Trading Scheme, at least three times more expensive than the predicted benchmark avoidance cost of €30/tonne CO_2.

From a purely economic standpoint, it appears to be much cheaper to reduce greenhouse gas emissions by improving the efficiencies of fossil fuel-powered generating stations than by substituting biodiesel for conventional diesel or bioethanol for gasoline* (figure 6.7).

The heavy current flowing in favor of biodiesel production may, however, be very difficult to reverse. Even in Brazil, where a national program was launched in 2002, biodiesel production is seen as a lever for new markets for agribusinesses, improving

---

* Brazilian sugarcane-derived ethanol may, depending on the sugarcane price, be cheaper to produce than conventional gasoline and represent a negative abatement cost (figure 6.7).

rural employment, and forms part of the government's policies to eliminate poverty; targets under "Probiodiesel" include 2% of all transport diesel to be biodiesel by 2008 (when all fuel distributors will be required to market biodiesel) and 5% by 2013.[80] As a chemical technology, biodiesel production is easily integrated with the existing infrastructure of heavy chemical industry, sharing sites and power costs: a BASF maleic anhydride plant at Feluy (Belgium) shares waste heat with a biodiesel production unit operated by Neochim.[81] Industrial plants that used to produce glycerol are now closing down to be replaced by others that use glycerol as a raw material, owing to the large surplus of glycerol formed as a coproduct during the production of biodiesel; in parallel, research efforts to find new applications of glycerol as a low-cost feedstock for functional derivatives have led to the introduction of a number of selective processes for converting glycerol into commercially valued products.[82]

Although the public face of soybean oil-derived biodiesel in Brazil and elsewhere remains that of sustainable symbiotic nitrogen fixation with the bacterium *Bradyrhizobium japonicum* and soybean cultivars selected to grow in the arid savannah in the hinterland state of Mato Grosso, environmentalist concerns about "deforestation diesel" remain acute: Brazil already exports 20 million tonnes of soybean annually and plans to increase this to 32 million tonnes by 2015.[83] To what extent soybean-plantation cultivation encroaches from savannah to adjacent rain forest will largely determine how "sustainable" this major source of biodiesel feedstock will prove.

6.2   FISCHER-TROPSCH DIESEL: CHEMICAL BIOMASS–TO–LIQUID FUEL TRANSFORMATIONS

6.2.1   THE RENASCENCE OF AN OLD CHEMISTRY FOR BIOMASS-BASED FUELS?

The generation of a combustible gas, synthesis gas ("syngas"), from biomass was discussed briefly in chapter 2 (section 2.1). Technologies for the conversion of coal and natural gas to liquid fuels were also included in chapter 5 (section 5.6) as part of a survey of different strategies for adapting to potentially dwindling crude oil reserves. The chemistry of gas-to-liquid fuel transformations was developed in the first quarter of the twentieth century and utilized extensively in Germany during World War II; further evolution led to commercial production processes being initiated for peacetime purposes in the 1990s.[84,85]

The essential step, known as the Fischer-Tropsch (FT) reaction, can be written as

$$nCO + 2nH_2 \rightarrow [CH_2]_n + nH_2O,$$

where $[CH_2]_n$ represents a range of hydrocarbons, ranging from low-molecular-weight gases ($n = 1$, methane), by way of gasoline ($n = 5–12$), diesel fuel ($n = 13–17$), and as far as solid waxes ($n > 17$). The reaction requires catalysts for realistic rates to be achieved, usually iron or cobalt (although transition metals will function effectively) at high temperatures (180–350°C) and high pressures; the higher the temperature, the higher the proportion of gas and liquid hydrocarbon products.

To date, no process has been commercialized from plant biomass feedstocks, and the FT technology could be described as "radical" or "*n*th" generation for biofuels were it not that the key elements of the chemistry and production options are reasonably well established in industrial processes with fossil inputs; in a climate of high crude oil prices, the environmental desirability of low-sulfur diesel, and the drive to commercialize otherwise unmarketable natural gas in remote locations are important synergies (table 6.6).[86] FT biomass-to-liquid fuel (FT-BtL) from lignocellulosic sources is particularly attractive because of the high CO_2 emission reduction potential (up to 90% when substituting conventional gasoline and diesel) and the ability to use woody materials from low-grade land, thus avoiding the pressures on land use in OECD countries contemplating agriculture-based bioethanol or biodiesel production on a large scale.[79] The principal barrier to large-scale biomass FT-BtL appears to be the suboptimal mixture of gases in syngas as prepared from plant materials: the lower the molar ratio of H_2:CO, the more the proportion of high-molecular-weight products formed in the FT reaction, but biomass gasification results in a wide range of H_2:CO ratios, often with an excess of CO, together with appreciable amounts of CO_2, methane, and higher hydrocarbons as well as smaller amounts of condensable tars and ammonia.[87]

The methane can be transformed to CO and H_2 by a number of different reactions, including the uncatalyzed (but again high-temperature and high-pressure) processes:[88]

$$CH_4 + O_2 \rightarrow CO_2 + 2H_2O \text{ and } CH_4 + H_2O \rightarrow CO + 3H_2$$

Partial removal of CO (and formation of additional H_2) is possible by the water-shift reaction:

$$CO + H_2O \rightarrow CO_2 + H_2$$

Finally, the physical removal (adsorption) of CO_2 (an inert gas for FT reactions) is relatively straightforward, but a higher-yielding process can be devised (at least, in principle) by including a catalytic reduction of the CO_2 to using multiple FT reactors in series with an intermediate water removal step:[89]

$$CO_2 + 3H_2 \rightarrow [-CH_2-] + 2H_2O$$

TABLE 6.6

Operating and Planned FT Plants Based on Methane (Natural Gas) Feedstock

Country	Companies	Production level (barrels per day)	Start-up year
South Africa	PetroSA	20,000	1992
Malaysia	Shell	15,000	1993
Qatar	Sasol, Qatar Petroleum, Chevron	34,000	2005
Nigeria	Chevron Nigeria, Nigeria National Petroleum	34,000	2007
Qatar	Shell, Qatar Petroleum	140,000	2009
Qatar	ExxonMobil, Qatar Petroleum	154,000	2011

Complete wood-based FT-BtL production involves, therefore, a multistage process, incorporating biomass pretreatment, syngas purification, and optional syngas recycling, plus gas turbine power generation for unused syngas and, for FT diesel, a hydrocracking step to generate a mixture of diesel, naphtha, and kerosene (figure 6.8).[87,88]

6.2.2　Economics and Environmental Impacts of FT Diesel

In comparison with natural gas-based FT syntheses, biomass requires more intensive engineering, and gas-cleaning technology has been slow to evolve for industrial purposes — although for the successful use of biomass, it is essential because of the sensitivity of FT catalysts to contaminants. In year 2000 U.S. doller terms, investment costs of $200–340 million would be required for an industrial facility, offering conversion efficiencies of 33–40% for atmospheric gasification systems and 42–50% for pressured systems, but the estimated production costs for FT diesel were high, more than 10 times those of conventional diesel.[89] Two years later,

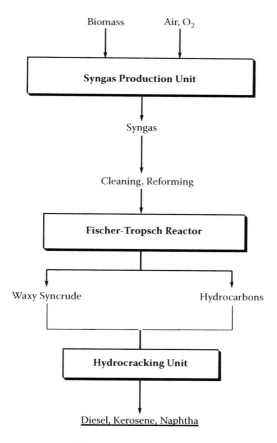

FIGURE 6.8　Outline scheme of FT diesel production from biomass. (Data from Tijmensen et al.[88] and Morales and Weckhuysen.[86])

the same research group from the Netherlands was predicting the same production costs and concluded that, unless the environmental benefits of FT diesel were valued in economic terms, the technology would only become viable if crude oil prices rose substantially.[90] This did, in the event, occur (figure 5.1), and the cost differential has undoubtedly narrowed — although with no signs of a surge in investor confidence.

If it could be produced economically, using an energy crop such as switchgrass as the substrate, FT diesel rates better than E85 (from corn-derived ethanol) as a biofuel in assessments performed by the Argonne National Laboratory (figure 6.9).[91,92] FT diesel greatly outperformed E85 for total fossil fuels savings and also exhibited much reduced emissions of total particulates, sulfur oxides, and nitric oxides — although it fared worse than E85 using the criterion of total CO. Compared with conventional diesel fuel, FT diesel had higher total emissions of volatile organic carbon, CO, and nitric oxides (figure 6.9).

An experimental biofuel conversion technology, presently explored only in the Netherlands, is that for HydroThermalUpgrading diesel.[93] At high temperature (300–350°C) and pressure, wet biomass feedstocks such as beet pulp, sludge, and bagasse can be converted to a hydrocarbon-containing liquid that, after suitable refining, can be blended with conventional diesel in any proportion without engine adjustments. A pilot plant remains the focus for further process optimization and development.

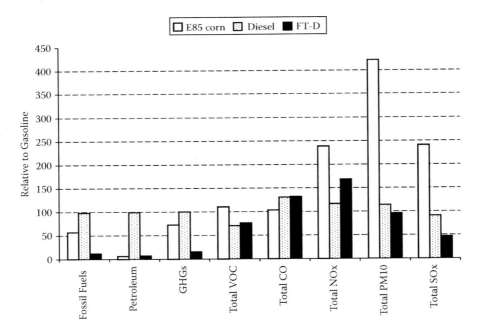

FIGURE 6.9 Well-to-wheel energy use and emissions for E85 from corn, diesel, and FT diesel produced from switchgrass, relative to conventional reformulated gasoline. (Data from Wu et al.[91])

6.3 METHANOL, GLYCEROL, BUTANOL, AND MIXED-PRODUCT "SOLVENTS"

6.3.1 METHANOL: THERMOCHEMICAL AND BIOLOGICAL ROUTES

The versatility of the FT process is that almost any hydrocarbon produced that can be derived from petroleum can be made from syngas, not only alkanes and alkenes but oxygenated compounds; the exact mixture of products obtained can be varied by choices of catalyst, pressure, and temperature, and straight-chain alcohols are produced in the "synthol" reaction at 400–450°C and 14-MPa pressure in the presence of an iron catalyst.[94] Industrial production from natural gas has, however, been dominated since the 1960s by a lower temperature and pressure process invented by Imperial Chemical Industries in which CO, CO_2, and H_2, derived by steam reforming, are reacted over a mixed $Cu/ZnO/Al_2O_3$ catalyst at 250°C and 50–10 MPa when two reactions occur:[95]

$$CO + 2H_2 \rightarrow CH_3OH$$

$$CO_2 + 3H_2 \rightarrow CH_3OH + H_2O$$

A recent development has been to combine syngas production from methane with the reduction of ZnO to metallic zinc in a metallurgical plant; the syngas has a H_2:CO ratio of approximately 2:1, highly suitable for methanol production.[96] A renewable-resource route for methanol (one of the largest bulk chemicals in the contemporary world) is, however, entirely feasible via biomass gasification as an intermediate step, and this would be entirely appropriate given methanol's older name of "wood alcohol," indicative of its historical provenance by incomplete combustion.

As an "energy carrier," methanol is inferior to ethanol, with an energy content of only 75% (on either a weight or a volume basis) compared with ethanol and approximately 50% compared with conventional gasoline.[97] Blends of methanol with conventional gasoline up to 20% can be tolerated without the need for engine modifications, that is, as a fuel extender; the corrosive effect of methanol on some engine materials limits the extent of this substitution.[98] Methanol would have been an excellent replacement for MTBE as a gasoline oxygenate additive (chapter 1, section 1.4), but its acute neurotoxicity is well known and a barrier to several potential uses. One notable exception, however, is as an at-site (or on-board) source of hydrogen for fuel cells (chapter 7, section 7.1): between 1983 and 2000, nearly 50 patents were granted for catalytic methanol "reforming" systems to automobile producers (such as General Motors, Daimler-Benz, DaimlerChrysler, and Honda), chemical multinationals (DuPont, BASF, etc.), and major oil companies (CONOCO, Standard Oil Company, etc.).[95] Combined reforming with liquid water and gaseous oxygen has been intensively investigated for use in mobile applications for transportation:

$$(s+p)CH_3OH\ (l) + sH_2O\ (l) + 0.5pO_2 \rightarrow (s+p)CO_2 + (3s+2p)H_2$$

since the composition of the reactant feed can be varied and the process carried out under a wide range of operating conditions.*

The first pilot plant for testing and evaluating the production process for "bio-methanol" was established in a program that commenced in 2000 between the Ministry of Agriculture, Forestry, and Fisheries of Japan and Mitsubishi Heavy Industries at Nagasaki (Japan); various feedstocks have been investigated, including wood, rice husks, rice bran, and rice straw.[99] The test plant consisted of

- A drier and grinder for the biomass input (crushed waste wood)
- A syngas generator
- A gas purifier
- A methanol synthesis vessel (with an unspecified catalyst)

The pilot plant was designed for a capacity of 240 kg/day, with a methanol yield (weight of methanol produced per unit dry weight of material) of 9–13%. A larger plant (100-tonne daily capacity) is predicted to have a much higher methanol yield (38–50%).

No economic analysis of the Japanese pilot facility has been published, but a theoretical study of methanol production via the syngas route suggested that methanol from biomass (by 2002) had production costs approximately twice those of conventional gasoline on an equal energy basis.[100] The surge in crude oil and gasoline refinery gate costs subsequently (figure 5.1) implies that methanol production costs from biomass sources would now (mid-2008) be competitive. This encouraging result is very timely because direct methanol fuel cell (DMFC) technology has reached the stage where Toshiba in Japan has announced the development of a micro-DMFC suitable for powering MP3 players.[101] A U.S. patent covering aspects of DMFC construction was also issued in March 2007 to Creare (Hanover, New Hampshire).[102] The Japanese invention utilizes a polymer electrolyte membrane device with the electrochemical reactions:

$$CH_3OH + H_2O \rightarrow CO_2 + 6H^+ + 6e^- \text{ (anode)}$$

$$1.5O_2 + 6H^+ + 6e^- \rightarrow 3H_2O \text{ (cathode)}$$

The inputs are concentrated methanol and air (O_2); the only outputs are water and CO_2 and electricity (100 mW) sufficient to power a portable device for 20 hours on a 2-cm^3 charge of solvent. The prospects for large DMFCs for heavier duty use are presently unclear.

Another thermochemical route has been explored to convert methanol to another biofuel, dimethylether (DME), $(CH_3)_2O$, a highly volatile liquid that is a suitable fuel for diesel engines because of its low self-ignition temperature and high cetane number.[103] Although bio-DME has only half the energy content of conventional diesel, diesel engines can easily be retrofitted for bio-DME use. Well-to-wheel analyses

---

* Ethanol is another candidate H_2 producer, but the reaction must be performed at higher temperatures and undesirable by-products are formed.[95]

showed that bio-DME was a little inferior to FT diesel for total fossil fuel substitution and pollutant emissions.[91,92]

6.3.2   GLYCEROL: FERMENTATION AND CHEMICAL SYNTHESIS ROUTES

Glycerol represents 10% by weight of typical triglycerides, and biodiesel production generates large amounts of this coproduct (figure 6.3). This obviously marginalizes any chemical routes to glycerol synthesis that are either cost-inefficient or otherwise suboptimal in a competitive manufacturing environment.[82] By any manufacturing route, glycerol production provides a good input to FT conversion to liquid alkanes because glycerol can be converted over platinum-based catalysts into syngas at relatively low temperatures, 225–350°C.[104] The gas mixture from glycerol conversion at 300–450°C has a molar excess of H_2 over CO (up to 1.83:1), a high ratio (up to 90:1) between CO and CO_2, and only traces of methane. With a subsequent FT step, the overall conversion of glycerol to hydrocarbons can be written as

$$25C_3O_3H_8 \rightarrow 7C_8H_8 + 19CO_2 + 37H_2O$$

and is a mildly exothermic process (enthalpy change −63 kJ/mol glycerol). High rates of conversion of glycerol into syngas were observed using glycerol concentrations of 20–30% (w/w).

Unlike ethanol, a major fermentation product of only a selected few microbial species, glycerol is ubiquitous because of its incorporation into the triglycerides that are essential components of cellular membranes, as well as being accumulated in vegetable oils (figure 6.1). Unlike methanol, which is toxic to microbial species as well as higher animals and has no biosynthetic pathway, glycerol is benign and has a well-characterized route from glucose and other sugars (figure 2.3). In fermentations for potable ethanol, the priority is to regulate glycerol accumulation as its formation is a waste of metabolic potential in fuel alcohol production — and much research effort has, therefore, been devoted to minimizing glycerol formation by yeasts by, for example, regulating the glucose feeding rate to maintain an optimal balance of CO_2 production and O_2 consumption.[105]

Conversely, maximizing glycerol production by yeasts is also straightforward, successful strategies including:[106]

- Adding bisulfite to trap acetaldehyde (an intermediate in the formation of ethanol), thus inhibiting ethanol production and forcing glycerol accumulation to restore the balance of intracellular redox cofactors
- Growing yeast cultures at much higher pH values (7 or above) than traditionally used for ethanol fermentations
- Using osmotolerant yeasts — glycerol is often accumulated inside yeast cells to counteract the adverse effects of high osmotic pressures

Suitable osmotolerant strains can accumulate 13% (w/v) glycerol within four to five days. Even more productive is the osmophilic yeast *Pichia farinosa* that was reported to produce glycerol at up to 30% (w/v) within 192 hours in a fed-batch fermentation with glucose as the carbon source as a molar yield of glycerol from glucose of 0.90.[107]

Fixing a maximum theoretical yield of glycerol from glucose is difficult because it is highly dependent on the totality of biochemical routes available to the producer cells. With *Saccharomyces cerevisiae* genetically manipulated to overproduce glycerol, the glycerol yield was 0.50 g/g of glucose, that is, a molar yield of 0.98.[108] With this genetic background, a yield of 1 mol of glycerol/mol of glucose consumed may be the maximum obtainable (figure 6.10). Although an osmotolerant *Saccharomyces* strain isolated from sugarcane molasses could accumulate higher levels of glycerol (up to 260 g/l), the molar yield was still 0.92.[109] An osmotolerant *Candida glycerinogenes* was, however, reported to produce glycerol with a molar yield of 1.25.[110] Breaking through the limitation imposed by redox cofactors can, therefore, be accomplished by natural biochemistry. Some yeasts harbor a well-characterized "short circuit" for NADH oxidation, bypassing an energy conservation mechanism via an alternative oxidase, a mitochondrial enzyme constitutively expressed in industrial strains of *Aspergillus niger* used in citric acid prorduction.[111–113] Similar enzymes are known in plants, fungi, and many types of yeast; overexpressing the alternative oxidase gene in another yeast of industrial relevance (*Pichia pastoris*) resulted in a small increase in growth rate.[114] Because the functioning of the alternative oxidase system is thought to circumvent any restrictions imposed on productivity by redox cofactor regeneration in citric acid producers, a bioconversion of each mole of glucose to produce nearly 2 mol of glycerol might be achievable.

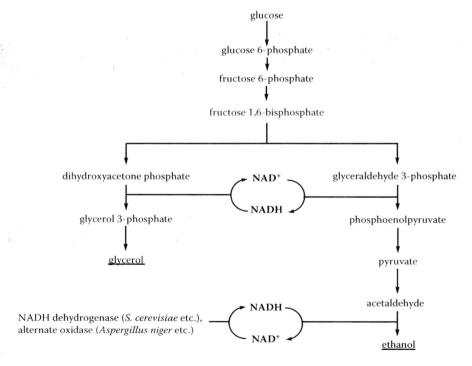

FIGURE 6.10   Redox balance in biosynthetic routes to glycerol and ethanol in *S. cerevisiae*. (Modified from Overkamp et al.[108])

Even with presently known levels of glycerol production, it is technically feasible to process a harvested filtered culture broth directly for syngas formation from glycerol.[104] Because 50% of the annual costs of FT fuels from biomass were considered to be capital-related, combined syngas/FT conversion of glycerol would be expected to entail markedly lower costs, eliminating the need for a biomass gasifier and gas-cleaning steps.[89,104]

6.3.3 ABE (ACETONE, BUTANOL, AND ETHANOL) AND "BIOBUTANOL"

Glycerol was produced on an industrial scale by fermentation in the first quarter of the twentieth century (especially during World War I) but then declined, unable to compete with chemical synthesis from petrochemical feedstocks.[106] A similar historical fate occurred with the ABE fermentation-producing "solvents," that is, acetone, butanol, and ethanol in various proportions. Beginning (as with fuel ethanol) with the oil crises of the 1970s, renewed interest was evinced in the technology, aided greatly by the accelerating advance of microbial physiology and genetics at that time.[115,116] The microbial species capable of this multiproduct biosynthesis are clostridia, which also have remarkable appetites for cellulosic and hemicellulosic polymers, able to metabolize hexose sugars and pentoses (usually, both xylose and arabinose).[117,118] This again parallels the drive to produce ethanol from lignocellulosic biomass substrates (chapter 3, section 3.3.2.5). It came as no surprise, therefore, when the neologism "biobutanol" (for n-butanol, C_4H_9OH) appeared. DuPont, Wilmington, Delaware, and British Petroleum are the companies most associated with the development of butanol as an advanced biofuel and which aim to market biobutanol by the end of 2007; according to the DuPont publicity material (www2.dupont.com), biobutanol's advantages are persuasive:

- Butanol has a higher energy content than ethanol and can be blended with gasoline at higher concentrations for use in standard vehicle engines (11.5% in the United States, with the potential to increase to 16%).
- Suitable for transport in pipelines, butanol has the potential to be introduced into gasoline easily and without additional supply infrastructure.
- Butanol/gasoline mixtures are less susceptible to separate in the presence of water than ethanol/gasoline blends, demanding no essential modifications to blending facilities, storage tanks, or retail station pumps.
- Butanol's low vapor pressure (lower than gasoline) means that vapor pressure specifications do not need to be compromised.
- Production routes from conventional agricultural feedstocks (corn, wheat, sugarcane, beet sugar, cassava, and sorghum) are all possible, supporting global implementation.
- Lignocellulosics from fast-growing energy crops (e.g., grasses) or agricultural "wastes" (e.g., corn stover) are also feasible feedstocks.

The principal hurdles to process optimization were in manipulating cultures and strains to improve product specificity (figure 6.11) and yield and in reducing the toxicity of butanol and O_2 (the fermentation must be strictly anaerobic)

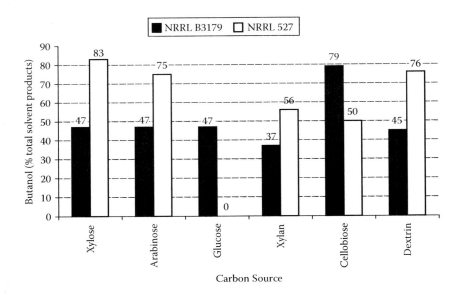

FIGURE 6.11 Variation in butanol production with two strains of *Clostridium acetobutylicum* grown on six different carbon sources. (Data from Singh and Mishra.[118])

to producing cells.[118,119] Notable among advances made in the last decade are the following:

- Isolation of hyperproducing strains — *Clostridium beijerinckii* BA101 expresses high activities of amylase when grown in starch-containing media, accumulating solvents up to 29 g/l and as high as 165 g/l when adapted to a fed-batch fermentation with product recovery by pervaporation using a silicone membrane.[120–122]
- Gas stripping has also been developed as a cost-effective means to remove butanol and reduce any product inhibition.[123]
- At the molecular level, the high product yields with hyperproducing strains can be ascribed to a defective glucose transport system exhibiting poor regulation and a more efficient use of glucose during the solventogenic stage.[124]
- The demonstration that the ABE fermentation can utilize corn fiber sugars (glucose, xylose, arabinose, and galactose) and is not inhibited by major sugar degradation products of pretreated lignocellulosic substrates.[125,126]
- Overexpression of a single clostridial gene to increase both solvent production and producer cell tolerance of product accumulation.[127]
- Improved understanding of the molecular events causing loss of productivity in solventogenic strains spontaneously or during repeated subculturing or continuous fermentation.[128,129]

A technoeconomic evaluation of a production facility with an annual capacity of 153,000 tonnes published in 2001 estimated production costs for butanol of $0.29/kg ($0.24/l, assuming a density of 0.8098 kg/l, or $0.89/gallon), assuming a conversion efficiency of 0.50 g products per gram of glucose and corn as the feedstock.[130] The calculations were noted to be very sensitive to the price paid for the corn, the worst-case scenario costs reaching $1.07/kg; with the best-case scenario, the production costs were probably competitive with conventional gasoline (at that time showing a high degree of price instability), allowing for a lower energy content (figure 5.1).

The downstream processing operations for the ABE fermentation are necessarily more complex than for fermentations with single product, for example, ethanol. Not only can the insoluble materials from the harvested fermentation be used as a source for animal feed production, but the fermentation broth must be efficiently fractionated to maximize the economic returns possible from three saleable solvent products. Detailed analysis of a conventional downstream process modeled solvent extraction (by 2-ethyl-1-hexanol), solvent stripping, and two distillation steps to recover 96% of the butanol from a butanol-dominated mix of products.[131] An optimal arrangement of these downstream steps could reduce the operating costs by 22%.

Advanced bioprocess options have included the following:

- A continuous two-stage fermentation design to maintain the producing cells in the solventogenic stage[132]
- Packed bed biofilm reactors with *C. acetobutylicum* and *C. beijerinckii*[133]
- A continuous production system with a high cell density obtained by cell recycling and capable of operation for more than 200 hours without strain degeneration or loss of productivity[134]
- Simultaneous saccharification and fermentation processes have been investigated by adding exogenous cellulase to poorly cellulolytic strains[135]

A novel feedstock for biobutanol production is sludge, that is, the waste product in activated sludge processes for wastewater treatments; this material is generated at 4×10^7 m³/year in Japan and most is discharged by dumping.[136] Adding glucose to the sludge supported growth and butanol production and a marked reduction in the content of suspended solids within 24 hours. In the Netherlands, domestic organic waste, that is, food residues, have been tested as substrates for the clostridial ABE fermentation, using chemical and enzymic pretreatments; growth and ABE formation were supported mainly by soluble sugars, and steam pretreatment produced inhibitors of either growth or solvent formation.[137,138]

Echoing the theme of recycling is the MixAlco process, developed at Texas A&M University, College Station, Texas; this can accept sewage and industrial sludges, manure, agricultural residues, or sorted municipal waste) as a feedstock, treated with lime and mixed with acid-forming organism from a saline environment to produce a mixture of alcohols that are subsequently thermally converted to ketones and hydrogenated to alcohols, predominately propanol but including higher alcohols.[139] This is another fermentation technology awaiting testing at a practical commercial scale.

6.4 ADVANCED BIOFUELS: A 30-YEAR TECHNOLOGY TRAIN

Of the biofuels discussed in this section, only biodiesel has reached full commercialization and not all of the others have even been tested in pilot plants and demonstration units as a prelude to scale up. Any conclusions as to their impact on fuel usage, the substitution of fossil fuels or reduction in greenhouse gas emissions is consequently highly speculative. Nevertheless, a European consortium of research institutes and university groups from five nations has commenced building advanced biofuels into a comprehensive model of biofuel chain options until 2030, commissioned by the European Union to identify a robust biofuels strategy to minimize costs and identify key technological, legislative, and policy developments that are required.[140] Among the key preliminary predictions are that:

- Biodiesel and cereal- and sugar-derived bioethanol use in Europe will peak at approximately 10% of total transportation fuels by 2020.
- By 2020, advanced biofuels (including lignocellulosic ethanol) will contribute an equal share of total transportation fuels but will peak at no more than 30% by 2050, the maximum being limited by the amount of land available for energy crops.
- The introduction of advanced biofuel options may meet considerable introductory cost barriers.
- Advanced biofuels may require "stepping-stone" strategies (e.g., the short-term development of lignocellulosic biomass supply chains for power generation by cofiring in power generating plants) or realistic synergies (e.g., coproduction of FT diesel and hydrogen for use in fuel cells).
- Commencing by 2020, the use of hydrogen-powered fuel cells is the only route to replacing fossil fuels for 50% or more of light-duty vehicle transportation needs.
- For heavy duty trucks, where fuel cells are unlikely to meet the demands for either high continuous loads or in long-distance transport, advanced biofuels may be the long-term market solution.

That biofuels — in particular, ethanol and biodiesel — might only be a transitional rather than a lasting solution for sustainable passenger transport is one challenging hypothesis but one with a rival: that globalization of biofuels will lead to a split between the North (the United States, Europe, and Japan) and the South (Latin America, Africa, and South and Southeastern Asia), the former increasingly focused on the production of advanced biofuels (including hydrogen-based systems), whereas the South develops a long-term economic strategy based on sugar ethanol and biodiesel, trading internationally and gradually supplanting OPEC.[141] Tropical countries "do it better," that is, can produce ethanol with positive energy gains, because they achieve at least twice the yield per hectare than can be demonstrated in temperate countries producing corn- or grain-derived ethanol; as locations for biodiesel crops, locales in the Southern Hemisphere simply cannot be bettered, for land availability, climate, the type of plant species grown or the yields extracted from them — table 6.7 lists recent investment programs in the Southern Hemisphere.[142]

In this analysis, the status of India and China is ambiguous, straddling the boundary between agricultural and industrial nations: both rank in the top 11 of nations with the least car ownership (6 and 11/1000 population, respectively), but both are in

TABLE 6.7

Investments in African and Other Biofuels Programs Announced in 2006

Country	Biofuel option	Program	Investor
Swaziland	Biodiesel	*Jatropha* plantations and processing facilities	D1 (UK)
Zambia	Biodiesel	*Jatropha* plantations and processing facilities	D1 (UK)
Liberia	Biodiesel	Palm oil crushing and processing	IBF (UK)
Nigeria	Bioethanol	Cassava plantations and processing	Nigerian National Petroleum Corporation
Senegal	Biodiesel	*Jatropha* plantations and processing facilities	Bioking (The Netherlands)
China	Bioethanol	To be finalized	Suedzucker (Germany)
Argentina	Bioethanol	Corn and sugarcane crops and processing	Private financier investment, $250-300 million

Source: Data from Matthews.[141]

the top 10 of the world's largest economies.[142] Like Brazil, India and China have the climate and rainfall (in some regions, at least, and seasonally) to establish large-scale bioethanol and biodiesel programs, and both could include lignocellulosic ethanol in the range of feedstocks. On the calculation basis that the creation of another 18 "Brazils" during the course of the next decade could produce sufficient biofuels to substitute for 20% of OECD gasoline needs by 2020, the countries could include, in a swathe across the tropics: Brazil (doubling its present output), Argentina, Colombia, Ecuador, and Chile; Central American and Caribbean countries; Nigeria, Togo, Democratic Republic of Congo, Angola, Namibia, Ghana, Senegal, Gambia, Guinea, Togo, Mali, Chad, Zimbabwe, Zambia, Kenya, Swaziland, Rwanda, Uganda, Tanzania, and South Africa; Malaysia, Thailand, the Philippines, Indonesia, and Vietnam; Indian Ocean islands (Mauritius etc.); China; and India.[141]

If, however, responsible cultivation fails to be maintained (or even established), if agriculture and ergoculture (energy crops) generate a dangerous level of competition and raise world prices unacceptably, and/or if global climate control needs solutions on a shorter time scale than envisioned by signers to the present Kyoto Agreement — or, most problematically, if President George W. Bush's "20 in 10" proposal announced in January 2007, that is, that 20% of U.S. gasoline use would be replaced with biofuels (domestic and imported) within ten years, cannot be realized and existing biofuels impose agricultural conflicts but fail to yield quick reductions in greenhouse gas emissions — what role has the North to play?

There are certainly minor improvements to established routes to biofuels to be optimized or new developments to be exploited with novel technological solutions, for example,

- The ability to utilize suboptimally grown or ripened crops, unsuitable for edible purposes, for biodiesel production[143]

- Innovative heterogeneous catalysts for transesterification reactions in biodiesel production[144]
- Producing very high quality FT diesel (from fossil sources) to be used in regions with very stringent specifications for diesel fuels and their emissions or to upgrade below specification diesels by blending[145]
- The utilization of (perhaps, unpurified) biomass-generated syngas in fermentations to support the production of either ethanol or major commodity chemicals[146,147]

The existing scientific base in OECD economies is, however, most likely to be best employed by successfully defining novel technologies to tackle the dual problems of dwindling oil reserves and out-of-control CO_2 emissions. Beyond the second-generation biofuels discussed in this chapter lies a raft of long-considered (sometimes, long-researched) topics, to which — to differing extents — the adjective "speculative" is often applied. All of these have strong biotechnological inputs and can be grouped as the radical options for biofuels.

REFERENCES

1. McDonnell, K., et al., Properties of rapeseed oil for use as a diesel fuel extender, *J. Am. Oil Chem. Soc.*, 76, 539, 1999.
2. *World Energy Outlook*, International Energy Agency, Paris, 2006, chap. 14.
3. *Commercial Biodiesel Production Plants June 7, 2007* and *Biodiesel Production Plants in Pre-construction (April 28, 2006)*, National Biodiesel Board, www.nbb.org.
4. Culshaw, F. and Butler, C., *A Review of the Potential of Biodiesel as a Transport Fuel*, Department of Trade and Industry Energy Technology Support Unit Report ETSU-R-71, Her Majesty's Stationery Office, London, 1993.
5. Mittelbach, M. and Enzelsberger, H., Transesterification of heated rapeseed oil for extending diesel fuel, *J. Am. Oil Chem. Soc.*, 76, 545, 1999.
6. Lide, D.R. (Ed.), *CRC Handbook of Chemistry and Physics*, 73rd edition, Boca Raton, section 7–29, 1992.
7. Lee, I., Johnson, A., and Hammond, E.G., Reducing the crystallization temperature of biodiesel by winterizing methyl soyate, *J. Am. Oil Chem. Soc.*, 73, 631, 1996.
8. Dunn, R.O., Thermal analysis of alternative diesel fuels from vegetable oils, *J. Am. Oil Chem. Soc.*, 76, 109, 1999.
9. Carroll, K.K., Erucic acid as the factor in rape oil affecting adrenal cholesterol in the rat, *J. Biol. Chem.*, 200, 287, 1953.
10. Mietkiewska, E. et al., Seed-specific heterologous expression of a nasturtium *FAE* gene in *Arabidopsis* results in a dramatic increase in the proportion of erucic acid, *Plant Physiol.*, 136, 2665, 2004.
11. Mittelbach, M. et al., Diesel fuel derived from vegetable oils: preparation and use of rape oil methyl ester, *Energy Agric.*, 2, 369, 1983.
12. Jeong, G.T. et al., Production of biodiesel fuel by transesterification of rapeseed oil, *Appl. Biochem. Biotechnol.*, 113–116, 747, 2004.
13. de Oliveira, D. et al., Optimization of alkaline transesterification of soybean oil and castor oil for biodiesel production, *Appl. Biochem. Biotechnol.*, 121–124, 553, 2005.
14. Karmee, S.K. and Chadha, A., Preparation of biodiesel from crude oil of *Pongamia pinnata*, *Bioresour. Technol.*, 96, 1425, 2005.
15. Felizardo, P. et al., Production of biodiesel from waste frying oils, *Waste Manag.*, 26, 487, 2006.

16. Jeong, G.T. and Park, D.H., Batch (one- and two-stage) production of biodiesel fuel from rapeseed oil, *Appl. Biochem. Biotechnol.*, 129–132, 668, 2006.
17. Vicente, G., Martínez, M., and Aracil, J., Optimisation of integrated biodiesel production. I. A study of the biodiesel purity and yield, *Bioresour. Technol.* 98, 1724, 2007.
18. Vicente, G., Martínez, M., and Aracil, J., Optimisation of integrated biodiesel production. II. A study of the material balance, *Bioresour. Technol.* 98, 1754, 2007.
19. Toda, M. et al., Green chemistry: biodiesel made with sugar catalyst, *Nature*, 438, 178, 2005.
20. Shibasaki-Kitakawa, N. et al., Biodiesel production using anionic ion-exchange resin as heterogeneous catalyst, *Bioresour. Technol.*, 98, 416, 2007.
21. da Silva, N.L. et al., Optimization of biodiesel production from castor oil, *Appl. Biochem. Biotechnol.*, 129–132, 405, 2006.
22. Kusdiana, D. and Saka, S., Two-step preparation for catalyst-free biodiesel fuel production: hydrolysis and methyl esterification, *Appl. Biochem. Biotechnol.*, 113–116, 781, 2004.
23. Warabi, Y., Kusdiana, D., and Saka, S., Biodiesel fuel from vegetable oil by various supercritical alcohols, *Appl. Biochem. Biotechnol.*, 113–116, 793, 2004.
24. Demirbas, A., Studies on cottonseed oil biodiesel prepared in non-catalytic SCF conditions, *Bioresour. Technol.*, in press, 2008.
25. Linko, Y.Y. et al., Biodegradable products by lipase biocatalysis, *J. Biotechnol.*, 66, 41, 1998.
26. Du, W., Xu, Y., and Liu, D., Lipase-catalysed transesterification of soya bean oil for biodiesel production during continuous batch operation, *Biotechnol. Appl. Biochem.*, 38, 103, 2003.
27. Hernández-Martín, E. and Otero, C., Different enzyme requirements for the synthesis of biodiesel: Novozym® 435 and Lipozyme® TL IM, *Bioresour. Technol.*, 99, 277, 2008.
28. Sanchez, F. and Vasudevan, P.T., Enzyme catalyzed production of biodiesel from olive oil, *Appl. Biochem. Biotechnol.*, 135, 1, 2006.
29. Modi, M.K. et al., Lipase-mediated conversion of vegetable oils into biodiesel using ethyl acetate as acyl acceptor, *Bioresour. Technol.*, 98, 1260, 2007.
30. Noureddini, H., Gao, X., and Philkana, R.S., Immobilized *Pseudomonas cepacia* lipase for biodiesel fuel production from soybean oil, *Bioresour. Technol.*, 96, 769, 2005.
31. Chen, G., Ying, M., and Li, W., Enzymatic conversion of waste cooking oils into alternative fuel: biodiesel, *Appl. Biochem. Biotechnol.*, 129–132, 911, 2006.
32. Chen, J.W. and Wu, W.T., Regeneration of immobilized *Candida antarctica* lipase for transesterification, *J. Biosci. Bioeng.*, 95, 466, 2003.
33. Fukuda, H., Kondo, A., and Noda, H., Biodiesel fuel production by transesterification of oils, *J. Biosci. Bioeng.*, 92, 405, 2001.
34. Matsumoto, T. et al., Yeast whole-cell biocatalyst constructed by intracellular overproduction of *Rhizopus oryzae* lipase is applicable to biodiesel fuel production, *Appl. Microbiol. Biotechnol.*, 57, 515, 2001.
35. Hama, S. et al., Lipase localization in *Rhizopus oryzae* cells immobilized within biomass support particles for use as whole-cell biocatalysts in biodiesel-fuel production, *J. Biosci. Bioeng.*, 101, 328, 2006.
36. Luo, Y. et al., A novel psychrophilic lipase from *Pseudomonas fluorescens* with unique property in chiral resolution and biodiesel production via transesterification, *Appl. Microbiol. Biotechnol.*, 73, 349, 2006.
37. Modi, M.K. et al., Lipase-mediated transformation of vegetable oils into biodiesel using propan-2-ol as acyl acceptor, *Biotechnol. Lett.*, 28, 637, 2006.
38. Salis, A. et al., Biodiesel production from triolein and short chain alcohols through biocatalysis, *J. Biotechnol.*, 119, 291, 2005.

39. Du., W. et al., Novozym 435-catalysed transesterification of crude soya bean oils for biodiesel production in a solvent-free medium, *Biotechnol. Appl. Biochem.*, 40, 187, 2004.

40. Berchmans, H.J. and Hirata, S., Biodiesel production from crude *Jatropha curcas* L. seed oil with a high content of free fatty acids, *Bioresour. Technol.*, in press, 2008.

41. Zullaikah, S. et al., A two-step acid-catalyzed process for the production of biodiesel from rice bran oil, *Bioresour. Technol.*, 96, 1889, 2005.

42. Kusidiana, D. and Saka, S., Methyl esterification of free fatty acids of rapeseed oil as treated in supercritical methanol, *J. Chem. Eng. Japan*, 34, 383, 2001.

43. Leevijit, T. et al., Performance test of a 6-stage continuous reactor for palm methyl ester production, *Bioresour. Technol.*, 99, 214, 2008.

44. Stavarache, C. et al., Ultrasonically driven continuous process for vegetable oil transesterification, *Ultrason. Sonochem.*, 14, 413, 2007.

45. Dubé, M. A, Tremblay, A.Y., and Liu, J., Biodiesel production using a membrane reactor, *Bioresour. Technol.*, 98, 639, 2007.

46. Xu, Y. et al., A novel enzymatic route for biodiesel production from renewable oils in a solvent-free medium, *Biotechnol. Lett.*, 25, 1239, 2003.

47. Baber, T.M. et al., Application of catalytic ozone chemistry for improving biodiesel product performance, *Biomacromolecules* 6, 1334, 2005.

48. Anonymous, Process incorporates renewables as part of refining operations, *Hydrocarbon Processing*, 85, 33, November 2006.

49. Pimentel, D. and Patzek, T.W., Ethanol production using corn, switchgrass, and wood; biodiesel production using soybean and sunflower, *Nat. Resour. Res.*, 14, 65, 2005.

50. Zhang, Y. et al., Biodiesel production from waste cooking oil. I. Process design and technological assessment, *Bioresour. Technol.*, 89, 1, 2003.

51. Zhang, Y. et al., Biodiesel production from waste cooking oil. II. Economic assessment and sensitivity analysis, *Bioresour. Technol.*, 90, 229, 2003.

52. Haas, M.J. et al., A process model to estimate biodiesel production costs, *Bioresour. Technol.*, 97, 671, 2006.

53. Canakci, M., The potential of restaurant waste lipids as biodiesel feedstocks, *Bioresour. Technol.*, 98, 183, 2007.

54. Kram, J.W., A hard row to how, *Biodiesel Magazine*, 2007, www.biodieselmagazine.com.

55. Sheehan, J. et al., *Life Cycle Inventory of Biodiesel and Petroleum Diesel for Use in an Urban Bus*, Final report, NREL/SR-580-24089 UC Category 1503, National Renewable Energy Laboratory, Golden, CO, May, 1998.

56. Van Gerpen, J. and Shrestha, D., *Biodiesel Energy Balance*, available from: http://uidaho.edu/bioenergy.

57. Wesseler, J., Opportunities (costs) matter: a comment on Pimentel and Patzek "Ethanol production using corn, switchgrass, and wood; biodiesel production using soybean and sunflower," *Energy Policy*, 35, 1414, 2007.

58. Thompson, J.C. and He, B., Characterization of crude glycerol from biodiesel production from multiple feedstocks, *Appl. Eng. Agri.*, 22, 261, 2006.

59. da Costa, R.E. et al., *The Energy Balance in the Production of Palm Oil Biodiesel — Two Case Studies: Brazil and Colombia*, Swedish Bioenergy association (available from: http://svebio.se/attachments/33/295.pdf).

60. Hill, J. et al., Environmental, economic, and energetic costs and benefits of biodiesel and ethanol biofuels, *Proc. Natl. Acad. Sci., USA*, 103, 11206, 2006.

61. Nabi, M.N., Akhter, M.S., and Zaglul Shahadat, M.M., Improvement of engine emissions with conventional diesel fuel and diesel-biodiesel blends, *Bioresour. Technol.*, 97, 372, 2006.

62. Canakci, M., Combustion characteristics of a turbocharged DI compression ignition engine fueled with petroleum diesel fuels and biodiesel, *Bioresour. Technol.*, 98, 1167, 2007.

63. Jeong, G-T., Oh, Y-T., and Park, D-H., Emission profile of rapeseed methyl ester and its blend in a diesel engine, *Appl. Biochem. Biotechnol.*, 129–132, 165, 2006.

64. Jung, H., Kittelson, D.B., and Zachariah, M.R., Characteristics of SME biodiesel-fueled diesel particle emissions and the kinetics of oxidation, *Environ. Sci. Technol.*, 40, 4949, 2006.

65. Lapuerta, M., Rodríguez-Fernández, J., and Agudelo, J.R., Diesel particulate emissions from used cooking oil biodiesel, *Bioresour. Technol.*, 99, 731, 2008.

66. Bünger, J. et al., Strong mutagenic effects of diesel engine emissions using vegetable oil as fuel, *Arch. Toxicol.*, 81, 599, 2007.

67. Yuan, C.S. et al., A new alternative fuel for reduction of polycyclic aromatic hydrocarbon and particulate matter emissions from diesel engines, *J. Air Waste Manag. Assoc.*, 57, 465, 2007.

68. Cardone, M. et al., *Brassica carinata* as alternative oil crop for the production of biodiesel in Italy: engine performance and regulated and unregulated exhaust emissions, *Environ. Sci. Technol.*, 36, 4656, 2002.

69. Bünger, J. et al., Mutagenic and cytotoxic effects of exhaust particulate matter of biodiesel compared to fossil diesel fuel, *Mutat. Res.*, 415, 13. 1998.

70. Bünger, J. et al., Mutagenicity of diesel exhaust particles from two fossil and two plant oil fuels, *Mutagenesis*, 15, 391, 2000.

71. Swanson, K.J., Madden, M.C., and Ghio, A.J., Biodiesel exhaust: the need for health effects research, *Environ. Health Perspect.*, 115, 496, 2007.

72. Liu, F. and Fang, B., Optimization of bio-hydrogen production from biodiesel wastes by *Klebsiella pneumoniae*, *Biotechnol. J.*, 2, 374, 2007.

73. Ito, T. et al., Hydrogen and ethanol production from glycerol-containing wastes discharged after biodiesel manufacturing process, *J. Biosci. Bioeng.*, 100, 260, 2005.

74. Suehara, K. et al., Biological treatment of wastewater discharged from biodiesel fuel production plant with alkali-catalyzed transesterification, *J. Biosci. Bioeng.*, 100, 437, 2005.

75. Henderson, K., Biofuels — growing energy solutions for the transport sector, *Energy World*, April 2007, 10.

76. Archer, G., Delivering sustainable biofuels, *Energy World*, April 2007, 13.

77. Dewulf, J., van Langenhove, H., and van de Velde, B., Exergy-based efficiency and renewability assessment of biofuel production, *Environ. Sci. Technol.*, 39, 3878, 2005.

78. Morris, G., *Strong land use is key to developing South African biofuels*, Biofuel Review of the Renewable Energy and Energy Efficiency Partnership, April 10, 2007, http://www.reeep.org/index.cfm?articleid=1628.

79. Frondel, M. and Peters, J., *Biodiesel: a new Oildorado?*, Discussion Paper 36, Rheinisch-Westfälisches Institut für Wirtschaftforschung, Essen (Germany), ISBN 3-936454-58-2, 2005.

80. Ramos, L.P. and Wilhelm, H.M., Current status of biodiesel development in Brazil, *Appl. Biochem. Biotechnol.*, 121–124, 807, 2005.

81. Note, in *Hydrocarbon Processing*, vol. 85, no. 11, November 2006, 39.

82. Pagliaro, M. et al., From glycerol to value-added products, *Angew. Chem. Int. Ed.*, 46, 4434, 2007.

83. Müller, S., Sunshine, sand and soybean king. Brazil's rise to agricultural superpower, *BioForum Europe*, vol. 5, part 11, May 2007, 17.

84. Schulz, H., Short history and present trends of FT synthesis, *Appl. Catalysis A*, 186, 3, 1999.

85. Sie, S.T. and Krishna, R., Fundamentals and selection of advanced FT-reactors, *Appl. Catalysis A*, 186, 55, 1999.

86. Morales, F. and Weckhuysen, B.M., Promotion effects in Co-based Fischer-Tropsch catalysis, *Catalysis*, 19, 1, 2006.

87. Campbell, I., *Biomass, Catalysts and Liquid Fuels*, Holt, Rinehart and Winston Ltd., London, 1983, chap. 5.
88. Tijmensen, M.J.A. et al., Exploration of the possibilities for production of Fischer Tropsch liquids and power via biomass gasification, *Biomass Bioenergy*, 23, 129, 2002.
89. Srinivas, S., Malik, R.K., and Mahajani, S.M., Fischer-Tropsch synthesis using bio-syngas and CO_2, *Advances in Energy Research (AER - 2006): Proceedings of the 1st National Conference on Advances in Energy Research*, Department of Energy Systems Engineering, Indian Institute of Technology, Bombay, 2006, 317.
90. Hamelinck, C.N. et al., Production of FT transportation fuels from biomass: technical options, process analysis and optimization, and development potential, *Energy*, 29, 1743, 2004.
91. Wu, M., Wu, Y., and Wang, M., *Mobility Chains Analysis of Technologies for Passenger Cars and Light-Duty Vehicles Fueled with Biofuels: Application of the GREET Model to the Role of Biomass in America's Future Energy (RBAFE) Project*, Argonne National Laboratory, Argonne, IL, May 2005.
92. Wu, M., Wu, Y., and Wang, M., Energy and emission benefits of alternative transportation liquid fuels derived from switchgrass: a fuel life cycle assessment, *Biotechnol. Prog.*, 22, 1012, 2006.
93. *HTU diesel*, http://www.refuel.eu/biofuels/htu-diesel/.
94. Schobert, H.H., *The Chemistry of Hydrocarbon Fuels*, Butterworth & Co., London, 1990, chap.11.
95. Agrell, J. et al., Catalytic hydrogen generation from methanol, *Catalysis*, 16, 67, 2002.
96. Ale Ebrahim, A. and Jamshidi, E., Synthesis gas production by zinc oxide reaction with methane: elimination of greenhouse gas emission from a metallurgical plant, *Energy Conversion Manag.*, 45, 345, 2004.
97. Campbell, I., *Biomass, Catalysts and Liquid Fuels*, Holt, Rinehart and Winston Ltd., London, 1983, chap.1.
98. *Biomethanol*, http://www.refuel.eu/biofuels/biomethanol/.
99. Nakagawa, H. et al., Biomethanol production and CO_2 emission reduction from forage grasses, trees, and crop residues, *Japan Agr. Res. Quart.*, 41, 173, 2007.
100. Hamelinck, C.N. and Faaij, A.P.C., Future prospects for production of methanol and hydrogen from biomass, *J. Power Sources*, 111, 1, 2002.
101. *Toshiba Announces World's Smallest Direct Methanol Fuel Cell with Energy Output of 100 Milliwatts*, press release, 24 June 2004, http://www.toshiba.co.jp/about/press/2004_06/pr2401.htm.
102. Izenson, M.G., Crowley, C.J., and Affleck, W.H., Lightweight direct methanol fuel cell, U.S. Patent 7,189,468, 2007.
103. *Bio-DME*, http://www.refuel.eu/biofuels/bio-dme/.
104. Soares, R.R., Simonetti, D.A., and Dumesic, J.A., Glycerol as a source for fuels and chemicals by low-temperature catalytic processing, *Angew. Chem. Int. Ed.*, 45, 3982, 2006.
105. Bideaux, C. et al., Minimization of glycerol production during high-performance fed-batch fermentation process in *Saccharomyces cerevisiae*, using a metabolic model as a prediction tool, *Appl. Environ. Microbiol.*, 72, 2134, 2006.
106. Wang, Z. et al., Glycerol production by microbial fermentation: a review, *Biotechnol. Adv.*, 19, 201, 2001.
107. Vijaikishore, P. and Karanth, N.G., Glycerol production by fermentation: a fed-batch approach, *Biotechnol. Bioeng.*, 30, 325, 1987.
108. Overkamp, K.M. et al., Metabolic engineering of glycerol production in *Saccharomyces cerevisiae*, *Appl. Environ. Microbiol.*, 68, 2814, 2002.
109. Gong, C.S. et al., Coproduction of ethanol and glycerol, *Appl. Biochem. Biotechnol.*, 84–86, 543, 2000.

110. Zhuge, J. et al., Glycerol production by a novel osmotolerant yeast *Candida glycerinogenes*, *Appl. Microbiol. Biotechnol.*, 55, 686, 2001.

111. Kirimura, K., Hirowatari, Y., and Usami, S., Alterations of respiratory systems in *Aspergillus niger* under conditions of citric acid fermentation, *Agric. Biol. Chem.*, 51, 1299, 1986.

112. Kirimura, K., Yoda, M., and Usami, S., Cloning and expression of the cDNA encoding an alternative oxidase gene from *Aspergillus niger* WU-2223L, *Curr. Genet.*, 34, 472, 1999.

113. Karaffa, L. and Kubicek, C.P., *Aspergillus niger* citric acid accumulation: do we understand this well working black box? *Appl. Microbiol. Biotechnol.*, 61, 189, 2003.

114. Kern, A. et al., *Pichia pastoris* "just in time" alternative respiration, *Microbiology*, 153, 1250, 2007.

115. Awang, G.M., Jones, G.A., and Ingledew, W.M., The acetone-butanol-ethanol fermentation, *Crit. Rev. Microbiol.*, 15, suppl. 1, S33, 1988.

116. Woods, D.R., The genetic engineering of microbial solvent production, *Trends Biotechnol.*, 13, 259, 1995.

117. Mitchell, W.J., Physiology of carbohydrate to solvent conversion by clostridia, *Adv. Microb. Physiol.*, 39, 31, 1998.

118. Singh, A. and Mishra, P., *Microbial Pentose Utilization. Current Applications in Biotechnology, Prog. Ind. Microbiol.*, 33, 1995, chap. 7.

119. Ezeji, T.C., Qureshi, N., and Blaschek, H.P., Bioproduction of butanol from biomass: from genes to reactors, *Curr. Opin. Biotechnol.*, 18, 220, 2007.

120. Qureshi, N. and Blaschek, H.P., Production of acetone-butanol-ethanol (ABE) by a hyper-producing mutant strain of *Clostridium beijerinckii* BA101 and recovery by pervaporation, *Biotechnol. Prog.*, 15, 594, 1999.

121. Qureshi, N. and Blaschek, H.P., Butanol production using *Clostridium beijerinckii* BA101 hyper-producing mutant strain and recovery by pervaporation, *Appl. Biochem. Biotechnol.*, 84–86, 225, 2000.

122. Qureshi, N. and Blaschek, H.P., Recent advances in ABE fermentation: hyper-butanol producing *Clostridium beijerinckii* BA101, *J. Ind. Microbiol. Biotechnol.*, 27, 287, 2001.

123. Ezeji, T.C., Qureshi, N., and Blaschek, H.P., Butanol fermentation research: upstream and downstream manipulations, *Chem. Rec.*, 4, 305, 2004.

124. Lee, J. et al., Evidence for the presence of an alternative glucose transport system in *Clostridium beijerinckii* NCIMP 8052 and the solvent-hyperproducing mutant BA101, *Appl. Environ. Microbiol.*, 71, 3384, 2005.

125. Qureshi, N. et al., Butanol production from corn fiber xylan using *Clostridium acetobutylicum*, *Biotechnol. Prog.*, 22, 673, 2006.

126. Ezeji, T.C., Qureshi, N., and Blaschek, H.P., Butanol production from agricultural residues: impact of degradation products on *Clostridium beijerinckii* growth and butanol fermentation, *Biotechnol. Bioeng.*, 97, 1460, 2007.

127. Tomas, C.A., Walker, N.E., and Papoutsakis, E.P., Overexpression of *groESL* in *Clostridium acetobutylicum* results in increased solvent production and tolerance, prolonged metabolism, and changes in the cell's transcriptional program, *Appl. Environ. Microbiol.*, 69, 4951, 2003.

128. Scotcher, M.C. and Bennett, G.N., *SpoIIE* regulates sporulation but does not directly affect solventogenesis in *Clostridium acetobutylicum* ATCC 824, *J. Bacteriol.*, 187, 1930, 2005.

129. Stim-Herndon, K.P. et al., Analysis of degenerate variants of *Clostridium acetobutylicum* ATCC 824, *Anaerobe*, 2, 11, 1996.

130. Qureshi, N. and Blaschek, H.P., ABE production from corn: a recent economic evaluation, *J. Ind. Microbiol. Biotechnol.*, 27, 292, 2001.

131. Liu, J. et al., Downstream process synthesis for biochemical production of butanol, ethanol, and acetone from grains: generation of optimal and near-optimal flowsheets with conventional operating units, *Biotechnol. Prog.*, 20, 1518, 2004.

132. Mutschlechner, O., Swoboda, H., and Gapes, J.R., Continuous two-stage ABE-fermentation using *Clostridium beijerinckii* NRRL B592 operating with a growth rate in the first stage vessel close to its maximal value, *J. Mol. Microbiol. Biotechnol.*, 2, 101, 2000.

133. Qureshi, N. et al., Biofilm reactors for industrial bioconversion processes: employing potential of enhanced reaction rates, *Microb. Cell Factories*, 4: 24, 2005.

134. Tashiro, Y. et al., High production of acetone-butanol-ethanol with high cell density culture by cell-recycling and bleeding, *J. Biotechnol.*, 120, 197, 2005.

135. Claassen, P.A.M. et al., Utilisation of biomass for the supply of energy carriers, *Appl. Microbiol. Biotechnol.*, 52, 741, 1999.

136. Kobayashi, G. et al., Utilization of excess sludge by acetone-butanol-ethanol fermentation employing *Clostridium saccharoperbutylacetonicum* NRR N1-4 (ATTC 13564), *J. Biosci. Bioeng.*, 99, 517, 2005.

137. Claassen, P.A.M. et al., Acetone, butanol and ethanol production from domestic organic waste by solventogenic clostridia, *J. Mol. Microbiol. Biotechnol.*, 2, 39, 2000.

138. Lopez-Contreras, A.M. et al., Utilisation of saccharides in extruded domestic organic waste by *Clostridium acetobutylicum* ATCC 824 for production of acetone, butanol and ethanol, *Appl. Microbiol. Biotechnol.*, 54, 162, 2000.

139. Holtzapple, M.T. et al., Biomass conversion to mixed alcohol fuels using the MixAlco process, *Appl. Biochem. Biotechnol.*, 77–79, 609, 1999.

140. Londo, M. et al., REFUEL: an EU road map for biofuels, http://www.refuel.eu/uploads/media/kaft_harvest_so_far_new.pdf.

141. Mathews, J.A., Biofuels: what a biopact between North and South could achieve, *Energy Policy*, 35, 3550, 2007.

142. *The Economist* Newspaper Ltd., *Pocket World in Figures*, 2007 Ed., Profile Books Ltd., London, 2006.

143. Kulkarni, M., Dalai A.K., and Bakhshi, N.N., Utilization of green seed canola oil for biodiesel production, *J. Chem. Technol. Biotechnol.*, 81, 1886, 2006.

144. Lotero, E. et al., The catalysis of biodiesel synthesis, *Catalysis*, 19, 41, 2006.

145. Dry, M.E., High quality diesel via the Fischer-Tropsch process — a review, *J. Chem. Technol. Biotechnol.*, 77, 43, 2002.

146. Datar, R.P. et al., Fermentation of biomass-generated producer gas to ethanol, *Biotechnol. Bioeng.*, 86, 587, 2004.

147. Sakai, S. et al., Acetate and ethanol production from H_2 and CO_2 by *Moorella* sp. using a repeated batch culture, *J. Biosci. Bioeng.*, 99, 252, 2005.

7 Radical Options for the Development of Biofuels

7.1 BIODIESEL FROM MICROALGAE AND MICROBES

7.1.1 MARINE AND AQUATIC BIOTECHNOLOGY

In mid-2007, a review in the journal *Biotechnology Advances* concluded that microalgae appeared to be the only source of renewable biodiesel capable of meeting the global demand for petrodiesel transportation fuels.[1] The main argument was that the oil productivity of selected microalgae greatly exceeds that of the best seed oil-producing terrestrial plants; although both life forms utilize sunlight as their ultimate energy source, microalgae do so far more efficiently than do crop plants. By then, three U.S. companies were developing commercial bioreactor technologies to produce biodiesels from "oilgae" (as the producing species have been termed): Greenfuel Technologies (Cambridge, Massachusetts), Solix Biofuels (Fort Collins, Colorado), and PetroSun (Scottsdale, Arkansas) via its subsidiary Algae Biofuels.

This is a very different scenario from that detailed in the 1998 close-out report on nearly decades of research funded by the U.S. DOE.[2] This program had set out to investigate the production of biodiesel from high-lipid algae grown in ponds and utilizing waste CO_2 from coal-fired power plants. The main achievements of the research were the following:

1. The establishment of a collection of 300 species (mostly green algae and diatoms), housed in Hawaii, that accumulated high levels of oils; some species were capable of growth under extreme conditions of temperature, pH, and salinity.
2. A much greater understanding of the physiology and biochemistry of intracellular oil accumulation — in particular, the complex relationships among nutrient starvation, cell growth rate, oil content, and overall oil productivity.
3. Significant advances in the molecular biology and genetics of algae,* including the first isolation from a photosynthetic organism of the gene encoding acetyl-CoA carboxylase, the first committed step in fatty acid biosynthesis.[3]
4. The development of large-surface-area (1000 m^2) pond systems capable of utilization of 90% of the injected CO_2.

---

* As a (possible) harbinger of the future, the most significant plant oil in the literature of science fiction was that elaborated by the triffids, a species whose true biological and predatory capabilities were not adequately understood before disaster struck (John Wyndham, *Day of the Triffids*, Michael Joseph, London, 1951).

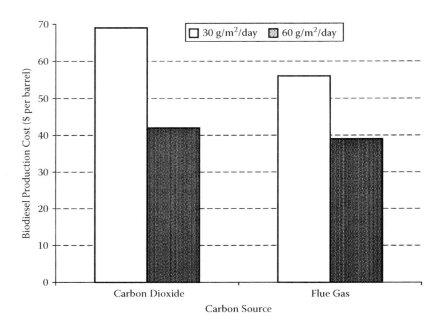

FIGURE 7.1 Estimated production costs of biodiesel from microalgae with two different carbon sources and at differing productivities. (Data from Sheehan et al.[2])

Although algal production routes had the enormous advantage of not encroaching on arable land or other agricultural resources for food crops, the perceived problem in the 1998 report was the high cost of algal biodiesel relative to conventional automotive fuels, up to $69/barrel in 1996 prices; the higher the biological productivity, the lower the production costs, whereas using flue gas was more economical than buying CO_2 supplies (figure 7.1). With crude oil prices then being $20/barrel or less, such production costs were disappointingly high (figure 1.3). Post 2000, oil prices three to four times higher allow a very different interpretation of the algal biodiesel option (figure 1.11).

The open-pond technology was not only the simplest but also the cheapest production choice. Closed-system production offered far more controllable growth environments for the algae, but the cost of even the simplest tubular photobioreactors were projected to have 10 times higher capital costs than open-pond designs. In addition, open-pond cultures had been commercialized for high-value algal chemical products — and any attempt at large-scale (>1 ton/year) closed-production systems had failed.[2] Choices of location and species had dramatically increased productivity during the lifetime of the program, from 50 to 300 tonnes/hectare/year, close to the calculated theoretical maximum for solar energy conversion (10%). The report concluded, therefore, that microalgal fuel production was not limited by engineering issues but by cultivation factors, including species control in large outdoor environments, harvesting methods, and overall lipid productivity. Encouragingly, the potential supply of industrial waste sources of CO_2 in the United States by 2010 was estimated to be as high as 2.25×10^6 tonnes/year, with Fischer-Tropsch conversion plants from fossil

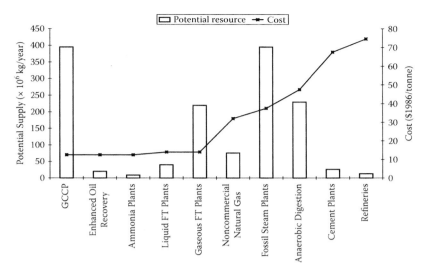

FIGURE 7.2 Potential supplies and costs of CO_2 for microalgal biodiesel production in the United States by 2010. (Data from Sheehan et al.[2])

fuels (chapter 6, section 6.2) and gasification/combined cycle power facilities offering the largest amounts of CO_2 at low cost prices (figure 7.2).

The era of high oil prices, interest in and research effort into algal sources of oils for biodiesel production has become more globally distributed. Typical of this recent change in scientific and technical priorities has been Chinese studies of *Chlorella prototothecoides* — but with the cells grown heterotrophically* (using chemical nutrients) rather than photosynthetically with a source of CO_2:

- Heterotrophically grown cells contained 57.9% oil, more than three times higher than in autotrophic (photosynthetically CO_2-fixing) cells; chemical pyrolysis yielded an oil with a lower oxygen content, a higher heating value, a lower density, and a lower viscosity than autotrophic cell bio-oil.[4]
- With a corn powder hydrolysate as carbon source (rather than glucose), a high cell concentration could be achieved; the oil (55.2%) could be efficiently extracted with hexane as a solvent and converted to biodiesel by transesterification with an acid catalyst.[5]
- Optimization of the transesterification defined a temperature of 30°C and a methanol:oil molar ratio of 56:1, resulting in a process time of 4 hours.[6]
- The process could be upscaled from 5 to 11,000 l, maintaining the lipid content; hexane-extracted oil could be transformed to methyl esters using an immobilized lipase and with a transesterification efficiency of more than 98% within 12 hours.[7]

---

* Heterotrophic cultivation was mentioned in descriptions of experiments in the 1998 review, but no data for production systems using this approach were included.[2]

As a very fast-growing "crop" (in comparison with terrestrial species), microalgae can also be viewed (even if grown autotrophically in external environments) as a lignin-less biomass source, potentially capable of being used as the substrate for ethanol production as well as biodiesel.[8]

7.1.2 "MICRODIESEL"

The cultivation of photosynthetic microalgae under dark conditions, supplied with organic carbon, closely resembles typical microbial fermentations. Because several bacterial species are well known as accumulators of triglycerides (oils) and esters of fatty acids with long-chain alcohols (waxes), the logical conclusion was to combine these biosynthetic abilities with that of ethanol formation to generate the precursors of triglycerides in microbial production systems, that is, "microdiesel" produced without any need for a chemically or enzymatically catalyzed transesterification.[9] The simple bacterium *Escherichia coli* was used as host for the *Zymomonas mobilis* pyruvate decarboxylase and alcohol dehydrogenase genes for ethanol production (chapter 3, section 3.3.2) together with the gene encoding an unspecific wax ester synthase/acyl-CoA: diacylglycerol transferase from a bacterial strain (*Acinetobacter baylyi*) known to accumulate lipid as an internal cell storage reserve. The resulting recombinant could accumulate ethyl esters of fatty acids at up to 26% of the cellular dry mass in fermentations fed with glucose. Insomuch as glucose is a fully renewable carbohydrate supply (via, e.g., cellulose or starch), microdiesel is a genuinely sustainable source of preformed transportation fuel — although the chemical engineering aspects of its extraction from bacterial cells and the economics of its production systems require further definition.

A refinement preliminary to industrial feasibility studies would be to transfer to a host capable of higher endogenous accumulation of lipids; many of these are Gram-negative species (like *E. coli*), and ethanol production in such species is a well-understood area of biochemistry.[10,11]

7.2 CHEMICAL ROUTES FOR THE PRODUCTION OF MONOOXYGENATED C6 LIQUID FUELS FROM BIOMASS CARBOHYDRATES

5-Hydroxymethylfurfural (HMF) was discussed in chapter 3 as a toxic product of acidic pretreatment techniques for biomass. The boiling point of HMF is too high (291°C) to be considered as a liquid fuel, but if HMF is subject to chemical hydrogenolysis of two of its C–O bonds, a more volatile product, 2,5-dimethylfuran (DMF) is formed (figure 7.3).[12] DMF has a boiling point of 93°C, 20°C higher than ethanol, and has a Research Octane Number of 119 — the by-product 2-methylfuran has an even higher RON (131) but is more water-soluble than DMF.

HMF is most readily formed by the dehydration of fructose, a naturally occurring sugar and a straightforward isomerization product of glucose; mineral acids such as hydrochloric acid (HCl) can be used to catalyze the reaction, 88% conversion being achieved at 180°C.[12] A solvent such as *n*-butanol (chapter 6, section 6.3.3) can then be employed to extract the HMF before hydrogenolysis over a mixed Cu-Ru catalyst at 220°C.

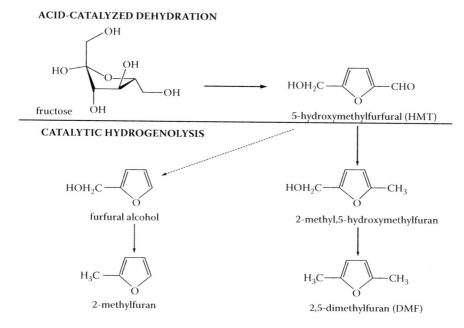

ACID-CATALYZED DEHYDRATION

fructose

5-hydroxymethylfurfural (HMT)

CATALYTIC HYDROGENOLYSIS

furfural alcohol

2-methyl,5-hydroxymethylfuran

2-methylfuran

2,5-dimethylfuran (DMF)

FIGURE 7.3 Thermochemical production of C6 furans and C5 by-products from fructose. (Modified from Román-Leshkov et al.[12])

Such production routes, beginning with enzymic conversion of glucose to fructose and proceeding via entirely thermochemical processes, have been described as "hybrid."[13] They have the advantage of avoiding reliance on large fermentation vessels for the production step(s), therefore being potentially much more rapid. Their economics could be similar to, or an improvement on, those for Fischer-Tropsch liquid fuels (chapter 6, section 6.2). The conversion of glucose to fructose, catalyzed by the enzyme glucose isomerase, has been a major industrial application of enzymology since the 1960s, the product (high-fructose corn syrup) being introduced as a substitute for Cuban sugar in the U.S. reduced-calorie sweetener market.[14] The enzyme technology has been continuously improved, evolving to immobilized forms of the enzyme; the potential of enzymes from hyperthermophilic microbes has now been explored, with a stability at 80°C rivaling that of conventional enzyme processes operated at 55–65°C.[15] Rapid and efficient processing of glucose solutions to high concentrations of fructose is feasible if the desirable biocatalytic and thermostability properties of suitable enzymes can be realized.

7.3 BIOHYDROGEN

7.3.1 THE HYDROGEN ECONOMY AND FUEL CELL TECHNOLOGIES

The International Energy Agency, in its 2006 review of world energy trends, forecasts that by 2030, hydrogen-powered vehicles may have begin to "decarbonize" transportation — if, that is, production from low- and zero-carbon sources develops, if there are breakthroughs in hydrogen storage and if the necessary infrastructure (requiring huge investments) develops.[16] The chemistry of hydrogen combustion entirely avoids greenhouse gas emissions:

$$2H_2 + O_2 \rightarrow 2H_2O$$

whether this occurs in thermal power generation or in any of the presently developed types of hydrogen fuel cell (table 7.1).[17]

In principle, generating H_2 from the most abundant potential source — water — is eminently straightforward, that is, the electrolysis of water, but the smallest amount of electricity that can produce 1 mol of H_2 from 1 mol of water is 237 kJ, whereas the amount of heat generated by the combustion of H_2 is 285.6 kJ.[18] Although fossil fuels are the main source of electric power generation, switching to the "hydrogen economy" will only be an inefficient means of reducing the emissions of CO_2 and other greenhouse gases because the net energy balance of this route to H_2 production is no more favorable than that often calculated for corn-derived ethanol (chapter 1, section 1.6.1). Other routes are known, for example the direct thermal decomposition of water, and thermochemical, photochemical, and photoelectrochemical technologies, but how "green" the resulting H_2 production is depends critically on the mix of fossil and nonfossil inputs used for power generation, locally or nationally.

The Division of Technology, Industry, and Economics of the United Nations Environment Programme noted in its 2006 review that publicly funded research into hydrogen technologies was intensive in OECD nations (figure 7.4).[19] Both OECD countries and a growing number of developing economies have active "hydrogen economy" targets:

- Japan was the first country to undertake an ambitious fuel cell program, 10 years of R&D funded at $165 million, completed in 2002; following on, the New Hydrogen Project focuses on commercialization, funding reaching $320 million in 2005 and with the aims of producing and supporting 50,000 fuel cell-powered vehicles by 2010 and 5 million by 2020 (with 4000 H_2 refueling stations by then), 2,200 MW of stationary fuel cell cogeneration systems by 2010, and 10,000 MW by 2020.

TABLE 7.1
Hydrogen-Fuel Cells: Types, Fuels, and Power Ranges

Fuel cell type	Operating temperature (ºC)	Electric efficiency (%)	Power range (kW)
Alkaline	60–120	35–55	<5
Proton exchange membrane	50–100	35–45	5–120
Phosphoric acid	approx. 220	40	200
Molten carbonate	approx. 650	>50	200-MW
Solid oxide	approx. 1000	>50	2-MW

Source: Data from Hoogers.[17]

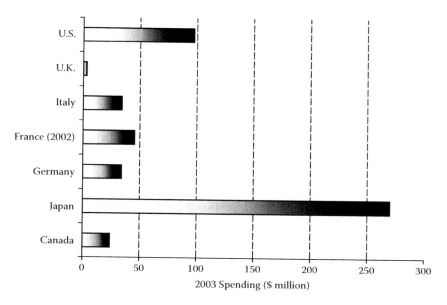

FIGURE 7.4   Publicly funded research into hydrogen technologies in the largest OECD economies, 2003. (Data from *The Hydrogen Economy.*[19])

- The transition to the hydrogen economy envisaged by the U.S. government (the Hydrogen Fuel Initiative) is set to proceed via four phases, of technology development, initial market penetration, infrastructural investment, and full realization to begin by 2025.
- Funding for the hydrogen economy in the European Union was provided by the Renewable Energy Sixth Framework program from 2003 to 2006, and subsequent plans are expected to generate combined public and private funding of approximately $2.8 billion by 2011.
- Canada's H_2 R&D focuses on the Ballard PEM fuel cell and the Hydrogenics alkaline water electrolyser, with public funding of more than $25 million/year.
- Korea has budgeted $586 million for hydrogen-related projects through to 2011, aiming at the introduction of 10,000 fuel cell vehicles, development of H_2 production from renewable resources, and development of a 370-MW capacity stationary fuel cell.
- India allocated $58 million from 2004 to 2007 for projects in universities and governmental research laboratories, with car manufacturers expected to contribute $116 million by 2010.
- Russia began to fund a joint project between the Russian Academy of Science and the Norlisk Nikel Company at $30 million in 2005 on fuel cell development.
- Brazil's Hydrogen Roadmap focuses on production from water electrolysis, reforming of natural gas, reforming or gasification of ethanol and other biofuels, storage technologies (including metal hydrides), and fuel cells.

In 2007, the first liquid H_2-dispensing fuel pump was installed in Norway as the first step in providing the "H_2 highway," a 360-mile route from Stavanger to Oslo that is expected to be complete by 2009, whereas California launched a H_2 highway network to include up to 200 fueling stations by 2010.

7.3.2 BIOPRODUCTION OF GASES: METHANE AND H_2 AS PRODUCTS OF ANAEROBIC DIGESTION

"Biogas," that is, a mixture of CH_4 and CO_2 prepared usually from the anaerobic digestion of waste materials by methanogenic bacterial species (*Methanosarcina*, *Methanosaeta*, *Methanobrevibacter*, etc.), is a technology applied globally and is ideally suited for local use in rural communities in developing economies as a cheap source of nonbiomass direct fuel.[20,21] As a low-technology but established approach to wastewater treatment, it is applicable on an industrial scale, its only disadvantages being the need to remove malodorous volatile sulfur compounds. Anaerobic digestion is also a relatively efficient means of capturing the energy present in biological materials (figure 2.2). As links to biofuels production, biogas production not only has a historical claim for practical implementation but is also an ideal means of purifying wastewater from bioethanol facilities for detoxification and recirculation, thus reducing production costs by generating locally an input for combined heat and power or steam generation.[22]

Much less widely known are the bacteria that can form H_2 as an end product of carbohydrate metabolism. Included in the vast number of species capable of some kind of biological fermentation (figure 2.3) are a wide array of microbes from anaerobic environments (including *Escherichia coli*) that were known as active research topics as far back as the 1920s (and some of which were even discovered by Louis Pasteur in the nineteenth century).[23,24] The ability of microorganisms isolated from the digestive tract to produce H_2 from cellulosic substrates is another scientific research subject with a surprisingly long history.[25]

7.3.2.1 Heterotrophic Microbes Producing H_2 by Hydrogenase Activity

The best taxonomically and physiologically characterized examples of H_2 producers are clostridia, but other genera (including bacilli), as well as a microbial flora from anoxic marsh sediments and other environments are known that are capable of the H_2 production and either the ABE "solvent" fermentation (chapter 6, section 6.3.3), the accumulation of one or more of the ABE trio, carboxylic acids (acetic, butyric, etc.) and/or other products (acetoin, 2,3-butanediol, etc.).[26]

A wide spectrum of carbon sources supports H_2 production at rates up to 1,000 ml/hr/g cells at a maximum yield of 4 mol of H_2/mol glucose with the stoichiometry:[26]

$$C_6H_{12}O_6 + 2H_2O \rightarrow 2CH_3COOH + 2CO_2 + 4H_2$$

This reaction is sufficiently exothermic (to support microbial growth). The yield of H_2 is, however, subject to feedback inhibition by H_2, requiring that the partial pressure of the gas be kept low to avoid problems with growth rate or a shift to acid production.

Another H_2-forming fermentation has butyric acid as its major acidic product:

$$C_6H_{12}O_6 \rightarrow C_3H_7COOH + 2CO_2 + 2H_2$$

although the molar production of H_2 is only half that of acetate-accumulating strains.

The key enzyme in heterotrophic H_2 producers is hydrogenase, an enzyme that catalyzes the reoxidation of reduced ferredoxin (Fd), an iron-containing protein reduced by ferredoxin-NAD and pyruvate-ferredoxin oxidoreductases, with the liberation of molecular hydrogen (figure 7.5):[27]

$$2Fd^{2+} + 2H^+ \rightarrow 2Fd^{3+} + H_2$$

A summary of hydrogenase-containing bacteria is given in table 7.2.

Hydrogenases are a diverse group of enzymes and are often cataloged on the basis of the metal ion they contain as an essential component of the active site.[28] The fastest H_2-evolving species under laboratory conditions, *Clostridium acetobutylicum*, produces two different hydrogenases:[29–31]

- An iron-containing enzyme, whose gene is located on the chromosome
- A dual-metal (nickel, iron) enzyme whose gene is located on a large plasmid

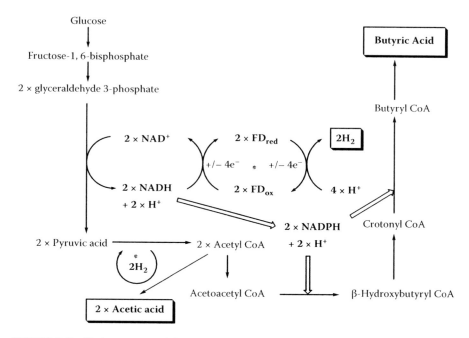

FIGURE 7.5 Hydrogenase and the reoxidation of redox cofactors in acid- and H_2-producing clostridial species. (Modified from Moat and Foster.[26])

TABLE 7.2
Biochemistry of Hydrogenase-Containing Bacteria

Microbe	Examplary species	Substrate	H_2 productivity (mole/mole)
Anaerobes			
Clostridia	*Clostridium butyricum. C. beijerincki*	Glucose	up to 4
Methylotrophs	*Methylomonas albus. M. trichosporium*	Formic acid	up to 2.45
Methanogens	*Methanobacterium soehgenii*	Formic acid	1
Rumen bacteria	*Ruminococcus albus*	Glucose	up to 0.59
Archaea	*Pyrococcus furiosus*	Glucose etc.	?
Facultative anaerobes			
Enterobacteria	*Escherichia coli. Enterobacter aerogenes*	Glucose	up to 0.9
Aerobes			
Alcaligens	*Alcaligenes eutrophus*	Gluconic acid, formic acid, etc.	?
Bacilli	*Bacillus licheniformis*	Glucose	up to 1.5

Source: Data from Nandi and Sengupta.[27]

The iron-dependent hydrogenase from *C. acetobutylicum* has a specific activity eightfold higher than similar enzymes from green algae even when all three enzymes are expressed in and purified from the clostridial host.[32] The active sites of iron-dependent hydrogenases may be the simplest such structures yet studied at the molecular level. Of enormous potential importance for the industrial development of hydrogenases is the finding that even simple complexes of iron sulfide and CO mimic hydrogenase action.[33] The crucial structure involves two iron atoms with different valency states at different stages of the reaction mechanism (figure 7.6). These findings raise the possibility of rational design of improved hydrogenases by the binding of novel metal complexes with existing protein scaffolds from known enzymes.

In contrast, nickel-iron bimetallic hydrogenases possess complex organometallic structures with CO and cyanide (CN^-) as additional components, the metal ions bound to the protein via multiple thiol groups of cysteine resides, and an important coupling between the active site and iron-sulfur clusters.[34,35] Multigene arrays are required for the biosynthesis of mature enzyme.[36] Nevertheless, progress has been impressive in synthesizing chemical mimics of the organometallic centers that contain elements of the stereochemistry and atomic properties of the active site.[37] The enzyme kinetics of nickel-iron hydrogenases remain challenging, and it is possible that more than one type of catalytic activation step is necessary for efficient functioning in vivo.[38,39]

Such advances in basic understanding will, however, open the door to replacing expensive metal catalysts (e.g., platinum) in hydrogen fuel cells by iron- or iron/nickel-based biocatalysts — the sensitivity of many hydrogenases to inhibition by O_2

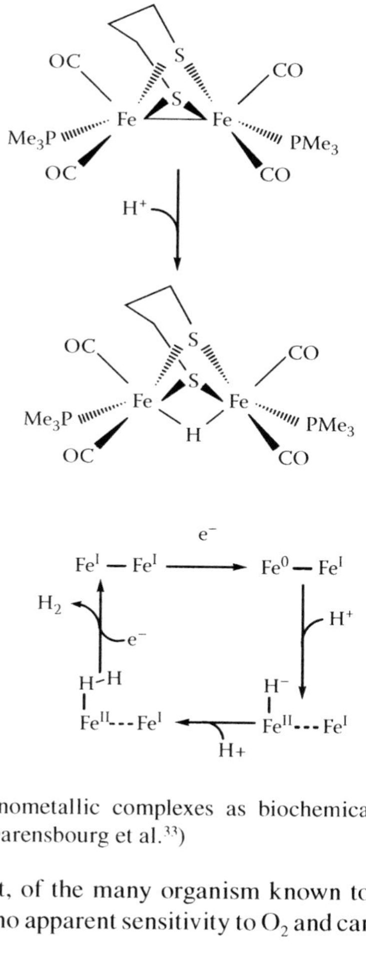

FIGURE 7.6 Simple organometallic complexes as biochemical mimics of hydrogenase enzymes. (Modified from Darensbourg et al.[33])

is a serious drawback but, of the many organism known to produce hydrogenases, some contain forms with no apparent sensitivity to O_2 and can function under ambient levels of the gas.[40]

7.3.2.2 Nitrogen-Fixing Microorganisms

The fixation of gaseous molecular nitrogen, N_2, to biologically utilizable nitrogen is performed as an essential part of the global nitrogen cycle by bacteria that may be either free living or exist in symbiotic associations with plants, bivalves, or marine diatoms.[21] The central enzymic reduction of nitrogen to ammonia is that catalyzed by nitrogenase:

$$N_2 + 8H^+ + 8e^- \rightarrow 2NH_3 + H_2$$

Molecular hydrogen is an obligatory product of the overall reaction — which, however, is so energy-expensive that most nitrogen fixers rarely evolve H_2, employing another hydrogenase ("uptake hydrogenase") to recycle the H_2 via oxidation by O_2 and conservation of part of the potentially released energy to support nitrogenase action. For this reason, nitrogenase-containing microbes are not viewed as likely sources of H_2 for biofuels.

7.3.2.3   Development of "Dark" H_2 Production Systems

The major challenge in developing commercial H_2 generation by "dark" biotechnology, that is, by fermentation processes, has been the low productivity of natural H_2-evolving microbes, and this can be resolved into two distinct limitations: the molar yield from a fermentable substrate and the expected low growth rates and cell densities of microbial producers under energy-poor environmental conditions.

On the basis of data presented in table 7.2, a maximum of 4 mol of H_2 per mole of fermentable glucose substrate equates (on a mass balance) to only 8 g/180 g of sugar consumed. In cell-free biotransformations, using mixtures of enzymes, a nearly threefold higher productivity has been demonstrated.[41] If the pentose phosphate pathway (figure 3.2) can be run in a cyclic manner to completely oxidize glucose 6-phosphate to CO_2 and H_2O, each mole of glucose consumed (in six "turns" of the cycle) can generate 12 mol of reduced cofactor:

$$6C_6H_{12}O_6 + 12\ NADP^+ \rightarrow 5C_6H_{12}O_6 + 6CO_2 + 12NADPH + 12H^+$$

Coupling the reoxidation of NADPH to the hydrogenase from *Pyrococcus furiosus* (table 7.2), one of only a few hydrogenases known to accept NADPH as a reducing agent, generated 116 mol H_2/mol of phosphorylated glucose oxidized.[41]

Japanese researchers genetically modified a strain of *E. coli* to overexpress the gene for formate hydrogen lyase, an enzyme catalyzing the reaction:

$$HCOOH \rightarrow CO_2 + H_2$$

By growing the cells to high cell densities under glucose-supported aerobic conditions before transfer to an anaerobic fermentor, a high catalytic potential for formate transformation to H_2 was established, reaching 300 L H_2/hr/l culture.[42] This rate of H_2 production could support a 1-kW fuel cell operating at 50% efficiency using only 2 l of culture medium maintained under continuous conditions by a feed of formic acid. Further strain construction (deleting lactate dehydrogenase and fumarate reductase genes) has improved the induction of the formate hydrogen lyase activity.[43] The same genetic manipulations have eliminated side reactions of (phosphoenol)pyruvic acid in glucose-grown *E. coli*, maximizing the transformation of pyruvate to formate via the pyruvate formate lyase-catalyzed step:

$$CH_3COCOOH + CoASH \rightarrow CH_3CO\text{-}SCoA + HCOOH$$

Such a genetic background (with formate hydrogen lyase as the next step) increases the production of H_2 from glucose as the fermentation substrate, although only rates of approximately 20 l/hr/l culture have been achieved.[44]

Thermophilic and hyperthermophilic microbes are obvious choices for production strains to be cultured at high temperatures to accelerate H_2 formation.

A strain of the bacterium *Klebsiella oxytoca* isolated from a hot spring in China could produce H_2 even in the presence of 10% O_2 in the gas phase but had a low molar yield (1 mol/mol glucose consumed).[45] The extreme thermophile *Caldicellulosiruptor saccharolyticus* shows up to 92% of the theoretical H_2 yield from glucose (4 mol/mol) at low growth rates at 72–73°C, indicating possible applications in long-term free or immobilized cultures.[46] This organism also can produce H_2 from hydrolyzed paper sludge industrial waste as the sole carbon source and is unusual in that it can utilize xylose faster than glucose.[47,48]

An advanced bioprocess option for H_2 production utilizes a membrane bioreactor to maintain the bacteria inside the reactor while allowing fluids to exit.[49] This design could be the optimal methodology to restrict the growth of methanogenic bacteria that consume H_2 and generate CH_4, a gas with only 42 percent of the energy content of H_2.[50] Restricting the residence time of materials in a continuous flow reactor system allows H_2 producers to outcompete the slower-growing methanogens. With a 12-hr residence time, glucose could be utilized as a substrate for H_2 production with an overall consumption of 98% and an efficiency (assuming 4 mol H_2/mol glucose) of 25%, accumulating H_2 at a concentration of 57–60% (volume basis) in the headspace.[49]

The production of H_2 need not be based on pure bacterial cultures, mixed cultures, even ones with only indirect evidence of the microbial flora present, being suitable for wastewater treatment or for local production sites in isolated rural localities. A clostridial population (on the basis of the spectrum of metabolites produced in parallel to H_2) provided a system capable of stable and prolonged production, with H_2 reaching 51 percent in the gas phase and with no methanogenesis observed.[51] Some process control is, however, unavoidable to maintain H_2-evolving capacity, particularly pH: maintaining a pH of 6.0 may inhibit the growth of lactobacilli in a mixed culture of *Clostridium* and *Coprothermobacter* species that could utilize untreated sludge and lake sediment material as substrates.[52] The choice of pH regulant may, however, be crucial, and the accumulation of sodium to toxic levels was a severe limitation in a continuous biohydrogen system from sucrose-supplemented anaerobic sewage sludge.[53] At a constant pH, the combination of substrate-material retention time and temperature (in the 30–37°C range) can have a marked effect on the balance between different clostridial species, the appearance of nonclostridial bacterial species and the overall molar yield of H_2 from carbohydrates.[54] In a molasses wastewater treatment plant in China, the H_2 production rate was highest in an ethanol-forming stage of the process, and at least six types of H_2-producing microbe were present, predominantly a novel species, *Ethanoligenes harbinese*.[55] Such a complex microbial ecology may be highly adaptable to differing types and compositions of carbon sources during production cycles or when seasonally available.

The use of advanced reactor types has been explored; for example, the fluidized bed reactor design has been explored with a mixed community that rapidly established H_2 production from *C. butyricum*; instability developed during the course of time as propionate producers gradually took over, and biofilm-type reactors may not be the optimal design because of the efficient adhesion of H_2-consuming microbial species to the carrier.[56] A trickle-bed reactor packed with glass beads inoculated with a pure culture of *C. acetobutylicum* certainly gave high H_2 gas concentrations but soon (60–72 hours) clogged because of bacterial growth.[57]

Irrespective of the long-term prospects for the industrial production of H_2, dark fermentations are very likely to be permanent features of wastewater treatment technologies and as an alternative to methane for local "biogas" production — indeed an obvious application of the trickle-bed reactor may be for the treatment of high-carbohydrate wastewaters, requiring no energy input for stirring a conventional mixed tank and producing H_2 as a recoverable fuel gas as well as a more dilute stream for conventional biogas production.[57] The use of water streams with lower organic loading may, however, be advantageous for H_2 production because supersaturation of the gas space inside bioreactors may feed back to inhibit H_2 synthesis.[58] Removing CO_2 (e.g., by the use of KOH to absorb the gas) is also beneficial to H_2 production, probably by minimizing the flow of utilizable substrates to acetogenic bacteria capable of synthesizing acetic acid from H_2 and CO_2.[59] Animal waste-contaminated water can also be made acceptable to biological H_2 producers if the ammonia concentration can be reduced and maintained below toxic concentrations in continuous flow systems, especially if the microbial community can be gradually adapted to increased ammonia levels.[60] Food processing aqueous streams with high chemical oxygen demands can support biohydrogen production at 100 times the rates possible with domestic wastewaters — often reaching commercially viable amounts of H_2 if used on-site as a heating fuel.[61]

As an excellent example of the opportunistic use of H_2-producing microbes in biofuels production, a strain of *Enterobacter aerogenes* was shown to be highly adept at producing both H_2 and ethanol from glycerol-containing wastewaters from biodiesel production; continuous production in a packed bed reactor using porous ceramic support material maximized the H_2 production rate.[62] But, as a final twist, H_2 producers may have an unexpected role in assisting a microbial community of methanogens to achieve full productivity, that is, a syntropic relationship may be established to provide the methanogens with a readily utilizable substrate; adding mesophilic or thermophilic H_2-producing cultures increases biogas production from animal manure slurry, and the added species persist for several months of semicontinuous operation.[63]

7.3.3   PRODUCTION OF H_2 BY PHOTOSYNTHETIC ORGANISMS

In comparison with fermentor-based H_2 production, the use of photosynthetic organisms has received wider publicity because envisaged bioprocesses have convincing environmental credentials, that is, the ability to produce a carbonless fuel using only water, light, and air (CO_2) as inputs. Hydrogen photobiology is, however, highly problematic because of the incompatibility of the two essential steps:

- In the first stage, water is split to produce O_2.
- In the second stage, the photoproduced electrons are combined with proteins to form H_2 by either a hydrogenase or a nitrogenase — and O_2 is a potent inhibitor of such an "anaerobic" system.

Nature provides two related solutions to this dilemma.[64] First, filamentous cyanobacteria (e.g., *Anabaena cylindrica*) that compartmentalize the two reactions into different types of cell: vegetative cells for generating O_2 from water and using the reducing power to fix CO_2 into organic carbon compounds that then pass to specialized nitrogenase-containing heterocyst cells that evolve H_2 when N_2 reduction is blocked by low ambient concentrations of N_2. The second scenario is that of non-heterocystous cyanobacteria that separate O_2 and H_2 evolution temporally (in day and night cycles), although the same overall effect could be achieved using separate light and dark reactors. With either type of nitrogen-fixing organism, however, the high energy requirement of nitrogenase would lower solar energy conversion efficiencies to unacceptably low levels.

Hydrogenase is the logical choice of biocatalyst for H_2 production, and nearly 35 years have now passed since the remarkable experimental demonstration that simply mixing chloroplasts isolated from spinach leaves with hydrogenase and ferredoxin isolated by cells of *Clostridium kluyveri* generated a laboratory system capable of direct photolysis of water and H_2 production.[65] The overall reaction sequence in that "hybrid" biochemical arrangement was

$$H_2O + light \rightarrow \tfrac{1}{2}O_2 + 2e^- + 2H^+ \rightarrow ferredoxin \rightarrow hydrogenase \rightarrow H_2$$

Light-induced photolysis of water produced electrons that traveled via the photosystems of the chloroplast preparations to reduce ferredoxin before hydrogenase catalyzing the reunion of electrons and protons to form molecular hydrogen. For over a quarter of a century, therefore, the nagging knowledge that direct photolytic H_2 production is technically feasible has both tantalized and spurred on research into solar energy conversion.

Thirty years ago, the abilities of some unicellular green (chlorophyll-containing) algae, that is, microalgae, to generate H_2 under unusual (O_2-free) conditions where hydrogenase was synthesized had already been defined.[66] Such microalgae can, when illuminated at low light intensities in thin films (5–20 cellular monolayers), show conversion efficiencies of up to 24% of the photosynthetically active radiation.[67]

What is the biological function of hydrogenase in such highly aerobic organisms? An induction period with darkness and anaerobiosis appears to be essential.[68] Photosynthetic H_2 production is also enhanced if the concentration of CO_2 is low, suggesting that the hydrogenase pathway is competitive with the normal CO_2-fixing activity of chloroplasts.[69] Because the electron transport via the hydrogenase pathway is still coupled to bioenergy conservation (photosynthetic phosphorylation), hydrogenase may represent an "emergency" strategy in response to adverse environmental conditions, for example, in normally above-ground plant parts subject to water logging and anaerobiosis where essential maintenance and cellular repair reactions can still operate with a continuing source of energy. It naturally follows that reintroduction of O_2 and CO_2 would render such a function of hydrogenase superfluous — and explains the inhibitory effect of O_2 on hydrogenase and the ability of even background levels of O_2 to act as an electron acceptor in direct competition with hydrogenase-mediated H_2 production.[70] Genetic manipulation and directed evolution of algal hydrogenases with reduced or (in the extreme case) no sensitivity to O_2 is, therefore, unavoidable if maximal and sustained rates of photohydrogen production can be achieved in microalgal systems.

Progress has begun to be made on the molecular biology of microalgal hydrogenases, including the isolation and cloning of the two genes for the homologous iron hydrogenases in the green alga *Chlamydomonas reinhardtii*.[71] Random mutagenesis of hydrogenase genes could rapidly isolate novel forms retaining activity in the presence of O_2 and/or improved hydrogenase kinetics. Screening mutants of *Chlamydomonas reinhardtii* has, however, revealed unexpected biochemical complexities, in particular the requirement for functional starch metabolism in H_2 photoproduction.[72] Several changes were indeed identified during the successful improvement of H_2 photoproduction by this alga:[73]

- There was rational selection of mutants with altered electron transport activities with maximized electron flow to hydrogenase.
- Isolates were then screened for increased H_2 production rates, leading to a mutant with reduced cellular O_2 concentrations, thus having less inhibition of hydrogenase activity.
- The most productive mutant also had large starch reserves.

Using the conventional representation of electron transport inside chloroplast membrane systems, the possible interactions of photohydrogen production and other photosynthetic activities can be visualized (figure 7.7).[74] Active endogenous metabolism could remove photoproduced O_2 by using O_2 as the terminal electron

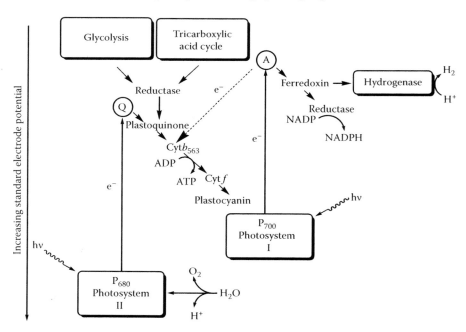

FIGURE 7.7 Z scheme of photosynthetic electron flow in green plants and algae showing links to carbon metabolism and hydrogen production: Q, A, primary electron acceptors in Photosystems II and I, respectively; dotted arrow signifies cyclic electron flow. (Modified from Melis and Happe.[74])

acceptor in mitochondria; the problem of different spatial sites for O_2 production and O_2 utilization still, however, requires a reduced sensitivity of hydrogenase to O_2 as the gas cannot be removed instantaneously — only in "test-tube" systems can O_2-removing chemicals be supplied, for example, as glucose plus glucose oxidase to form gluconic acid by reaction between glucose and O_2.[65]

The obvious implication of the redox chemistry of figure 7.7 is that the normal processes of photosynthesis, involving reduction of NADP for the subsequent reduction of CO_2 to sugars, can be separated in time, with light-dependent O_2 evolution and dark H_2 production or if H_2 production can proceed with inhibited O_2 evolution, that is, "indirect biophotolysis."[64] The particular advantage of this arrangement is that the light-dependent stage can be operated in open pools to maximize productivity at minimal cost. Sustained H_2 production could be achieved over approximately 100 hours after transfer of light-grown C. *reinhardtii* cells to a medium deficient in sulfur; these conditions reversibly inactivated Photosystem II and O_2 evolution, whereas oxidative respiration *in the continued light* depleted O_2, thus inducing hydrogenase.[75] The subsequent H_2 production only occurred in the light and was probably a means of generating energy by Photosystem I activity (figure 7.7). Starch and protein were consumed while a small amount of acetic acid was accumulated.

This was the first reported account of a single-organism, two-stage photobiological production process for H_2, although a prototype light/dark device using three stages (one light and two dark) with a marine microalga and a marine photosynthetic bacterium was tested in Japan in the 1990s.[76] How much H_2 could a microalgae-based approach produce? With C. *reinhardtii* cells given an average irradiance of 50 mol photons/m²/day (a possible value in temperate latitudes, although highly variable on a day-to-day and seasonal basis), the maximum H_2 production would be 20 g/m²/day, equivalent to 80 kg/acre/day (or 200 kg/hectare/day) — but the likely value, allowing for low yields of H_2 production measured under laboratory conditions, the far from complete absorption of incident light, and other factors, is only 10% of this.[74] In a further refinement of this approach, the sulfate-limited microalgae were shown to form a stable process for 4000 hours: two automated photobioreactors were coupled to first grow the cells aerobically before being continuously delivered to the second, anaerobic stage.[77] Until all the biological and physical limitations can be overcome, however, large infrastructural investments in high and predictable sunlight regions would be required, and the capital costs for such solar power stations would be high, but the technical complexity may only approximate that of installing extensive photovoltaic cell banks for the direct production of electricity, an option vigorously advocated by critics of biofuels programs.[78,79]

Cyanobacteria ("blue-green algae") are prokaryotes but share with higher photosynthetic organisms the basic electron transport chains of Photosystems I and II (figure 7.7). The molecular biology and biochemistry of hydrogenases in cyanobacteria is well understood, the complete genomes of several such organisms have been sequenced, and interspecies gene transfer is established.[80] Much of the research has unfortunately concentrated on nitrogenase as a source of H_2, but many cyanobacteria contain hydrogenases catalyzing the reversible formation of H_2, a route with far more biotechnological potential for commercial H_2 generation, and protein engineering has begun to reduce the O_2 sensitivities of cyanobacterial hydrogenases.[81] The physiological role of hydrogenase in cyanobacteria has been debated for decades; recent results suggest some kind of safety valve function under low O_2 condition when a light-to-dark transition occurs, and inactivating quinol oxidase (an enzyme with a similar hypothetical function) and nitrate reductase (a third electron "sink") increase photohydrogen evolution rates.[82]

Thermophilic cyanobacteria are known to be capable of H_2 photoproduction at up to 50°C in open-air cultures maintained for more than 3 weeks.[83,84] If a fermentable carbon source is supplied, a sustained photoevolution of H_2 can be achieved, with photolysis of water (a Photosystem II activity — see figure 7.7), whereas carbohydrate-mediated reduction of the plastoquinone pool continues independently.[85] This H_2 production system has been termed "photofermentation"; in principle, relatively little light energy is required to drive the reaction because of the energy input from the fermentable substrate.[64] The green alga C. reinhardtii shares this pattern of metabolism with cyanobacteria, behaving under photofermentative conditions much like an enteric bacterium such as E. coli, exhibiting pyruvate formate lyase activity and accumulating formate, ethanol, acetate, CO_2, and H_2 as well as glycerol and lactate.[86]

The overlapping molecular structures of cyanobacteria and nonphotosynthetic bacteria were exemplified by the coupling (both in vivo and in vitro) between cyanobacterial photosynthetic electron transport components with clostridial hydrogenase; even more remarkable was the expression in a Synechococcus strain of the hydrogenase gene from C. pasteurianum, the enzyme being active in the cyanobacterial host.[87] As a possible pointer to the future of designing an improved photosynthetic organism for H_2 production, the "hard wiring" of a bacterial hydrogenase with a peripheral subunit of a Photosystem I subunit of the cyanobacterium Thermosynechococcus elongatus resulted in a fusion protein that could associate functionally with the rest of the Photosystem I complex in the cyanobacterium and display light-driven H_2 evolution.[88]

Photosynthetic bacteria differ from other photosynthetic organisms in using bacteriochlorophyll rather than chlorophyll as the central pigment for light-induced electron transport; they also lack Photosystem II (figure 7.7) and perform anoxygenic photosynthesis and require electron donors more reduced than water, including reduced sulfur and organic compounds.[89] Being able to fix gaseous nitrogen, the photosynthetic bacteria contain nitrogenase in addition to hydrogenase and occur globally in widely different habitats, including fresh, brackish, and sea waters, hot sulfur springs, paddy fields, wastewaters, and even in Antarctica. Hydrogen can be photoproduced in the presence of an organic substrate, sometimes with high efficiencies deduced from the maximum theoretical H_2 production on a molar basis (table 7.3). Both free and immobilized cells have been used to produce H_2 during extended periods (table 7.4). All photosynthetic bacteria can use H_2 as a reductant for the fixation of CO_2 into organic carbon, and considerable reengineering of the molecular biochemistry is unavoidable if the cells are to be evolved into biological H_2 producers.[27] Photofermentations are also known, and Rhodobacter capsulatus has been used as a test organism to evaluate photobioreactor designs potentially reaching 3.7% conversion efficiency of absorbed light energy into H_2 fuel energy.[90]

Photosynthetic bacteria may have the additional capability of catalyzing the "water shift" reaction (chapter 6, section 6.2.1):

$$CO + H_2O \rightarrow CO_2 + H_2$$

TABLE 7.3
Photofermentative Hydrogen Production by Photosynthetic Bacteria

Species	Substrate	Stoichiometry	Conversion efficiency (%)
Rhodobacter capsulatus	Acetate	$C_2H_4O_2 + 2H_2O = 2CO_2 + 4H_2$	57–100
Rhodobacter capsulatus	Butyrate	$C_4H_8O_2 + 6H_2O = 4CO_2 + 10H_2$	23–80
Rhodopseudomonas sp.7	Ethanol	$C_2H_5O + 3H_2O = 2CO_2 + 6H_2$	45
Rhodobacter sphaeroides	Glucose	$C_6H_{12}O_6 + 6H_2O = 6CO_2 + 12H_2$	99
Rhodomicrobium vannielii	Lactate	$C_3H_6O_3 + 3H_2O = 3CO_2 + 6H_2$	78–100
Rhodobacter sphaeroides	Malate	$C_4H_6O_5 + 3H_2O = 4CO_2 + 6H_2$	57–100
Rhodopseudomonas sp.7	Propanol	$C_3H_8O + 5H_2O = 3CO_2 + 9H_2$	36
Rhodopseudomonas palustris	Pyruvate	$C_3H_4O_3 + 3H_2O = 3CO_2 + 5H_2$	52
Rhodobacter capsulatus	Succinate	$C_4H_6O_4 + 4H_2O = 4CO_2 + 7H_2$	72

Source: Data from Sasikala et al.[89]

TABLE 7.4
Photofermentative Hydrogen Production by Immobilized Cells

Species	Electron donor	Immobilization method	H_2 Evolution rate (ml/hr/g dry weight)
Rhodobacter sphaeroides	Malate	Alginate	16.2
Rhodospirillum rubrum	Acetate	Alginate	9.0
Rhodospirillum rubrum	Lactate	Alginate	30.7
Rhodopseudomonas sp.7	Starch	Alginate	80.0
Rhodospirillum rubrum	Lactate	Agar beads	57.3
Rhodospirillum molischianum	Wastewater	Agar blocks	139.0
Rhodopseudomonas palustris	Malate	Agar blocks	41.0
Rhodobacter capsulatus	Lactate	Carrageenan	111.0
Rhodospirillum rubrum	Acetate	Agar cellulose fiber	15.6
Rhodospirillum rubrum	Lactate	Agarose	22.9
Rhodospirillum rubrum	Lactate	Pectin	21.0

Source: Data from Sasikala et al.[89]

but, unlike the thermochemical process, at moderate temperatures and without multiple passages of gases through the reaction vessel.[64] A continuous process was devised for *Rhodospirillum rubrum* with illumination supplied by a tungsten light.[91] With biomass as the substrate for gasification, a substantially (if not entirely) biological process for H_2 production can be envisaged. A National Renewable Energy Laboratory report concluded that a biological reactor would be larger and slower but could achieve comparable efficiencies of heat recovery in integrated systems; the most likely niche market use would occur in facilities where the water gas shift was an option occasionally (but gainfully) employed but where the start-up time for a thermal catalytic step would be undesirable.[92]

Patents describing processes for H_2 production using photosynthetic microbes cover at least 23 years, and include topics as diverse as their basic biology, molecular and enzymic components, and analytical methodologies. In the last 5 years, however, several patents have also appeared focusing on biohydrogen production by fermentative organisms (table 7.5).

TABLE 7.5
U.S. Patents Covering Photosynthetic and Fermentative H_2 Production

Date	Title	Assignee/Inventor(s)	Patent
Photoproduction			
10/4/1984	Method for producing hydrogen and oxygen by use of algae	U.S. Department of Energy	US 4,442,211
1/30/1985	Process for producing hydrogen by alga in alternating light/dark cycle …	Miura et al.	US 4,532,210
3/31/1992	Measurement of gas production of algal clones	Gas Research Institute, Chicago, IL	US 5,100,781
2/16/1999	Process for selection of oxygen-tolerant algal mutants that produce H_2 under aerobic conditions	Midwest Research Institute, Kansas City, MO	US 5,871,952
6/25/2002	Molecular hydrogen production by direct electron transfer	McTavish, H.	US 6,410,258
1/24/2006	Hydrogen production using hydrogenase-containing oxygenic photosynthetic organisms	Midwest Research Institute, Kansas City, MO	US 6,989,252
2/13/2007	Modulation of sulfate permease for photosynthetic hydrogen production	University of California, Oakland, CA	US 7,176,005
6/12/2007	Fluorescence techniques for on-line monitoring of state of hydrogen-producing microorganisms	Midwest Research Institute, Kansas City, MO	US 7,229,785
Fermentation			
9/10/2002	System for rapid biohydrogen phentotypic screening of microorganisms using chemochromic sensor	Midwest Research Institute, Kansas City, MO	US 6,448,068
3/1/2005	Method of producing hydrogen gas by using hydrogen bacteria	Japan Science and Technology Corporation	US 6,860,996
5/3/2005	Method and apparatus for hydrogen production from organic wastes and manure	Gas Technology Institute, Des Plaines, IL	US 6,887,692
8/1/2006	Method for hydrogen production from organic wastes using a two-phase bioreactor system	Gas Technology Institute, Des Plaines, IL	US 7,083,956
6/19/2007	Process for enhancing anaerobic biohydrogen production	Feng Chia University, Taiwan	US 7,232,669

7.3.4 EMERGENCE OF THE HYDROGEN ECONOMY

It is highly doubtful that industrial biohydrogen processes will be the entry points for the widespread use of H_2 as a fuel. Despite a number of major national and international initiatives and research programs, fossil fuel-based and alternative energy processes are widely considered to be essential before 2030, or even as late as 2050. Of these nonbiological technologies, H_2 production by coal gasification is clearly the worst alternative in terms of fossil energy use and greenhouse gas emissions (figure 7.8).[93] Nevertheless, gasification and electricity-powered electrolytic routes to H_2 offer the promise of production costs rivaling or even less than those of conventional gasoline for use in fuel cell-powered vehicles with an anticipated fuel economy approximately twice that of conventional internal combustion engines (figure 7.9). As a carbonless production route, the internationally accepted "route map" is the sulfur-iodine cycle based on the three reactions:

$$H_2SO_4 \rightarrow SO_2 + H_2O + \frac{1}{2}O_2 \ [850°C]$$

$$I_2 + SO_2 + 2H_2O \rightarrow 2HI + H_2SO_4 \ [120°C]$$

$$2HI \rightarrow H_2 + I_2 \ [220–330°C]$$

The high temperatures required for the first reaction have prompted research programs investigating solar-furnace splitting of sulfuric acid, for example, in the five-nation project HYTHEC (HYdrogen THErmochemical Cycles), involving research teams from France, Germany, Spain, Italy, and the United Kingdom in the "search for a long-term massive hydrogen production route" that would be sustainable and independent of fossil fuel reserves (www.hythec.org).

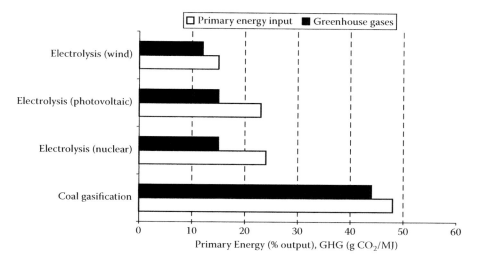

FIGURE 7.8 Alternative nonbiological production routes for H_2: primary (fossil) energy inputs and associated greenhouse gas emissions. (Data from Mason.[93])

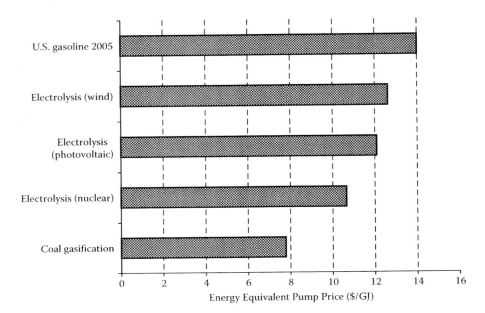

FIGURE 7.9 Predicted retail (pump) prices for H_2 and gasoline on an equal energy basis and assuming fuel cell vehicles are more efficient than conventional internal combustion engine vehicles. (Data from Mason.[93])

The enormous added bonus of biohydrogen would be the use of other highly renewable resources as well as avoiding undue reliance on nuclear technology (an alternative means of providing the power for very-high-temperature reactors), highly persuasive rationales for the continuing interest in biohydrogen energy in the twenty-first century as exemplified by the International Energy Agency's Hydrogen Implementing Agreement whose Task 15 involves Canada, Japan, Norway, Sweden, the Netherlands, the United Kingdom, and the United States in four R&D areas:[94]

- Light-driven H_2 production by microalgae
- Maximizing photosynthetic efficiencies
- H_2 fermentations
- Improving photobioreactors for H_2 production

In Japan, all the major automobile manufacturers are active in the development of fuel cell-powered vehicles: Toyota, Honda, Nissan, Mazda, Daihatsu, Mitsubishi, and Suzuki.[94]

In Europe, HYVOLUTION is a program with partners from 11 European Union countries, Russia, and Turkey, funded by approximately $9.5 million, and aiming to establish decentralized H_2 production from biomass, maximize the number and diversity of H_2 production routes, and increase energy security of supply at both local and regional levels (www. biohydrogen.nl/hyvolution). The approach is based on combined bioprocesses with thermophilic and phototrophic bacteria to provide H_2

production with high efficiencies in small-scale, cost-effective industries to reduce H_2 production costs to \$10/GJ by 2020 — with production costs in the \$5–7/GJ range, biomass-derived H_2 would be highly competitive with conventional fuels or biofuels.[95] Principal subobjectives for HYVOLUTION include the following:

- Pretreatment technologies to optimize biodegradation of energy crops
- Maximized conversion of biomass to H_2
- Assessment of installations for optimal gas cleaning
- Minimum energy demand and maximal product output
- Identification of market opportunities for a broad feedstock range

Based in Sweden, the SOLAR H program links molecular genetics and biomimetic chemistry to explore radically innovative approaches to renewable H_2 production, including artificial photosynthesis in manmade systems (www.fotmol.uu.se). Japanese research has already explored aspects of this interface between industrial chemistry and photobiology, for example, incorporating an artificial chlorophyll (with a zinc ion replacing the green plant choice of magnesium) in a laboratory system with sucrose, the enzymes invertase and glucose oxidase, together with a platinum colloid to photoevolve H_2.[96]

The size of the investment required to bring the hydrogen economy to fruition remains, however, daunting: from several billion to a few trillion dollars for several decades.[97] The International Energy Agency also estimates that H_2 production costs must be reduced by three- to tenfold and fuel cell costs by ten- to fiftyfold. Stationary fuel cells could represent 2–3% of global generating capacity by 2050, and total H_2 use could reach 15.7 EJ by then. There are some appreciated risks in these prognostications, with governments holding back from imposing fuel taxes on H_2 but imposing high CO_2 penalties being strongly positive for increasing the possible use of H_2, whereas high fuel cell prices for automobiles will be equally negative (figure 7.10).

7.4 MICROBIAL FUEL CELLS: ELIMINATING THE MIDDLEMEN OF ENERGY CARRIERS

Hydrogen ions (protons, H^+) can accept reducing equivalents (conventionally represented as electrons, e^-) generated either photosynthetically or by the oxidation of organic and inorganic substrates inside microbial cells:

$$2e^- + 2H^+ \rightarrow H_2$$

The terminal electron donor (e.g., reduced ferredoxin) could donate electrons to the anode of a battery. Protons could then, in the presence of O_2, complete the electric circuit at the cathode by the reaction:

$$O_2 + 4e^- + 4H^+ \rightarrow 2H_2O$$

thus forming a highly environmentally friendly source of electric power (a battery), fueled by microbial metabolic activity. That, in essence, is the definition of a microbial fuel cell (MFC).[98–100]

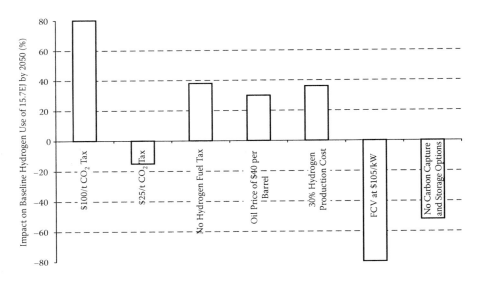

FIGURE 7.10 Sensitivity analysis of predicted H_2 use by 2050. (Data from *Prospects for Hydrogen and Fuel Cells.*[97])

At its simplest, an MFC is a dual-chamber device with an electrolyte, a cation exchange membrane to separate anodic and cathodic compartments, a supply of O_2 for the cathode, and an optional sparge of inert gas for the anode (figure 7.11). The transfer of electrons to the anode may be either direct (via unknown terminal electron donors on the cell surface) or employing redox-active "mediators" that can be reduced by the cells and reoxidized at the anode (e.g., neutral red reduced by hydrogenase).[101,102] A wide spectrum of microbial species have been tested in MFC environments, usually anaerobes or facultative anaerobes chosen to function in the O_2-deficient anode compartment. Recent examples include

- Immobilized cells of the yeast *Hansenula anomala*[103]
- A mixed microbial community of *Proteobacterium, Azoarcus,* and *Desulfuromonas* species with ethanol as the fuel source[104]
- *Desulfitobacterium hafniense* with humic acids or the humate analog anthraquinone-2,6-disulfonate added as an electron-carrying mediator with formic acid, H_2, lactate, pyruvate, or ethanol as the fuel[105]
- *E. coli* in MFCs as power sources for implantable electronic devices[106]

The first use of the term "microbial fuel cell" appears to date from the early 1960s in studies of hydrocarbon-metabolizing *Nocardia* bacteria by research scientists of the Mobil Oil Company, but the basic concepts may date back 30 or even 50 years earlier.[107] Developments of MFCs as commercial and industrial functionalities are methods of water treatment and as power sources for environmental sensors; the power produced by these systems is currently limited, primarily by high internal (ohmic) resistance, but improvements in system architecture might result in power generation that is more dependent on the bioenergetic capabilities of the

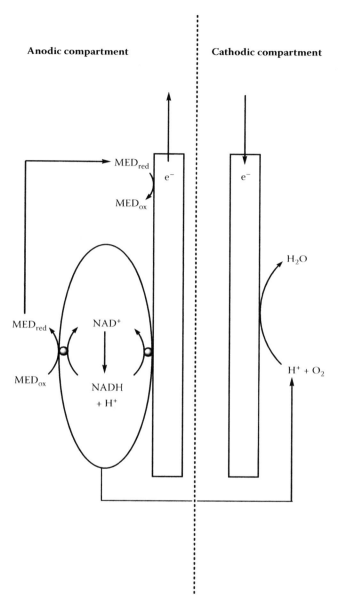

Anodic compartment

Cathodic compartment

FIGURE 7.11 Redox reactions occurring in an MFC: MED is the soluble mediator reduced by the microbial terminal electron donor at the microbial cell surface.

microorganisms.[108] They are close to devices mobilized for the sets of science fiction films that call on "bioelectricity" in all its many conceptual forms (e.g., in the *Matrix* trilogy) and can easily be imagined in self-reliant ecosystems in deep space travel (*Silent Running*). But are they meaningful additions to the biofuels armory on planet Earth?

Creating a scalable architecture for MFCs is essential to provide large surface areas for oxygen reduction at the cathode and bacteria growth on the anode; a tubular ultrafiltration membrane with a conductive graphite coating and a nonprecious metal catalyst can be used to produce power in an MFC and is a promising architecture that is intrinsically scalable for creating larger systems.[109] For the anodes, highly conductive noncorrosive materials are needed that have a high specific surface area (i.e., surface area per volume) and an open structure to avoid biofouling; graphite fiber brush anodes have high surface areas and a porous structure can produce high power densities, qualities that make them ideal for scaling up MFC systems.[110] The technology has long existed for very-large-scale microbial fermentations for bulk chemicals (e.g., citric acid and amino acids), with stirred tank volumes greater than 500,000 l being widely used; such structures could, in principle, be the "anodic" compartments of stationary MFC generators. Because rumen bacteria have been shown to generate electricity from cellulosic materials, potentially immense substrate supplies could be available for MFC arrays.[111] Even greater flexibility can be designed, for example, coproducing H_2 and ethanol production from glycerol-containing wastes discharged from a biodiesel fuel production plant with *Enterobacter aerogenes* in bioelectrochemical cells with thionine as the exogenous electron transfer mediator.[112]

7.5   BIOFUELS OR A BIOBASED COMMODITY CHEMICAL INDUSTRY?

Technical advances will define to what extent agricultural and wastewater resources can be transformed into quantities of electricity and liquid and gaseous fuels as energy carriers and to what extent these will contribute to local and national power demands, but certain industries could benefit greatly from considering their wastes as sources of bioenergy, either for immediate use or as a saleable commodity. To illustrate this, a bilateral study between Japan and Malaysia of the waste streams from the Malaysian palm oil industry (42.5 million tons in 2001) demonstrated that, with *Enterobacter* and *Clostridium* strains evolving H_2 or producing acetone, butanol, and ethanol (chapter 6, section 6.3.3), more than 62,000 tons of oil equivalent in total energy could be generated.[113]

Biohydrogen (however it is produced by living cells) has the potential to marginalize all other biofuels, from ethanol and biodiesel to all presently contemplated "bio" options for mass transportation — if, that is, onboard fuel cells achieve their cost and vehicle range targets. In one possible future, mariculture units growing cyanobacteria in coastal waters will be the sources of renewable energy from water and sunlight, providing the H_2 as the most environmentally friendly energy carrier.[114] In another (Chinese) model, cultures of *Propionibacterium*, *Clostridium*, and *Bacteriodes* species relentlessly ferment whatever substrates can be made available to them to form H_2 and ethanol and/or mixtures of acids, quite independently of sunlight, and in processes that can be managed by simply adjusting the pH to determine the product stream.[115]

Even the hydrogen economy has its critics, however, and in another possible future, hybrid gasoline-electric vehicles will dominate the highways by as early as 2020, reducing gasoline and greenhouse gas emissions by 30–50%, with no major investments in fuel infrastructure; they may even be dually hybrid, being able to run on gasoline-biofuel blends, traveling 500 miles on a gallon of gasoline mixed with five gallons of cellulosic ethanol.[116]

When, in 1975, Ballard occupied a derelict motel in southern Arizona at the beginning of the quest to develop a viable technology to power an electric vehicle and so reduce dependency on fossil fuels, optimism may have been tempered with the realization that a long and uncertain journey had just commenced, but optimism was certainly rampant by the late 1990s.[117] As an "entirely unauthorized" biography of Ballard (the early pacesetters in fuel cell technology for transportation) noted, the starting pistol for the race to develop a marketable fuel cell-powered automobile was fired in April 1997 when Daimler-Benz paid nearly $200 million for a 25% share in Ballard and committed itself to invest a further $300 million.[118] A decade later, progress continues worldwide, but projections about who crosses the winning line first and the eventual date of mass use of such ecofriendly vehicles tends to relentlessly slip back into "decades away" — inventors have, necessarily, to be ever hopeful about the futures of their brain children but underestimate production costs and the full sequence of engineering and other events that lead to commercialization.

Even if developments in the next two to three decades render both fuel ethanol and biodiesel obsolete, however, the many advances made in the biotechnology of the bioproduction of biofuels will not prove to be wasted. To return once more to the preface, biomass as the main supply of chemical feedstocks may be unavoidable in the twenty-first century as increased demand for gasoline in the rapidly developing economies of Asia and South America applies a price tourniquet to petrochemicals, particularly rapidly if oil reserves prove smaller than estimated or if an accelerating CO_2-dominated climate change forces political action to restrain CO_2-producing industries.[118] Can agricultural sources ever be justified as substrates for the production of transportation fuels? Or solely for automobile fuels?

Or can biofuels gain global approval as part of the mix of products emanating from biorefineries, in a flexible output that could replace petrochemicals, provide biofuels for blends according to market demands, and provide fuels for multiple types of fuel cells? The next and final chapter explores how the biorefinery concept emerged in the 1990s to be the beacon of a radically different vision of how biotechnology and commodity chemical production can merge in another blueprint for a sustainable mobile and industrial society.

REFERENCES

1. Chisti, Y., Biodiesel from microalgae, *Biotechnol. Adv.*, 25, 294, 2007.
2. Sheehan, J. et al., *A Look Back at the U.S. Department of Energy's Aquatic Species Program, Biodiesel from Algae*, NREL/TP-580-24190, National Renewable Energy Laboratory, Golden, CO, July 1998.
3. Roessler, P.G. et al., Characteristics of the gene that encodes acetyl-CoA carboxylase in the diatom *Cyclotella cryptica*, *Ann. N.Y. Acad. Sci.*, 721, 250, 1994.
4. Miao, X. and Wu, Q., High yield bio-oil production from fast pyrolysis by metabolic controlling of *Chlorella protothecoides*, *J. Biotechnol.*, 110, 85, 2004.
5. Xu, H., Miao, X., and Wu, Q., High quality biodiesel production from a microalga *Chlorella protothecoides* by heterotrophic growth in fermenters, *J. Biotechnol.*, 126, 499, 2006.

6. Miao, X. and Wu, Q., Biodiesel production from heterotrophic microalgal oil, *Bioresour. Technol.*, 97, 841, 2006.
7. Li, X., Xu, H., and Wu, Q., Large-scale biodiesel production from microalgal *Chlorella protothecoides* through heterotrophic cultivation in fermenters, *Biotechnol. Bioeng.*, 98, 764, 2007.
8. Schubert, C., Can biofuels finally take center stage?, *Nat. Biotechnol.*, 24, 777, 2006.
9. Kalscheuer, R., Stölting, T., and Steinbüchel, A., Microdiesel: *Escherichia coli* engineered for fuel production, *Microbiology*, 152, 2529, 2006.
10. Alvarez, H.M. and Steinbüchel, A., Triacylglycerols in prokaryotic microorganisms, *Appl. Microbiol. Biotechnol.*, 60, 367, 2002.
11. Ingram, L.O. et al., Genetic engineering of ethanol production in *Escherichia coli*, *Appl. Environ. Microbiol.*, 53, 2420, 1987.
12. Román-Leshkov, Y. et al., Production of dimethylfuran for liquid fuels from biomass-derived carbohydrates, *Nature*, 447, 982, 2007.
13. Schmidt, L.D. and Dauenhauer, P.J., Hybrid routes to biofuels, *Nature*, 447, 914, 2007.
14. Bhosale, S.H., Rao, M.B., and Deshpande, V.V., Molecular and industrial aspects of glucose isomerase, *Microbiol. Rev.*, 60, 280, 1996.
15. Bandish, R.K. et al., Glucose-to-fructose conversion at high temperature with xylose (glucose) isomerases from *Streptomyces murinus* and two hyperthermophilic *Thermotoga* species, *Biotechnol. Bioeng.*, 80, 185, 2002.
16. *World Energy Outlook*, International Energy Agency, Paris, 2006, chap. 10.
17. Hoogers, G., *Introduction*, in Hoogers, G. (Ed.), *Fuel Cell Technology Handbook*, CRC Press LLC, Boca Raton, 2003, chap. 1-1.
18. Takahashi, T., *Water Electrolysis*, in Ohta, T. (Ed.), *Solar-Hydrogen Energy Systems*, Pergamon Press, Oxford, 1979, chap. 3.
19. *The Hydrogen Economy: A Non-Technical Review*, United Nations Environment Programme, Paris, 2006.
20. Lettinga, G., Sustainable integrated biological wastewater treatment, *Water Sci. Technol.*, 33, 85, 1996.
21. Claassen, P.A.M. et al., Utilisation of biomass for the supply of energy carriers, *Appl. Microbiol. Biotechnol.*, 52, 741, 1999.
22. Torry-Smith, M., Sommer, P., and Ahring, B.K., Purification of bioethanol effluent in an USAB reactor system with simultaneous biogas formation, *Biotechnol. Bioeng.*, 84, 7, 2003.
23. Thimann, K.V., *The Life of Bacteria*, 2nd ed., The Macmillan Company, New York, 1963, chap. 14.
24. Thimann, K.V., *The Life of Bacteria*, 2nd ed., The Macmillan Company, New York, 1963, chap. 15.
25. Thimann, K.V., *The Life of Bacteria*, 2nd ed., The Macmillan Company, New York, 1963, chap. 17.
26. Moat, A.G. and Foster, J.W., *Microbial Physiology*, 3rd ed., John Wiley-Liss, Inc., New York, 1995, chap. 7.
27. Nandi, R. and Sengupta, S., Microbial production of hydrogen: an overview, *Crit. Rev. Microbiol.*, 24, 61, 1998.
28. Vignais, P.M., Billoud, B., and Meyer, J., Classification and phylogeny of hydrogenases, *FEMS Microbiol. Rev.*, 25, 455, 2001.
29. Soni, B.K., Soucaille, P., and Goma, G., Continuous acetone butanol fermentation: influence of vitamins on the metabolic activity of *Clostridium acetobutylicum*, *Appl. Microbiol. Biotechnol.*, 27, 1, 1987.
30. Gorwa, M.F., Croux, C., and Soucaille, P., Molecular characterization and transcriptional analysis of the putative hydrogenase gene of *Clostridium acetobutylicum* ATCC 824, *J. Bacteriol.*, 178, 2668, 1996.

31. Nölling, J. et al., Genome sequence and comparative analysis of the solvent-producing bacterium *Clostridium acetobutylicum*. *J. Bacteriol.*, 183, 4823, 2001.
32. Girbal, L. et al., Homologous and heterologous overexpression in *Clostridium aceto-butylicum* and characterization of purified clostridial and algal Fe-only hydrogenases with high specific activities, *Appl. Environ. Microbiol.*, 71, 2777, 2005.
33. Darensbourg, M.Y. et al., The organometallic active site of [Fe]hydrogenase: models and entatic states, *Proc. Natl. Acad. Sci., USA*, 100, 3683, 2003.
34. Bagley, K.A. et al., Infrared-detectable groups sense changes in charge density on the nickel center in hydrogenase from *Chromatium vinosum*, *Biochemistry*, 34, 5527, 1995.
35. Volbeda, A. et al., Structure of the [NiFe] hydrogenase active site: evidence for biologically uncommon Fe ligands, *J. Am. Chem. Soc.*, 118, 12989, 1996.
36. Lenz, O. et al., Requirements for heterologous production of a complex metalloenzyme: the membrane-bound [NiFe] hydrogenase, *J. Bacteriol.*, 187, 6590, 2005.
37. Zhu, W. et al., Modulation of the electronic structure and the Ni–Fe distance in heterobimetallic models for the active site in [NiFe]hydrogenase, *Proc. Natl. Acad. Sci., USA*, 102, 18280, 2005.
38. Armstrong, F.A., Hydrogenases: active site puzzles and progress, *Curr. Opin. Chem. Biol.*, 8, 133, 2004.
39. Ösz, J. et al., Theoretical calculations on hydrogenase kinetics: explanation of the lag phase and the enzyme concentration dependence of the activity of hydrogenase uptake, *Biophys. J.*, 89, 1957, 2005.
40. Tye, J.W. et al., Better than platinum? Fuel cells energized by enzymes, *Proc. Natl. Acad. Sci, USA*, 102, 16911, 2005.
41. Woodward, J. et al., Enzymatic production of biohydrogen, *Nature*, 405, 1014, 2000.
42. Yoshida, A. et al., Enhanced hydrogen production from formic acid by formate hydrogen lyase-overexpressing *Escherichia coli* strains, *Appl. Environ. Microbiol.*, 71, 6762, 2005.
43. Yoshida, A. et al., Efficient induction of formate hydrogen lyase of aerobically grown *Escherichia coli* in a three-step biohydrogen production process, *Appl. Microbiol. Biotechnol.*, 74, 754, 2007.
44. Yoshida, A. et al., Enhanced hydrogen production from glucose using *ldh*- and *frd*-inactivated *Escherichia coli* strains, *Appl. Microbiol. Biotechnol.*, 73, 67, 2006.
45. Minnan, L. et al., Isolation and characterization of a high H_2-producing strain *Klebsiella oxytoca* HP1 from a hot spring, *Res. Microbiol.*, 156, 76, 2005.
46. de Vrije, T. et al., Glycolytic pathway and hydrogen yield studies of the extreme thermophile *Caldicellulosiruptor saccharolyticus*, *Appl. Microbiol. Biotechnol.*, 74, 1358, 2007.
47. Kadar, Z. et al., Hydrogen production from paper sludge hydrolysate, *Appl. Biochem. Biotechnol.*, 105–108, 557, 2003.
48. Kadar, Z. et al., Yields from glucose, xylose, and paper sludge hydrolysate during hydrogen production by the extreme thermophile *Caldicellulosiruptor saccharolyticus*, *Appl. Biochem. Biotechnol.*, 113–116, 497, 2004.
49. Oh, S.-E. et al., Biological hydrogen production using a membrane bioreactor, *Biotechnol. Bioeng.*, 87, 119, 2004.
50. Liu, H. and Fang, H.P., Hydrogen production from wastewater by acidogenic granular sludge, *Water Sci. Technol.*, 47, 153, 2002.
51. Cheong, D.Y., Hansen, C.L., and Stevens, D.K., Production of bio-hydrogen by mesophilic anaerobic fermentation in an acid-phase sequencing batch bioreactor, *Biotechnol. Bioeng.*, 96, 421, 2007.
52. Kawagoshi, Y. et al., Effect of inoculum conditioning on hydrogen fermentation and pH effect on bacterial community relevant to hydrogen production, *J. Biosci. Bioeng.*, 100, 524, 2005.

53. Kyazze, G. et al., Performance characteristics of a two-stage dark fermentative system producing hydrogen and methane continuously, *Biotechnol. Bioeng.*, 97, 759, 2007.

54. Iyer, P. et al., H_2-producing bacterial communities from a heat-treated soil inoculum, *Appl. Microbiol. Biotechnol.*, 66, 166, 2004.

55. Ren, N. et al., Microbial community structure of ethanol type fermentation in biohydrogen production, *Environ. Microbiol.*, 9, 1112, 2007.

56. Koskinen, P.E.P., Kaksonen, A.H., and Puhakka, J.A., The relationship between instability of H_2 production and compositions of bacterial communities within a dark fermentation fluidized-bed reactor, *Biotechnol. Bioeng.*, 97, 742, 2007.

57. Zhang, H., Bruns, M.A., and Logan, B.E., Biological hydrogen production by *Clostridium acetobutylicum* in an unsaturated flow reactor, *Water Res.*, 40, 728, 2006.

58. Van Ginkel, S.W. and Logan, B.E., Increased biological hydrogen production with reduced organic loading, *Water Res.*, 39, 3819, 2005.

59. Park, W. et al., Removal of headspace CO_2 increases biological hydrogen production, *Environ. Sci. Technol.*, 39, 4416, 2005.

60. Salerno, M.B. et al., Inhibition of biohydrogen production by ammonia, *Water Res.*, 40, 1167, 2006.

61. Van Ginkel, S.W., Oh, S.-E., and Logan, B.E., Biohydrogen gas production from food processing and domestic wastewaters, *Int. J. Hyd. Energy*, 30, 1535, 2005.

62. Ito, T. et al., Hydrogen and ethanol production from glycerol-containing wastes discharged after biodiesel manufacturing process, *J. Biosci. Bioeng.*, 100, 260, 2005.

63. Bagi, Z. et al., Biotechnological intensification of biogas production, *Appl. Microbiol. Biotechnol.*, 76, 473, 2007.

64. Benemann, J., Hydrogen biotechnology: progress and prospects, *Nature Biotechnol.*, 14, 1101, 1996.

65. Benemann, J. et al., Hydrogen evolution by a chloroplast-ferredoxin-hydrogenase system, *Proc. Natl. Acad. Sci., USA*, 70, 2317, 1973.

66. Gaffron, H. and Rubin, J., Fermentative and photochemical production of hydrogen in algae, *J. Gen. Physiol.*, 26, 219, 1942.

67. Greenbaum, E., Energetic efficiency of hydrogen photoevolution by algal water splitting, *Biophys. J.*, 54, 365, 1988.

68. Roessler, P.G. and Lien, S., Activation and *de novo* synthesis of hydrogenase in *Chlamydomonas*, *Plant Physiol.*, 76, 1086, 1984.

69. Kessler, E., Effect of anaerobiosis on photosynthetic reactions and nitrogen metabolism in the green alga *Chlamydomonas reinhardtii*, *Arch. Microbiol.*, 93, 91, 1973.

70. Lee, J.W. and Greenbaum, E., A new oxygen sensitivity and its potential application in photosynthetic H_2 production, *Appl. Biochem. Biotechnol.*, 105–108, 303, 2003.

71. Melis, A., Seibert, M., and Happe, T., Genomics of green algal hydrogen research, *Photosynth. Res.*, 82, 277, 2004.

72. Posewitz, M.C. et al., Hydrogen photoproduction is attenuated by disruption of an isoamylase gene in *Chlamydomonas reinhardtii*, *Plant Cell*, 16, 2151, 2004.

73. Kruse, O. et al., Improved photobiological H_2 production in engineered green algal cells, *J. Biol. Chem.*, 280, 34170, 2005.

74. Melis, A. and Happe, T., Hydrogen production. Green algae as a source of energy, *Plant Physiol.*, 127, 740, 2001.

75. Melis, A. et al., Sustained photobiological hydrogen gas production upon reversible inactivation of oxygen evolution in the green alga *Chlamydomonas reinhardtii*, *Plant Physiol.*, 122, 127, 2000.

76. Akano, T. et al., Hydrogen production by photosynthetic microorganisms, *Appl. Biochem. Biotechnol.*, 57–58, 677, 1996.

77. Fedorov, A.S. et al., Continuous hydrogen production by *Chlamydomonas reinhardtii* using a novel two-stage, sulfate-limited chemostat system, *Appl. Biochem. Biotechnol.*, 121–124, 403, 2005.

78. Patzek, T.W., Thermodynamics of the corn-ethanol biofuel cycle, *Crit. Rev. Plant. Sci.*, 23, 519, 2004.

79. Patzek, T.W. and Pimentel, D., Thermodynamics of energy production from biomass, *Crit. Rev. Plant. Sci.*, 24, 329, 2006.

80. Tamagnini, P. et al., Hydrogenases and hydrogen metabolism of cyanobacteria, *Microbiol. Mol. Biol. Rev.*, 66, 1, 2002.

81. McTavish, H., Sayavedra-Soto, L.A., and Arp, D.J. Substitution of *Azotobacter vinelandii* hydrogenase small-subunit cysteines by serines can create insensitivity to inhibition by O_2 and preferentially damages H_2 oxidation over H_2 evolution, *J. Bacteriol.*, 177, 3960, 1995.

82. Gutthann, F. et al., Inhibition of respiration and nitrate assimilation enhances photohydrogen evolution under low oxygen concentrations in *Synechocystis* sp. PCC 6803, *Biochim. Biophys. Acta*, 1767, 161, 2007.

83. Miyamoto, K., Hallenbeck, P.C., and Benemann, J.R., Nitrogen fixation by thermophilic blue-green algae (cyanobacteria): temperature characteristics and potential use in biophotolysis, *Appl. Environ. Microbiol.*, 37, 454, 1979.

84. Miyamoto, K., Hallenbeck, P.C., and Benemann, J.R., Hydrogen production by the thermophilic alga *Mastigocladus laminosus*: effects of nitrogen, temperature, and inhibition of photosynthesis, *Appl. Environ. Microbiol.*, 38, 440, 1979.

85. Cournac, L. et al., Sustained photoevolution of molecular hydrogen in a mutant of *Synechocystis* sp. strain PCC 6803 deficient in the type I NADPH-dehydrogenase complex, *J. Bacteriol.*, 186, 1737, 2004.

86. Hemschemeier, A. and Happe, T., The exceptional photofermentative hydrogen metabolism of the green alga *Chlamydomonas reinhardtii*, *Biochem. Soc. Trans.*, 33, 39, 2005.

87. Asada, Y. and Miyake, J., Photobiological hydrogen production, *J. Biosci. Bioeng.*, 88, 1, 1999.

88. Ihara, M. et al., Light-driven hydrogen production by a hybrid complex of a [NiFe]-hydrogenase and the cyanobacterial photosystem I, *Photochem. Photobiol.*, 82, 676, 2006.

89. Sasikala, K. et al., Anoxygenic phototrophic bacteria: physiology and advances in hydrogen production technology, *Adv. Appl. Microbiol.*, 38, 211, 1993.

90. Hoekema, S. et al., Controlling light-use by *Rhodobacter capsulatus* continuous cultures in a flat-panel photobioreactor, *Biotechnol. Bioeng.*, 95, 613, 2006.

91. Najafpour, G. et al., Hydrogen as a clean fuel via continuous fermentation by an aerobic photosynthetic bacteria, *Rhodospirillum rubrum*, *Afr. J. Biotechnol.*, 3, 503, 2004.

92. Amos, W.A., *Biological Water-Gas Shift Conversion of Carbon Monoxide to Hydrogen. Milestone Completion Report*, NREL-MP-560-35592, National Renewable Energy Laboratory, Golden, CO, January 2004.

93. Mason, J.E., World energy analysis: H_2 now or later?, *Energy Policy*, 35, 1315, 2007.

94. *Hydrogen and Fuel Cells: Review of National R & D Programs*, International Energy Agency, Paris, 2004.

95. Hamelinck, C.N. and Faaij, A.P.C., Future prospects for production of methanol and hydrogen from biomass, *J. Power Sources*, 111, 1, 2002.

96. Takeuchi, Y. and Amao, Y., Biohydrogen production from sucrose using the light-harvesting function of zinc chlorophyll-*a*, *Bull. Chem. Soc. Japan*, 78, 622, 2005.

97. *Prospects for Hydrogen and Fuel Cells*, International Energy Agency Paris, 2005.

98. Tayhas, G. and Palmore, R., Bioelectric power generation, *Trends. Biotechnol.*, 22, 99, 2004.

99. Trabaey, K. and Verstraete, W., Microbial fuel cells: novel biotechnology for energy generation, *Trends. Biotechnol.*, 23, 291, 2005.

100. Logan, B.E. and Regan, J.M., Microbial fuel cells: challenges and applications, *Environ. Sci. Technol.*, 40, 5172, 2006.

101. Chaudhuri, S.K. and Lovley, D.R., Electricity generation by direct oxidation of glucose in mediatorless microbial fuel cells, *Nature Biotech.*, 21, 1229, 2003.

102. McKinlay, J.B. and Zeikus, J.G., Extracellular iron reduction is mediated in part by Neutral Red and hydrogenase in *Escherichia coli*, *Appl. Environ. Microbiol.*, 70, 3467, 2004.

103. Prasad, D. et al., Direct electron transfer with yeast cells and construction of a mediatorless microbial fuel cell, *Biosens. Bioelectron.*, 22, 2604, 2007.

104. Kim, J.R. et al., Electricity generation and microbial community analysis of alcohol powered microbial fuel cells, *Bioresour. Technol.*, 98, 2568, 2007.

105. Milliken, C.E. and May, H.D., Sustained generation of electricity by the spore-forming, Gram-positive, *Desulfitobacterium hafniense* strain DCB2, *Appl. Microbiol. Biotechnol.*, 73, 1180, 2007.

106. Justin, G. et al., Biofuel cells: a possible power source for implantable electronic devices, *Conf. Proc. IEEE Eng. Med. Biol. Soc.*, 6, 4096, 2004.

107. Davis, J.B. and Yarborough, H.F., Preliminary experiments on a microbial fuel cell, *Science*, 137, 615, 1962.

108. Logan, B.E. and Regan, J.M., Electricity-producing bacterial communities in microbial fuel cells, *Trends Microbiol.*, 14, 512, 2006.

109. Zuo, Y. et al., Tubular membrane cathodes for scalable power generation in microbial fuel cells, *Environ. Sci. Technol.*, 41, 3347, 2007.

110. Logan, B. et al., Graphite fiber brush anodes for increased power production in air-cathode microbial fuel cells, *Environ. Sci. Technol.*, 41, 3341, 2007.

111. Rismani-Yazdi, H. et al., Electricity generation from cellulose by rumen microorganisms in microbial fuel cells, *Biotechnol. Bioeng.*, 97, 1398, 2007.

112. Sakai, S. and Yagishita, T., Microbial production of hydrogen and ethanol from glycerol-containing wastes discharged from a biodiesel fuel production plant in a bioelectrochemical reactor with thionine, *Biotechnol. Bioeng.*, 98, 340, 2007.

113. Ngan, M.A. et al., *Development of the Conversion Technology of Biomass into Bioenergy*, www.nedo.go.jp/english/archives/17017/pdf/H-20Y_E.pdf.

114. Sakurai, H. and Masukawa, H., Promoting R & D in photobiological hydrogen production utilizing mariculture-raised cyanobacteria, *J. Marine Biotechnol.*, 9, 1436, 2007.

115. Ren, N.Q. et al., Assessing the optimal fermentation type for bio-hydrogen production in continuous-flow acidogenic reactors, *Bioresour. Technol.*, 98, 1774, 2007.

116. Romm, J., The car and fuel of the future, *Energy Policy*, 34, 2609, 2006.

117. Koppel, T., *Powering the Future. The Ballard Fuel Cell and the Race to Change the World*, John Wiley & Sons Canada, Ltd., Toronto, 1999.

118. Hansen, J. et al., Climate change and trace gases, *Philos. Trans. R. Soc., Ser. A*, 365, 1925, 2007.

8 Biofuels as Products of Integrated Bioprocesses

8.1 THE BIOREFINERY CONCEPT

As a neologism, "biorefinery" was probably coined in the early 1990s by Charles A. Abbas of the Archers Daniel Midland Company, Decatur, Illinois, extrapolating the practices implicit in the fractionation of corn and soybean — the wet milling process was an excellent example of a protobiorefinery (figure 1.20). Certainly, by the late 1990s, the word (or, in an occasional variant usage "biomass refinery") was becoming increasingly popular.[1] The concept has carried different meanings according to the user, but the central proposition has been that of a comparison with the petrochemical refinery that produces not only gasoline and other conventional fuels but also petrochemical feedstock compounds for the chemical industry: from a biorefinery, on this formal analogy, the fuels would include ethanol, biodiesel, biohydrogen, and/or syngas products, whereas the range of fine chemicals is potentially enormous, reflecting the spectrum of materials that bacterial metabolism can fashion from carbohydrates and other monomers present in plant polysaccharides, proteins, and other macromolecules (figure 8.1).

The capacity to process biomass material through to a mixture of products (including biofuels) for resale distinguishes a biorefinery from, for example, a "traditional" fermentation facility manufacturing acids, amino acids, enzymes, or antibiotics, industrial sites that may use plant-derived inputs (corn steep liquor, soybean oil, soy protein, etc.) or from either of the two modern polymer processes producing any one output from biomass resources that are often discussed in the context of biorefineries:

- Cargill Dow's patented process for polylactic acid ("Natureworks PLA"), pioneered at a site in Blair, Nebraska; this was the first commodity plastic to incorporate the principles of reduced energy consumption, waste generation, and emission of greenhouse gases and was awarded the 2002 Presidential Green Chemistry award.[2]
- 1,3-Propanediol (1,3-PD) produced from glucose by highly genetically engineered *Escherichia coli* carrying genes from baker's yeast and *Klebsiella pneumoniae* in a process developed by a DuPont/Tate & Lye joint venture; 1,3-PD is a building block for the polymethylene terphthalate polymers used in textile manufacture.[3]

However good are these example of the use of modern biotechnology to support the bulk chemistry industry, they center on single-product fermentations (for lactic acid and 1,3-PD, respectively) that are not significantly different from many earlier bacterial bioprocesses — in particular, lactic acid has a very long history as a microbial ingredient of yogurts and is used in the food industry to control pH, add flavor, and

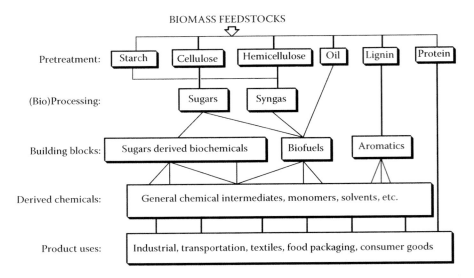

FIGURE 8.1   Material flow in an idealized biorefinery.

control microbial growth in products as diverse as alcoholic beverages, frozen desserts, and processed meat; the lactic acid production sector has major manufacturers in China, the United States, and Europe that utilize lactobaccili, bacilli, or *Rhizopus* molds in large-scale fermentations.

The following three definitions for biorefineries focus on the multiproduct (usually) biofuel-associated nature of the envisaged successors to fossil-based units:

1. The U.S. Department of Energy: "A biorefinery is an overall concept of a processing plant where biomass feedstocks are converted and extracted into a spectrum of valuable products."[4]
2. The National Renewable Energy Laboratory: "A biorefinery is a facility that integrates biomass conversion processes and equipment to produce fuels, power, and chemicals from biomass. The biomass concept is analogous to today's petroleum refineries, which produce multiple fuels and products from petroleum."[5]
3. "Third generation (generation-III) and more advanced biorefineries … will use agricultural or forest biomass to produce multiple product streams, for example ethanol for fuels, chemicals, and plastics."[1]

In early 2008, no such biorefineries exist, but the concept provides a fascinating insight into how biofuel-production facilities could develop as stepping stones toward the global production of chemical intermediates from biomass resources if lignocellulosic ethanol fails to meet commercial targets or if other developments (e.g., the successful emergence of a global hydrogen economy) render liquid biofuels such as bioethanol and biodiesel short-lived experiments in industrial innovation.*

---

* In July 2007, Honda unveiled its FCX Concept hydrogen fuel cell car capable of 100 mph and with a range of 350 miles.

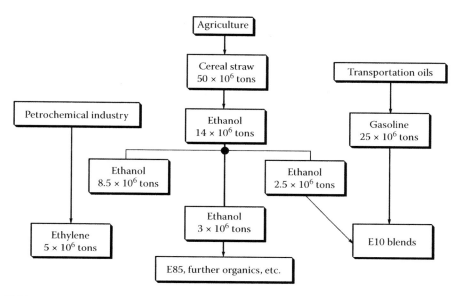

FIGURE 8.2 A possible biobased organic chemical industry in Germany: annual material flows. (Data from Kamm et al.[1])

The sheer scale of chemical endeavor possible from biomass resources is moreover extremely persuasive:[6]

- By 2040, a world population of 10 billion could be supported by 2 billion hectares of land for food production leaving 800 million hectares for nonfoods.
- With a "modest" increase in agricultural productivity to 40 tonnes/hectare/year, this land surplus to food production could yield 32 billion tonnes/year.
- Adding in 12 billion tonnes annually from forests and other agricultural waste streams yields 50 billion tonnes.

Of this total, only 1 billion tonnes would be required to generate all the organics* required as chemical feedstocks — leaving the rest for biofuels, including traditional biomass as a direct source of power and heat.

Calculations prepared from and for German industry show that agricultural waste only, that is, cereal straw, could match the total demand for E10:gasoline blends as well as all the ethylene manufactured for the national chemical and plastics industries plus surplus ethanol for use in E85 blends and other chemical uses (figure 8.2).[7]

---

* *Green chemistry* was coined in the mid-1990s by the U.S. Environmental Protection Agency; of the twelve principles of green chemistry, the ninth is that raw materials should be renewable; inorganic materials appear difficult to fit into this vision but could be accessed by recycling.[6]

Included in the range of roles proposed for biorefineries, as codified by biorefinery.nl, the umbrella organization in the Netherlands tasked with developing strategic aspects of biorefineries (www.biorefinery.nl), are the following:

1. Primary processing units for waste streams from existing agricultural endeavors
2. Essential technologies for ensuring that biomass-derived ethanol and other biofuels can be produced at costs competitive with conventional fuels
3. New additions to be integrated with the infrastructure of agricultural processing — these might include (in Europe) beet sugar refineries

As corollaries and (probably) axiomatic truths, biorefineries will only become "interesting" (as players in the industrial economy) when they reach large scales of operation and contribute significant amounts of materials to widely used and/or specialist chemistry platforms while being driven not essentially or solely as means to reduce greenhouse gas emissions* but by considerations of the future depletion of fossil fuel reserves and the desire to broaden the substrate base, with governments being instrumental in catalyzing these developments by favorable taxation regimes and economic subsidies.

8.2   BIOMASS GASIFICATION AS A BIOREFINERY ENTRY POINT

The potential of biorefineries to generate a wider collection of chemical feedstocks presently derived from petroleum is best visualized with biomethanol as the starting point (figure 8.3). Methanol obtained from the gasification of biomass (chapter 6, section 6.3.1) can be transformed by well-known purely chemical reactions to form, among many other chemicals:[8]

- Formaldehyde (CH_2O) is used in the production of resins, textiles, cosmetics, fungicides, and others; formaldehyde accounted for 35% of the total worldwide production of methanol in the mid-1990s and is prepared by the oxidation of methanol with atmospheric O_2 using a variety of catalysts.
- Acetic acid (CH_3COOH) is a major acid in the food industry and as feedstock for manufacturing syntheses, in the production of some plastics, fibers, and others; acetic acid is manufactured by the carbonylation of methanol with CO.
- Formic acid (HCOOH) is a preservative.
- Methyl esters of organic and inorganic acids are used as solvents and methylation reagents and in the production of explosives and insecticides.
- Methylamines are precursors for pharmaceuticals.
- Trimethylphosphine is used in the preparation of pharmaceuticals, vitamins, fine chemicals, and fragrances.
- Sodium methoxide is an organic intermediate and catalyst.
- Methyl halides are solvents, organic intermediates, and propellants.
- Ethylene is used for plastics and as an organic intermediate.

---

* With this motivation, simply burning biomass to produce energy and replace coal (the highest specific CO_2 generator) would be the strategic option of choice for Europe, if not elsewhere.

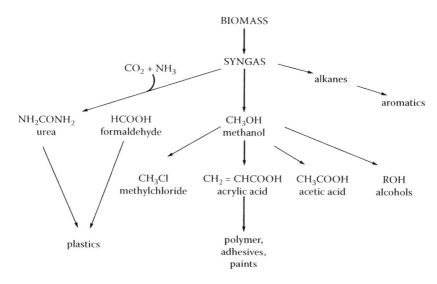

FIGURE 8.3 Chemical production routes for industrial feedstocks from methanol. (Modified from Kamm et al.[1])

8.3 FERMENTATION BIOFUELS AS BIOREFINERY PIVOTAL PRODUCTS

Ethanol is readily dehydrated by chemical reactions to ethylene (ethene):

$$C_2H_5OH \rightarrow C_2H_4 + H_2O$$

in a reversal of the chemistry used to manufacture "industrial" alcohol from petrochemical sources. Ethylene per se is a key intermediate in organic chemistry for plastics (polyethylenes); in 2005, its worldwide production was estimated to be 113 million tonnes.[9] From ethylene (a compound with no direct end uses), the vast "hinterland" of petrochemical production rapidly opens up by way of

- Ethylene oxide (C_2H_4O), a starting material for the manufacture of acrylonitrile, nonionic surfactants, and others, and a ripening agent for fruits
- Ethylene glycol ($C_2H_5O_2$), a solvent as well as an intermediate in the synthesis of synthetic fibers
- Ethylene chlorohydrin (C_2H_5ClO), another solvent and an intermediate in the production of agrochemicals
- Ethyl bromide (C_2H_5Br), an ethylating agent in organic syntheses

Approximately 0.9 million tonnes of glycerol is synthesized via chemical routes annually; two-thirds of this could be isolated from bioethanol fermentation broths as glycerol-rich stillage.[3] Biodiesel wastes and direct fermentations could, however, supply far more glycerol (chapter 6, section 6.3.2) — in fact, so much glycerol is presently being produced as a coproduct of biodiesel production that, although not a biological process, biodiesel manufacture is an imminent development of the biorefinery concept (see section 8.3.3). Important derivatives of glycerol prepared chemically include[10]

- Glycerol trinitrate, that is, nitroglycerin for explosives
- Epichlorohydrin, the most important material for the production of epoxy resins (and historically the immediate precursor of glycerol in chemical manufacture)[11]
- Oxidation products as intermediates for pharmaceutical synthesis, including glyceric, tartronic, hydroxypyruvic, and mesoxalic acids and dihydroxyacetone
- *Tert*butyl ethers formed by the reaction between glycerol and alkenes such as 2-butene

Approximately 36% of global glycerol production is directed toward such chemical intermediates and products, most of the remainder finding uses in the food and cosmetic industries.

Butanol (chapter 6, section 6.3.3) is a major solvent and finds numerous applications during the manufacture of plastics and textiles.[11] Butanol could be catalytically dehydrated to yield l-butene, one of four butene isomers (1, *cis*-2, *trans*-2, and *iso* forms); none of these highly volatile hydrocarbons exist naturally but are produced from petroleum refineries; the complexity of the composition of the butene-containing C4 streams from crude oil "cracking" has limited the chemical exploitation of butanes as intermediates and feedstocks.[12] The availability of pure 1-butene in feedstock quantities may herald new horizons for industrial synthetic chemistry.

These examples demonstrate one fundamental feature of biorefineries, that is, the ability to replace petroleum refineries as sources of both liquid fuels (ethanol, biobutanol, etc.) and (by catalytic transformation of these compounds) feedstock chemicals. This model implies a short linear chain of sequential biochemical and chemical processes:

biomass substrate → ethanol, methanol, glycerol, butanol → feedstock chemicals

This could operate on an either/or basis, either producing biofuels for transportation uses or proceeding straight to the production of bulk chemicals. The development of a full infrastructure for biofuels distribution would bring in parallel a means of transporting the liquid biofuels to existing chemical industrial sites, thus reducing the total investment cost associated with the transition from oil dependency to a biobased commodity economy, and, although the economic analysis of biorefineries is poorly developed, the price tag will be high: one estimate puts the cost of a biorefinery capable of processing 2000 tonnes/day as $500 million.[13] Approximately 500 such facilities will be required to process the "billion tons" annually required for the mass production of lignocellulosic ethanol in the United States (chapter 2, section 2.7), thus representing a $200 billion total investment.

A more elaborate model for biorefineries entails the dual bioproduction of biofuels and other fermentation products.[14] The logistic basis for this design is the multiple nature of lignocellulosic and whole-plant carbohydrate streams (chapter 2, section 2.3), that is, hemicellulose-derived pentoses, cellulose-derived glucose and oligoglucans, and starch-derived glucose if grains are processed. Given the wide spectrum of naturally occurring and genetically engineered ethanologens (chapter 3), ethanol could be produced entirely from one of these carbohydrate streams, leaving the others as substrates for different types of fermentations — potentially as wide a choice as that of fermentations known or already used for industrial production. To substantially narrow the field, twelve building block chemicals that can be produced from sugars via biological or chemical conversions and that can be subsequently converted to a number of high-value bio-based chemicals or materials have been identified in a report prepared for the U.S. DOE.[15] These building block chemicals are molecules with multiple functional groups that can transformed into new families of useful molecules:

- 1,4-Diacids (succinic, fumaric, and malic), all intermediates of the tricarboxylic acid cycle and easily bioproduced by microbes
- 2,5-Furan dicarboxylic acid, chemically produced by the oxidative dehydration of C6 sugars
- 3-Hydroxypropionic acid, a microbial product
- Aspartic acid, an amino acid biosynthesized by all living organisms
- Glutamic acid, another amino acid and one of the major products of industrial fermentations for fine chemicals (as monosodium glutamate, MSG)
- Glucaric acid, chemically produced by the nitric acid oxidation of starch
- Itaconic acid, a tricarboxylic acid manufactured on an industrial scale by the fungus *Aspergillus oryzae*
- Levulinic acid, chemically produced by the acid-catalyzed dehydration of sugars
- 3-Hydroxybutyrolactone, chemically produced by the oxidative degradation of starch by hydrogen peroxide
- Glycerol (chapter 6, section 6.3.2)
- Sorbitol, a sugar alcohol derived from glucose (chemically by hydrogenation) but also known as an enzyme-catalyzed product of glucose metabolism
- Xylitol/arabinitol (chapter 3, section 3.2)

These compounds were chosen by consideration of the potential markets for "building blocks" and their derivatives and the technical complexity of their (bio)synthetic pathways. A second-tier group of building blocks was also identified as viable candidates, most of which are microbial products: gluconic acid, lactic acid, malonic acid, propionic acid, the two triacids, citric and aconitic, xylonic acid, acetoin, furfural, levoglucosan, and the three amino acids, lysine, serine, and threonine (table 8.1).

The building block compounds were clearly differentiated from two other potential biorefinery products: direct product replacements and novel products. The former might include acrylic acid manufactured from lactic acid rather than from fossil-derived propylene; markets already existed for the compound and the cost structures and growth potential of these markets were well understood, thus reducing the risks involved in devising novel production routes. On the other hand, novel products such as polylactic acid (section 8.1) had no competing routes from fossil reserves, had unique properties (thus rendering cost issues less crucial), and were intended to meet new markets.

TABLE 8.1
Microbial Routes Known for Future Biobased Chemical Building Blocks

Building block	Aerobic fermentation	Anaerobic fermentation
Three-carbon compounds		
Glycerol	Yeast/fungal and bacterial	Yeast/fungal and bacterial
Lactic acid	Yeast/fungal	Commercial bacterial process
Propionic acid	None	Bacterial
Malonic acid	Yeast/fungal	None
3-Hydroxypropionic acid	Yeast/fungal and bacterial	None
Serine	Commercial bacterial process	None
Four-carbon compounds		
3-Hydroxybutyrolactone	None	None
Acetoin	Yeast/fungal and bacterial	Bacterial
Aspartic acid	Yeast/fungal and bacterial	None
Fumaric acid	Yeast/fungal and bacterial	None
Malic acid	Yeast/fungal and bacterial	None
Succinic acid	Yeast/fungal and bacterial	Bacterial
Threonine	Commercial bacterial process	None
Five-carbon compounds		
Arabitol	Yeast/fungal	Yeast/fungal
Xylitol	Yeast/fungal	Yeast/fungal and bacterial
Furfural	None	None
Glutamic acid	Commercial bacterial process	None
Itaconic acid	Commercial fungal process	None
Levulinic acid	None	None
Six-carbon compounds		
2,5-Furan dicarboxylic acid	None	None
Aconitic acid	Yeast/fungal	None
Citric acid	Commercial fungal process	None
Glucaric acid	Yeast/fungal and bacterial	None
Gluconic acid	Commercial fungal process	None
Levoglucosan	None	None
Lysine	Commercial fungal process	None
Sorbitol	Yeast/fungal and bacterial	None

Source: Data from Werpy and Petersen.[15]

In contrast, the building block compounds were envisaged as being the starting points for diverse portfolios of products, both replacing existing fossil-based compounds and offering novel intermediates for chemical syntheses. This combination has three advantages:

- The market potential is expanded.
- Multiple possible markets can reduce risks.
- Capital investment can be spread across different industrial sectors.

To illustrate these points, the 12 "finalists" could give rise to many derivatives that would find immediate or short-term uses in fields as diverse as transportation (polymers for automobile components and fittings, anticorrosion agents, and oxygenates), recreation (footgear, golf equipment, and boats), and health and hygiene (plastic eyeglasses, suntan lotions, and disinfectants).[15] Three of them will now be considered in greater detail to explore possible key features and likely variables in the development of biorefineries.

8.3.1 Succinic Acid

This C4 dicarboxylic acid is one of the key intermediates of glucose catabolism in aerobic organisms (including *Homo sapiens*) but can also be formed anaerobically in fermentative microbes (figure 8.4). In either case, CO_2 is required to be "fixed" into organic chemicals; in classical microbial texts, this is described as "anaplerosis," acting to replenish the pool of dicarboxylic and tricarboxylic acids when individual compounds (including the major industrial products citric, itaconic, and glutamic acids) are abstracted from the intracellular cycle of reactions and accumulated in the extracellular medium. Under anaerobic conditions, and given the correct balance of fermentation

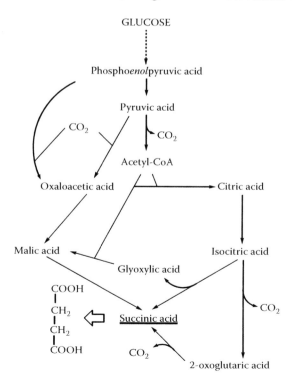

FIGURE 8.4 Succinic acid: known biosynthetic routes from glucose.

products, a net "dark" fixation of CO_2 can occur, and it is this biological option that has been most exploited in the development of modern biosynthetic routes:

- The rumen bacterium *Actinobacillus succinogenes* was discovered at Michigan State University and commercialized by MBI International, Lansing, Michigan.[16,17] Succinate yields as high as 110 g/l have been achieved from glucose.
- At the Argonne National Laboratory, Argonne, Illinois, a mutant of *E. coli* unable to ferment glucose because of inactivation of the genes encoding lactate dehydrogenase and pyruvate formate lyase spontaneously gave rise to a chromosomal mutation that reestablished glucose fermentative capacity but with an unusual spectrum of products: 1 mol succinate and 0.5 mol each of acetate and ethanol per mole of glucose consumed.[18,19] The second mutation was later mapped to a glucose uptake protein that, when inactivated or impaired, led to slow glucose transport into the cells and avoided any repression of genes involved in this novel fermentation.[20] The result is a curious fermentation in which redox equivalents are balanced by a partition of carbon between the routes to succinate and that to acetate and ethanol (in equal measures), pyruvate being "oxidatively" decarboxylated rather than being split by pyruvate formate lyase activity or reduced to lactic acid, both routes lost from wild-type *E. coli* biochemistry in the parental strain (figure 8.5). The maximum conversion of glucose to succinate by this route is 1 mol/mol, a carbon conversion of 67%; succinate titers have

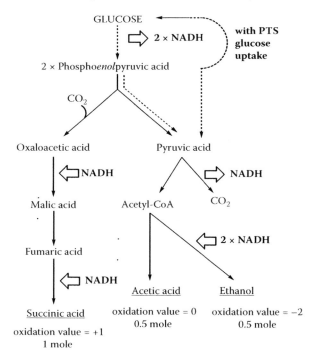

FIGURE 8.5   Redox balance in the fermentation of glucose to succinic acid by *Escherichia coli*.

reached 75 g/l. Because *E. coli* is only facultatively anaerobic, biomass in the fermentation can be generated rapidly and to a high level under aerobic conditions, O_2 entry then being restricted to transform the process to one of anaerobic metabolism.[21] The same organism can successfully utilize both glucose and xylose in acid hydrolysates of corn straw and generate succinate as a fermentation product.[22]

- In wild-type *E. coli*, glucose fermentations produce complex mixtures of acid and nonacidic products, in which succinate may be only a minor component (chapter 2, section 2.2). Nevertheless, the succinate titer can be greatly increased by process optimization, and Indian researchers achieved more than 24 g/l within 30 hours with laboratory media and 17 g/l in 30 hours in a fermentor with an economical medium based on corn steep liquor and cane sugar molasses.[23,24] The same group at the University of Delhi have enhanced succinate productivity with *Bacteroides fragilis*, another inhabitant of the human gut and intestine but an obligately anaerobic species.[25,26]

- In complete contrast, an *aerobic* system for succinate production was designed with a highly genetically modified *E. coli*, using the same glucose transport inactivation described above but also inactivating possible competing pathways and expressing a heterologous (*Sorghum vulgare*) gene encoding PEP carboxylase, another route for anaplerosis.[27,28] A succinate yield of 1 mol/mol glucose consumed was demonstrated, with a high productivity (58 g/l in 59 hours) under fed-batch aerobic reactor conditions. The biochemistry involved in this production route entails directing carbon flow via anaplerotic reactions to run "backward" through the tricarboxylic acid cycle, that is, in the sequence:

$$PEP \rightarrow oxaloacetate \rightarrow malate \rightarrow fumarate \rightarrow succinate$$

- At the same time, succinate is produced in the "forward" direction by blocking the normal workings of the cyclic pathway (with a necessary loss of carbon as CO_2) and the activation of a pathway (the "glyoxylate shunt") normally only functioning when *E. coli* grows on acetate as a carbon source:

$$citrate \rightarrow isocitrate \rightarrow succinate + glyoxylate$$

- The enzymes catalyzing the final two steps in the pathway from PEP, malate dehydratase and fumarate reductase, can be overexpressed in bacterial species and are the subjects of two recent patent applications from Japan and Korea.[29,30]

- To return to anaerobic rumen bacteria, *Anaerobiospirillum succiniciproducens* had a short but intense history as a candidate succinate producer.[31–33] Fermenting glucose in a medium containing corn steep liquor as a cost effective source of nitrogen and inorganic nutrients, succinate titers reached 18 g/l from 20.2 g/l of glucose, equivalent to a conversion efficiency of 1.35 mol/mol.[34]

- The same research group at the Korean Advanced Institute of Science and Technology, Daejeon, Republic of Korea, then isolated a novel rumen bacterial species, *Mannheimia succiniciproducens*, and has determined its complete genomic sequence as well as constructing a detailed metabolic network for the organism.[35–37] Mutants of this microbe can produce succinate with much reduced amounts of other acids and can anaerobically ferment xylose and wood hydrolysate to succinate.[38,39]

There are grounds to predict that overexpressing genes for anaplerotic pathway enzymes would enhance succinate production (and, in other genotypes or fermentation conditions, the accumulation of other acids of the tricarboxylic cycle); experimental evidence amply confirms this prediction.[40–45] With the capabilities to perform metabolic computer-aided pathway analysis with known gene arrays, comparison of succinate producers and nonproducers and between different species would be expected to greatly accelerate progress toward constructing the "ideal" microbial cell factory. Comparison of *E. coli* and *M. succiniciproducens* suggested five target genes for inactivation but combinatorial inactivation did not result in succinate overproduction in *E. coli*; two of the identified genes — *ptsG* (the glucose transport system) and *pykF* (encoding pyruvate kinase, the enzyme interconverting PEP and pyruvic acid), together with the second pyruvate kinase gene (*pykA*) — increased succinate accumulation by more than sevenfold, although succinate was still greatly outweighed by the other fermentation products (formate, acetate, etc.).[46] Eliminating the glycolytic pathway below PEP will clearly aid succinate production, but the producing cells will by then be highly dependent on organic nutrients (including many amino acids) for growth and maintenance.

The combination of several related technologies is particularly appealing for commercial production of succinic acid:

- Optimization of the Argonne National Laboratory strains for succinate production by the Oak Ridge National Laboratory, Oak Ridge, Tennessee
- Innovations for succinate recovery from the fermentation broth at Argonne National Laboratory
- An improved succinic acid purification process[47]
- The development of catalytic methods for converting succinic acid to 1,4-butanediol and other key derivatives at Pacific Northwest National Laboratory, Richmond, Washington[48]

These advances have moved biologically derived succinic acid close to commercialization as a component of the first genuine biorefinery.

8.3.2 Xylitol and "Rare" Sugars as Fine Chemicals

Xylitol was a significant biochemical feature of the metabolic routes for the xylose presented to ethanologenic cultures in hydrolysates of the hemicelluloses from lignocellulosic biomass (chapter 2, section 2.3, and chapter 3, section 3.2). With an organoleptic sweetness to human taste approximately equivalent to sucrose, however, it is a fine chemical product in its own right as a low-calorie sweetener — xylose sugars are not metabolized by the human consumers of xylitol-containing chewing gums (figure 3.2). Other envisaged uses include[15]

- Production of anhydrosugars (as chemical intermediates) and unsaturated polyester resins
- Manufacture of propylene and ethylene glycols as antifreeze agents and unsaturated polyester resins
- Oxidation to xylonic and xylaric acids to produce novel polymers (polyesters and nylon-type structures)

The production of xylitol for use as a building block for derivatives essentially requires no technical development, and if the xylose feedstock is inexpensive (as a product of biomass processing), then the production of xylitol could be done for very low cost.

The accumulation of xylitol during ethanologenesis from lignocellulosic substrates is, of course, unwanted and quite undesirable — for process as well as for economic reason (chapter 3, section 3.2). Viewed as an economically valuable product, xylitol formation and production acquire a different biotechnological perspective, and patenting activity has recently been intense (table 8.2). Biochemical efforts also continue to locate and exploit enzymes for bioprocessing hemicelluloses and hemicellulosic waste streams, for example:

- An L-xylulose reductase identified from the genome sequence of the filamentous mold *Neurospora crassa* has been heterologously produced in *E. coli* for the production of xylitol.[49]

TABLE 8.2
Recent Patents for Xylitol Production Technologies

Date	Title	Assignee	Patent
8/7/2001	Yeast strains for the production of xylitol	Xyrofin Oy, Finland	US 6,271,007
4/20/2004	Manufacture of xylitol using recombinant microbial hosts	Xyrofin Oy, Finland	US 6,723,540
1/25/2005	Process for the simultaneous production of xylitol and ethanol	Xyrofin Oy, Finland	US 6,846,657
4/17/2006	Fermentation process for production of xylitol from *Pichia* sp.	Council of Scientific and Industrial Research, India	US 6,893,849
5/17/2005	Process for the production of xylitol	Danisco Sweeteners Oy, Finland	US 6,894,199
6/28/2005	Process for the production of xylitol	Danisco Sweeteners Oy, Finland	US 6,911,565
8/2/2005	Xylitol dehydrogenase of acetic acid bacteria and gene thereof	Ajinomoto Co., Inc., Japan	US 6,924,131
9/19/2005	Process for the simultaneous production of xylitol and ethanol	Danisco Sweeteners Oy, Finland	US 7,109,005

- A β-xylosidase from *Taloromyces emersonii* has been shown to be superior to the enzyme from the industrial fungus *Hypocrea jecorina* in releasing xylose from vinasse, the solid-waste material from ethanol fermentations.[50]
- Xylan residues in hemicelluloses can be variably esterified with acetyl, feruloyl, and *p*-coumaryl residues; hemicellulose deacetylating esterases have been characterized from fungal species and shown to only be highly effective when mixed in multiplexes of enzymes capable of using all the possible structures as substrates for enzyme action.[51]
- Feruloyl esterases have recently been designed as novel chimeric forms with cellulose and hemicellulose binding proteins to improve their efficiencies with plant polymeric substrates.[52,53]
- Ferulic acid has a number of potential commercial applications as an antioxidant, food preservative, anti-inflammatory agent, photoprotectant, and food flavor precursor; major sources — brewer's spent grain, wheat bran, sugarbeet pulp, and corn cobs — make up 1–2% of the daily output of the global food industry, and ferulic acid can be released from brewer's grain and wheat bran by feruloyl esterases from the thermophilic fungus *Humicola insolens*.[54]

Once acquired, a hemicellulose hydrolysate contains a variety of hexoses and pentoses; yeast highly suitable for the production of xylitol from the xylose present in the mix may, however, preferentially utilize glucose. One solution to this problem is to remove the rapidly utilized hexoses before removing the cells and replacing them with a purposefully xylose-grown cell batch, as was demonstrated with dilute acid hydrolysates of corn fiber.[55] With *Saccharomyces cerevisiae* expressing the *Pichia stipitis* gene for xylose reductase, the presence of glucose inhibited xylose uptake and biased the culture toward ethanol production; controlling the glucose concentration by feeding the fermentation to maintain a high xylose:glucose ratio resulted in a near-quantitative conversion of xylose to xylitol, reaching a titer of 105 g/l xylitol.[56] With a double-recombinant strain of *S. cerevisiae* carrying the xylose reductase genes from both *P. stipitis* and *Candida shehatae*, quite different microbial biochemistry occurred in a more process-friendly formation (at close to theoretical levels) of xylitol from a mixture of xylose (the major carbon source) and glucose, galactose, or mannose as the cosubstrate — indeed, the presence of the cosubstrate was mandatory for continued metabolism of the pentose sugar.[57] In a gene-disrupted mutant of *C. tropicalis*, with no measurable xylitol dehydrogenase activity, glycerol proved to be the best cosubstrate, allowing cofactor regeneration and redox balancing, with a xylitol yield that was 98% of the maximum possible.[58]

Bioprocess engineering for xylitol production appears straightforward; aerated cultures with pH regulation and operated at 30°C appear common. With Brazilian sugarcane bagasse as the source of the hemicellulosic sugars, xylitol from the harvested culture broth was recovered at up to 94% purity crystallization from a clarified and concentrated broth.[59]

Xylitol dehydrogenase enzymes catalyze reversible reactions. The catabolism of xylitol proceeds via the formation of D-xylulose (figure 3.2). Fungal pathways of L-arabinose utilization can include L-xylulose as an intermediate, a much less common pentulose sugar.[60–63] Such "rare" sugars are of increasing interest to metabolic biochemists because it has long been appreciated that many naturally occurring antibiotics and other bioactives contain highly unusual sugar residues, sometimes highly modified hexoses derived by lengthy biosynthetic pathways (e.g., erythromycins); many secondary metabolites elaborated by microbes may only exhibit weak antibiotic activity but are or can easily be converted into chemicals with a wide range of biologically important effects: antitumor, antiviral, immunosuppressive, anticholesterolemic, cytotoxic, insecticidal, or herbicidal.[64] To the synthetic chemist, therefore, the availability of such novel carbohydrates in large quantities offers new horizons in development novel therapeutic agents: for example, L-xylulose offers a promising route to inhibiting the glycosylation of proteins, including those of viruses.[65]

Equally, such unconventional chemicals may have very easily exploited properties as novel fine chemicals. The hexose D-tagatose is an isomeric form of the commonly occurring sugar D-galactose (a hexose present in hemicelluloses, figure 1.23) and has attracted interest for commercial development.[66] Because of its very rare occurrence in the natural world, a structure such as D-tagatose presents a "tooth-friendly" metabolically intractable sugar to human biochemistry. Fortuitously, the enzyme L-arabinose isomerase includes among its spectrum of possible substrates D-galactose; the enzyme can be found in common bacteria with advanced molecular genetics and biotechnologies, including *E. coli*, *Bacillus subtilis*, and *Salmonella typhimurium* and, when expressed in suitable hosts, can convert the hexose into D-tagatose with a 95% yield.[67] Even more promising for industrial use is the efficient bioconversion of D-galactose to D-tagatose using the immobilized enzyme, more active than free L-arabinose isomerase and stable for at least 7 days.[68] Both the enzyme and recombinant L-arabinose isomerase-expressing cells can be used in packed-bed bioreactors, the cells being particular adaptable to long production cycles.[69,70]

Synthetic chemistry offers only expensive and low-yielding routes to the rare sugars, but uncommon tetroses, pentoses, and hexoses can all be manufactured with whole cells or extracted enzymes acting on cheap and plentiful carbon sources, including hemicellulose sugars; bioproduction strategies use an expanding toolkit of enzymes, including D-tagatose 4-epimerase, aldose isomerase, and aldose reductase.[71–73] Together, the rare sugars offer new markets for sugars and sugar derivatives of at least the same magnitude as that for high-fructose syrups manufactured with xylose (glucose) isomerase.

8.3.3 GLYCEROL — A BIOREFINERY MODEL BASED ON BIODIESEL

Glycerol is an obligatory coproduct of biodiesel production, glycerol forming 10% by weight of the triglycerides that act as substrates for chemical or enzymic transesterification (figure 6.1). The waste stream is, however, highly impure, with the glycerol mixed with alkali, unreacted methanol, free fatty acids, and chemical degradation products. The glycerol can, at the cost of considerable expenditure of energy in gas sparging and flash distillation, be recovered at more than 99.5% purity.[74] Much simpler is to mix the crude glycerol with methanol and substitute waste petroleum oil and heavy fuel oil as a direct fuel.[75]

Glycerol represents a valuable chemical resource as a potential feedstock.[15] Research into value-added utilization options for biodiesel-derived glycerol has, therefore, ranged widely in the search for commercial applications using both chemical and biotechnological methods (table 8.3).[76] For the production of 1,3-PD for polymer manufacture, glycerol as source via microbial fermentation was (5–6 years ago) more expensive than either glucose or chemical routes from ethylene oxide or acolein.[3] Those economics have significantly altered now, and glycerol represents the shortest, most direct route for bioproduction of 1,3-PD, a two-reaction sequence comprising an enzyme-catalyzed dehydration followed by a reduction:

$$\text{glycerol} \rightarrow \text{3-hydroxypropionaldehyde} \rightarrow \text{1,3-PD}$$

Microbial studies have focused on the fermentative production of 1,3-propanediol by clostridial species, but this is greatly complicated by the multiplicity of other products, including n-butanol, ethanol, and acids.[77] Bioproduction of a nutraceutical fatty acid derivative (marketed as a dietary supplement) by microalgae is a route to a higher-value product than bulk chemicals.[78] As examples of chemical engineering, recent industrial patent documents have disclosed methods for converting the biodiesel waste glycerol into dichloropropanol and an alcohol-permeable membrane to generate a purified mixture of esters, glycerol, and unreacted alcohol.[79,80]

For many years, glycerol was not considered fermentable by *E. coli* but only by a limited number of related bacterial species; a landmark publication in 2006, however, reported that *E. coli* could efficiently ferment glycerol to ethanol (and a small amount of succinic acid) provided high pH in the culture was avoided — this is a crucial point because growth from glycerol requires an anaplerotic step (section 8.3.1), the CO_2 being generated by the pyruvate formate lyase reaction whose activity may be much reduced by high pH in the growth medium (figure 8.6).[81] The application of genetically engineered *E. coli* with superior ethanologenic potential (chapter 3,

TABLE 8.3
Chemical and Biotechnological Transformations of Biodiesel-Derived Glycerol

Product formed from glycerol	Chemical route	Fermentation route
1,3-Propanediol	Selective hydroxylation	*Clostridium butyricum, Klebsiella pneumoniae*
1,2-Propanediol	Hydrogenolysis	None
Dihydroxyacetone	Selective catalytic oxidation	*Gluconobacter oxydans*
Succinic acid	None	*Anaerobiospirillum succiniciproducens*
Hydrogen	Catalytic reforming	*Enterobacter aerogenes*
Polyesters	Catalyzed esterification with acids	None
Polyglycerols	Selective etherification	None
Polyhydroxyalkanoates	None	Various osmophilic microbial species

Source: Data from Pachauri and He.[76]

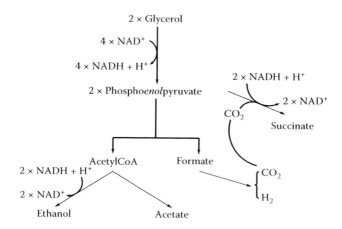

FIGURE 8.6 The anaerobic fermentation of glycerol to ethanol and succinic acid by *Escherichia coli*. (Modified from Dharmadi et al.[81])

section 3.3.2.1), combined with further manipulations to eliminate competing pathways of acid accumulation, could result in ethanol productivity approaching the theoretical molar production (mole per mole) from glycerol.

8.4 THE STRATEGIC INTEGRATION OF BIOREFINERIES WITH THE TWENTY-FIRST CENTURY FERMENTATION INDUSTRY

Producing low-volume, high-cost products or, at least, middle-volume and middle-price coproducts from a biorefinery will be essential in establishing crop- and biomass-dependent biorefineries as components of the industrial landscape. The presence of up to 500 biorefineries in North America and 500–1000 across Europe brings the option of large numbers of small- to medium-capacity sugar streams in widely scattered locations, often in agricultural or afforested regions. Each site could house a fermentation facility for fine chemicals as well as being a site for biofuels — the costs of transporting biomass and of high-volume substrate solutions long distances by road or rail are unlikely to appeal either to industry or to environmentalists. The implication is that large numbers of production sites will arise where many commercial products for the chemical, food, and other industries will be synthesized. In turn, this suggests a reversal of the process by which large-scale fermentation production for antibiotics, enzymes, vitamins, food flavors, and acids has been exported from the United States and Europe to areas with lower labor and construction costs (India, China, etc.). This has been described as "restructuring the traditional fermentation industry into viable biorefineries."[82] Such a vision will almost certainly challenge the imaginations of industrial fermentation companies and demand that much closer attention is paid to the practical economics of the biorefinery concept(s).

Nevertheless, long-term strategic research around the world has begun to explore what might be achieved with such an abundant supply of pentose sugars from lignocellulosic biomass and of the metabolic and biosynthetic uses that might result.

Corynebacterium glutamicum has, as its name suggests, been much used for the industrial production of amino acids such as glutamic acid and lysine for the food and agricultural feed sectors. To broaden its substrate utilization range to include xylose, a two-gene xylose catabolic pathway was constructed using the *E. coli xylA* gene (encoding xylose isomerase) with either the *E. coli xylB* gene (for xylulokinase) or a corynebacterial gene for this enzyme; recombinants could grow in minimal media with xylose as the sole carbon source under aerobic conditions or, when O_2-limited, utilize xylose alone or in combination with glucose.[83]

An unusual *Lactobacillus* sp. strain MONT4, isolated from a high-temperature fermenting grape must, is uniquely capable of fermenting L-arabinose to a mixture of D- and L-lactic acids; the organism contains two separate genes encoding lactate dehydrogenase with differing stereochemistries.[84] D-lactic acid, along with "unnatural" D-acids and amino acids, is contained in a series of bioactives developed as antiworming agents.[85] Chemical transformations of lactic acids can yield a variety of chemical intermediates and feedstocks, notably acrylic acid and propylene glycol, compounds with major existing petrochemical-derived markets.[86]

Few (if any) major industrial-scale processes could fail to utilize individual sugars or mixtures of the carbohydrates emanating from lignocellulosic biomass processing or biodiesel production — even as atypical a microbe as *Streptomyces clavuligerus* (unable to metabolize glucose) can use glycerol for fermentations for the medically important β-lactam inhibitor clavulanic acid.[87] The extensive experience of adapting microbes to growing and functioning in lignocellulosic hydrolysates (chapters 3 and 4) should be readily transferable to industrial strains that already are expected to produce large amounts of high-value products in extremely concentrated media with sugars, oligosaccharides, or plant oils as substrates and with vegetable proteins or high concentrations of ammonium salts as nitrogen sources. Conversely, the genomes of already adapted ethanologens could (with removal of genes for ethanol formation) provide platform hosts for the expression of other biosynthetic pathways.

Only biopharmaceuticals, that is, recombinant proteins expressed in and produced by animal cell cultures, yeast, or *E. coli*, are impossible to merge with biorefineries because of regulatory restrictions and the absolute requirement for sterile cleanliness at the production facility; with the rise of biogeneric products, the production of biopharmaceuticals, already becoming global, will have completely left its early exclusive bases in southern California and western Europe.[88] Less highly regulated bioprocesses, including all fermentation-derived chemicals not intended for biomedical use, are likely to migrate to sites of cheap and abundant carbon, nitrogen, and mineral nutrients, probably assisted by grants and incentives to provide employment in rural or isolated areas.

As an example of a tightly closed circular biorefinery process, consider biodiesel production from a vegetable oil — or, as work in Brazil indicates, intact oil-bearing seeds[89] — using enzyme-catalyzed transesterification rather than alkaline hydrolysis/methanolysis (chapter 6, section 6.1.2). The glycerol-containing effluent could be a carbon source for the methylotrophic yeast *Pichia pastoris* (able also to catabolize any contaminating methanol) that is widely used for expressing heterologous proteins, including enzymes; if the esterase used in the biodiesel process were to be produced on-site by a *Pichia* fermentation, the facility could be self-sufficient biotechnologically (figure 8.7). More ambitiously, any solid materials from the fermentation together with any "waste" plant material could provide the substrates for solid-state fermentations producing bioinsecticides, biopesticides, and biofertilizers for application to the farmland used to grow the oil crop itself.[90]

Microbial enzymes and single-cell proteins (as well as xylitol, lactic acid, and other fermentation products) have become targets for future innovations in utilizing sugarcane bagasse in Brazil.[91] The bioproduction of bacterial polyhydroxyalkanoate polymers to replace petrochemical-based plastics is another goal of projects aiming to define mass-market uses of agricultural biomass resources in Brazil and elsewhere.[92] Polyhydroxybutyrate and related polymers have had a long but unsuccessful record of searching for commercialization on a large scale on account of their uncompetitive economics and not always entirely industry-friendly chemical and physical properties, but high oil prices will increasingly sideline the former, whereas continued research into the vast number of possible biosynthetic structures may alight on unsuspected properties and uses — this also illustrates an important distinction made by the founders of "biocommodity engineering," that is, that replacement of a fossil resource-derived product by a biomass-derived compound of identical composition is a conservative strategy, whereas a more radical one is that of substituting the existing chemical product with a biochemical with equivalent functionalities but with a distinct composition, which involves a more protracted transition but could be more promising in the long term.[93]

Using blue-green algae, not only to generate photobiohydrogen, but as microorganisms capable of much wider (and mostly unexplored) metabolic and biosynthetic capabilities, also offers innovative polymeric products, that is, exopolysaccharides whose massive cellular production has elicited interest in exploring their properties

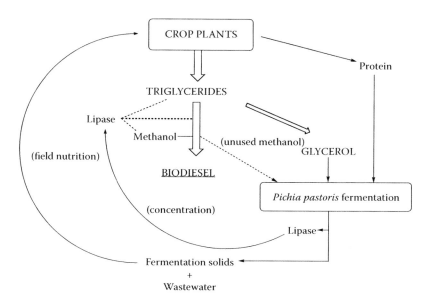

FIGURE 8.7 A "closed loop" integrated biotechnological process for biodiesel production.

and uses as industrial gums, bioflocculents, soil conditioners, emulsifiers and stabilizers, and vehicles for the removal and recovery of dissolved heavy metals.[94]

The biggest and most challenging leap for feedstocks for biorefineries, however, is that of using known microbial biochemistry to adsorb excess CO_2 in the atmosphere. This is one of the two means — the other being the genetic manipulation of higher plants to increase their photosynthetic efficiencies (chapter 4, section 4.7) — that biotechnology could make decisive contributions to the global campaign to reduce atmospheric CO_2 levels. Although plants use the Calvin-Benson cycle to fix CO_2 into organic carbon compounds (initially sugar phosphates), CO_2 fixation by "dark" metabolic processes are considerably less well publicized. The most recently discovered pathway was defined as recently as 1989 in an archeon, a type of single-celled microbe that is bacteria-like but evolved as an ancient line quite separately from the eubacteria and blue-green algae and whose members usually inhabit extreme environments (figure 8.8).[95] The extensive cultivation of photobioreactors anywhere but in climates and locations with long guaranteed daily hours of intense sunlight is inefficient except in thin films, energy-requiring (e.g., to maintain an optimum temperature of 24°C or more*), and always limited by variations in the light/dark cycle.[96] Light-independent bacterial bioprocesses avoid such limitations; if combined with biosynthetic pathways for high-value chemicals, bioreactors supplied with CO_2 pumped into underground storage or with chemically adsorbed and released CO_2 could be a primary technology for the later twenty-first century, whereas known and planned biofuels (other than H_2) may prove disappointing in the extent to which their production and use over their full life cycles actually reduces transportation greenhouse gas emissions (chapter 1, section 1.6.2, and chapter 6, section 6.1.4), applying biotechnology to use accumulated CO_2 as a process input would be a more widely applauded achievement.

Chemical routes to carbon capture and sequestration have already begun to establish themselves, especially for the most pertinent application of removing (as far as possible) CO_2 emissions from power stations; trapping CO_2 with amine-based systems has, however, been questioned as to its economics and its use of energy.[97,98] Pumping trapped CO_2 into sandstone deposits underneath the North Sea is practiced by the Norwegian oil industry but is of doubtful legality as well as requiring constant monitoring — it is, in many ways, a Faustian bargain with uncertain implications for the future despite its technical feasibility.[99]

8.5   POSTSCRIPT: WHAT BIOTECHNOLOGY
COULD BRING ABOUT BY 2030

Vast caverns of CO_2-absorbing bacterial fermentations producing high carbon-content products with immediate human use — including bacterial cellulose as fiber, single-cell protein, or bioplastics — may be industrial realities for the later years of the present century but perhaps a more compressed timescale should occupy a high priority on the biofuels and climate change agenda. Over the coming 25 years, "hard

---

* The example quoted was for a site in northwestern Europe (the Netherlands).

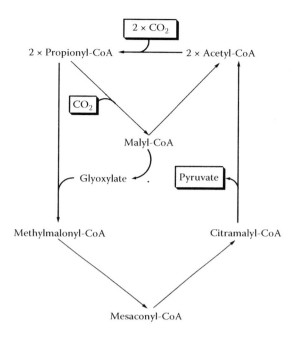

FIGURE 8.8 Schematic of a bicyclic CO_2 fixation pathway in *Chloroflexus aurantiacus* and other archeons. (Modified from Ishii et al.[95])

truths" about the global energy future will be (or are) unavoidable, and the role of biomass and other renewables in the emerging technological mix is a key issue.[100]

Any individual's ranking of the immediate challenges that could be met by biotechnology is biased and partial but a useful departure point may be the priority list of discussion items in a major international conference on biofuels held in 2007.

8.5.1 Chicago, Illinois, October 16–18, 2007

The conference, entitled Cellulosic Ethanol and 2nd Generation Biofuels, was held over three days in October 2007 and focused on scientific and economic issues beyond those of starch- or sucrose-derived fuel ethanol. Itemized separately, the presentations and panel discussions were

- Economic and commercial outlook for advanced biofuels — see chapter 6 (after section 6.1)
- Developing butanol as a transportation fuel — see chapter 6, section 6.3.3
- Commercialization of second-generation biofuels — see chapters 6 and 8
- The role of the federal government in supporting cellulosic ethanol development — see chapter 5, in particular, sections 5.2.2 and 5.2.7
- California's commitment to advanced biofuels — see chapter 1, section 1.6.2
- Design and engineering challenges for cellulosic ethanol plants — see chapter 4

- Development of cellulosic ethanol — perspective from the European Union — see chapter 5, section 5.2.7
- Commercial development of cellulosic ethanol — steps toward industrial-scale production — see chapters 4 and 5
- Tapping the potential of forest products as a feedstock for cellulosic ethanol — see chapter 4, section 4.2.6, and chapter 5, section 5.4
- Logistics of switchgrass as a cellulosic feedstock — see chapter 4, section 4.2.2, and chapter 5, section 5.4
- Overcoming resource and transportation constraints for cellulosic ethanol — see chapter 4, section 4.2, and chapter 5, section 5.4
- Feedstock procurement and logistics for cellulosic ethanol — see chapter 4, section 4.2, and chapter 5, section 5.4
- Advances in pretreatment and cellulosic ethanol production — see chapters 2 and 4
- Enzymes and ethanologen challenges for cellulosic ethanol — see chapter 2, sections 2.5 and 2.6, and chapter 3
- A different approach to developing yeasts for biomass conversion — see chapter 3, sections 3.1, 3.2, and 3.4
- Siting and financing a cellulosic ethanol plant — see chapter 5
- Dealing with legal and policy risks for ethanol and advanced biofuels — see chapter 5, especially section 5.2.2
- Investor perspectives on developing and financing cellulosic ethanol — see chapter 5

From this agenda, it is evident that, unlike starch- and sucrose-derived ethanol industries (very much still in the "boom" phase of development worldwide), cellulosic ethanol still remains on the fringes even in the developed economy (the United States) with the most overt public support of cellulosic ethanol while investors fret about fiscal uncertainties and technologists scrabble to convince potential manufacturers that the major scientific hurdles have been overcome. In New Zealand, the Genesis Research and Development sold (in early November 2007) BioJoule, its biofuels business that aimed at establishing a lignocellulosic ethanol and biorefinery facility, citing the difficulties in raising capital and grant funding to pursue such an aim.

Other than cellulosic ethanol, other advanced biofuels (with the exception of "biobutanol") remain in need of serious industrial partners or timelines. The obvious danger to many environmentalists is that the partial vacuum in the supply and demand cycle will inevitably be met by ecologically unstable developments in tropical regions to establish monoculture plantations of "energy crops" to further harden the international trade in "off-the-shelf" biofuels.

In the short term, it can readily be concluded that all the elements needed to establish a cellulosic ethanol industry do now exist — and have existed for a decade (and possibly longer) — and that the biotechnologists could now leave the stage to determined commercial interests to finance, set up, and maintain industrial production facilities. Casualties are inevitable: some second-generation companies will undoubtedly fail financially, but this will be a part of whatever learning curve the new biobased industry will encounter, and as a safety net, in the United States and other OECD nations, enough government support is likely to remain on offer to smooth the transition as long as crude oil prices are sufficiently high for biofuels to be pump price competitive with gasoline.

8.5.2 Biotechnology and Strategic Energy Targets Beyond 2020

It is not surprising that critics of biofuels are skeptical about (or hostile to) further research into bioproduction routes, especially from food crop sources, because the proponents of biofuels have often confounded two quite separate issues:

1. Plant biomass as a supply of feedstocks for fuels for automobiles owned and operated by private drivers
2. Plant biomass as a source of industrial chemicals to replace petrochemicals "eventually" or "sooner or later"

Biofuels need not be manufactured from food crops, but both biofuels and biorefineries must utilize lignocellulosic substrates to satisfy the massive demand that is presently met worldwide by a very large manufacturing sector, that is, the oil, gas, and petrochemical industries, based on an unsustainable basis (fossil fuels).

In addition, biofuels may be a poor choice in the long term when compared with photovoltaic cells to capture solar energy and the plethora of "renewable" (wind, wave, geothermal, etc.) energy sources, alternatives that have many dedicated publicists.

Further scientific research from a biotechnological perspective is unlikely to stumble on arguments to convince the critics of biofuels to radically revise their position or dubious investors to reassess the risks. Based on the material presented in this and the preceding seven chapters, however, a priority list can be assembled for bioenergy aims that are achievable within the next three to four decades (chapter 5, section 5.6), that is, before oil depletion becomes acute — viewed not purely as the narrow biofuels agenda but as part of the unavoidable need to develop a viable biocommodity production system within that time.

First, biohydrogen is the sole means of breaking any dependence or "addiction" to CO_2 cycles in the industrial world. Is either photoproduction or "dark" bacterial fermentation energetically sufficient and cost-efficient to seriously rival solar-powered or chemical routes to generate the enormous quantities required for fuel cell technologies on a global scale? Critics of fuel cell-powered mass transportation point to the cost of the units and the practical problems inherent in supplying inflammable H_2. In October 2007, however, a report appeared of an elegant solution to both issues: a fuel cell designed for automobiles without any expensive platinum catalyst and capable of using hydrazine (N_2H_4) as a convenient liquid form of H_2.[101] The production of hydrazine has a total energy efficiency of 79% and the refilling energy efficiency of hydrazine is higher than of H_2 because energy is required to compress gaseous H_2. Whether this direct hydrazine fuel cell proves to be the final form of a vehicle-compatible fuel cell, a similar design is likely to evolved within the next 20–50 years. A viable technology to bioproduce H_2 will, therefore, be highly desirable on an industrial scale to eliminate reliance on natural gas as the major long-term source of H_2 for this supposedly zero-carbon fuel.

Second, lignocellulosic biomass is the main biological source of fuel ethanol (and/or other biofuels) on a truly mass-sufficient basis to displace (or replace) gasoline. Can bioprocesses achieve the volumes of product required to displace up to 30% of gasoline demand before 2030, with or without tax incentives, as a nascent biocommodity sector seeking funding, with multiple, rival microbial catalysts and possible substrates, and with significant governmental guidance (or interference) to ensure continual development despite any short-term fluctuations in oil price or availability? Lignocelluloses remain refractory — this is particularly obvious when compared with the ease of processing of sugarcane and corn grains — and biotechnology might profitably be applied to optimizing the use of clostridial microbes that have evolved during a billion years or more precisely to utilize the food resource that cellulose presents. Clostridial cellulases are a much-undervalued source of novel biocatalysts.[102,103] Fungal cellulases can, as a further innovation, be incorporated into bacterial cellulosomic structures with enhanced activity toward cellulosic substrates.[104] Clostridia can even perform the ultimate in resource-efficient biofuels production, transforming domestic organic waste to ethanol and butanol.[105,106] As arguments rage over the use of food crops and agriculturally valuable land for biofuels production, exploiting what the global ecosphere offers in efficient microbial biocatalysts adept at recycling the mountains of waste produced by human communities remains not only environmentally attractive but could, as the twenty-first century progresses, make overwhelming economic sense.

Third — and crucial to the success of biofuels production and biorefineries — have the best sources of the optimal choice of carbohydrate polymer- and lignin-degrading enzymes been located? The identification, gene mining, characterization, and successful manufacture of complex mixtures of enzymes to pretreat efficiently, rapidly, and without generating any unwanted or toxic products and the full range of mechanically disintegrated biomass feedstocks that could be available (on a seasonal or serendipitous or opportunistic basis) are essential to the rise of the much-anticipated biobased economy. Unconventional and little-researched microorganisms are still fertile grounds for exploring even as intensively investigated enzymes as cellulases but artificially designed, multiple enzyme mixtures still appear to mimic the natural microbial world in offering the highest bioprocessing capabilities.[107]

Fourth, lignins cannot be eliminated (as essential components of terrestrial plant structure), but their carbon is underused and undervalued. Can laccases and other enzymes or whole wood-rotting cells be developed to liberate the benzenoid structures that are presently petrochemical products? Lignins can also be inputs for hydroprocessed high-octane fuels.[108] Could they be enzymically processed at sufficiently rapid rates and with high carbon recoveries to enter the list of biorefinery resources? Broadening the question of the likely feedstocks for the hypothetical future biorefineries, can new procedures (based on enzymes or mild chemical treatments or a mixture of both) define better ways of separating and fractionating plant biomass constituents, that is, are presently operated feedstock pretreatment techniques for bioethanol production (some of which have been known and discussed for decades) only crude approximations to what is required for fully developed biorefinery operations?[109–111]

Fifth, optimizing the interface between biorefineries and "traditional" fermentation manufacture will smooth the transition to the biobased economy. Can existing fermentation processes operated at commercial scales of production be adapted to at least comparable yields and economic cost with novel sources of ingredients for microbial media presented by refineries?[112,113] The industrial fermentation sector evolved under conditions where the supply of inputs usually mirrored products for the food industry (glucose syrups, soybean oil, soybean protein, etc.). Locating the manufacture of fine chemicals alongside the primary processing facilities for biorefineries allows economies of scale and shared use but will require the rethinking and reformulation of media, nutrient supply, and feeding strategies to efficiently utilize the sugar units and other plant-derived product streams that will be available on an increasingly large scale.

Finally, harmonizing agronomy and biotechnology will accelerate the approach to the biorefinery model of the future commodity chemicals market. Plant biotechnologists view the woefully low efficiency of capture of solar energy as a prime target for genetic therapy, and this issue may still limit greatly how much of the world's present lavish energy use could be met by biomass as the ultimate bioenergy source.[114] At the same time, the debate over the ecological desirability of monoculture plantations as sustainably supplying biomass for biofuels production remains a pointed argument — grassland pastures may simply outperform any dedicated "energy crops."[115]

In the prelude to the establishment of a biorefineries for advanced biofuels, commodity chemicals, or both, however, there is still time to resolve an equally pertinent conundrum: should reliable crop as biorefinery inputs be (quickly) defined, relevant processing technologies fixed, products and coproducts agreed (nationally, if not internationally), and the various interdependent operations of a biorefinery be designed around those agronomic choices?[116,117] Or is the clearly most energy-efficient route of choice for biomass use that of thermochemical decomposition (domestically and in localized industry via wood-burning stoves of high energy conversions)? Biomass gasification dovetails very well into the existing oil and natural gas sectors and their chemistries, and in the first half of the twenty-first century Fischer-Tropsch liquid fuels remain an obvious short-term choice. Syngas is highly versatile, but can expanding the knowledge base fully exploit this versatile carbon and energy source? In particular, can syngas-based fermentations be the optimum biomass conversion methodology for bioethanol production?[118]

8.5.3 DO BIOFUELS NEED — RATHER THAN BIOTECHNOLOGY — THE PETROCHEMICAL INDUSTRY?

There is little doubt that biotechnology has — or will — produce viable solutions to all the manifold problems inherent in processing lignocellulosic biomass with maximum efficiency into ethanol and other advanced biofuels. But are these techniques and methodologies simply too elaborate and overtechnologized for the task in hand? Although academic groups insert increasingly complex and inclusive arrays of genes into microorganisms to metabolize the whole gamut of sugars present in woody feedstocks, is this the realistic technology toolbox for the massive amounts of primary chemical building blocks and biofuels that will (eventually) be required?

Of the three established economically significant biofuels presently produced, two (starch- and sugar-derived ethanol) have required only modest advances in carbohydrate-processing enzymes as essential inputs from industrial biotechnology, whereas the third (biodiesel) has evolved as an entirely chemical operation. Recent applications of chemical catalysis to the production of fuels and other value-added chemicals from biomass-derived oxygenated feedstocks have now opened up routes involving well-defined reactions of organic chemistry (dehydration/dehydrogenation, aldol condensation, etc.).[119] To the process chemist's eyes, biorefineries are entirely analogous to petrochemical refineries in that a limited range of major feedstocks — biomass hydrolysates, vegetable oils, pyrolytic bio-oils, lignin, and others (in biorefineries) and oils and diesel fuel (in a petrochemical refinery) — are convertible by catalytic cracking, hydrotreating, and hydrocracking to a range of fuels (gasoline, diesel, aviation fuel, and liquid petroleum gas) and chemicals (olefins, etc.).[120] The logical extrapolation is that biomass-derived and fossil fuel inputs to refineries could be mixed — and trials of that concept have indeed already begun, with oil companies exploring cofeeding of biomass and petroleum feedstocks, the production of biofuels in petroleum refineries, and the direct production of diesel fuel from vegetable oils by a hydrotreatment process. Eventually (and perhaps within the next 20–60 years), the massive infrastructure of oil refineries may find their conventional feedstocks dwindling; the massive accumulated experience of chemical engineering could easily accommodate new feedstock types and modified chemical catalytic processes. Although biotechnologists have focused for the past decade on refining ethanologens and other biofuels producers to utilize more of the available carbon in highly processed lignocellulosic substrates, the chemists can offer pyrolytic and thermochemical technologies that can transform all the available carbon in crude biomass to forms that can either be catalytically modified to value-added chemicals or act as inputs to Fischer-Tropsch fuel production streams. At high temperatures and pressures, given well-designed catalysts and predictable downstream operations, no material is "refractory." If the investment in the chemical hardware has already been made, why not ease into a new feedstock regime in a form of "reeducation," gradually restructuring the industry to a sustainable future?

Between 20 and 60 years: either limit is allowed by present versions of the "Hubbert curve" before the question of how to replace dwindling oil reserves in the production of the myriad chemicals on which Western industrialized societies rely becomes acute.[121] What biotechnology now makes very clear is that understanding the metabolism of even as ancient a "domesticated" microbe as the common ethanologenic yeast cannot be achieved purely in terms of genes and gene function but must include fermentation media and operating engineering in any attempt to map the biochemistry and molecular biology accompanying ethanol formation and accumulation in real-world and real-time fermentors.[122,123]

That the two final scientific references in this volume explore the computational re-creation of metabolism inside the fungus whose secretion of cotton-destroying enzymes initiated modern scientific exploration of cellulases and in ethanol-forming yeast cells is entirely appropriate. Nine millennia have passed since ethanol-forming microbes were first consciously — and (possibly) wisely — used by *Homo sapiens*; in 2007, we learned how to construct computer simulations of yeast metabolic networks and perform dynamic analyses and were able to conclude that genetic information alone is unable to optimize any biochemical pathway or maximize the yield of ethanol or any other fermentation product. Mathematical modeling, genetic manipulations, and the application of advanced chemical engineering can combine to achieve optima but only on the clear understanding that any conclusion may not be portable to simple change in circumstances (with glucose and xylose as cosubstrates rather than when individually consumed) or to each and every running of a complex series of events, that is, the biologically variable growth and metabolism of a microorganism under multiple physiological stresses in concentrated media and driven to maximal productivity.

Twenty, 40, 50, or 100 years: yeast cells (or at least, those in laboratory strains) have had little more than 20 years of deliberate, goal-driven molecular evolution, and this despite an incomplete understanding of gene expression and its endogenous regulation. "Scientific" studies of beer and wine yeasts, their inheritance patterns, stability, and metabolic capabilities may extend to a century. Pasteur was defining modern concepts of fermentation little more 150 years ago. The present generation of research scientists may be unique: it is the first to work with the sense of a severe time limit, not racing to launch a product or file a patent, but with a deadline, the realization that — however imprecisely computed is that future date — novel technologies must be in place before petrochemicals become increasingly scarce and all ready-at-hand energy sources become valuable commodities to be bartered, in other words, an ideal time in which to put to good use all the knowledge gained over nine millennia of evolving biotechnological craft and science.

REFERENCES

1. Kamm, B. et al., Biorefinery systems — an overview, in Kamm, B., Gruber, P.R., and Kamm, M. (Eds.), *Biorefineries: Industrial Processes and Products*, Volume 1, Wiley-VCH Verlag GmbH & Co. KGaA, Weinheim, 2007, chap. 1.
2. Gruber, P., Henton, D.E., and Starr, J., Polylactic acid from renewable resources, in Kamm, B., Gruber, P.R., and Kamm, M. (Eds.), *Biorefineries: Industrial Processes and Products*, Volume 2, Wiley-VCH Verlag GmbH & Co. KGaA, Weinheim, 2007, chap. 14.
3. Zeng, A.P. and Biebl, H., Bulk chemicals from biotechnology: the case of 1,3-propanediol production and the new trends, *Adv. Biochem. Eng./Biotechnol.*, 74, 239, 2002.
4. *Energy, Environment, and Economics (E3) Handbook*, 1st ed., September, 1997. U.S. Department of Energy, http://www.oit.doe.gov/e3handbook.
5. National Renewable Energy Laboratory (NREL), http://www.nrel.gov/biomass/biorefinery.html.
6. Clark, J., Introduction, in Clark, J. and MacQuarrie, D.J. (Eds.), *Handbook of Green Chemistry and Technology*, Blackwell Science Ltd., Oxford, 2002, chap. 1.
7. Kamm, B., Gruber, P.R., and Kamm, M., Biorefineries — industrial processes and products, in *Ullmann's Encyclopedia of Industrial Chemistry*, e-edition, Wiley-VCH Verlag GmbH & Co. KGaA, Weinheim, 2007.
8. Fiedler, E. et al., Methanol, in *Ullmann's Encyclopedia of Industrial Chemistry*, e-edition, Wiley-VCH Verlag GmbH & Co. KGaA, Weinheim, 2005.
9. Zimmermann, H. and Walzl, R., Ethylene, in *Ullmann's Encyclopedia of Industrial Chemistry*, e-edition, Wiley-VCH Verlag GmbH & Co. KGaA, Weinheim, 2007.
10. Christoph, R. et al., Glycerol, in *Ullmann's Encyclopedia of Industrial Chemistry*, e-edition, Wiley-VCH Verlag GmbH & Co. KGaA, Weinheim, 2006.
11. Sienel, G., Rieth, R., and Rowbottom, K.T., Epoxides, in *Ullmann's Encyclopedia of Industrial Chemistry*, e-edition, Wiley-VCH Verlag GmbH & Co. KGaA, Weinheim, 2005.
12. Hahn, H.-D., Dämbkes, G., and Ruppich, N. et al., Butanols, in *Ullmann's Encyclopedia of Industrial Chemistry*, e-edition, Wiley-VCH Verlag GmbH & Co. KGaA, Weinheim, 2005.
13. Obenhaus, F., Droste, W., and Neumeister, J., Butenes, in *Ullmann's Encyclopedia of Industrial Chemistry*, e-edition, Wiley-VCH Verlag GmbH & Co. KGaA, Weinheim, 2005.

14. Clements, L.D. and Van Dyne, D.L., The lignocellulosic biorefinery — a strategy for returning to a sustainable source of fuels and industrial organic chemicals, in Kamm, B., Gruber, P.R., and Kamm, M. (Eds.), *Biorefineries: Industrial Processes and Products*, Volume 1, Wiley-VCH Verlag GmbH & Co. KGaA, Weinheim, 2007, chap. 5.

15. Werpy, T. and Petersen, G. (Eds.), *Top Value Added Chemicals from Biomass*. Volume 10. *Results of Screening for Potential Candidates from Sugars and Synthesis Gas*, Pacific Northwest National Laboratory and National Renewable Energy Laboratory, August 2004, http://www.osti.gov/bridge.

16. Guettler, M.V., Jain, M.K., and Soni, B.K., Process for making succinic acid, microorganisms for use in the process and methods of obtaining the microorganisms, U.S. Patent 5,723,322, March 3, 1998.

17. Zeikus, J.G., Jain, M.K., and Elankovan, P., Biotechnology of succinic acid production and markets for derived industrial products, *Appl. Microbiol. Biotechnol.*, 51, 545, 1999.

18. Donnelly, M.I. et al., A novel fermentation pathway in an *Escherichia coli* mutant producing succinic acid, acetic acid, and ethanol, *Appl. Biochem. Biotechnol.*, 70–72, 187, 1998.

19. Donnelly, M.I., Millard, C.S., and Stols, L., Mutant *E. coli* strain with increased succinic acid production, U.S. Patent 5,770,435, June 23, 1998.

20. Chatterjee, R. et al., Mutation of the *ptsG* gene results in increased production of succinic acid in fermentation of glucose by *Escherichia coli*, *Appl. Environ. Microbiol.*, 67, 148, 2001.

21. Nghiem, N.P. et al., Method for the production of dicarboxylic acids, U.S. Patent 5,869,301, February 9, 1999.

22. Donnelly, M.I., Sanville-Millard, C.S., and Nghiem, N.P., Method to produce succinic acid from raw hydrolysates, U.S. Patent 6,743,610, June 1, 2004.

23. Isar, J. et al., A statistical method for enhancing the production of succinic acid from *Escherichia coli* under anaerobic conditions, *Bioresour. Technol.*, 97, 1443, 2006.

24. Agarwal, L. et al., A cost effective fermentative production of succinic acid from cane molasses and corn steep liquor by *Escherichia coli*, *J. Appl. Microbiol.*, 100, 1348, 2006.

25. Isar, J. et al., Succinic acid production from *Bacteroides fragilis*: process optimization and scale up in a bioreactor, *Anaerobe*, 12, 231, 2006.

26. Isar, J. et al., A statistical approach to study the interactive effects of process parameters on succinic acid production from *Bacteroides fragilis*, *Anaerobe*, 13, 50, 2007.

27. Lin, H., Bennett, G.N., and San, K.-Y., Fed-batch culture of a metabolically engineered *Escherichia coli* strain designed for high-level succinate production and yield under aerobic conditions, *Biotechnol. Bioeng.*, 90, 775, 2005.

28. San, K.-Y., Bennett, G.N., and Sanchez, A., Mutant *E. coli* strain with increased succinic acid production, U.S. Patent Application 2006/0046288, March 2, 2006.

29. Lee, S.Y. et al., Novel gene encoding fumarate dehydratase C and method for preparing succinic acid using the same, U.S. Patent Application 2007/0042477, February 22, 2007.

30. Murase, M. et al., Process for producing succinic acid, U.S. Patent Application 2006/0205048, September 14, 2006.

31. Lee, P.C. et al., Succinic acid production by *Anaerobiospirillum succiniciproducens*: effects of the H_2/CO_2 supply and glucose concentration, *Enz. Microb. Technol.*, 24, 549, 1999.

32. Lee, P.C. et al., Effects of medium components on the growth of *Anaerobiospirillum succinproducens* and succinic acid production, *Proc. Biochem.*, 35, 49, 1999.

33. Lee, P.C. et al., Batch and continuous fermentation of succinic acid from whey by *Anaerobiospirillum succiniciproducens*, *Appl. Microbiol. Biotechnol.*, 54, 23, 2000.

34. Lee, P.C. et al., Fermentative fermentation of succinic acid from glucose and corn steep liquor by *Anaerobiospirillum succiniciproducens*, *Biotechnol. Bioproc. Eng.*, 5, 379, 2000.

35. Lee, P.C. et al., Isolation and characterization of a new succinic acid-producing bacterium, *Mannheimia succiniciproducens* from bovine rumen, *Appl. Microbiol. Biotechnol.*, 58, 663, 2002.

36. Hong, S.H. et al., The genome sequence of the capnophilic rumen bacterium *Mannheimia succiniciproducens*, *Nat. Biotechnol.*, 22, 1275, 2004.
37. Kim, T.Y. et al., Genome-scale analysis of *Mannheimia succiniciproducens* metabolism, *Biotechnol. Bioeng.*, 97, 657, 2007.
38. Lee, S.Y. and Lee, S.J., Novel rumen bacteria variants and process for preparing succinic acid employing the same, International Patent Application WO 2005/052135, June 9, 2005.
39. Kim, D.Y. et al., Batch and continuous fermentations of succinic acid from wood hydrolysate by *Mannheimia succiniciproducens* MBEL55E, *Enz. Microb. Technol.*, 35, 648, 2004.
40. Millard, C.S. et al., Enhanced production of succinic acid by overexpression of phosphoenolpyruvate carboxylase in *Escherichia coli*, *Appl. Environ. Microbiol.*, 62, 1808, 1996.
41. Stols, L. and Donnelly, M., Production of succinic acid through overexpression of NAD⁺-dependent malic enzyme in an *Escherichia coli* mutant, *Appl. Environ. Microbiol.*, 63, 2695, 1997.
42. Stols, L. et al., Expression of *Ascaris suum* malic enzyme in a mutant *Escherichia coli* allows production of succinic acid from glucose, *Appl. Biochem. Biotechnol.*, 63–65, 153, 1997.
43. Hong, S.H. and Lee, S.Y., Metabolic flux analysis for succinic acid production by recombinant *Escherichia coli* with amplified malic enzyme activity, *Biotechnol. Bioeng.*, 74, 89, 2001.
44. Kim, P. et al., Effect of overexpression of *Actinobacillus succinogenes* phosphoenolpyruvate carboxylase on succinate production in *Escherichia coli*, *Appl. Environ. Microbiol.*, 70, 1238, 2004.
45. Lin, H., Bennett, G.N., and San, K.-Y., Metabolic engineering of aerobic succinate production systems in *Escherichia coli* to improve process productivity and achieve the maximum theoretical succinate yield, *Metab. Eng.*, 7, 116, 2005.
46. Lee, S.Y. et al., Metabolic engineering of *Escherichia coli* for enhanced production of succinic acid based on genome comparison and *in silico* gene knockout simulation, *Appl. Environ. Microbiol.*, 71, 7880, 2005.
47. Yedur, S., Berglund, K.A., and Dunuwila, D.D., Succinic acid production and purification, U.S. Patent 6,265,190, July 24, 2001.
48. *Production of Chemicals from Biologically Derived Succinic Acid (BDSA)*, www.pnl.gov/biobased/docs/succinic.pdf.
49. Nair, N. and Zhao, H., Biochemical characterization of an L-xylulose reductase from *Neurospora crassa*, *Appl. Environ. Microbiol.*, 73, 2001, 2007.
50. Rasmussen, L.E. et al., Mode of action and properties of the β-xylosidases from *Talaromyces emersonii* and *Trichoderma reesei*, *Biotechnol. Bioeng.*, 94, 869, 2006.
51. Tenkanen, M., Action of *Trichoderma reesei* and *Aspergillus oryzae* esterases in the deacetylation of hemicelluloses, *Biotechnol. Appl. Biochem.*, 27, 19, 1998.
52. Lavasseur, A. et al., Design and production in *Aspergillus niger* of a chimeric protein associating a fungal feruloyl esterase and a clostridial dockerin domain, *Appl. Environ. Microbiol.*, 70, 6984, 2004.
53. Lavasseur, A. et al., Production of a chimeric enzyme tool associating the *Trichoderma reesei* swollenin with the *Aspergillus niger* feruloyl esterase A for release of ferulic acid, *Appl. Microbiol. Biotechnol.*, 73, 872, 2006.
54. Faulds, C.H. et al., Arabinoxylan and mono- and dimeric ferulic acid release from brewer's grain and wheat bran by feruloyl esterases and glycosyl hydrolases from *Humicola insolens*, *Appl. Microbiol. Biotechnol.*, 64, 644, 2004.
55. Leathers, T.D. and Dien, B.S., Xylitol production from corn fibre hydrolysates by a two-stage fermentation process, *Proc. Biochem.*, 35, 765, 2000.

56. Lee, W.-J., Ryu, Y.-W., and Seo, J.-H., Characterization of two-substrate fermentation processes for xylitol production using recombinant *Saccharomyces cerevisiae* containing xylose reductase gene, *Proc. Biochem.*, 35, 1199, 2000.

57. Govinden, R. et al., Xylitol production by recombinant *Saccharomyces cerevisiae* expressing the *Pichia stipitis* and *Candida shehatae XYL*-1 genes, *Appl. Microbiol. Biotechnol.*, 55, 76, 2001.

58. Ko, B.S., Kim, J., and Kim, J.H., Production of xylitol from D-xylose by a xylitol dehydrogenase gene-disrupted mutant of *Candida tropicalis*, *Appl. Environ. Microbiol.*, 72, 4207, 2006.

59. Martinez, E.A. et al., Downstream process for xylitol produced from fermented hydrolysate, *Enz. Microb. Technol.*, 40, 1193, 2007.

60. Verho, R. et al., A novel NADH-linked L-xylulose reductase in the L-arabinose catabolic pathway of yeast, *J. Biol. Chem.*, 279, 14746, 2004.

61. Poonperm, W. et al., Production of L-xylulose from xylitol by a newly isolated strain of *Bacillus pallidus* Y25 and characterization of its relevant enzyme xylitol dehydrogenase, *Enz. Microb. Technol.*, 40, 1206, 2007.

62. Suzuki, T. et al., Cloning and expression of NAD$^+$-dependent L-arabinitol 4-dehydrogenase gene (*ladA*) of *Aspergillus oryzae*, *J. Biosci. Bioeng.*, 100, 472, 2005.

63. Aarnikunnas, J.S. et al., Cloning and expression of a xylitol-4-dehydrogenase gene from *Pantoea ananatis*, *Appl. Environ. Microbiol.*, 72, 368, 2006.

64. Strohl, W.R., Industrial antibiotics: today and the future, in Strohl, W.R. (Ed.), *Biotechnology of Antibiotics*, Marcel Dekker, Inc., New York, 1997, chap. 1.

65. Muniruzzaman, S. et al., Inhibition of glycoprotein processing by L-fructose and L-xylulose, *Glycobiology*, 8, 795, 1996.

66. Kim, P., Current studies on biological tagatose production using L-arabinose isomerase: a review and future perspective, *Appl. Microbiol. Biotechnol.*, 65, 243, 2004.

67. Roh, H.J. et al., Bioconversion of D-galactose into D-tagatose by expression of L-arabinose isomerase, *Biotechnol. Appl. Biochem.*, 31, 1, 2000.

68. Kim, P. et al., High production of D-tagatose, a potential sugar substitute, using immobilized L-arabinose isomerase, *Biotechnol. Prog.*, 17, 208, 2001.

69. Kim, H.J. et al., A feasible enzymatic process for D-tagatose production by an immobilized thermostable L-arabinose isomerase in a packed-bed bioreactor, *Biotechnol. Prog.*, 19, 400, 2003.

70. Jung, E.S., Kim, H.J., and Oh, D.K., Tagatose production by immobilized recombinant *Escherichia coli* cells containing *Geobacillus stearothermophilus* L-arabinose isomerase mutant in a packed-bed bioreactor, *Biotechnol. Prog.*, 21, 1335, 2005.

71. Doten, R.C. and Mortlock, R.P., Production of D- and L-xylulose by mutants of *Klebsiella pneumoniae* and *Erwinia uredovora*, *Appl. Environ. Microbiol.*, 49, 158, 1985.

72. Granström, T.B. et al., Izumoring: a novel and complete strategy for bioproduction of rare sugars, *J. Biosci. Bioeng.*, 97, 89, 2004.

73. Menavuvu, B.T. et al., Novel substrate specificity of D-arabinose isomerase from *Klebsiella pneumoniae* and its application to production of D-altrose and D-psicose, *J. Biosci. Bieng.*, 102, 436, 2006.

74. Aitken, J.E., Purification of glycerin, U.S. Patent 7,126,032, October 24, 2006.

75. Pech, C.W. and Hopp, J.R., Method for manufacture and use of the waste stream from biodiesel production crude (glycerin) as a commercial fuel, U.S. Patent Application 2007/0113465, May 24, 2007.

76. Pachauri, N. and He, B., Value-added utilization of crude glycerol from biodiesel production: a survey of current research activities, 2006 ASABE Annual International Meeting, paper number 066223, 2006.

77. Biebl, H., Fermentation of glycerol by *Clostridium pasteurianum* — batch and continuous fermentations, *J. Ind. Microbiol. Biotechnol.*, 27, 18, 2001.

78. Pyle, D. and Wen, Z., Production of omega-3 polyunsaturated fatty acid from biodiesel waste glycerol by microalgal fermentation, 2007 ASABE Annual International Meeting, paper number 077028, 2007.
79. Frafft, P. et al., Process for producing dichloropropanol from glycerol, the glycerol coming eventually from the conversion of animal fats in the manufacture of biodiesel, U.S. Patent Application 2007/0112224, May 17, 2007.
80. Bournay, L. and Baudot, A., Process for producing fatty acid alkyl esters and glycerol of high-purity, U.S. Patent 7,138,536, November 21, 2006.
81. Dharmadi, Y., Murarka, A., and Gonzalez, R., Anaerobic fermentation of glycerol by *Escherichia coli*: a new platform for metabolic engineering, *Biotechnol. Bioeng.*, 94, 821, 2006.
82. Koutinas, A.A. et al., Cereal-based biorefinery development: integrated enzyme production for cereal flour hydrolysis, *Biotechnol. Bioeng.*, 97, 61, 2007.
83. Kawaguchi, H. et al., Engineering of a xylose metabolic pathway in *Corynebacterium glutamicum*, *Appl. Environ. Microbiol.*, 72, 3418, 2006.
84. Weekes, J. and Yüksel, G.U., Molecular characterization of two lactate dehydrogenase genes with a novel structural organization on the genome of *Lactobacillus* sp. strain MONT4, *Appl. Environ. Microbiol.*, 70, 6290, 2004.
85. Yanai, K. et al., *Para*-position derivatives of fungal anthelmintic cyclodepsipeptides engineered with *Streptomyces venezuelae* antibiotic resistance genes, *Nat. Biotechnol.*, 22, 848, 2004.
86. Varadarajan, S. and Miller, D.J., Catalytic upgrading of fermentation-derived organic acids, *Biotechnol. Prog.*, 15, 845, 1999.
87. Mousdale, D.M., Metabolic analysis and optimisation of microbial and animal cell bioprocesses, in El-Mansi, E.M.T. et al. (Eds.), *Fermentation Microbiology and Biotechnology*, 2nd ed., CRC Press, Boca Raton, 2006, chap. 6.
88. Mousdale, D.M., Bioprocess expression and production technologies in India, in *Advances in Biopharmaceutical Technology in India*, BioPlan Associates and Society for Industrial Microbiology, 2008, chap. 13.
89. Khalil, C.N. and Leite, L.C.F., Process for producing biodiesel fuel using triglyceride-rich oleagineous seed directly in a transesterification reaction in the presence of an alkaline alkoxide catalyst, U.S. Patent 7,112,229, September 26, 2006.
90. Pandey, A., Soccol, C.R., and Mitchell, D., New developments in solid state fermentation. I. Bioprocesses and products, *Proc. Biochem.*, 35, 1153, 2000.
91. Pessoa, A. et al., Perspectives on bioenergy and biotechnology in Brazil, *Appl. Biochem. Biotechnol.*, 121–124, 59, 2005.
92. Solaiman, D.K. et al., Conversion of agricultural feedstocks and co-products into poly(hydroxyalkanoates), *Appl. Microbiol. Biotechnol.*, 71, 783, 2006.
93. Lynd, L.R., Wyman, C.E., and Gerngross, T.U., Biocommodity engineering, *Biotechnol. Prog.*, 15, 777, 1999.
94. Li, P., Harding, S.E., and Liu, Z., Cyanobacterial exopolysaccharides: their nature and potential biotechnological applications, *Biotechnol. Genet. Eng. Rev.*, 18, 404, 2001.
95. Ishii, M. et al., Occurrence, biochemistry and possible biotechnological application of the 3-hydroxypropionate cycle, *Appl. Microbiol. Biotechnol.*, 64, 605, 2004.
96. Bosma, R., Prediction of volumetric productivity of an outdoor photobioreactor, *Biotechnol. Bioeng.*, 97, 1108, 2007.
97. Rao, A.B. et al., Evaluation of potential cost reductions from improved amine-based CO_2 capture systems, *Energy Policy*, 34, 3765, 2006.
98. Rubin, E.S., Chen, C., and Rao, A.B., Cost and performance of fossil fuel power plants with CO_2 capture and storage, *Energy Policy*, 35, 4444, 2007.
99. Spreng, D., Marland, G., and Weinberg, A.M., CO_2 capture and storage: another Faustian bargain?, *Energy Policy*, 35, 850, 2007.
100. *Facing the Hard Truths about Energy: A Comprehensive View to 2030 of Global Oil and Natural Gas*, National Petroleum Council, July 2007, www.npc.org.

101. Asazwa, K. et al., A platinum-free zero-carbon-emission easy fuelling direct hydrazine fuel cell for vehicles, *Angew. Chem. Int. Ed.*, 46, 8024, 2007.
102. López-Contreras, A.M. et al., Production by *Clostridium acetobutylicum* ATCC 824 of CelG, a cellulosomal glycoside hydrolase belonging to family 9, *Appl. Environ. Microbiol.*, 69, 869, 2003.
103. López-Contreras, A.M. et al., Substrate-induced production and secretion of cellulases by *Clostridium acetobutylicum*, *Appl. Environ. Microbiol.*, 70, 5238, 2004.
104. Mingardon, F. et al., Incorporation of fungal cellulases in bacterial minicellulosomes yields viable, synergistically acting cellulolytic complexes, *Appl. Environ. Microbiol.*, 73, 3822, 2007.
105. López-Contreras, A.M. et al., Utilisation of saccharides in extruded domestic organic waste by *Clostridium acetobutylicum* ATCC 824 for production of acetone, butanol and ethanol, *Appl. Microbiol. Biotechnol.*, 54, 162, 2000.
106. Claassen, P.A., Budde, M.A., and López-Contreras, A.M., Acetone, butanol and ethanol production by solventogenic clostridia, *J. Mol. Microbiol. Biotechnol.*, 2, 39, 2000.
107. Gusakov, A.V. et al., Design of highly efficient cellulase mixtures for enzymatic hydrolysis of cellulose, *Biotechnol. Bioeng.*, 97, 1028, 2007.
108. Shabtail, J.S. et al., Process for converting lignins into a high octane blending component, U.S. Patent Application 2003/0115792, June 26, 2003.
109. Montross, M.D. and Crofcheck, C.L., Effect of stover fraction and storage method on glucose production during enzymatic hydrolysis, *Bioresour. Technol.*, 92, 269, 2004.
110. Bootsma, J.A. and Shanks, B.H., Hydrolysis characteristics of tissue fractions resulting from mechanical separations of corn stover, *Appl. Biochem. Biotechnol.*, 125, 27, 2005.
111. Akin, D.E. et al., Corn stover fractions and bioenergy: chemical composition, structure, and response to enzyme pretreatment, *Appl. Biochem. Biotechnol.*, 129–132, 104, 2006.
112. Zverlov, V.V. et al., Bacterial acetone and butanol production by industrial fermentation in the Soviet Union: use of hydrolyzed agricultural waste for biorefinery, *Appl. Microbiol. Biotechnol.*, 71, 587, 2006.
113. Thomsen, M.H., Complex media from processing of agricultural crops for microbial fermentation, *Appl. Microbiol. Biotechnol.*, 68, 598, 2005.
114. Pimentel, D. and Lal, R., Letter, *Science*, 317, 897, 2007.
115. Palmer, M.W., Letter, *Science*, 317, 897, 2007.
116. Koutinas, A.A., Wang, R., and Webb, C., Restructuring upstream bioprocessing: technological and economical aspects for production of a generic microbial feedstock from wheat, *Biotechnol. Bioeng.*, 85, 524, 2004.
117. Koutinas, A.A. et al., Development of an oat-based biorefinery for the production of L(+)-lactic acid by *Rhizopus oryzae* and various value-added coproducts, *J. Agric. Food Chem.*, 55, 1755, 2007.
118. Inokuma, K. et al., Characterization of enzymes involved in the ethanol production of *Moorella* sp. HUC22-1, *Arch. Microbiol.*, 188, 37, 2007.
119. Chheda, J.N., Huber, G.W., and Dumesic, J.A., Liquid-phase catalytic processing of biomass-derived oxygenated hydrocarbons to fuels and chemicals, *Angew. Chem. Int. Ed.*, 46, 7164, 2007.
120. Huber, G.W. and Corma, A., Synergies between bio- and oil refineries for the production of fuels from biomass, *Angew. Chem. Int. Ed.*, 46, 7184, 2007.
121. Hubbert, M.K., *Nuclear Energy and the Fossil Fuels*, Publication no. 95, Shell Development Company, Exploration and Production Research Division, Houston, TX, June 1956.
122. Druzhinina, I.S. et al., Global carbon utilization profiles of wild-type, mutant, and transformant strains of *Hypocrea jecorina*, *Appl. Environ. Microbiol.*, 72, 2126, 2006.
123. Hjerstad, J.L., Henson, M.A., and Mahadevan, R., Genome-scale analysis of *S. cerevisiae* metabolism and ethanol production in fed-batch culture, *Biotechnol. Bioeng.*, 97, 1190, 2007.

Index